Additive Manufacturing of Ceramics

Online at: https://doi.org/10.1088/978-0-7503-4831-7

Additive Manufacturing of Ceramics

Ling Bing Kong
College of New Materials and New Energies, Shenzhen Technology University, Shenzhen 518118, Guangdong, People's Republic of China

Zhuohao Xiao
School of Materials Science and Engineering, Jingdezhen Ceramic University, Jingdezhen 333403, Jiangxi, People's Republic of China

Bin He
College of New Materials and New Energies, Shenzhen Technology University, Shenzhen 518118, Guangdong, People's Republic of China

Yin Liu
School of Pharmacy and Materials, Huainan Union University, Huainan 232038, Anhui, People's Republic of China

IOP Publishing, Bristol, UK

ISBN 978-0-7503-4831-7 (ebook)
ISBN 978-0-7503-4829-4 (print)
ISBN 978-0-7503-4832-4 (myPrint)
ISBN 978-0-7503-4830-0 (mobi)

DOI 10.1088/978-0-7503-4831-7

Version: 20250901

IOP ebooks

British Library Cataloguing-in-Publication Data: A catalogue record for this book is available from the British Library.

Published by IOP Publishing, wholly owned by The Institute of Physics, London

IOP Publishing, No.2 The Distillery, Glassfields, Avon Street, Bristol, BS2 0GR, UK

US Office: IOP Publishing, Inc., 190 North Independence Mall West, Suite 601, Philadelphia, PA 19106, USA

Contents

Preface

3D printing, also known as additive manufacturing, is a revolutionary technology that constructs objects by building them up through the precise deposition or solidification of materials guided by digital models. Unlike conventional subtractive manufacturing, in which materials are removed from bulks to create designed structures, the incremental nature of 3D printing offers unparalleled flexibility in design. Significantly, it has enabled the fabrication of parts with complex geometries that were previously impossible or extremely difficult to achieve using traditional subtractive methods, thereby supporting customizability across various industries.

The development of 3D printing has brought significant value to industrial sectors, driving disruption and remolding manufacturing and design practices. The technology provided a platform to achieve near-net-shape fabrication, pushing the boundaries toward zero waste in manufacturing. Furthermore, 3D printing enabled rapid prototyping, accelerating development cycles and playing a critical role in shaping Industry 5.0 products and services through design flexibility and responsiveness to emerging needs. Its impact spans across a wide range of industries. In the healthcare sector, it has continued to offer new possibilities for fabricating customized prosthetics, surgical models and even bio-printed tissues. In the aerospace and automotive sectors, it is especially useful in manufacturing components with lightweight characteristics without sacrificing mechanical strength, improving production efficiencies, and enhancing performances. In architecture, 3D printing enables the construction of unique building structures and features with flowing or organic shapes, facilitating seamless realization of artistic visions.

Comparatively, the application of 3D printing has experienced broader success in metallic and polymeric materials than ceramics. This book aims to provide an overview on the progress in the fabrication and development of 3D printed ceramics and ceramic materials. There are seven chapters, each covering different printing strategies, such as stereolithography (SLA), jet printing, selective laser sintering (SLS), and extrusion freeforming (EFF), as well as their potential applications. It can serve as reference for senior undergraduate students, postgraduate students, researchers, and engineers in materials science and engineering, biomaterials engineering, applied physics, applied chemistry, chemical engineering, mechanical engineering, intelligent manufacturing, and related fields, in both academia and industry.

Acknowledgements

This work was financially supported by the projects of Shenzhen Key Laboratory of Applied Technologies of Super-Diamond and Functional Crystals (ZDSYS20230626091303007).

Keywords

3D printing, additive manufacturing, ceramics, stereolithography (SLA), jet printing, selective laser sintering (SLS), extrusion freeforming (EFF), laminated object manufacturing (LOM), layer-by-layer deposition, tape-casting, slurry, suspension, inks, organic binder, dispersant, plasticizer, solid loading, biomedical materials, piezoelectric ceramics, microwave materials, transparent ceramics, ceramic matrix composites, eutectic ceramics, glass-ceramics

IOP Publishing

Ling Bing Kong, Zhuohao Xiao, Bin He and Yin Liu

Chapter 1

Introduction

This chapter is to present a brief introduction to 3D printing and its applications in the production of ceramics and ceramic materials, together with arrangement of the book.

Materials can be broadly classified into four types, including metals, polymers, ceramics, and composites. Metals are known for their excellent mechanical strengths, electrical and thermal conductivities, ductility, and machinability. Polymers, such as plastics and rubbers, are lightweight, highly flexible, and chemically stable. Ceramics are characterized by their high hardness and exceptional temperature resistance, while exhibiting low electrical conductivity and high brittleness. Composites are materials made of two or more components to achieve specific properties, such as the combined advantages of their constituent materials or even the development of entirely new properties that are not present in the individual components. These characteristics give each type of material its own unique pros and cons, and distinct suitability for different applications.

3D printing, also known as additive manufacturing (AM), enables new approaches to address the challenges and limitations of the traditional subtractive processing methods. Over the years, various specific strategies have been developed for the 3D printing of metallic, polymeric, ceramic, and composite materials [1–8]. In particular, ceramic 3D printing techniques have garnered considerable research attention as it offers a means to fabricate ceramic components of high complexity and eliminate the need for costly mechanical post-processing, along with the failure-inducing flaws inherent to traditional methods [9].

Significant progress has been made in the fabrication and development of ceramics and ceramic materials through the use of 3D printing technology, with intensive reviews published in the open literature, providing insights into the different techniques [9, 10–13]. Notably, the fabrication of ceramics by direct laser sintering/melting has been well summarized, including materials such as silica (SiO_2) and zirconia (ZrO_2), and ceramic-reinforced metal matrix composites [10].

doi:10.1088/978-0-7503-4831-7ch1

Challenges surrounding this technique were thoroughly reviewed and evaluated. The design, processing, and characterization of lithography-based 3D printed ceramics have also been comprehensively reviewed [9]. Building on this, the challenges and opportunities for enhancing mechanical performance of the printed ceramics were extensively discussed. Another review on porous ceramic structures achieved through 3D printing was published, with methods covering binder jetting, selective laser sintering, direct ink writing, stereolithography, and laminated object manufacturing [11]. Therefore, it is crucial to take a holistic look at the applications of 3D printing for ceramics and ceramic materials, synthesizing the insights from the various reviews to understand the broader implications.

Ceramic 3D printing techniques can be classified into three categories according to the form of ceramic feedstock employed for printing: (i) slurry-based, (ii) powder-based, and (iii) solid-based [14]. For slurry-based techniques, colloids are used, consisting of ceramic powders combined with polymers that function as binders, dispersants, plasticizers, and other additives [15–19]. Powder-based techniques utilises loose ceramic particles directly in powder beds [20]. In solid-based techniques, sheets or filaments composed primarily of ceramic powders are used.

This book aims to provide an overview of the progress in ceramics and ceramic materials, realized through well-established 3D printing technology across the different categories, for various applications. Slurry-based techniques, namely stereolithography (SLA), jet printing, and extrusion freeforming (EFF), are presented in chapters 2–4, respectively. Chapter 5 covers selective laser sintering (SLS), a powder-based technique, while chapter 6 focuses on laminated object manufacturing (LOM), a solid-based technique. The last chapter briefly describes the potential applications in several aspects. It is worth noting that given the large variations due to the combination of the different types of ceramic materials and the 3D processes, detailed content should be referred to the original literatures.

References

[1] Barui S, Mandal S and Basu B 2017 Thermal inkjet 3D powder printing of metals and alloys: current status and challenges *Curr. Opin. Biomed. Eng.* **2** 116–23

[2] Chohan J S, Kumar R, Singh S, Sharma S and Ilyas R A 2022 A comprehensive review on applications of 3D printing in natural fibers polymer composites for biomedical applications *Funct. Compos. Struct.* **4** 034001

[3] Dall'Ava L, Hothi H, Di Laura A, Henckel J and Hart A 2019 3D printed acetabular cups for total hip arthroplasty: a review article *Metals* **9** 729

[4] McMullan R, Golbang A, Salma-Ancane K, Ward J, Rodzen K and Boyd A R 2025 Review of 3D printing of polyaryletherketone/apatite composites for lattice structures for orthopedic implants *Appl. Sci.-Basel* **15** 1804

[5] Okolie O, Stachurek I, Kandasubramanian B and Njuguna J 2020 3D printing for hip implant applications: a review *Polymers* **12** 2682

[6] Ravichandran D, Xu W H, Jambhulkar S, Zhu Y, Kakarla M, Bawareth M and Song K 2021 Intrinsic field-induced nanoparticle assembly in three-dimensional (3D) printing polymeric composites *ACS Appl. Mater. Interfaces* **13** 52274–94

[7] Ucak N, Cicek A and Aslantas K 2022 Machinability of 3D printed metallic materials fabricated by selective laser melting and electron beam melting: a review *J. Manuf. Processes* **80** 414–57
[8] Zhang C, Ouyang D, Pauly S and Liu L 2021 3D printing of bulk metallic glasses *Mater. Sci. Eng. R-Rep.* **145** 100625
[9] Lube T, Staudacher M, Hofer A K, Schlacher J and Bermejo R 2023 Stereolithographic 3D printing of ceramics: challenges and opportunities for structural integrity *Adv. Eng. Mater.* **25** 202200520
[10] Sing S L, Yeong W Y, Wiria F E *et al* 2017 Direct selective laser sintering and melting of ceramics: a review *Rapid Prototyp. J.* **23** 611–23
[11] Zhang F, Li Z A, Xu M J, Wang S Y, Li N and Yang J Q 2022 A review of 3D printed porous ceramics *J. Eur. Ceram. Soc.* **42** 3351–73
[12] Hwa L C, Rajoo S, Noor A M, Ahmad N and Uday M B 2017 Recent advances in 3D printing of porous ceramics: a review *Curr. Opin. Solid State Mater. Sci.* **21** 323–47
[13] Wang L K, Feng J Z, Jiang Y G, Li L J and Feng J 2023 Direct-ink-writing 3D printing of ceramic-based porous structures: a review *J. Inorg. Mater.* **38** 1133–48
[14] Chen Z W, Li Z Y, Li J J, Liu C B, Liu C, Yang L, Wang P, Yi H, Lao C S and Yuelong F 2019 3D printing of ceramics: a review *J. Eur. Ceram. Soc.* **39** 661–87
[15] Hoffmann M, Schubert N H, Günster J, Stawarczyk B and Zocca A 2025 Additive manufacturing of glass-ceramic dental restorations by layerwise slurry deposition (LSD-print) *J. Eur. Ceram. Soc.* **45** 117235
[16] Su C Y, Wang J C, Chen D S, Chuang C C and Lin C K 2020 Additive manufacturing of dental prosthesis using pristine and recycled zirconia solvent-based slurry stereolithography *Ceram. Int.* **46** 28701–9
[17] Wang J W, Shaw L L and Cameron T B 2006 Solid freeform fabrication of permanent dental restorations via slurry micro-extrusion *J. Am. Ceram. Soc.* **89** 346–9
[18] Zhang G, Chen H, Yang S B, Guo Y, Li N, Zhou H and Cao Y 2018 Frozen slurry-based laminated object manufacturing to fabricate porous ceramic with oriented lamellar structure *J. Eur. Ceram. Soc.* **38** 4014–9
[19] Zhang G R, Carloni D and Wu Y Q 2020 3D printing of transparent YAG ceramics using copolymer-assisted slurry *Ceram. Int.* **46** 17130–4
[20] Verga F, Borlaf M, Conti L *et al* 2020 Laser-based powder bed fusion of alumina toughened zirconia *Addit. Manuf.* **31** 100959

Ling Bing Kong, Zhuohao Xiao, Bin He and Yin Liu

Chapter 2

Stereolithographic (SLA) approaches

Stereolithography (SLA) has emerged as one of the fastest-evolving 3D printing technologies for the fabrication of ceramics and ceramic materials. In this technique, light-induced polymerization is utilized to build green bodies of ceramics from slurries, combined with precise control over the curing process, resulting in structures with complicated geometries. This chapter serves to offer an overview of the process in the manufacturing of various ceramics by using the SLA 3D printing technology.

2.1 Brief introduction

Stereolithography is defined as an approach that can be used to fabricate products in a three-dimensional way (stereo) by printing designed structures in a layer-by-layer way (lithography). The structures are designed by using a commercial software before printing. The printing process is applied to photosensitized monomer resin pastes or solutions, during which the printed layers are selectively solidified through photocuring by exposing to light at given wavelengths (e.g., UV). The monomers are polymerized under the irradiation. Therefore, the products are actually assembled with multilayer structures.

When using stereolithographic approach to fabricate ceramics, there are two processes, i.e., (i) direct and (ii) indirect processes. In the direct process, ceramic powders are first mixed with photocuring liquids at given concentrations, followed by the printing process. The printed items are then subject to burning and sintering to obtain final ceramic products. In this process, the polymers act as a support for the ceramics. Indirect process is essentially similar to the original polymer printing process, after which the printed items with predesigned dimensions and shapes are used as molds for ceramic parts. In both processes, the sintering step to obtain final ceramics products is near the same. Currently, stereolithographic approach has been employed to develop ceramics for various applications, such as biomedical devices,

doi:10.1088/978-0-7503-4831-7ch2

microelectron-mechanical systems (MEMS), sensors, piezoelectric actuators, turbine blades [1–7].

2.2 Working principles of stereolithographic approaches

As the photocuring liquid is selectively exposed to light irradiation, solid cure tracks are generated. The profile of the cure track can be determined by the cure depth (C_d) and line width (L_W). Specifically, C_d can be derived from the Jacobs formula, which is given by:

$$C_d = D_P \ln\left(\frac{E_{\max}}{E_c}\right) \tag{2.1}$$

where D_P, $E_{\max}$ (in J m^{-2}) and E_c are the penetration depth of light, peak energy density of the light exposure according to the Lambert–Beer law and minimum critical exposure level required to trigger the polymerization of the photocuring liquid, respectively.

By assuming that the intensity profile of the incident light follows Gaussian distribution, L_W can be expressed as [8–10]:

$$L_W = \sqrt{2}\, W_0 \sqrt{\ln\left(\frac{E_{\max}}{E_C}\right)} \tag{2.2}$$

where W_0 is the full width at half maximum (FWHM) in the Gaussian profile of the light. This approximation applies well to cases, such as photocuring of monomers that are irradiated with laser beams.

The monomers in the main component of photocuring liquids are either monomer resins (e.g., derivatives of vinyl ether, acrylate, epoxy and oxytane) or aqueous solutions with monomers (e.g., acrylamide-based derivatives) [11]. The photo-polymerization reactions are triggered with appropriate photoinitiators, which have concentrations to be $\leqslant$10 wt% in the photocuring solutions. The reactions could be either radical or ionic photopolymerization, depending on the types of the monomers [9, 10].

Each type of polymerization has its own limitations. For example, during the radical polymerization reactions, oxygen molecules tend to diffuse into the reaction system, leading to instability. Ionic photopolymerization is maintained for a certain time duration, after the light irradiation is switched off [8].

Since both radical and ionic polymerization are exothermic reactions, heat is released during the reactions, so that the printed parts would experience deformation in structure. The dimensional accuracy could be maintained by accurately controlling the irradiation energy of the light according to careful estimation of the heat release of the reaction. It has been confirmed that the increase in temperature inside the printed parts is proportionally increased with increasing irradiation energy of the light and decreased with increasing distance from the center of the light beam and increasing scanning rate [12].

For printing of ceramics, ceramic powders should be put in the monomer solutions, which would alter their response to the light irradiation because ceramic particles are most likely inactive. As a consequence, the irradiation energy would be reduced, owing to the light scattering from the ceramic particles. The effect can be described by using the Kubelka–Munk model [13]. In this model, the effect of ceramic particles on optical properties, including reflection, absorption and transmission, of the reaction systems in terms of the light response, is well elaborated. The irradiation energy (E) of the light at a given location (z) in the reaction system containing ceramic particles with certain concentrations can be derived as [13]:

$$E = E_{\max} \exp\left[-z\sqrt{K(K+2S)}\right] \tag{2.3}$$

where $E_{\max}$, S and K are peak energy density, specific scattering coefficient, specific absorption coefficient of a printed layer, respectively. The parameters are dependent on the optical thickness and physical density of the printed layer. Because the penetration depth (Dp) is inversely proportional to the value of reflectance, S can be replaced by reflective indices of the reaction systems (n_0) and the ceramic particles (n_p). Furthermore, the density of the systems can be described with the average particle size of the ceramic phase (d_{50}), the wavelength of the irradiation light (l) and the spacing intensity in between the ceramic particles (ϕ) [14–17]:

$$D_P \propto \left(\frac{2}{3}\right)\frac{d_{50}\lambda n_0}{\varnothing(\Delta n^2)} \tag{2.4}$$

where $\Delta n = n_p - n_0$, which is the difference in the reflective indices between the ceramic particles and the reaction systems. In a more recent study, Gentry and Halloran investigated the contribution of ceramic particles to the profile of the cure tracks of photosensitive liquids [18–20]. The scattering caused by ceramic particles led to broadening of the curing cone-shape like tracks, so that the penetration depth of the light is reduced. In this case, the cured track profiles could be evaluated more accurately by taking the penetration width and the critical exposure in the lateral direction.

The printed structures are subject to subsequent sintering process to obtain final ceramic products. Under certain sintering conditions, density of the final ceramics is closely related to properties of the green bodies, in which the degree of polymerization has a critical role. Currently, suspensions of SLA process have been made of ceramic powders with particle sizes in the range from nanometer to submicron scales. Generally, the smaller the particles, the stronger the light scattering will be caused inside the photocuring suspensions [21].

2.3 Fabrication and characterization of stereolithographically printed ceramics

2.3.1 General stereolithographic approach

Badev *et al* used real time infrared spectroscopy (RTIR) to study the photoinitiated polymerization of commercial polyether acrylate oligomers, in which

2,2-dimethoxy-1,2-phenyl acetophenone (DMPA) was used as the radical photoinitiator [21]. Their study focused on the effects of intensity of light irradiation, concentration of photoinitiator and dilute agent (1,6-hexanediol diacrylate or HDDA) on the efficiency of the polymerization. The optimized conversion rate was achieved, as the concentration of DMPA was 0.5 wt% and that of HDDA was in the range of 10–15 vol% without the presence of ceramic phases. Ceramic particles, including SiO_2, Al_2O_3, ZrO_2 and SiC, were mixed with the acrylate oligomer, forming slurries to develop ceramic green bodies with stereolithography. It was found that both the reaction kinetics and the conversion rate of the polymerization reactions were dependent on the type, particle size and content of the ceramic powders. The weight ratio of the ceramic phase over the polymeric one and viscosity of the reaction systems played an important role in determining the efficiency of the polymerization process.

Photopolymerization kinetics of the polyether acrylate (PEAAM) with different concentrations of DMPA, in the range of 0.25–1.0 wt%, were evaluated, with results that are illustrated in figure 2.1(a). As the concentration of the photoinitiator reached 0.7 wt%, the conversion rate started to decline. Such a reduction in the conversion rate was ascribed to two factors [22–25]. On one hand, the reactions could be terminated due to the increased number of macroradicals from the primary radicals in the photoinitiator. On the other hand, the primary radicals could recombine so as to result in termination of the polymerization reaction.

When the concentration of DMPA was fixed at 0.5 wt%, the degree of the photopolymerization of PEAAM increased with increasing intensity of the light irradiation, as demonstrated in figure 2.1(b). This is because the content of the primary radicals was increased as the light intensity was increased. As a result, the polymerization rate was enhanced and hence the conversion rate was increased accordingly [26–28].

It is well known that the viscosity of a slurry can be reduced by introducing a dilute agent, so that the content of ceramic phase can be increased. A promising dilute agent should have at least three requirements, i.e., (i) strong wetting capability for the ceramic phase, (ii) high miscibility with the monomers or oligomers to be used and (iii) high sensitivity to the irradiation of UV light to ensure high conversion rates.

The viscosity of HDDA is much lower than that of FEAAM viscosity, while the two have similar reaction characteristics. As the concentration of DMPA was kept unchanged at 0.5 wt% with respect to the acrylate and the irradiation intensity of UV light was fixed at 5.3 $mW \cdot cm^{-2}$, their conversion rates were nearly identical, which were about 80%. Figure 2.2 shows rheological characteristics of the samples with PEAAM/HDDA ratios in the range of 5–50 vol%. They are of Newtonian rheological behavior, with viscosity of the system to be decreased with increasing content of HDDA.

As a consequence, the reactivity of the systems was dependent on the content of HDDA. According to conversion rate versus the content of HDDA, there was a maximum at about 15 vol% HDDA, since the presence of HDDA had two opposite effects. This is likely due to two antagonistic effects induced by the introduction of

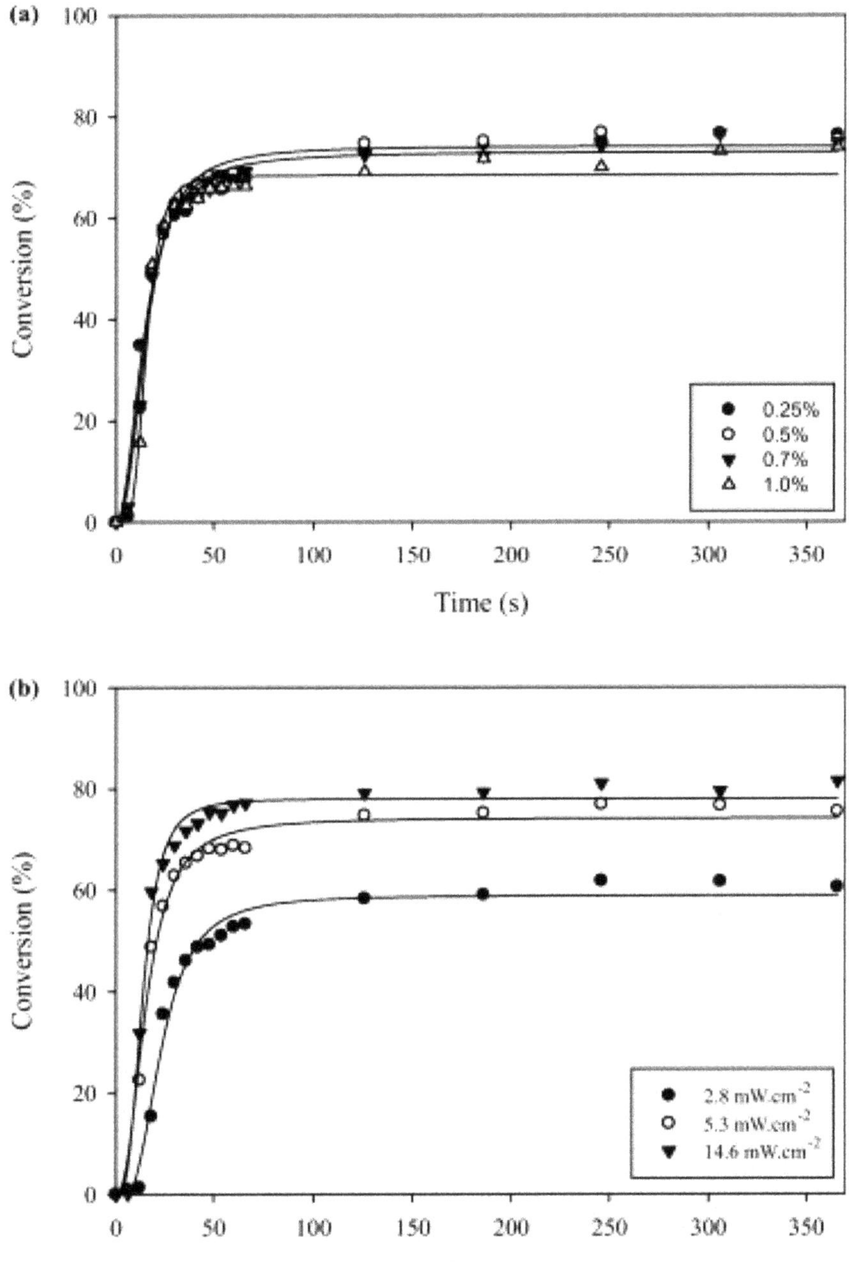

Figure 2.1. Double bond conversion rates of acrylate as a function of time: (a) systems with different contents of DMPA at photoinitiator concentration of $I_0 = 5.3\ \mathrm{mW \cdot cm^{-2}}$ and (b) reactions irradiated with different incident intensities of UV light at DMPA concentration of 0.5 wt%. Reprinted from [21], Copyright (2011), with permission from Elsevier.

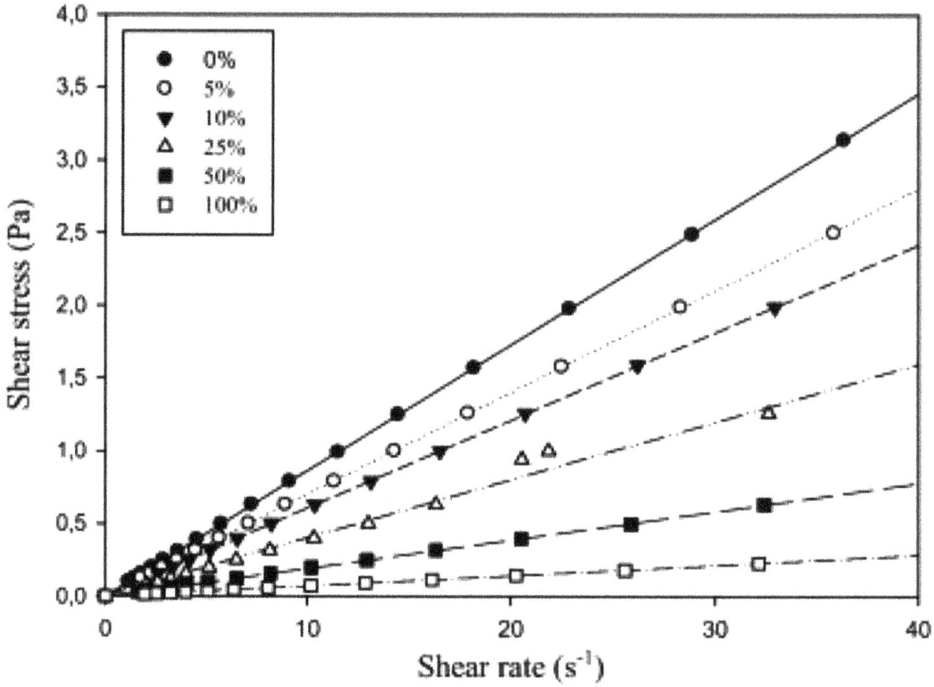

Figure 2.2. Shear stresses of the reaction systems with different PEAAM/HDDA compositions at 20 °C (HDDA vol%). Reprinted from [21], Copyright (2011), with permission from Elsevier.

HDDA. The mixing of HDDA effectively reduced the viscosity of the reaction system, thus accelerating the motion of the reactants and hence raising the polymerization efficiency [23, 29]. In addition, PEAAM is a tri-functional resin and HDDA is a di-functional one. The presence of excessive HDDA would reduce the concentration of function groups of the reactants, so that the network density was lowered. Therefore, the polymerization efficiency was decreased accordingly.

The effect of the addition of the ceramic phase on polymerization kinetic characteristics of PEAAM was evaluated, as the layer thickness was fixed at 150 ± 10 μm, while the light irradiation intensity was set at 5.3 mW · cm^{-2}. The presence of ceramic particles would alter the photocuring behavior of the polymerization systems by influencing their viscosity and optical characteristics. The effect was dependent on type, content, size and morphology of the ceramic particles.

Rheological behaviors of the PEAAM systems with SiO_2 loadings of 10–40 vol% and the polymerization efficiencies as a function of the concentration of different ceramic phases have been studied. In general, the addition of ceramic particles would increase viscosity of the reaction systems. However, it was observed that the addition of 10 vol% SiO_2 had no negative effect on polymerization efficiency of PEAAM, although the viscosity of the reaction system was increased. This observation implies that the polymerization reaction process was independent of the viscosity of the system within the concentration range of the ceramic phase.

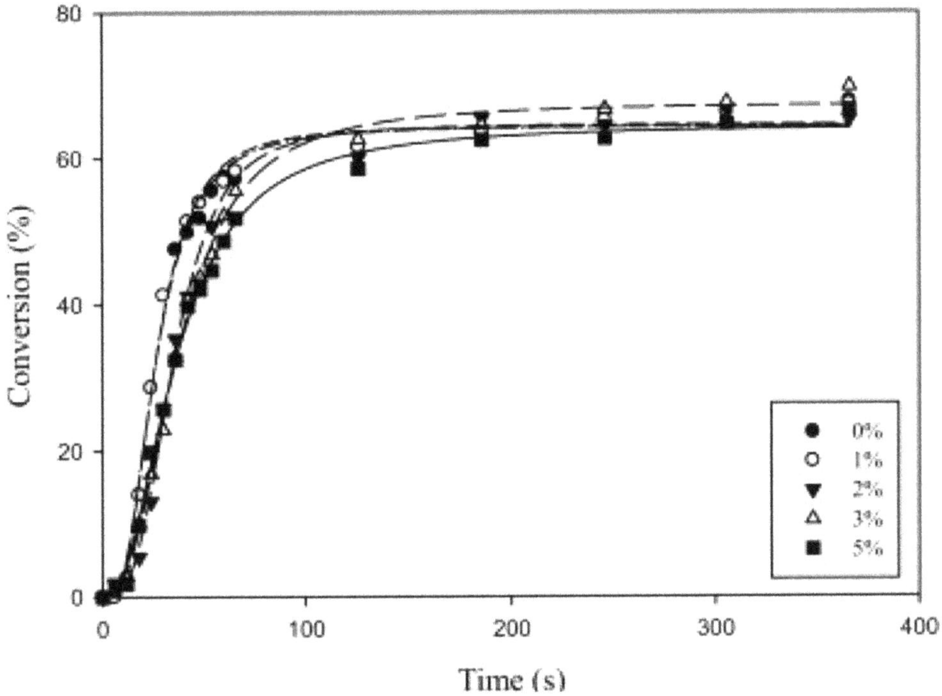

Figure 2.3. Conversion rates of the systems with PEAAM + 0.5 wt% DMPA containing 40 vol% SiO_2, having different weight percentages of dispersant, at light irradiation intensity of 5.3 mW · cm^{-2}. Reprinted from [21], Copyright (2011), with permission from Elsevier.

To verify such a hypothesis, the SiO_2/PEAAM system with 40 vol% SiO_2 was incorporated with phosphate ester as a dilute agent, which altered the interacting behavior of the ceramic particles inside the suspensions. The phosphate ester had a strong effect on viscosity of the reaction systems, displaying shear thickening characteristics that accelerated with increasing content of the ester. Nevertheless, the concentration of the phosphate ester should be $\leqslant$1 wt%. Otherwise, the molecular chains of the phosphate ester on the ceramic particles would interact, thus significantly increasing the viscosity of the reaction systems [30]. Figure 2.3 shows conversion rates of the systems, consisting of PEAAM + 0.5 wt% DMPA, with 40 vol% SiO_2 ceramic phase, having different weight percentages of dispersant, at light irradiation intensity of 5.3 mW · cm^{-2}. It is confirmed that the kinetics of the polymerization process is almost independent on the content of ceramic particles. Similar effects were observed when the ceramic particles are Al_2O_3, ZrO_2 and SiC.

The reactivity of the systems with ceramic phases was further evaluated by varying viscosity through the introduction of dilute agent (HDDA), with concentrations in the range of 5–25 vol%. At the same time, the content of the ceramic phases was fixed at 40 vol%. Based on the rheological experimental results of the suspensions with 40 vol% SiO_2, it was found that viscosity of the sample was increased to a maximum after the addition of 5 vol% HDDA. However, further increase in the concentration of HDDA led to a reduction in viscosity. Such an observation was out of expectation and not well understood. However, it is believed that the interactions among the three components in the systems would have played a critical role. Nevertheless, the conversion rate of the polymerization reaction was increased with increasing content of HDDA till 15 vol% and then declined above that.

For different ceramic phases, the optical properties influence the polymerization rate of the reaction systems. As expected, the closer the value of index of the ceramic phase to that of the resin, the weaker the effect would be [31]. For instance, since the index ratio of SiO_2 and resin is as low as 1.05, the presence of 10 vol% SiO_2 had nearly no effect on polymerization reactivity of the system. The reduction in conversion rate of the samples with higher concentrations of SiO_2 was attributed to the increase in the degree of scattering caused by the ceramic particles. For Al_2O_3 and ZrO_2, the index ratios are 1.20 and 1.51, respectively, so that their contents should be $\leqslant$10 vol%. In contrast, a small content of SiC could have a much stronger negative effect on the polymerization rate of the system, because of the light absorption of the ceramic particles [15].

Particle size of the ceramic phases also affects the conversion rate of PEAAM. Al_2O_3 powders with mean sizes of $d_{50} = 0.5$, 1.4 and 2.3 μm were examined. For the sample with Al_2O_3 powder of $d_{50} = 0.5$ μm, as the content of the ceramic phase was increased from 10 to 40 vol%, the conversion rate was decreased by about 10%. In comparison, as the particle size of the ceramic phase was $d_{50} = 2.3$ μm, the reduction of the conversion rate was 10%. This is simply because the scattering behavior of the ceramic particles is increased with decreasing particle size [32].

Wozniak *et al* studied viscose behaviors of transparent SiO_2 suspensions for stereolithography applications by using stereolithography techniques through curation with UV irradiation. The SiO_2 suspensions had solid loading levels of >40 vol%, which was realized by minimizing the van der Waals interactions and maximizing the affinity of functional groups of monomers to the surface of SiO_2 particles [9]. The suspensions have high optical transparency because of the matching in refractive index of the ceramic phase and the acrylate monomer, leading to a curing depth of as large as 10 mm.

Commercial SiO_2 nanosized powder (Aerosil OX50), with a specific surface area of 54 $m^2 \cdot g^{-1}$, was used. Figure 2.4 shows a presentative TEM image of the SiO_2 nanosized powder, with particles of a spherical shape and sizes in the range of 40–80 nm. Five monomers (4HBA, HEA, butyl acrylate, M200 and M282-PEGDA) were evaluated for dispersion of the ceramic phase, while three photoinitiators (Genomer ITX, Genocure TPO and Genocure LTM) were examined for the UV curation efficiency.

The monomers were used because their refractive indices were comparable with that of SiO_2, so that the van der Waals force could be effectively reduced and the ceramic-monomer suspensions had sufficiently high optical transparency. The di-functional M200 and PEGDA have dual functional groups, which facilitated strong cross-linking reactions, resulting in stronger of the polymerization product. In comparison, 4HBA and HEA contain a single functional polar OH group, similar

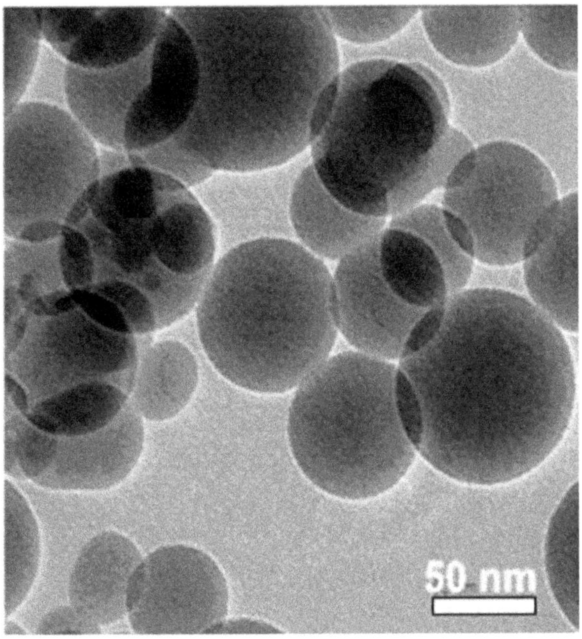

Figure 2.4. Representative TEM image of the SiO_2 nanosized powder (OX50). Reprinted from [9], Copyright (2009), with permission from Elsevier.

to the termination of butyl acrylate, allowing us to compare the polarity in influencing the dispersion efficiency. The photoinitiators have close UV absorption characteristics to the monomers and hence were selected for the study.

In fact, the loading level of the ceramic phase in the suspensions is not only determined by the matching in refractive index. It is also dependent on other factors. For instance, although polyethylene glycol 200 diacrylate (PEGDA) and 1,6-hexanediol diacrylate (Miramer M200) have index matching with SiO_2, the viscosity of the suspensions containing 15–18 vol% ceramic phase was higher than the limit of SLA processing, which is 5 Pa · s at 30 s^{-1} [33]. Therefore, it is also necessary to pay attention to the hydrophobic and hydrophilic characteristics of both the solvent and the ceramic phase to be dispersed. In this regard, 4HBA and HEA are more promising, because they have OH groups that can be attached onto the surface of SiO_2, thus achieving high loadings with relatively low viscosity.

It was found that the polymerization efficiency is not sufficient to ensure mechanical strength of the final products, if monomers with a single functional group are used, owing to the low density of cross-linking. To address this problem, a mixture of monomers with different functional groups was proposed. For instance, when 4HBA or HEA was mixed with 7 vol% PEGDA, optimal results could be achieved. Figure 2.5 shows viscosities of the reaction systems consisting of the mixed monomers and SiO_2. In both systems, the viscosity is increased with increasing content of SiO_2, with the one of HEA to be slower. For example, viscosity values of the 4HBA and HEA suspensions with 40 vol% SiO_2 were 1.75 and 0.5 Pa · s,

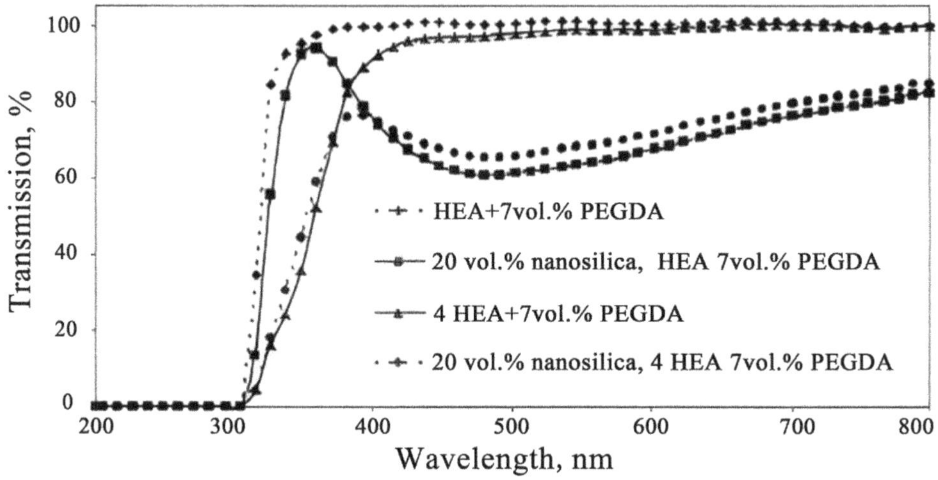

Figure 2.5. UV–vis spectra of the SiO_2 suspension and solutions of monomers. Reprinted from [9], Copyright (2009), with permission from Elsevier.

respectively. If butyl acrylate was used to disperse SiO_2, the concentration of the ceramic phase could not be over 10 vol%, to maintain a sufficiently low viscosity for SLA process. It is therefore concluded that the dispersion efficiency of ceramic phase could be enhanced through modification with high polarity functional groups or shortening of hydrocarbon spacers.

According to optical transmittance experimental results in the UV–vis regime, transparencies of the monomer mixtures (HEA/4HBA + 7% PEGDA) were both >90%, with cut-off wavelengths of 320 and 380 nm, respectively. Correspondingly, the suspension with 20 vol% SiO_2 nanosized particles exhibited optical transparency that was lower than those of the monomer mixtures in the UV regime by just 5%–20%.

Various samples were derived from suspensions with SiO_2 and 5% photoinitiator in the mixtures of HEA/PEGDA and 4HBA/PEGDA, with curing times of 15–25 and 40 s for those from the suspensions of HEA and 4HBA, respectively. Also, samples from the suspensions with TPO and LTM photoinitiators were mechanically much stronger than those made with ITX.

Dilatometry and TG/DTA were employed to characterize the printed green bodies from the suspensions with 40 vol% SiO_2 in 4HBA and PEGDA. There was an expansion of about 2% near 300 °C, which was ascribed to the thermal behavior of the polymers in the green bodies. Three weight losses were observed in the TGA curves. The first one was about 10%, corresponding to the expansion due to the decomposition of the polymers. The second weight loss was as high as 25%, over 250 °C–450 °C. It was accompanied by a shrinkage of about 7%. The last weight loss was finished at about 570 °C. The total weight loss was 40.5%, which was just slightly lower than the theoretical value of 42%. The shrinkage above 550 °C was attributed to the sintering of the samples. In addition, after sintering at 1250 °C, the relative density approached 99%.

Hinczewski *et al* used SLA to fabricate Al_2O_3 ceramic parts with complicated shapes from alumina suspensions with ceramic loading of 53 vol% [34]. The green bodies with complicated shapes were developed through layer-by-layer laser polymerization of the monomer in the suspensions with an Ar ionized laser. It was found that the components of the suspensions all had influence on their rheological properties. The viscosity of the slurries could be properly reduced by adding dispersants and diluents and increasing temperature, so that the SLA printing process could be suitably conducted. The suspensions displayed shear thinning behaviors well meeting the requirements of layer casting. The curing depth was about 200 μm and high scanning rates were allowed during the SLA process.

Acrylamide is an environmentally friendly constituent for 3D printing, which has attracted strong attention in additive manufacturing of polymers. Naturally, it has been widely used as a matrix for 3D printing of ceramic materials. For instance, Wang *et al* used acrylamide suspension of SiO_2 to make SiO_2 ceramics [35]. The authors examined the effects of type of dispersant, content of solid and mixing time on physical properties of the SiO_2 acrylamide suspensions. The dependence of curing behaviors of the suspensions on concentration of photoinitiator and the ratio of monomer and cross-linker were elucidated. SiO_2 ceramics were finally obtained after sintering at 1200 °C, exhibiting a density of $1.42\ g \cdot cm^{-3}$ and bending strength of about 13 MP.

Specifically, the monomer and cross-linker were dissolved in a mixed solution of pure water and glycerol to form a UV curable solution, to which SiO_2 powder with an average diameter of about 2 μm and specific surface area (SSA) of $5\ m^2 \cdot g^{-1}$ was added at a volumetric concentration of about 57%, resulting in 3D printing suspensions through ball milling. Four types of dispersants were used, including ammonium polyacrylate, sodium polyacrytale, polyvinyl pyrrolidone and polyacrylic acid. Ball milling time on performances of the suspensions was optimized.

Figure 2.6 depicts a schematic diagram to describe the principle for the 3D printing process, via a computer-aided design (CAD) model with slicing processing, by projecting onto the vat bottom with a digital micromirror device (DMD). The SiO_2 suspension was cast onto the vat in films, which were solidified owing to the strong bonding caused by the cross-linked polymer matrix through polymerization of the monomer after UV irradiation. The 3D printing process was conducted in a layer-by-layer way, to obtain SiO_2 green bodies, according to the predesign of the final ceramic parts.

To minimize the defects and anisotropic stress of the green bodies, it is necessary to optimize the properties of the suspensions with suitable viscosity by adjusting the solid content of the ceramic powders. As observed in figure 2.7, viscosity of the suspensions increased in an exponential manner, with increasing content of the SiO_2 powder, except for the ammonium polyacrylate. As a result, ammonium polyacrylate was selected for the experiment to ensure sufficiently high solid loading level. In addition, viscosity of the suspensions was initially reduced as the rotating speed was increased, followed by final stabilization, due to the effect of shearing thinning fluid.

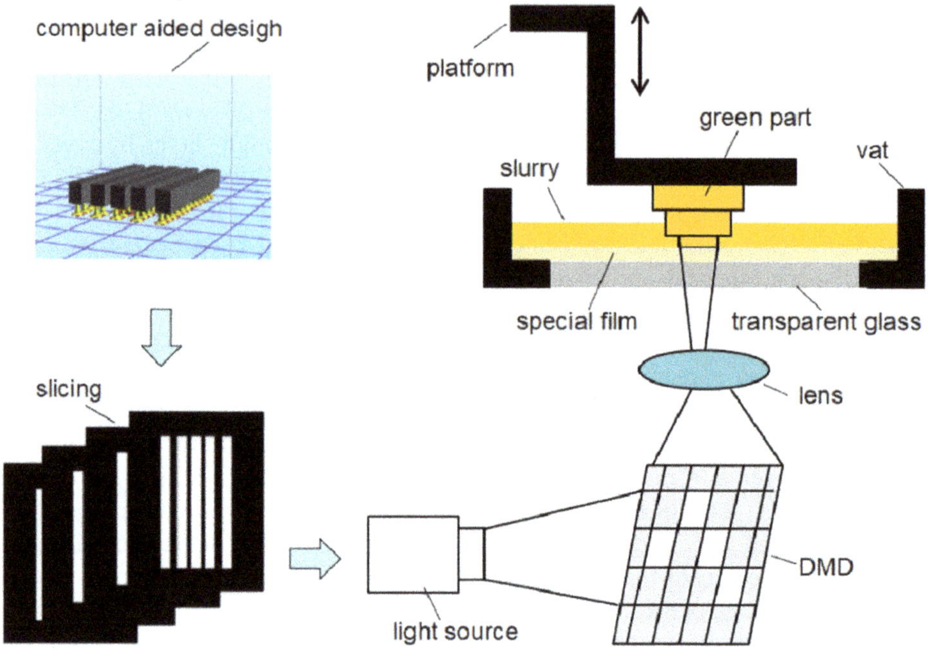

Figure 2.6. Schematic diagram showing working principles of the 3D printing process. Reprinted from [35], Copyright (2019), with permission from Elsevier.

More interestingly, the particle of the SiO_2 powder was reduced to about 1.2 μm from the original 2 μm, which is beneficial for the penetration depth of UV radiation [36].

Figure 2.8 shows viscosity of the SiO_2 suspensions versus ball milling time duration. The viscosity exhibits a decrease trend first, followed by a slight increase, with ball milling time. Without ball milling for sufficiently long time, the SiO_2 particles were relatively large in size, along with possible aggregation, thus leading to high suspension viscosity. With ammonium polyacrylate dispersant, the suspension has the lowest viscosity after ball milling for 20 h, which were used as optimal experimental conditions.

Suspensions with 42 vol% SiO_2 were prepared, by controlling the concentrations of ammonium polyacrylate in the range of 0–2.5 wt%, to optimize its content of ammonium polyacrylate. Initially, with increasing concentration of ammonium polyacrylate, viscosity of the suspensions was almost linearly decreased, reaching the minimum point at 1 wt% and then rising. Therefore, the optimal concentration of ammonium polyacrylate is 1 wt% in this case.

The content of photoinitiator also exerted an impact on the properties of the suspensions. It was found that 2 wt% photoinitiator is sufficient for complete curing of the printed layers, under the UV irradiation at 405 nm. This is simply because excessive photoinitiator would be accumulated on the surface of printed layers to hinder the absorption of UV light by the materials. As a result, the optimal

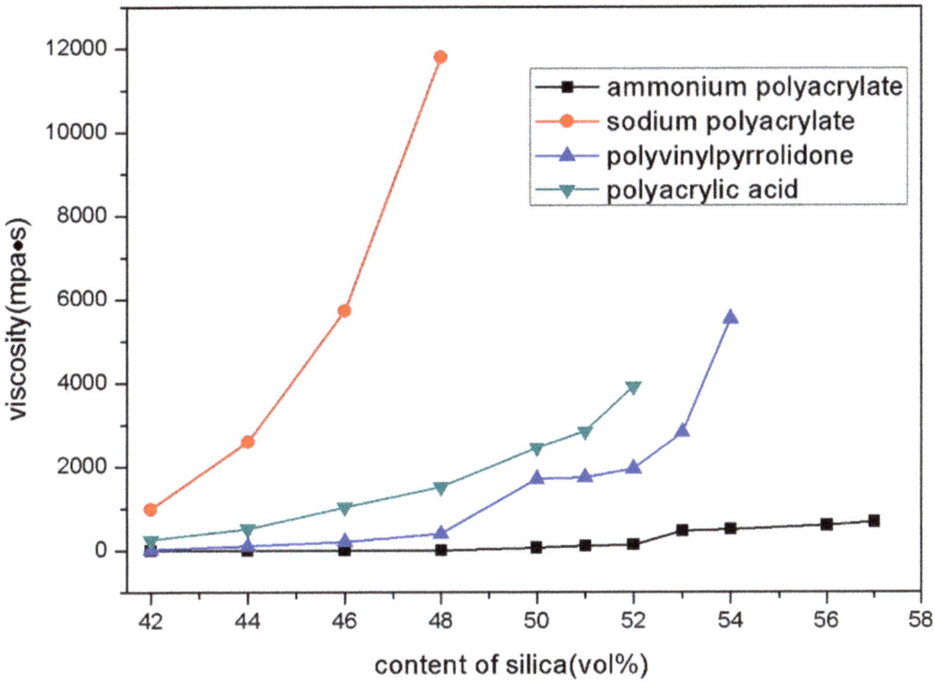

Figure 2.7. Relationships between viscosity of the suspensions and content of the silica. Reprinted from [35], Copyright (2019), with permission from Elsevier.

concentration of photoinitiator was 2 wt%. All other parameters were also optimized from the point of view of suspension performances.

Dependances of densification of the 3D-printed SiO_2 green bodies and porosity of the final SiO_2 ceramics on sintering temperature were systematically studied. Additionally, the sintering temperature also had an influence on phase composition of the SiO_2 ceramics. Single phase quartz is present in the samples after sintering at temperatures of $\leqslant$1200 °C. In the sample sintered at 1250 °C, phase transition from quartz to cristobalite occurred. As the sintering temperature was increased to 1250 °C, the samples were of single phase cristobalite.

Manière *et al* used stereolithography printed SiO_2 as an example to establish a model describing the anisotropic sintering behaviors of 3D-printed ceramic materials [37]. The modeling was based on experimental data. For instance, SEM results indicated that the samples sintered at different temperatures exhibited anisotropic characteristics, with different non-ideal porosity distribution in different printed layers, as shown in figure 2.9. Three aspects have been identified. Firstly, the building direction (*z*-axis) displayed larger shrinkage during sintering than others. Secondly, the anisotropic behavior was gradually leveled off, as the pores were eliminated, at certain temperatures, after which isotropic densification behavior was present, at the final stage of the sintering process. Lastly, coalescence took place for the pores at the layer interface. Once large pores are formed, it is difficult to remove by using pressureless sintering technology.

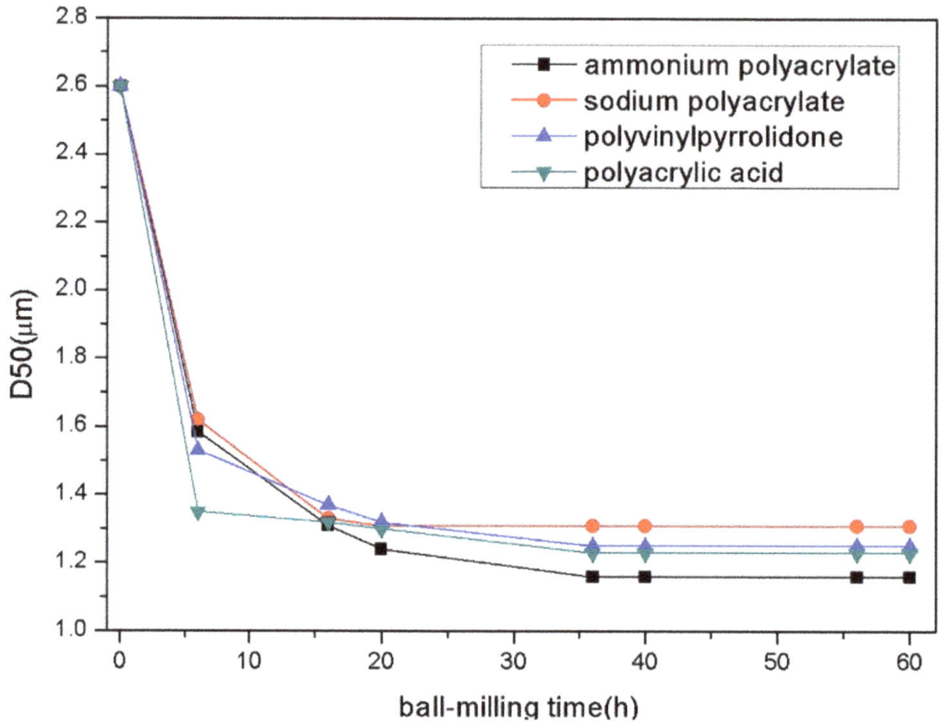

Figure 2.8. Variation in silica particle size with different ball milling time durations. Reprinted from [35], Copyright (2019), with permission from Elsevier.

Zhou *et al* examined the effects of composition and laser exposure intensity on curing performance of aqueous SiO_2 suspensions with relatively high concentrations [38]. The curing behaviors of the aqueous ceramic suspensions were characterized by evaluating curing depth and width of the basic building unit-printed lines. The depth of the lines was increased, but the width was decreased with increasing solid loading level of the silica suspensions. As expected, both the depth and width of the lines were decreased with increasing the scanning rate of the laser.

Two SiO_2 powders were used, with average particle sizes of 1.55 and 9.34 μm. The monomer was acrylamide (AM) and cross-linker was methylenebisacrylamide (MBAM), which were employed as a solute, while the mixture of SiO_2 sol and glycerol were utilized as solvent. The dispersant was sodium polyacrylate, while the photoinitiator was photocure-1173. The volumetric concentration of SiO_2 in the suspensions was 50%.

The laser used for the printing experiment had a maximum power of 270 mW, with a beam diameter of 140 μm. Since aqueous silica suspensions were insensitive to the shining of the laser beam in terms of curing, single curing lines tended to be distorted, when they were taken out from the SiO_2 suspensions and washed with DI-water. To tackle this issue, a single line mesh was proposed, which had a reinforced boundary. In this way, the integral strength of the mesh layer was enhanced and

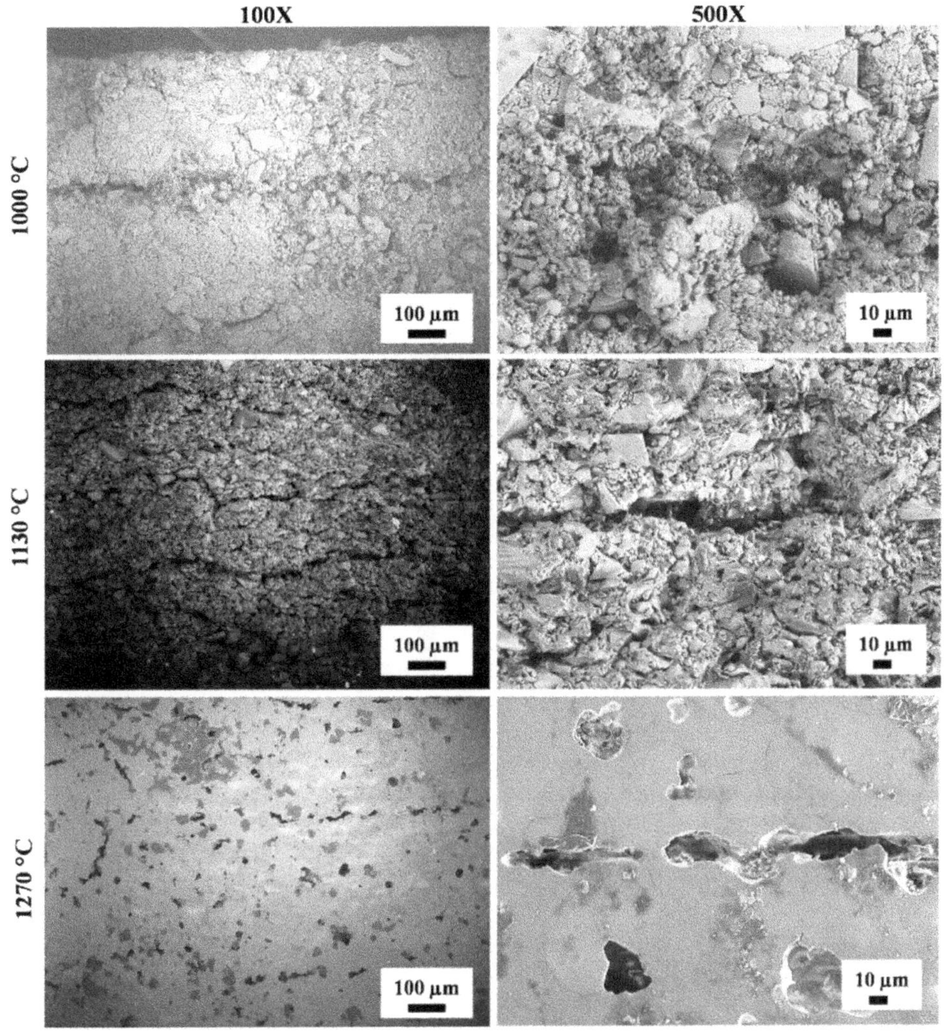

Figure 2.9. SEM images of the cubic samples sintered at different temperatures at a heating rate of 3 °C · min^{-1}. for all images, with building direction (*z*-axis) to be from bottom to top. Reprinted from [37], Copyright (2020), with permission from Elsevier.

distortion of the single lines was effectively prevented. The mesh layers were then removed after finishing the printing process.

Figure 2.10 shows the schematic diagram of the printing process, which involves four steps, including design of a three-dimensional structure with a concave structure with CAD software, addition of support for the structure, curing of the layer in the three-dimensional structure and taking out of the mesh layer. It was experimentally observed that the structure of the curing lines was well retained due to the reinforcement effect of the mesh.

Figure 2.11 shows curing effects of two printed lines, i.e., photosensitive resin and silica suspension. The line from the silica suspension was smoother than that from

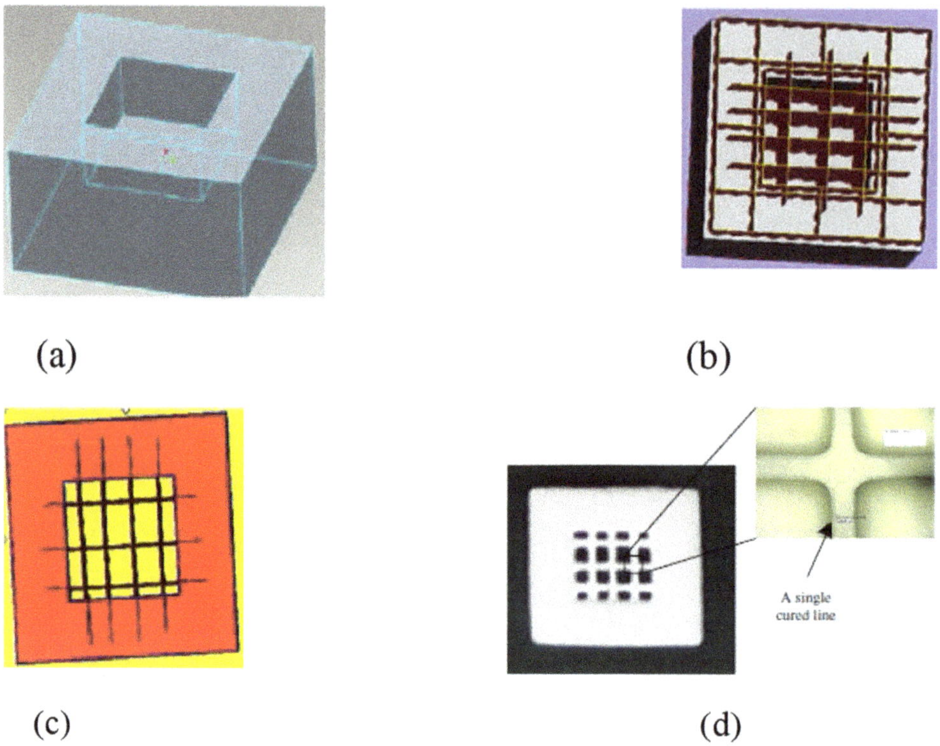

Figure 2.10. Fabricating process of the single curing lines: (a) design of a three-dimensional structure with an inner concave, (b) addition of supports, (c) a layer of the mesh structure for curing and (d) rinsing and drying the single curing line mesh. Reproduced from [38], with permission from Springer Nature.

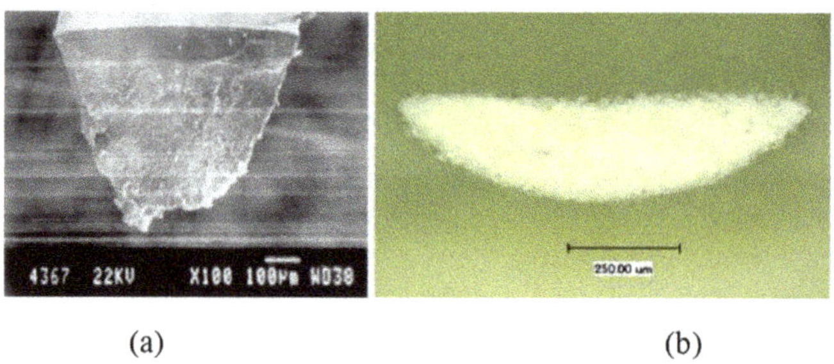

Figure 2.11. Cross-sectional view of a single curing line: (a) photosensitive resin and (b) silica suspensions. Reproduced from [38], with permission from Springer Nature.

the photosensitive resin. The difference in curing effect between the two lines could be explained by considering laser scattering of SiO_2 particles [14]. The propagation direction of the laser beam was altered, owing to the scattering effect of the SiO_2 particles, thus attenuating the intensity of the irradiation laser. As a consequence, the curing in the width direction was increased, while that in the depth direction was comparatively reduced. It is well known that the curing effect of conventional photosensitive resin is only dependent on phososensitivity of the resin and the intensity of the laser irradiation, without the scattering of ceramic particles.

The scattering characteristics of ceramic particles in a suspension are related to several factors, including volumetric concentration of the suspensions, diameter of the ceramic particles, concentration of monomers, intensity of laser irradiation and so on [39]. The classical scattering theories are most likely applicable to suspensions with quite low solid loading levels. For ceramic suspensions with high solid contents, both multiple scatterings and correlated scatterings should be taken into account. Therefore, the tradition scattering theories would not be validated. In factor, the scattering behaviors of the ceramic particles in concentrated suspensions have not been well elucidated till now [40–42].

Zheng *et al* prepared SiO_2 ceramics complexed with Al_2O_3 by using stereolithography, with the effect of content of Al_2O_3 on mechanical properties of the SiO_2 ceramics to be studied [43]. With the presence of Al_2O_3, the mechanical properties of the SiO_2 ceramics were enhanced, because the crystallization behavior of fused SiO_2 phase to cristobalite was promoted, owing to the seeding effect of Al_2O_3. In addition, with increasing content of Al_2O_3, linear shrinkage of the green bodies in the sintering process was first decreased and then increased, while both the room temperature and high temperature flexural strengths were first increased and then decreased. Since the presence of Al_2O_3 would hinder the viscous flow of SiO_2, the sample with 1 vol% Al_2O_3 had a linear shrinkage of only 1.6%. Accordingly, the room-temperature and the high-temperature flexural strengths were raised to 20.38 and 21.43 MPa, respectively. Such Al_2O_3 enhanced SiO_2 ceramics could be used as ceramic cores of hollow blades.

SiO_2 powder with nearly spherical particles and average particle size of 2 μm and Al_2O_3 powder with an average particle size of 3 μm were used to prepare ceramic suspensions. Photosensitive resin HDDA, polymer dispersant 41000 and photoinitiator diphenyl (2,4,6-trimethyl benzoyl) phosphine oxide were employed as polymer matrix and additives. Solid loading levels of the ceramic powders were 45% in volume. The concentrations of Al_2O_3 were up to 2 vol%. To obtain ceramic suspensions, the mixtures were ball milled for 30 min. Green bodies were printed with an optimal laser power of 0.75 W, the scanning rage of 3000 mm $\cdot$ s^{-1} and printed layer thickness of 50 μm. The printed green bodies were sintered at 1200 °C for about 6 h.

Figure 2.12 shows SEM images of the sintered SiO_2 ceramics with contents of Al_2O_3. All samples exhibited porous structure with relatively large grain sizes, independent on the content of Al_2O_3, as presented in figures 2.12(a)–(e). EDS spectrum of the ceramic sample with 1.5 vol% is shown in figure 2.12(f), demonstrating the presence of Al_2O_3. Additionally, the amorphous SiO_2 was crystallized forming a 3D network structure, simply because the crystallization temperature of fused silica is lower than the sintering temperature [44–46]. In addition, the introduction of Al_2O_3 slightly hindered the connection of the SiO_2 grains. Besides the crystallization and phase transition from β-cristobalite to α-cristobalite, mullite ($3Al_2O_3 \cdot 2SiO_2$) phase was formed due to the reaction of SiO_2 and Al_2O_3 [47, 48].

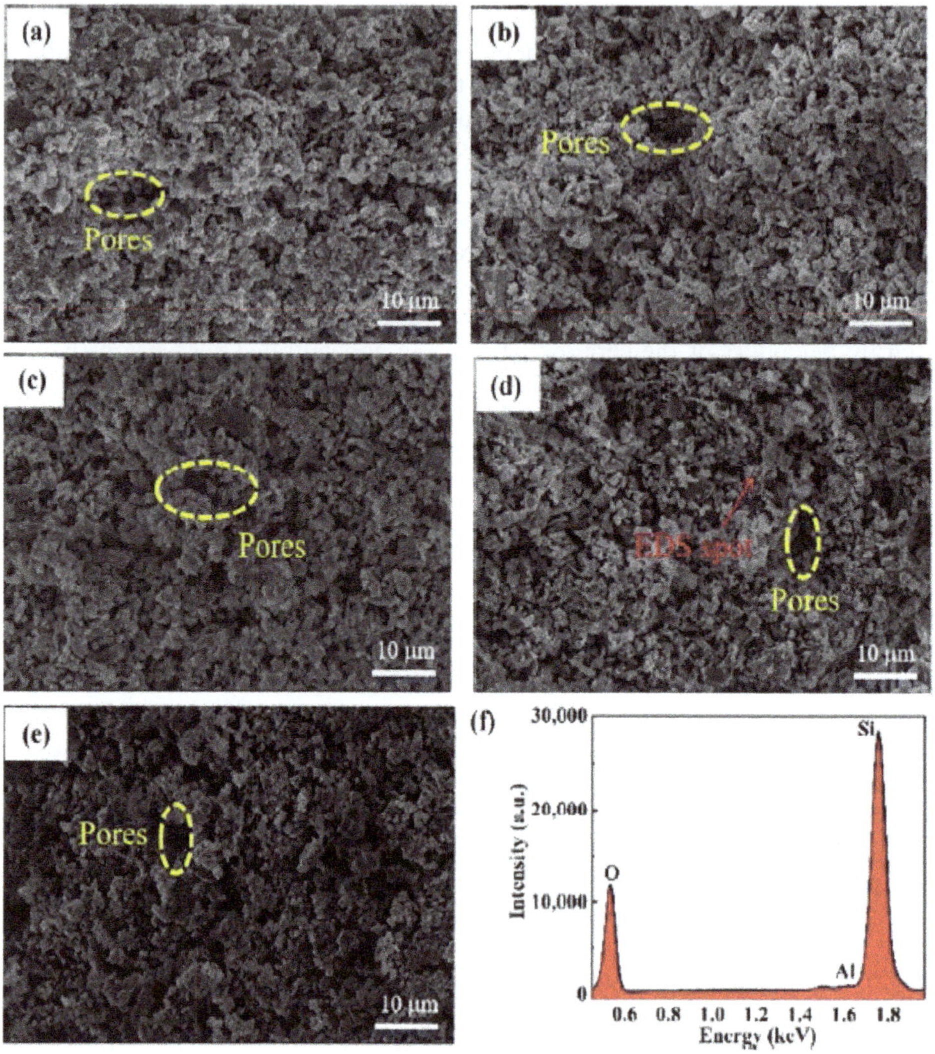

Figure 2.12. SEM images of the SiO_2 ceramics with different volumetric concentrations of Al_2O_3: (a) 0, (b) 0.5, (c) 1.0, (d) 1.5 and (e) 2.0, (f) EDS spectrum of the sample with 1.5 vol% Al_2O_3. Reproduced from [43].

Linear shrinkage rates of the SiO_2 ceramics with different concentrations of Al_2O_3 are depicted in figure 2.13(a). The linear shrinkage decreased from 2.6% to 1.6%, as 1.0 vol% Al_2O_3 was used. The decrease in linear shrinkage of the green bodies was attributed to the effect of the crystallization of the fused silica phases at high temperatures, induced by the crystalline phases, i.e., Al_2O_3 and cristobalite SiO_2. Also, because of the occurrence of the phase crystallization, the viscous flow induced densification was slowed down, which was the driving force of the sintering process [49–51]. After further increase in the content of Al_2O_3, the linear shrinkage was increased, which is ascribed to the promoted grain growth of the Al_2O_3 ceramics.

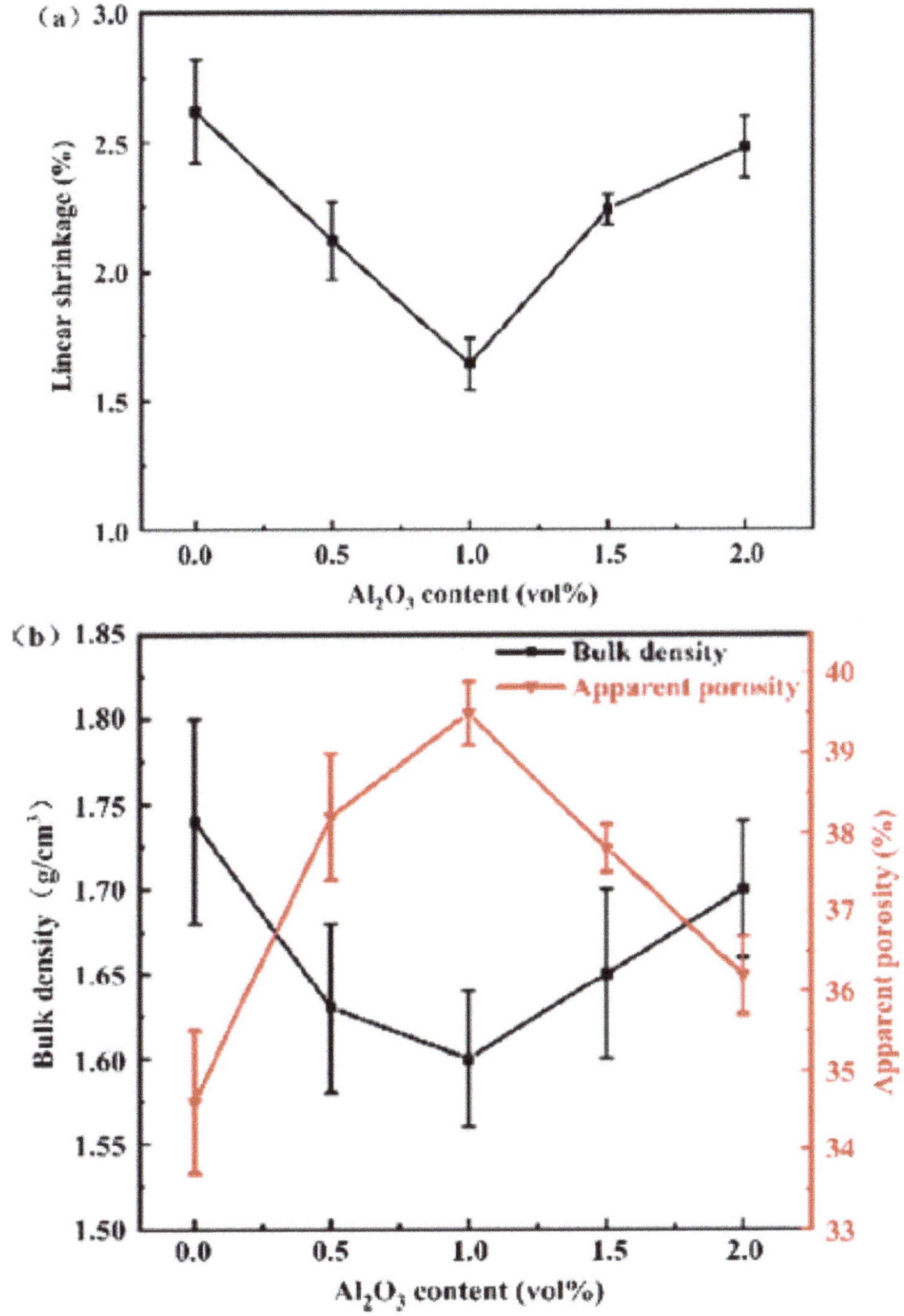

Figure 2.13. Densification and physical properties of the SiO_2 ceramics as a function of the content of Al_2O_3: (a) linear shrinkage and (b) density and porosity. Reproduced from [43]. CC BY 4.0.

Density and the porosity of the SiO_2 ceramics after sintering at 1200 °C are illustrated in figure 2.13(b). With the addition of 1 vol% Al_2O_3, β-cristobalite with cubic structure was present during the high temperature sintering process, which was transferred to α-cristobalite at about 250 °C during the cooling process [52–54]. The phase transformation resulted in a decrease by 5% in volume, which was possibly caused by the propagation of microcracks in the SiO_2 ceramics. The current effect of the phase transition and crystallization led to increase in porosity and decrease in density. The trend in porosity is inverse to that of the density.

Figure 2.14 shows flexural strengths of the SiO_2 ceramics measured at room temperature and 1550 °C, as a function of the content of Al_2O_3. The trend of flexural strength of the ceramics is opposite to that of the linear shrinkage of the green bodies. The room temperature flexural strength increased from 11.6 to 20.4 MPa, corresponding to the increase in the content of Al_2O_3 to 1 vol%. On one hand, noting that the cristobalite is stronger than amorphous silica, it is not surprising that the crystallization of the fused silica from amorphous state to cristobalite phase would result in increase in flexural strength [55–57]. On the other hand, the room temperature flexural strength of the ceramics decreased once the content of Al_2O_3 exceeded 1.5 vol%. This implies that excessive crystalline phase of cristobalite would bring down mechanical strength of the SiO_2 ceramics. It has been reported that as the content of the cristobalite is over 30%, the flexural strength of the fused silica

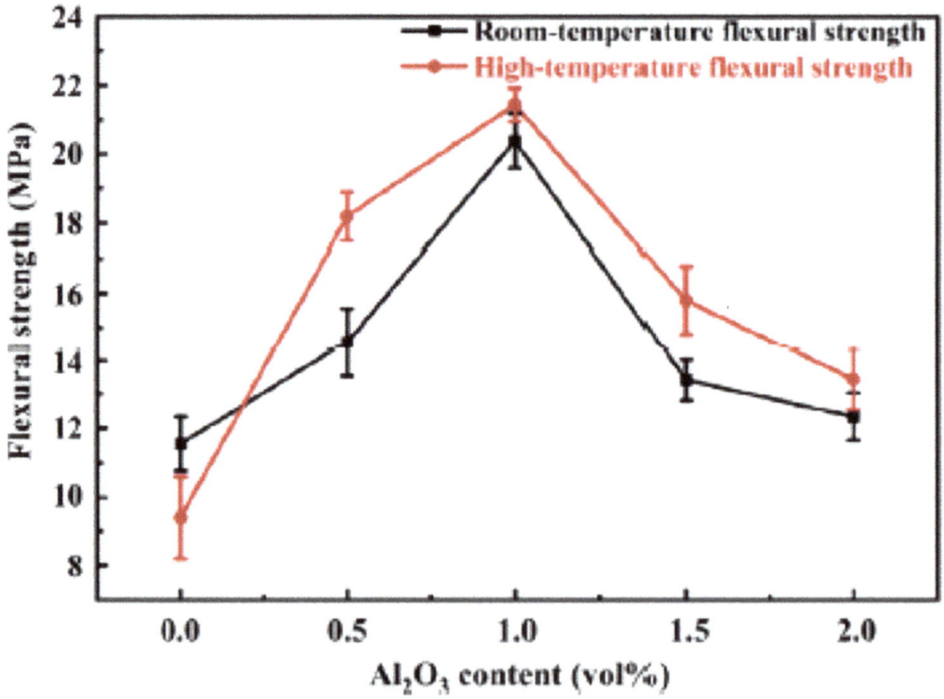

Figure 2.14. Room-temperature and high-temperature flexural strength of SiO_2 ceramics versus the concentration of Al_2O_3. Reproduced from [43]. CC BY 4.0.

ceramics tends to decline [55]. Therefore, overcrystallization should be avoided. In other words, the doping concentration of Al_2O_3 should be optimized. The variation trend in high-temperature mechanical strength is similar. Correspondingly, the high-temperature flexural strength was raised from 9.4 to 21.4 MPa as the content of Al_2O_3 was increased up to 1 vol%.

Yin *et al* fabricated green bodies of SiO_2-based ceramic cores, with the addition of ZrO_2 nanoparticles to enhance densification and mechanical strength, by using SLA 3D printing technology [58]. The effects of burying process of printing and heating rate of debinding on densification behavior and microstructures of SiO_2-based ceramic cores were explored. The burying powders acted as supports and offered stable thermal fields for the printed ceramic cores in the processes of both debinding and sintering. As the heating rate was 0.5 °C · min^{-1}, the SiO_2 ceramic cores exhibited relatively low shrinkage and high porosity during the process of burying. With the presence of the ZrO_2 nanoparticles, the densification of the green bodies and mechanical performances of the final ceramics were enhanced. Moreover, cracks were induced on the surface of the ceramics. The content of the ZrO_2 nanoparticles was optimized in the range of 1.5–2.0 wt%, corresponding to shrinkage rates of 2.9%–3.2%, apparent porosity of 27.1%–27.97%. Meanwhile, high bending strengths at room temperature and high temperature were 24.51–24.65 and 27.81–29.69 MPa, respectively.

The SiO_2 powder and ZrO_2 nanopowder used to prepare the slurries for printing the ceramic green bodies had D_{50} = 10 μm and D_{50} = 50 nm, respectively. Photosensitive resin, Solsperse 41000 dispersant and TPO photoinitiator were employed as matrix and additives for the slurries. Figure 2.15 shows a schematic diagram illustrating fabrication process of the ceramic green bodies by using SLA 3D printing. Solid loading of the ceramic slurries was 78% in weight percentage. The content of the ZrO_2 nanoparticles was in the range of 0.5–2.5 wt%. The laser power, the scanning speed and the layer thickness were 150 mW, 3000 mm · s^{-1} and 50 μm, respectively. The heating rates were set to be 0.1, 0.5, 1.0 and 2.0 °C · min^{-1}.

Cross-sectional SEM images of the ceramics after sintering at different heating rates are shown in figure 2.16. The sintered ceramics had relatively high porosity.

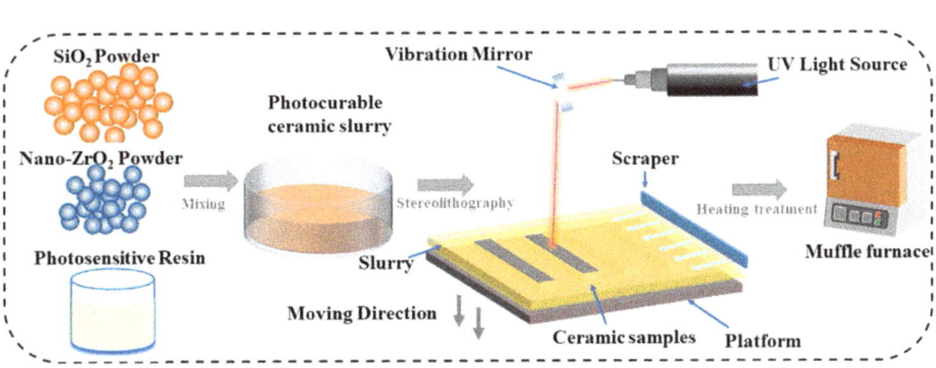

Figure 2.15. Flow chart of preparation of the SiO_2-based cores by using SLA 3D printing process. Reprinted from [58], Copyright (2023), with permission from Elsevier.

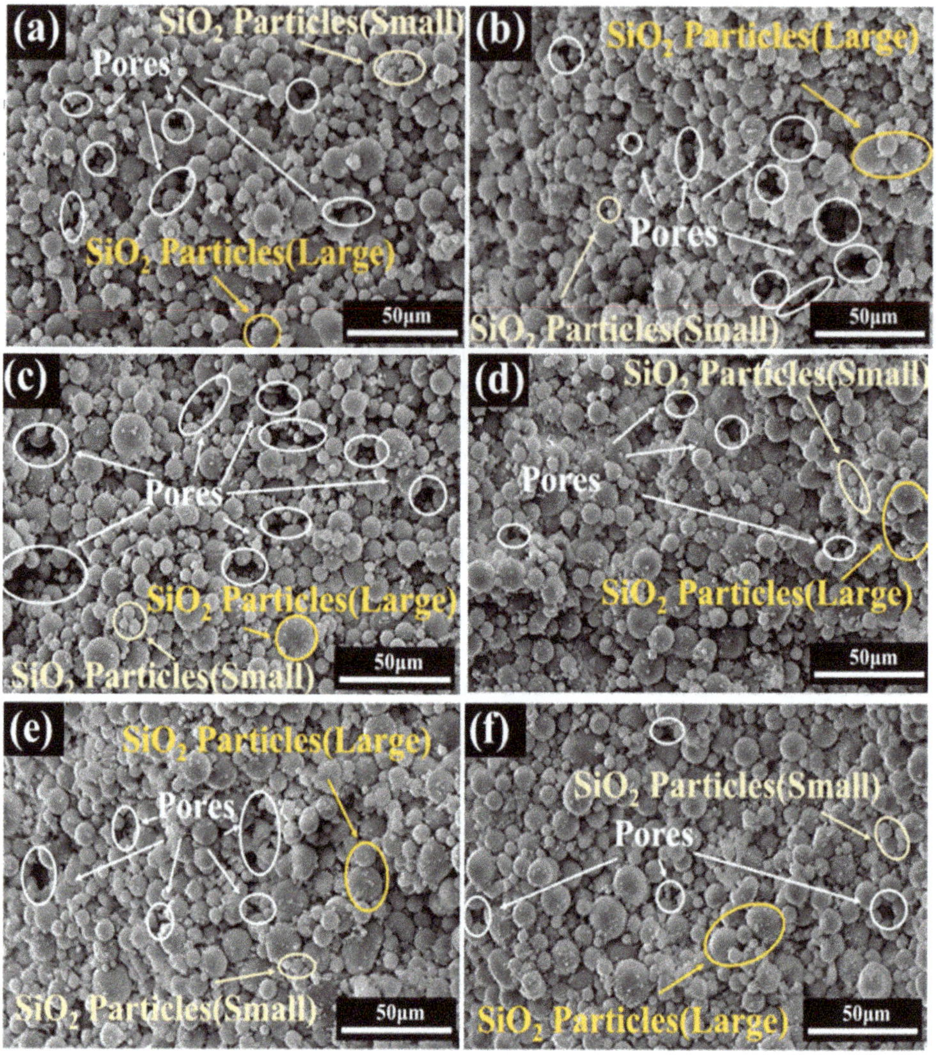

Figure 2.16. Cross-sectional SEM images of the SiO_2 ceramics processed at different heating rates with (a, b, c) and without (d, e, f) burying: (a, d) 0.1 °C · min^{-1}, (b, e) 0.5 °C · min^{-1} and (c, f) 1.0 °C · min^{-1}. Reprinted from [58], Copyright (2023), with permission from Elsevier.

The grains of different sizes were tightly compacted. After sintering at the heating rate of 1.0 °C · min^{-1}, the densification was optimized. The linear shrinkage in the z-direction, as the building direction in the SLA printing process, was larger than those in the x- and y-directions, since intralayer bonding was relatively weak [59, 60]. Independent of heating rate, all the ceramic samples for the burying process experienced smaller linear shrinkage than those for the non-burying process.

For instance, after sintering at 0.5 °C · min^{-1}, the linear shrinkage in the z-direction of the buried sample was 2.7%, whereas that of the non-buried sample was 4.0%. The linear shrinkage of the buried samples displayed relatively slight variation as the heating rates of <1.0 °C · min^{-1}, while it was increased to 3.2%, once the heating rate was raised to 2 °C · min^{-1}. However, the linear shrinkage of the non-buried samples was increased to 4.9%, at the heating rate of 1.0 °C · min^{-1}. However, it was slightly reduced to about 4.5%, as the heating rate was increased to 2 °C · min^{-1}.

The buried samples had higher porosity than the non-buried ones, disregarding the heating rate. With increasing heating rate, the porosity of the non-buried samples was decreased to 26% at the heating rate of 1.0 °C · min^{-1}, while it was increased to a relatively low extend. As the heating rate was lower than 1.0 °C · min^{-1}, the porosity of the buried samples was higher than 30%, while it was arrived at a maximum level of 30.7%, at the heating rate of 0.5 °C · min^{-1}. Density of the samples exhibited a variation trend that was opposite to that of their porosity. The density of the non-buried samples was slightly larger than that of the buried samples, regardless of the heating rate.

At a heating rate of <1.0 °C · min^{-1}, the non-buried samples possessed higher flexural strength than the buried ones. This is because the non-buried ceramics had denser microstructure, thus having stronger load resistance. After sintering at a heating rate of 1.0 °C · min^{-1}, flexural strength of the non-buried sample was maximized to 19.6 MPa, which was decreased to about 3.6 MPa thereafter. The sample processed at 2 °C · min^{-1} had the lowest flexural strength, mainly because of the absence of the supports and the rapid rise in temperature during the sintering process. As for the buried samples, the room temperature flexural strength increased from the lowest value of 11.1 MPa for the heating rate of 0.5 °C · min^{-1} to 12.8 MPa for 1.0 °C · min^{-1}.

The buried samples had smoother surfaces than the non-buried ones. In addition, the buried ceramics exhibited more microcracks, owing to the connected pores on their surfaces. Comparatively, the surface of the non-buried ceramics had hill-like undulations. Accordingly, the buried samples exhibited smaller surface roughness than the non-buried ones, regardless of the heating rate. With increasing heating rate, the top surface and side surface roughness values of the buried samples increased from 1.1 to 5.7 μm and from 3.02 to 7.7 μm, respectively. In contrast, the roughness of the non-buried ceramics remained almost unchanged. The side surface roughness was larger than that of the top surface, because of the characteristics of the printing layers.

According to the previous results, the effect of ZrO_2 nanoparticles on properties of the SiO_2 ceramics was conducted by focusing on the samples sintered at the heating rate of 0.5 °C · min^{-1}. After the introduction of the ZrO_2 nanoparticles, the ceramics exhibited microstructures with large grains to be densely compacted and small grains to be attached to large ones. Meanwhile, porosity of the ceramics was significantly decreased.

Figure 2.17 shows surface SEM images of the SiO_2 ceramics with different contents of ZrO_2 nanoparticles, indicating that their surfaces are relatively smooth, although small cracks are present. In figure 2.17(d), with the addition of 2.5 wt% ZrO_2 nanoparticles, the densification of the green bodies was greatly enhanced, thus

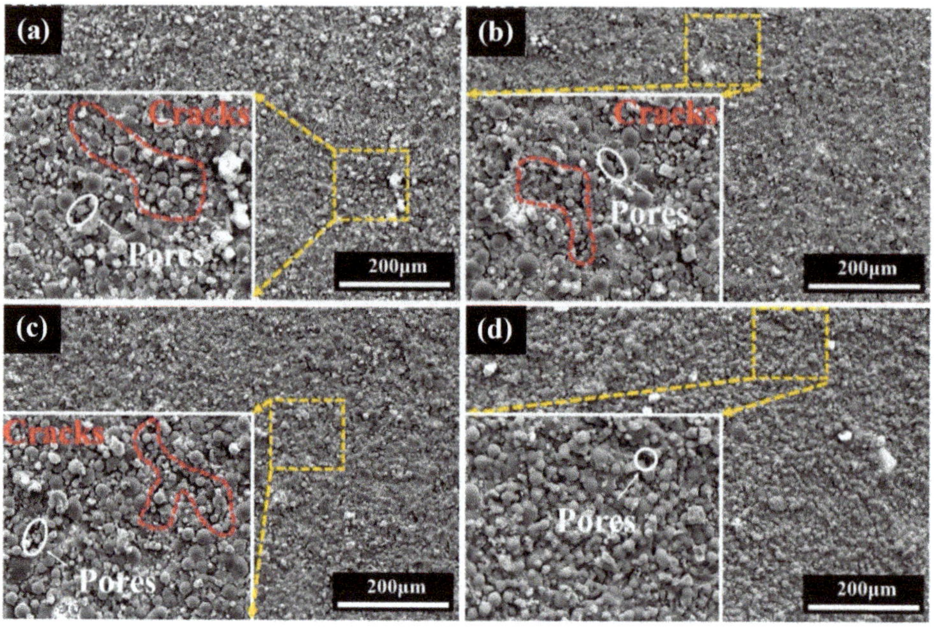

Figure 2.17. Surface SEM images of the SiO_2 ceramics with different contents of nano-ZrO_2: (a) 1.0 wt%, (b) 1.5 wt%, (c) 2.0 wt% and (d) 2.5 wt%. Reprinted from [58], Copyright (2023), with permission from Elsevier.

resulting in compact microstructure of the surface. In this case, the pores are distributed in a relatively isolated way, while all the pores were not interconnected.

The linear shrinkage of the SiO_2 ceramic green bodies increased with increasing concentration of the ZrO_2 nanoparticles. After the introduction of 2 wt% ZrO_2 nanoparticles, the linear shrinkage in the *z*-direction was increased from 2.7% for the sample without ZrO_2 to 3.2%. Accordingly, the porosity of the ceramic samples decreased with increasing content of ZrO_2. In comparison with the pure SiO_2 ceramics, the sample with 2.5 wt% ZrO_2 exhibited a reduction in porosity from 30.7% to 26.4%. As the level of the ZrO_2 nanoparticles increased from 0 to 2.5 wt%, the density of the SiO_2 ceramics increased from 1.47 to 1.64 $g \cdot cm^{-3}$. On one hand, because the density (5.89 $g \cdot cm^{-3}$) of ZrO_2 is higher than that (2.2 $g \cdot cm^{-3}$) of fused silica, the addition of ZrO_2 nanoparticles would result in an increase in density. On the other hand, the β-cristobalite to α-cristobalite phase transition was accompanied by a decrease in volume by about 5%, which was also responsible for the increase in porosity and decrease in density [53]. However, the concurrent effect led to decrease in porosity and increase in density of the final ceramics.

With the addition of 2.5 wt% ZrO_2 nanoparticles, the room temperature flexural strength of the ceramics increased from 11.1 MPa for the sample without ZrO_2 to 26.5 MPa. The samples with ZrO_2 nanoparticles in the range of 1.5–2.0 wt% had the room temperature flexural strength in the range of 24.5–24.7 MPa, which is superior to that (~12 MPa) reported in open literature [61]. It has been demonstrated that room temperature mechanical strengths of silica-based ceramic cores are closely related to sintering behavior and the content of cristobalite phase [53, 62]. In sintering process at high temperatures, fused silica transferred to β-cristobalite, which would transform to α-cristobalite, during the cooling process. It was experimentally illustrated that the SiO_2 ceramics after sintering at 1200 °C contained quartz, cristobalite and ZrO_2. The enhancement in the room temperature flexural strength of the SiO_2 ceramics was mainly ascribed to the improvement in the densification behavior, which in turn resulted in increment in mechanical strength. In addition, the increase in the high temperature flexural strength was more significant than the room temperature ones.

Zhao *et al* studied creep behavior of SiO_2 ceramic cores for investment casting by using SLA 3D printing process [63]. The SiO_2 ceramics were doped with different contents of $ZrSiO_4$ at stresses of 5–20 MPa over temperature range of 1400 °C–1500 °C. The creep rate of the ceramic samples was first increased and then decreased with increasing content of $ZrSiO_4$ up to 25%, arriving at the maximum value in the sample with 15% $ZrSiO_4$. The incorporation of $ZrSiO_4$ addition hindered the viscous flow behavior of the fused silica during the sintering process at high temperatures, thus weakening the delamination and hence increasing the creep rate of the ceramics. At certain levels of $ZrSiO_4$, interlayer densification could be effectively promoted, thus leading to enhancement in the creep resistance.

The fused silica powder used in this study had $D_{50} = 39$ μm, while $ZrSiO_4$ powder had $D_{50} = 10.5$ μm, with doping levels of 0, 5, 15 and 25 wt%. The two powders were mixed and dispersed with photosensitive resin (10D-S-02), with a solid loading of 58 vol%. 3D printing slurries were obtained by using ball milling at 400 rpm for 12 h, followed by defoaming in vacuum for 5 min. For SLA printing, the thickness of the forming layers was 100 μm, while the energy intensity of the laser was 5 $mW \cdot cm^{-2}$. The printing was in the direction along the height of samples. After printing, the samples were washed with ethanol and dried. The dried green bodies were debinded at 400 °C for 2 h, with a heating rate of 2 $°C \cdot min^{-1}$. Then, the samples were sintered at 1200 °C for 4 h, at a heating rate of 5 $°C \cdot min^{-1}$. They were then subject to creep testing at room temperature.

As an early example, Licciulli *et al* reported ZrO_2 reinforced Al_2O_3 ceramics by using an SLA 3D printing process, from alumina powder with acrylic modified zircon [33]. The zircon compound also served as organic photoactivated resin and acted as a dispersant to ensure high contents of Al_2O_3 powder without negatively affecting the viscosity. As a liquid ceramic precursor, the zircon compound was converted to oxide during the debinding process. Due to the homogeneous dispersion of the Al_2O_3 powder in the zircon acrylate, ZrO_2 submicron particles with uniform distribution were achieved after the thermal pyrolysis. The ZrO_2 particles were located at the Al_2O_3 grain boundaries.

The liquid reactive phase of the suspension is the zircon acrylate NZ39, neopentyl (diallyl)oxy, triacryl zirconate from KEN REACT. The Al_2O_3 powder has an average particle size of 1.8 μm and a specific surface area of 0.42 $m^2 \cdot g^{-1}$. After the green body was sintered at 1600 °C, the Al_2O_3 ceramics had a density of 3.77 $g \cdot cm^{-3}$, with a linear shrinkage of 16%. In the designed compositions, 1-hydroxy-cyclohexyl-phenyl-khetone Irgacure 184 was used as photoinitiator, at a level of 3 wt% with respect to the zirconate. The volume concentrations of the resin were controlled in the range of 40%–60%.

Light curing was conducted by using a He-Cd laser, with a specific power of 16 mW $\cdot$ mm^{-2}, a beam diameter of 0.2 mm and irradiation wavelength of 325 nm. A scanning mirror was employed to drive the laser beam on the surface of suspension. The system was controlled by using a software to construct rectangular samples through scanning the vat surface along parallel lines. The parameters to be controlled include the scanning direction, the dimension of the sample, the scan speed and number of scans per unit width. The last two parameters were set to control the energy per unit area.

A moving platform, which was controlled in a manual manner, was utilized to move downward to construct a new layer after every re-filling of the suspension. A recoating unit was kept at a fixed height, so that the resin level would move in the plane parallel to the surface of the resin. The shear rate applied to the suspension was dependent on the speed of the recoater. The laser exposure of each layer was set at an energy dose of 10 mJ $\cdot$ mm^{-2}. The thickness of the single printed layer was controlled to be about 0.1 mm to ensure a balance among curing depth, processing time and shape configuration.

The as-printed green samples were heated at 150 °C for 1 h, at a heating rate of 2 °C $\cdot$ min^{-1} to postcure the acrylates that might not be completely reacted during the laser irradiation. They were then debinded to remove the organic components, by calcining at 500 °C for 1 h, at a heating rate of about 3.3 °C $\cdot$ min^{-1}. Finally, Al_2O_3 ceramics were obtained after sintering at 1550 °C for 0.5 h, at a heating rate of 5 °C h.

A representative cross-sectional SEM image of the printed samples is shown in figure 2.18(a). The fractured surface is highly compacted, without obvious delamination, having only small spacing near the upper right corner. Interestingly, the ZrO_2 particles are notably attached on the Al_2O_3 grains, as seen in figures 2.18(b) and (c).

Wu *et al* fabricated complex-shaped alumina-based ceramic cores with complex shapes by using SLA 3D printing combined with gelcasting [64]. SLA process was first used to prepare integral sacrificial molds of resin, while the gelcasting process was employed to develop wet green bodies of ceramic cores, through polymerization of aqueous ceramic slurries. A freeze-drying process was utilized to dry the wet green bodies surrounded by the resin molds. After the resin molds were removed through debinding and calcination, complex-shaped Al_2O_3-based ceramic cores were derived. In addition, sintering behavior of the green bodies was enhanced by introducing MgO as a sintering aid.

The resin molds were made of Somos14120 resin with an integrated structure, which was then removed through post-thermal decomposition. The process allowed for very complicated structures within the resin molds [65]. To prepare the resin molds, a CAD model was established first, which was negative to the desired ceramic cores, by using PRO/E software. Then, they were transferred to an STL format, which was sliced into layers that are 0.05 mm in thickness, by using data-processing software. Eventually, the resin molds were obtained by using a SL apparatus, as

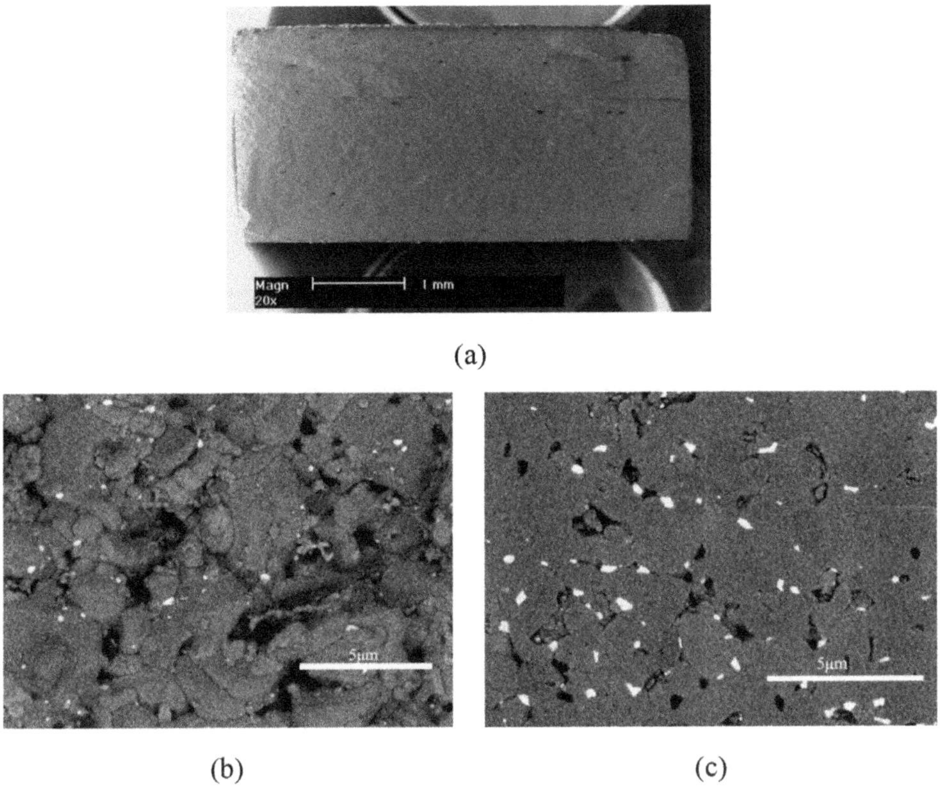

Figure 2.18. Representative SEM images illustrating microstructures of the printed parts: (a) fracture surface low magnification image, (b) unpolished cross-sectional image and (c) polished cross-sectional image. Reprinted from [33], Copyright (2005), with permission from Elsevier.

depicted in figure 2.19. It is extremely challenging to develop such complicated structures by using traditional manufacturing technologies. Moreover, a bottom gating structure was used to fill the ceramic slurries into the molds.

Two types of Al_2O_3 powders, with $D_{50} = 5$, 25 μm and purity of 99%, were used to fabricate ceramic cores, while MgO power ($D_{50} = 2$ μm and purity of 99.99%) and Y_2O_3 power ($D_{50} = 2$ μm and purity of 99.99%) were employed as sintering aids. Deionized water was utilized to disperse the oxide powders as the solvent, while other additives included acrylamide (CH_3CONH_2, AM) organic monomer, N,N-methylene diacrylamide ($C_7H_{10}N_2O_2$, MBAM) cross-linker, ammonium persulfate initiator, N,N,N′,N′-tetramethyl ethylenediamine ($C_6H_{16}N_2$, TEMED) catalyst and sodium polyacrylate dispersing agent.

For Al_2O_3 aqueous gelcasting, AM and MBAM, with a ratio of 24:1, were dissolved in deionized water to form a solution with a concentration of 25% [66–68]. After the addition of sodium polyacrylate (2.5 wt% of solid powers), the solution was adjusted to pH value of about 11 with ammonia solution. Oxide powder mixture, consisting of 75% coarser Al_2O_3, 16% finer Al_2O_3, 4% MgO and 5% Y_2O_3,

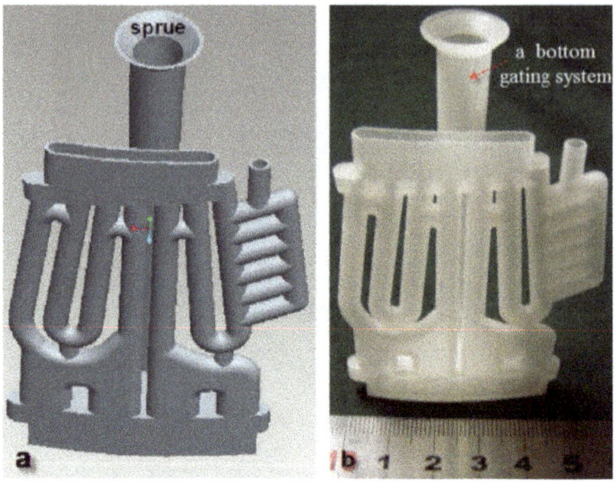

Figure 2.19. Sacrificial mold: (a) CAD model and (b) the resin mold fabricated by using SLA. Reprinted from [64], Copyright (2009), with permission from Elsevier.

were dispersed in the solution. After ball milling for 2–3 h, ceramic slurries were obtained, with sufficiently low viscosity of 0.735 Pa · s and high solid loading levels of up to 55 vol%.

It is well known that slurries should be degassed to reduce the bubbles and hence increase strength of the green bodies [69, 70]. After the addition of 1.0 wt% initiator and 0.2 wt% catalyst, the slurries were stirred in vacuum for 3–5 min to minimize the number of air bubbles, which were then cast into the resin molds. Meanwhile, the slurries were vibrated at amplitudes of 1–3 mm and frequencies of 30–60 Hz. With the presence of the initiator and catalyst, polymerization occurred in the slurries, resulting in green bodies. After that, the green bodies together with the resin molds were subject to freeze drying to remove the deionized water, with pre-freezing time and the sublimation time dependent on the size of the green bodies.

When the green bodies were freeze dried, the deionized water was first crystallized to small crystals, which were then sublimated. In this case, the solid–liquid interface tension force, as a main driving force for densification, was absent. As a result, a large number of pores was produced to minimize the shrinkage of the green bodies. In comparison, the air dried counterpart experienced much larger shrinkage. Furthermore, the pores were connected to form channels, allowing water molecules to be removed from the green bodies in an easier way.

Excess shrinkage during the sintering process should also be avoided to prevent cracking of the ceramic cores. As shown in figure 2.20(a), the sample was partly broken, due to the relatively large shrinkage of about 2%. It was found that the sintering shrinkage of the SiO_2 ceramics could be further reduced by doping MgO. With 2 wt% MgO, the shrinkage decreased from about 2% to 0.6%. As the content of MgO was raised to 4 wt%, the shrinkage was further reduced to 0.25%. The reduction in sintering shrinkage was ascribed to the formation of spinel $MgAl_2O_4$,

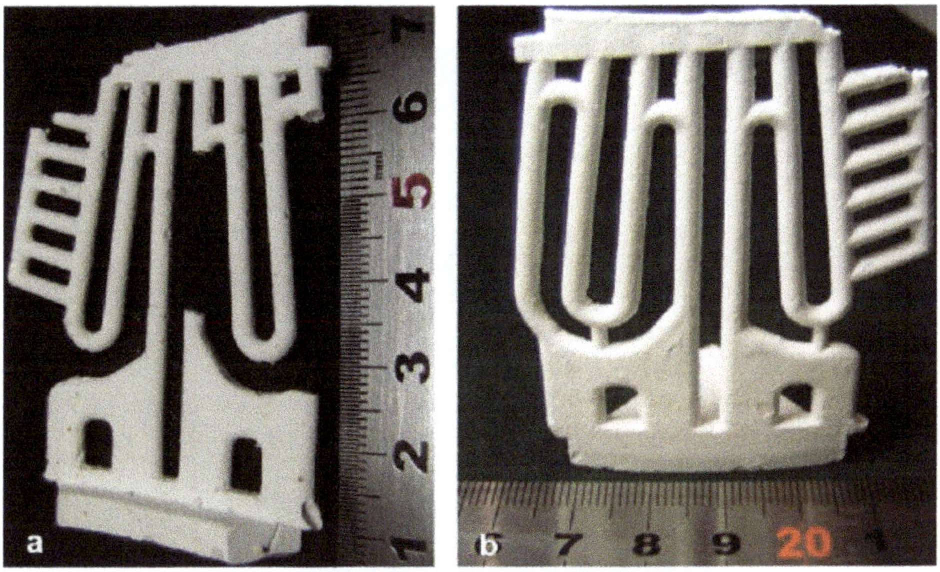

Figure 2.20. Complex-shaped ceramic cores: (a) failure sample and (b) integrated sample. Reprinted from [64], Copyright (2009), with permission from Elsevier.

owing to the reaction between Al_2O_3 and MgO during the high temperature sintering process. The formation of the spinel phase was accompanied by a volume expansion of about 2.5% [71]. An integrated sample is shown in figure 2.20(b). In other words, the optimal content of MgO is 4 wt%. In addition, the optimal sintering parameters included a sintering temperature of 1550 °C, time duration of 4 h and heating rate of 80 $°C \cdot min^{-1}$.

Hu *et al* reported 3D diamond electromagnetic band-gap (EBG) structures fabricated by using a combined process of SLA printing and gelcasting [72]. The EBG structures had a lattice constant of 12 mm, while the band-gap in the (110) direction was in the frequency range of 9–12 GHz. In this case, the SLA process was used to prepare epoxy resin molds for the inverse diamond EBG structures, as shown representatively in figure 2.21. After that, alumina slurries with 55 vol% Al_2O_3 solid loading were cast into the resin molds through a gelcasting process. Figure 2.22 shows an example of the whole EBG structure.

The EBG structures were designed by using a computerized CAD program, according to the electromagnetic theory of wave scattering. Then, the inverse diamond structures were derived with the Boolean operation, as seen in figure 2.21 (a). This inverse diamond EBG structures were converted into a prototype format as STL file, which sliced into sections with a thickness of 0.1 mm. Stereolithography printing was carried out with a computer-aided design/computer-assisted modeling (CAD/CAM) process. The system enabled the formation of two-dimensional layers of epoxy resin, which was photopolymerized with scanning of an ultraviolet (UV) laser at the wavelength of 325 nm. Finally, 3D structures were obtained through the

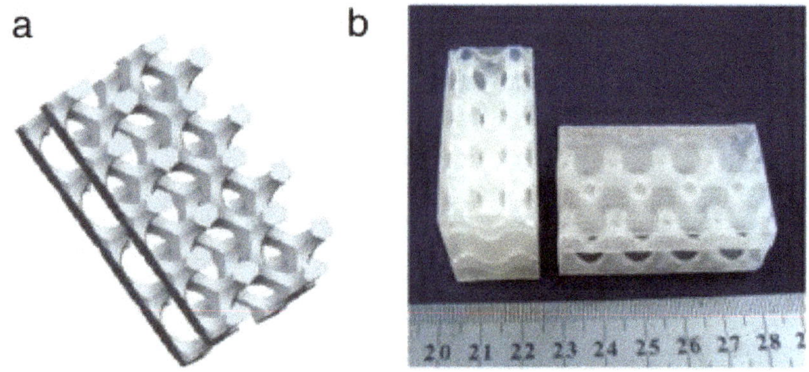

Figure 2.21. Three-dimensional CAD/CAM models and RP models of the inverse diamond EBG structures designed to achieve band-gap performances in the microwave frequency range. Reprinted from [72], Copyright (2012), with permission from Elsevier.

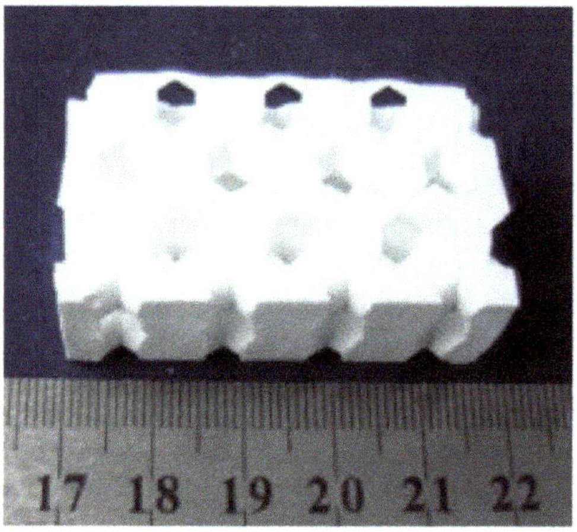

Figure 2.22. Photographs of the three-dimensional diamond EBG parts after sintering. Reprinted from [72], Copyright (2012), with permission from Elsevier.

layer-by-layer stacking process, as illustrated in figure 2.21(b). A typical EBG structure had a dimension of 50 mm × 30 mm × 20 mm. The laser beam used in the stereolithography system had a spot diameter of 100 μm, leading to single layer thickness of 100 μm. The scanning speed was set to be 90 mm · s^{-1}, with a dimensional accuracy of about 0.1%.

To prepare gelcasting slurries, two types of Al_2O_3 powders, with average particle sizes of 2 and 25 μm, were selected as the ceramic materials. Al_2O_3 powders with two particle sizes were used to achieve green bodies with high compact. Meanwhile, acrylamide (AM, CH_3CONH_2) was used as the monomer, while N,N-methylene diacrylamide (MBAM, $C_7H_{10}N_2O_2$) was employed as the cross-linking agent. The AM and MBAM were dissolved in deionized water to form a solution. Then, the

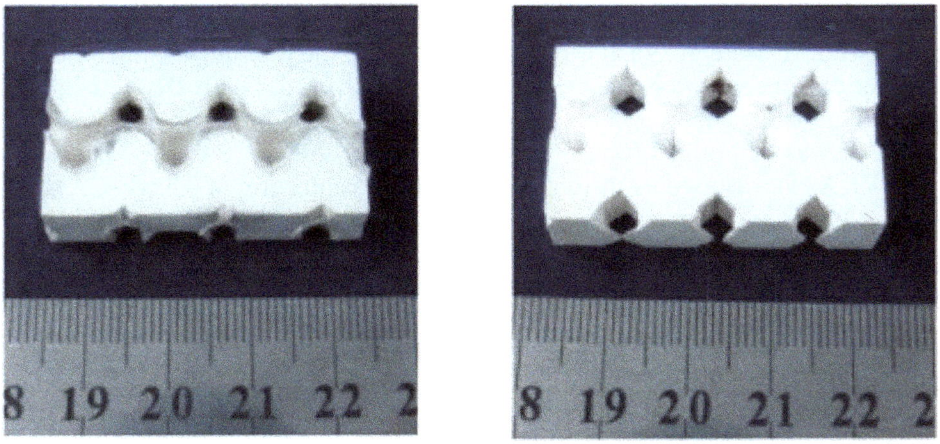

(a) (b)

Figure 2.23. Photographs of the three-dimensional diamond EBG samples after cutting in the (1, 1, 0) direction before (a) and after (b) sintering. Reprinted from [72], Copyright (2012), with permission from Elsevier.

two Al_2O_3 powders were dispersed in AM/MBAM solution, thus having alumina slurries with a solid loading level of 55 vol%. Additionally, 5 wt% aqueous solution of ammonium peroxydisulfate (APS) was added as the initiator (1.0 wt% of the slurry), followed by ball milling for 3 h. Prior to the gelcasting process, tetramethyl ethylenediamine (TMEDA) as a catalyst was introduced into the slurries. After thorough mixing, the slurries were quickly cast into the resin molds at a constant rate with the aid of vibration.

The green bodies were solidified through the gelation reaction triggered by the catalyst. After the gelation reaction was completed, the parts could be unmolded, followed by freeze drying to remove water. The dried green bodies were debinded and then sintered at 1550 °C. The samples exhibited a slight shrinkage of $<10\%$. The sintered structures had a dielectric constant of about 9, which is in a good agreement with that of Al_2O_3. The dielectric constant of the samples without sintering was 4, owing to the presence of pores and the organic additives. Figure 2.23 shows the three-dimensional diamond EBG samples after cutting in the (1, 1, 0) direction, for electromagnetic measurement.

Zhou *et al* developed fully dense defect-free Al_2O_3 ceramic cutting tools by using the SLA printing process [73]. Both the drying and debinding processes were optimized to achieve high quality products. As compared with the natural drying process, PEG400 liquid desiccant drying resulted in green bodies with relatively low deformation. A two-step debinding process was developed, consisting of a vacuum pyrolysis step and an air debinding step, which ensured the pyrolysis rate and suppressed the formation of defects in the green bodies. With optimal processing parameters, the final Al_2O_3 ceramic cutting tools had a relative density of 99.3% and Vickers hardness of $\sim$17.5 GPa.

To prepare aqueous acrylamide solution, acrylamide and methylene-bis-acrylamide (MBAM) were mixed at a ratio of 19:1, while the mixture was then dissolved in water at a concentration of 25 wt%. Then, Al_2O_3 powder with $D_{50} = 0.2$ μm was dispersed in the aqueous acrylamide solution at a solid loading level of 30 vol%. The suspension was ball milled for 18 h to form printing slurries. Polyvinyl pyrrolidone K15 (PVP K15) was used as the dispersant at a concentration of 0.4 wt%. After ball milling, the ceramic suspensions were vacuum degassed to remove the bubbles. The ceramic suspensions were made to be UV curable, by mixing with photoinitiator 1173, at a content of 1 wt%, with respect to the concentration of the aqueous acrylamide solution.

3D model was generated by using UG software, while the Magics software was used to produce the supporting structures and sliced parts. Al_2O_3 green bodies were printed with the ceramic suspensions. Figure 2.24 shows a schematic diagram for the SLA 3D printing process. The patterned layers were selectively cured with a UV laser. To avoid deformation, the green bodies after printing were subject to a suitable drying process. An effective drying process was developed, by using PEG 400 as the liquid desiccant.

The green bodies would be deformed if they were dried without any interference. This was explained by considering the anisotropy in evaporation rate of the water molecules, because of the inhomogeneity in the air flow on different surfaces of the printed parts, as demonstrated in figure 2.25(a). In comparison, more uniform water extraction rates were observed when using the liquid desiccant, thus leading to much more homogeneous shrinkages in different directions, as illustrated in figures 2.25(c) and (d). In addition, the PEG drying process also resulted in Al_2O_3 ceramics with higher density and homogeneous microstructure, as revealed in figure 2.26.

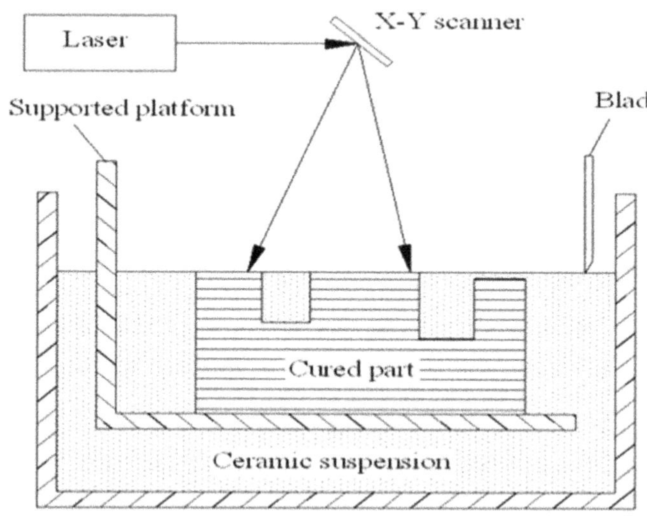

Figure 2.24. Schematic diagram of the SLA 3D printing process. Reprinted from [73], Copyright (2016), with permission from Elsevier.

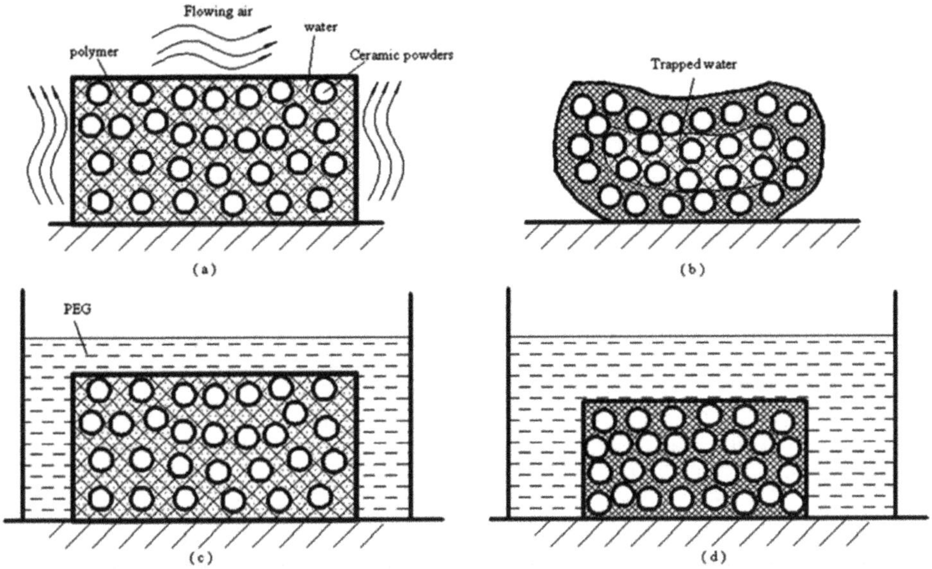

Figure 2.25. Schematic diagram showing the two drying processes: (a, b) natural drying and (c, d) PEG-assisted drying. Reprinted from [73], Copyright (2016), with permission from Elsevier.

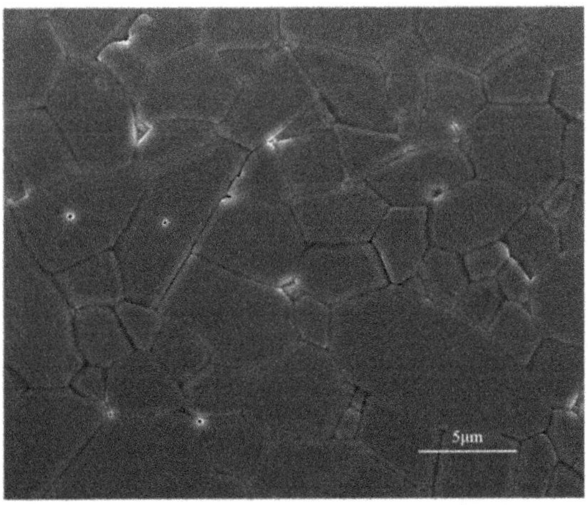

Figure 2.26. Representative polished surface SEM image of the Al_2O_3 ceramics sintered at 1650 °C through the combination of liquid desiccant drying and a two-step debinding process. Reprinted from [73], Copyright (2016), with permission from Elsevier.

An *et al* fabricated Al_2O_3 ceramics with complex shapes and structures, by using the SLA 3D printing technique [74]. Rheological properties and thermal behaviors of the Al_2O_3 slurries were characterized to identify the optimal molding and debinding processing parameters. Then, sintering and densification behaviors of the printed green bodies were evaluated. The green bodies after debinding exhibited a uniform particle packing state and a relatively narrow distribution of pore sizes. It was found that dimensions of the Al_2O_3 green bodies shrunk anisotropically when different processing steps were used. Significant reduction in pore size and enhancement in grain growth were present at sintering temperatures of over 1450 °C. The final ceramics possessed homogeneous microstructures, without an obvious interface between the adjacent printed layers. After sintering at 1650 °C, the Al_2O_3 ceramics displayed a relative density of 99.1% and a Vickers hardness of 17.9 GPa.

To prepare UV curable ceramic slurries, methylenebisacrylamide (MBAM), acrylamide (AM) and glycerol were mixed. Al_2O_3 powder with $D_{50} = 0.18$ μm was dispersed into the organic mixture, with a solid loading of 50 vol%. 0.3 wt% sodium polyacrylate was added as a dispersant. Al_2O_3 slurries were obtained after ball milling for 6 h. After that, a photoinitiator was added into the suspensions, at a concentration of 2 wt%. Room temperature rheological behaviors of the slurries were characterized. The viscosity decreased quickly with an increasing shear rate and then levelled off. The slurries with suitable flow properties for the SLA printing step had a viscosity of 10 Pa · s.

The SLA 3D printing process is schematically shown in figure 2.27(A). The slurries were poured into the vat in a given quantity, followed by spreading to form thin layers with a wiper blade, while the vat was spun with proper speed. After that, the 3D model was loaded to the software and then sliced into a 2D image. UV light irradiation was conducted with a computer to selectively scan on the surface of the slurry layer according to the 2D images. During the layer-by-layer scanning, photocurable resin in the slurry was photopolymerized. The Al_2O_3 particles were

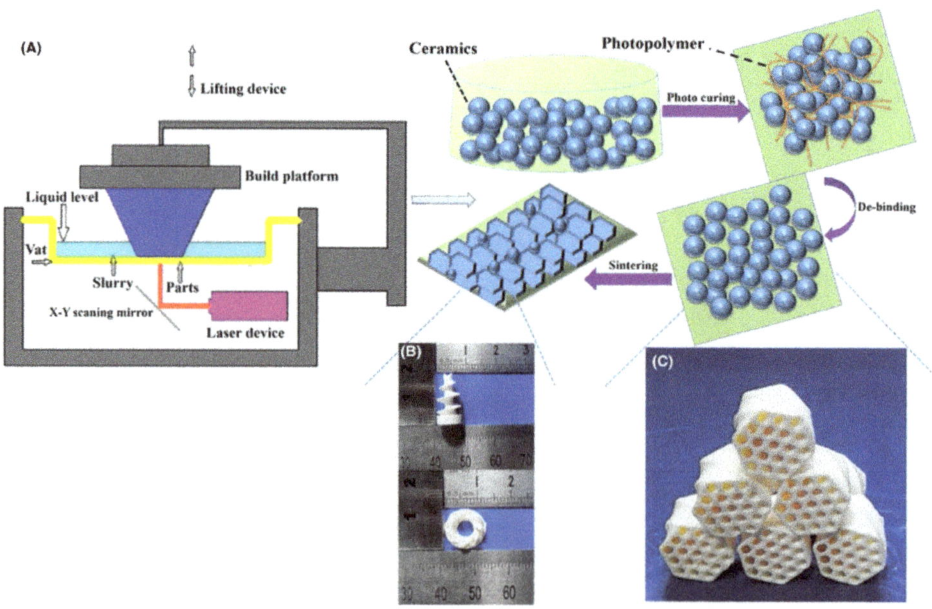

Figure 2.27. (A) Schematic diagram of the SLA system. (B, C) Photographs of the green bodies fabricated with the SLA system. [74] John Wiley & Sons. © 2017 The American Ceramic Society.

confined by the polymeric networks after a cross-linking reaction. Representative Al_2O_3 green bodies are shown in figures 2.27(B) and (C).

To prevent delamination and cracking, the green bodies were debound at 600 °C for 3 h and then sintered at temperatures of 1000 °C–1650 °C. After debinding, the samples experienced average shrinkages in the directions of length, width and height of 2.65%, 2.61% and 2.14%, respectively. Such a difference was related to the compact nature of the green bodies. In the height direction, the particles were relatively loosely packed, together with small gaps between every two adjacent layers, thus resulting in smaller shrinkage. In comparison, along the directions of width and length, the ceramic particles were packed in a more dense way was relatively compact [75]. However, after sintering at 1650 °C, the shrinkages of the samples in the directions of height, length and width were 19.4%, 19.6% and 22.6%, where the shrinkage in the height direction was larger than those in the directions of length and width, owing to the same reasons.

Figure 2.28 shows SEM images of the Al_2O_3 ceramic samples after sintering at different temperatures. The sintering process was likely divided into three stages, according to the microstructural evolution, corresponding to three temperature ranges, i.e., 1200 °C–1400 °C, 1400 °C–1600 °C and >1600 °C. At 1200 °C, grain rearrangement occurred in the green bodies, which is known as the initial stage of sintering, characterized by slight necking among the ceramic particles. At this stage, there was almost no grain growth, while pores between the Al_2O_3 grains were clearly present. After sintering temperatures of ⩾1400 °C, grain growth took place, due to the occurrence of atomic diffusion in the materials. Here, the densification process was closely linked to the particle rearrangement, while the increasing contribution of interfacial atomic diffusion became dominant [76]. Above 1600 °C, the densification was completed, with the sintering mechanism corresponding to solid-state diffusion.

Liu *et al* examined printability of Al_2O_3 ceramic slurries and mechanical performance of the final ceramics made with SLA 3D printing [77]. To evaluate

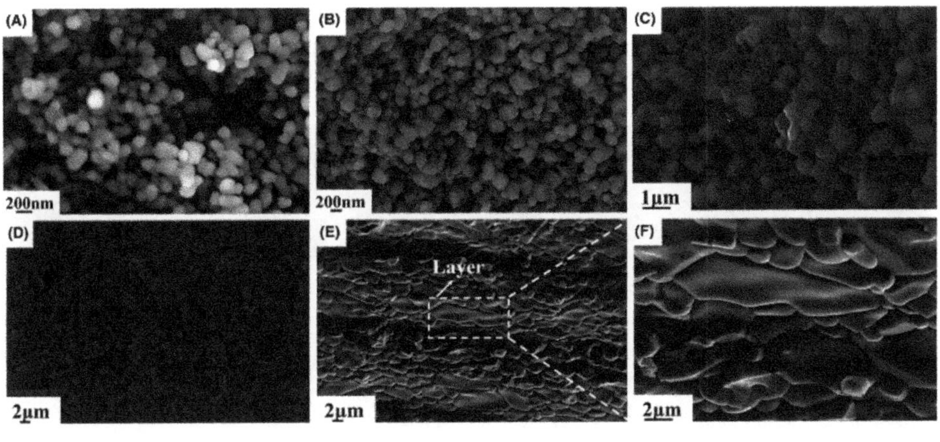

Figure 2.28. SEM images of the Al_2O_3 ceramics sintered at different temperatures: (A) 1200 °C, (B) 1400 °C, (C) 1500 °C, (D) 1600 °C and (E, F) interlayered structure at 1600 °C. [74] John Wiley & Sons. © 2017 The American Ceramic Society.

the effect of solid loading, samples with 48, 50, 52, and 55 volume fractions were studied. The Al_2O_3 powder had a specific surface area of 6.89 $m^2 \cdot g^{-1}$, with particle sizes of $D_{10} = 0.152$ μm, $D_{50} = 0.549$ μm and $D_{90} = 1.303$ μm. Acrylate monomers (1,6-hexanediol diacrylate, M1) and ethoxylated (5) pentaerythritol tetraacrylate (M2) were mixed, together with alkylamine dispersant, plasticizer (polyethylene glycol-400) and photoinitiator (Irgacure 184). The Al_2O_3 powder was introduced into the solution, followed by ball milling at 300 rpm for 4 h. Finally, Al_2O_3 slurries were obtained after ball milling for another 24 h.

To fabricate Al_2O_3 3D green bodies, the recoating layer thickness was controlled to be 60 μm. For experimental characterization, rectangular samples with dimensions of 25 mm × 5 mm × 4.2 mm were printed, while cylinder samples had dimensions of φ 20 mm × 12 mm. All green bodies were debound by heating to 330 °C at a heating rate of 0.5 °C · min^{-1} and then to 600 °C at 0.2 °C · min^{-1}. The samples were kept for 2 h at 70 °C, 330 °C and 600 °C, to ensure complete removal of all the organic components. After that, the samples were sintered 1580 °C for 2 h, at a heating rate of 5 °C · min^{-1}.

Figure 2.29 shows photographs of the four samples with different solid loading levels. Self-holding ability of the ceramic slurries are highly dependent on the solid loading level. Promising stability was maintained in the slurries with solid loading of up to 52 vol%. The 52 vol% slurry thus exhibited outstanding recoating performance. Although the ceramic slurries had sufficiently smooth flowing capability, particle agglomeration could not be avoided if the solid loading was too high. On one hand, the rheological behavior of the slurries could be different. On the other hand, to guarantee the surface quality of the green bodies, the slurries should be certain self-holding ability.

Figure 2.30 shows SEM images of the Al_2O_3 green bodies derived from the slurries with different solid loading levels. As seen in figure 2.30(a), when the solid loading was 48 vol%, the ceramic particles were wrapped by the cured resin, which served as plasticizer. For the samples with 50 and 52 vol% Al_2O_3 powder, the cross-sectional surfaces were homogeneous, suggesting uniform distribution of the ceramic particles in the cured polymeric matrix. When the solid loading level approached

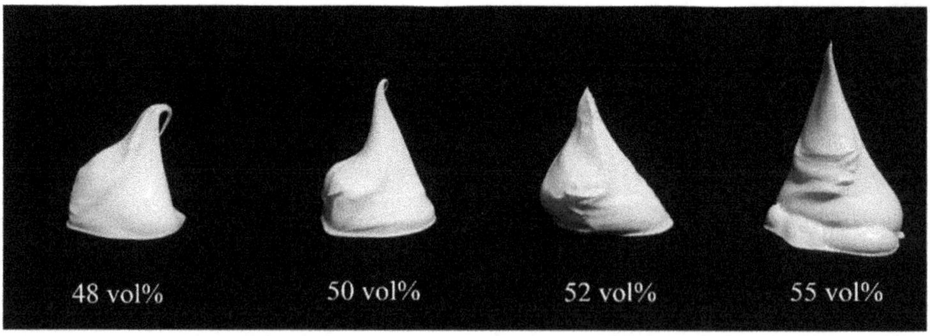

Figure 2.29. Photographs of the Al_2O_3 ceramic slurries that are different in self-holding ability. Reprinted from [77], Copyright (2020), with permission from Elsevier.

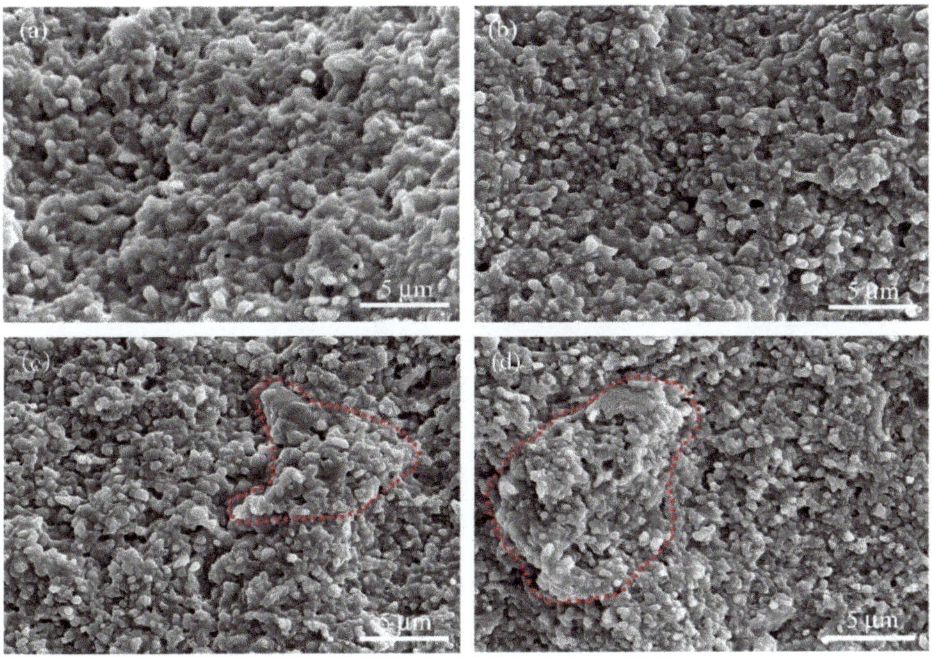

Figure 2.30. Cross-sectional SEM images of the green bodies with different solid loading levels: (a) 48 vol%, (b) 50 vol%, (c) 52 vol% and (d) 55 vol%. Reprinted from [77], Copyright (2020), with permission from Elsevier.

55 vol%, some ceramic particles were disconnected, as observed in figure 2.30(d). In addition, the agglomeration of the Al_2O_3 particles was more and more pronounced, with an increasing solid loading level.

Cross-sectional SEM images of the ceramic parts after sintering are shown in figure 2.31. The samples with lower solid loading levels, pores are present at the grain boundaries, as seen in figure 2.31(a). Those with higher solid loadings exhibited compact triple junctions, as revealed in figures 2.31(c) and (d). No external defects or delamination in the column height direction and along the circumference were observed in all the sintered ceramics from the SEM images, while flaws were indeed present in the computed tomography images. Nevertheless, the defects could be minimized by adjusting the solid loading level of the ceramic slurries. In this case, 52 vol% ceramic powder was the optimal solid loading value.

Qian *et al* used SLA 3D printing technology to develop Al_2O_3 ceramics with complex structures, focusing on the effects of particle size distribution on densification behaviors, microstructural development and mechanical properties [78]. Seven groups of samples with different particle size distributions were examined, in terms of μm/μm, including 30/5, 20/3, 10/2, 5/2, 5/0.8, 3/0.5 and 2/0.3, with a coarse/fine particle ratio of 6:4. The effect of particle size distribution on shrinkage, porosity and flexural strength was stronger than sintering temperature, as the difference in particle size approached tenfold. The presence of coarse ceramic particles had an influence on the accuracy of small-sized samples. When the particle

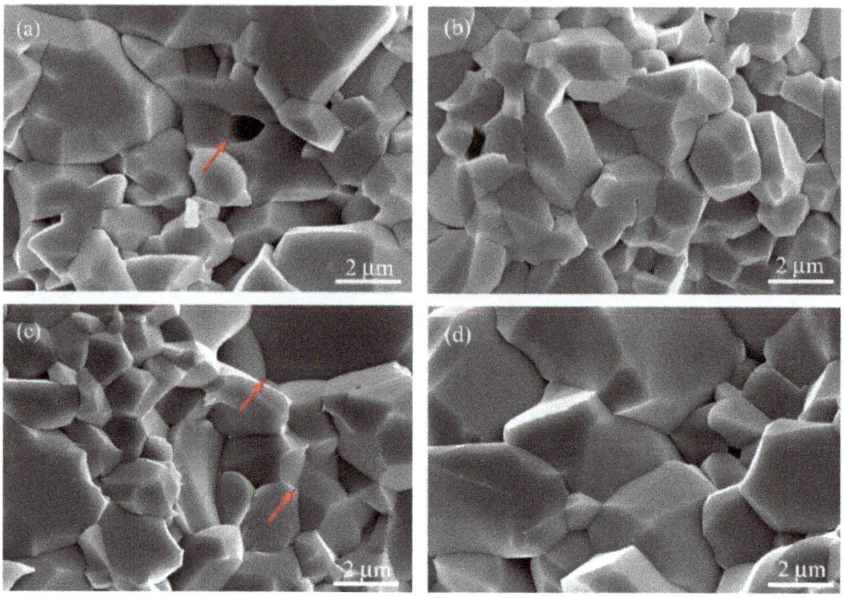

Figure 2.31. Cross-sectional SEM images of sintered Al_2O_3 ceramics with different solid loading levels: (a) 48 vol%, (b) 50 vol%, (c) 52 vol% and (d) 55 vol%. Reprinted from [77], Copyright (2020), with permission from Elsevier.

size was close to the width of the samples, such as 30 µm/5 µm and 5 mm, the shrinkage in width was comparable with that in height. If the particle size was smaller than the width of the samples, such as 2 µm/0.3 µm and 5 mm, the shrinkage in width was close to that in length.

To prepare ceramic slurries, 1,6-hexanediol diacrylate (HDDA) and ethoxylated pentaerythritol tetraacrylate (PPTTA) were used as monomers. 2,4,6-trimethyl benzoyl diphenylphosphine oxide (TPO) was used as a photoinitiator. Al_2O_3 powder with a purity of 99.99% was loaded with a volume concentration of 56%. To make a trade-off between printing precision and particle size, the thickness of each printed layer was set to be 100 µm, while the coarsest particle should have sizes of $D_{90} <$ 100 µm. Therefore, the Al_2O_3 powder met this requirement.

The Al_2O_3 powder was dispersed in resin by using ball milling for 12 h to form homogeneous slurries, which were subject to degass treatment in vacuum for 2 min, for the SLA printing process. A UV light source with a wavelength of 405 nm was used for photopolymerization. The green bodies had dimensions of height:width: height = 50 mm:5 mm:4 mm, with the height in the direction of printing, as shown schematically in figure 2.32.

The green bodies were calcined at 200 °C, 400 °C and 600 °C, for 3, 3 and 2 h, respectively, at an average heating rate of about 1 $°C \cdot min^{-1}$. Then, the temperature was raised to 900 °C, at a heating rate of 7.5 $°C \cdot min^{-1}$, while the time duration was 40 min. Finally, all the samples were sintered at temperatures of 1350 °C–1650 °C

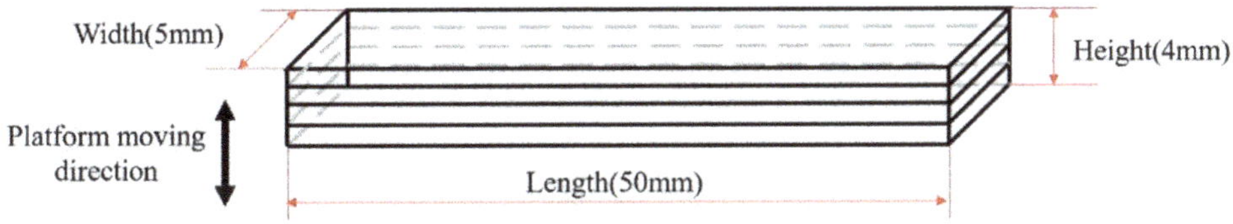

Figure 2.32. Schematic diagram of samples with designed dimension. Reprinted from [78], Copyright (2022), with permission from Elsevier.

for 2 h, at a heating rate of 5 °C · min^{-1}. After sintering, the samples were cooled to 800 °C at a cooling rate of 6 °C · min^{-1}, followed by cooling to room temperature.

Generally, the flexural strength of the ceramic samples with different groups of particle size showed an increase trend with increasing sintering temperature, except for the one with the particle size distribution of 2/0.3. For this group, the flexural strength decreased after sintering at 1650 °C. At given sintering temperatures, the flexural strength was increased with decreasing particle size.

Figure 2.33 shows SEM images of the samples sintered at different temperatures, demonstrating four sintering stages. In the early stage of particle rearrangement, the particles retained the original morphology. When the small particles were fused with the large ones, sintering necks began to form. During this process, the pores were eliminated, while the grain size had a relatively narrow distribution. Gradually, the grain growth occurred and porosity was reduced. However, an excessively high sintering temperature resulted in abnormal grain growth, which is harmful to mechanical properties of the final ceramics.

The samples from powder groups of 30/5 and 20/3 could only reach secondary sintering stage, because of their relatively low sinterability. The samples from the groups of 10/2, 5/2 and 5/0.8 exhibited similar densification behaviors. They had average grain sizes of 3–5 μm. The densification rate of the 2/0.3 group was much higher than that of the 3/0.5 group, especially at the sintering temperature of 1550 °C. At this sintering temperature, the 2/0.3 group was at the pore discharge stage, while the 3/0.5 group was still at the stage of particle fusion. This is reason why flexural strength of the 2/0.3 group was higher than that of the 3/0.5 group, as the sintering temperature was over 1500 °C.

The interlayer interface led to the content of resin in the height direction that was more than that between two adjacent layers, while the spacing related to the shrinkage after debinding was also larger. Additionally, the difference in curing conditions with respect to height, width and length were responsible for their difference in shrinkage. In fact, this trend was not observed across all particle size groups.

The shrinkage in length, width and height all increased with decreasing particle size, while the difference was very small between the two samples from the 30/5 and 20/3 groups. As the size of the coarse particles was smaller than 10 μm, the samples exhibited strong densification behaviors. The shrinkage in the height was slightly different between the samples from the 10/2 and 5/2 groups. The sample from the

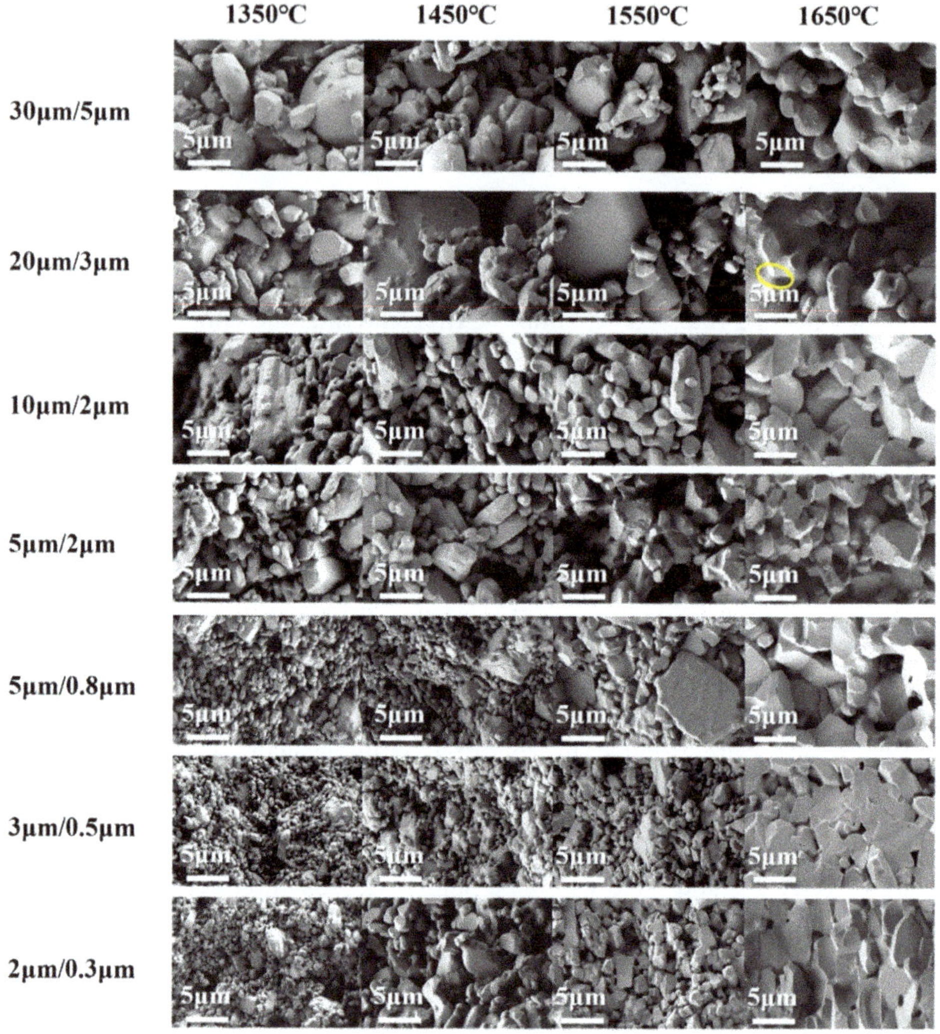

Figure 2.33. SEM images of the Al_2O_3 ceramics with different particle size distributions after sintering at different temperatures. The yellow circle is used to indicate a sintering neck. Reprinted from [78], Copyright (2022), with permission from Elsevier.

5/0.8 group had a larger shrinkage than that from the 5/2 group. Moreover, the sample from the 2/0.3 group displayed the highest shrinkage.

The ratio of the shrinkage in height direction to that in the length direction was about 1.5, for the samples from all particle size groups, suggesting that the effects of sintering temperature and particle size distribution on the shrinkage ratio were very weak. The ratio of shrinkage in the height direction to that in the width direction was close to 1 for the samples from the 30/5 and 20/3 groups. With decreasing particle size, the shrinkage in the width direction became closer and closer to that in the length direction. Because shrinkage ratio in the height direction to the length direction was nearly unchanged, it was implied that the resin present in between the printed layers could have a positive effect on densification.

For the shrinkage in width direction, the samples with coarse particles on the outer edges experienced relatively low densification rate. A resin layer with a certain thickness of the edges was observed in the specimens from the 20/3 and 30/5 groups. Such resin layers had strong influence on densification behaviors of the green bodies. Figure 2.34 shows the schematic diagram of the model used to describing this observation. In this case, the particles that were cured to be less than 2/3 in terms of their radius would be cleaned out during washing.

Cui *et al* made attempts to clarify the mechanism of crack generation in the degassing process of Al_2O_3 green bodies fabricated with SLA 3D printing technique [79]. The green bodies were calcined at temperatures of up to 700 °C, in air or Ar. SP-RH series resin and Al_2O_3 powder with an average particle size of 1 μm were used to prepare ceramic slurries for 3D printing. The resin was mixed with dispersant KOS 110 at a volume ratio of 10:1, which served as monomer. The ceramic slurries had a solid loading of 80 wt%.

In experiment, the Al_2O_3 powder and the monomer were thoroughly mixed through strong stirring for 2 h at 2000 r min^{-1} with a disperser, leading to homogeneous ceramic slurries. Meanwhile, the 3D model was sliced with the software of Magics, with proper support to be added. The thickness of the sliced layers was set to be 50 μm. The slurries were cast onto the platform with a squeegee, while the laser emitted UV light at a wavelength of 355 nm was used to solidify the printed layers. The Al_2O_3 green bodies were calcined in air or Ar at temperatures in the range of 200 °C–700 °C, for time duration of 2 h, at a heating rate of 1 $°C \cdot min^{-1}$.

In the as-printed green bodies, the Al_2O_3 particles were tightly wrapped by the polymer. After calcining at 200 °C, the resin began to melt, while the Al_2O_3 particles tended to fall off from the organic matrix. When the calcination temperature increased, decompose of the organic components continued, resulting in agglomeration for both the polymer and the Al_2O_3 particles. As the calcination temperature was raised to 500 °C, a membrane-like organic layer covered the surface of the green bodies, because diffusion of the resin from interior of the green bodies towards the

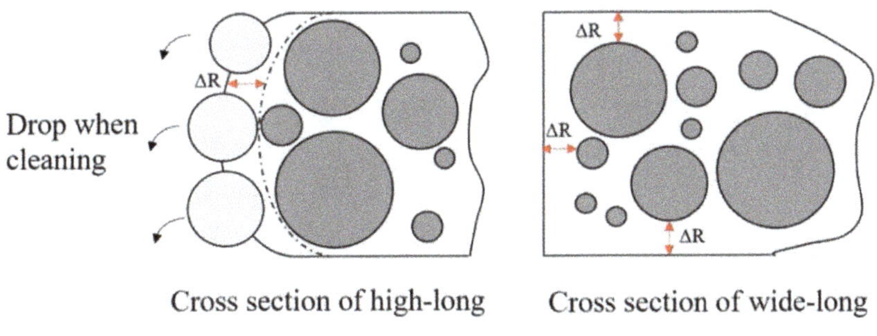

Figure 2.34. Schematic diagram of the model showing the resin layer at the edge of the sample, which was formed during the SLA printing process. Reprinted from [78], Copyright (2022), with permission from Elsevier.

surfaces occurred, owing to the pressure caused by the gas generated during the degrassing process and capillary force.

Once the resin in the green bodies was eliminated, the Al_2O_3 particles were exposed on the sample surface. At the same time, pores were present on the surface of the green bodies, with the porosity was increased with increasing calcination temperature. The pores were produced due to the rapid overflow of the products related to the decomposition of the organic components. The green bodies calcined in Ar exhibited a similar trend in microstructural evolution.

Figure 2.35 shows morphological profiles and crack microstructure of the Al_2O_3 ceramic green bodies calcined in air at different temperatures. After calcining at 400 °C, the organic components were molten near the cracks, with the Al_2O_3 particles densely aggregated, as seen in figure 2.35(B). An obvious crack was present in the sample, as illustrated in figure 2.35(A), implying that the thermally induced cracks propagated across the sample. After calcining at 500 °C, thermal expansion

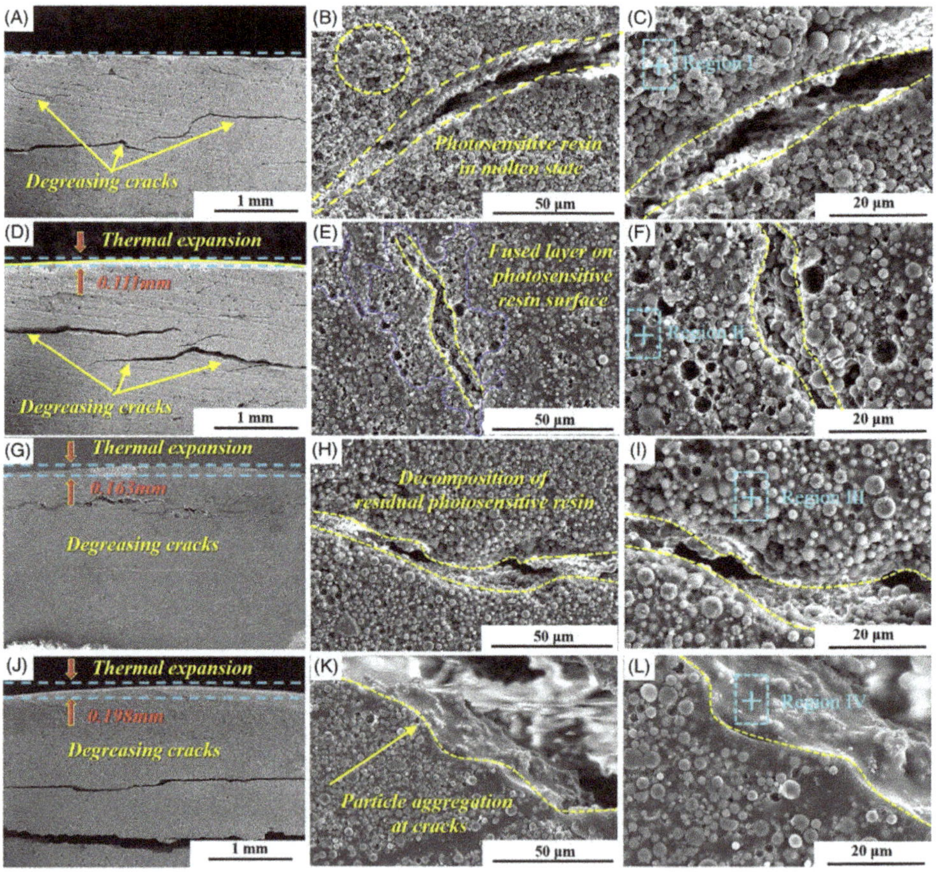

Figure 2.35. Microstructure and morphology of cracks in the green bodies after calcining at different temperatures in air: (A–C) 400 °C, (D-F) 500 °C, (G–I) 600 °C and (J–L) 700 °C. [79] John Wiley & Sons. © 2023 The American Ceramic Society.

deformation was induced, with a magnitude of 0.111 mm, while the crack was widened. Meanwhile, the resin covered the surface of the green body like a film, as revealed in figure 2.35(E).

No polymer component was observed near the crack, and the temperature inside the green body was slightly lower than that at the surface during the calcination process. For a similar reason, membrane-like resin was not formed near the crack. As the calcination temperature increased to 600 °C, the deformation related to thermal expansion increased to 0.163 mm, while tiny white particles were attached on the Al_2O_3 particles, which were derived from the polymer near complete decomposition, as observed in figures 2.35(G)–(I). With further increase in the calcination temperature to 700 °C, the deformation was about 0.2 mm, whereas the polymer components were entirely absent, due to the complete decomposition, as depicted in figures 2.35(J)–(L).

Figure 2.36 shows morphological and microstructural characteristics of cracks in the green bodies after calcining in Ar. Overall, the density of cracks in this group of samples was much smaller than that in the samples calcined in air. For instance, as the calcination temperature was 400 °C, cracks were present inside the green bodies, while the resin flow aggregated with a cluster-like distribution, as observed in figure 2.36(B). No thermal expansion occurred at the surface of the sample calcined at 500 °C, whereas pores and cracks were present in the interior of the green body. The cracks were generated in the direction perpendicular to that of the 3D printing, as indicated in figure 2.36(D).

At the same time, a film-like polymer layer was similarly present on the surface of the sample, as displayed in figure 2.36(E). No cracks appeared near the polymer layer. After calcination at 600 °C, the thermal expansion deformation was 0.115 mm, which was smaller than that by about 29%, as compared with that observed in the air calcined counter parts, as revealed in figures 2.36(G)–(I), When the degreasing temperature was 700 °C, the thermal expansion related deformation was 0.83 mm, while the organic components were completely decomposed.

Figure 2.37 shows a schematic diagram of the mechanism governing the formation of cracks and microstructure development of the green bodies, after calcination at different temperatures. The organic components were molten and evaporated, as depicted in figure 2.36(A). The water vapor was formed and CO_2 gas was produced, which escaped from the interior to the surface of the green bodies through diffusion or capillary effect. At the calcination temperature of 300 °C, the polymer components collapsed from the Al_2O_3 particles, the molten components experienced decomposition, resulting in acrylic acid with a relatively low boiling point, while partial carbonization of the organic components took place, as schematically described in figure 2.37(B).

Correspondingly, the green bodies exhibited a variation in color for the four stages, i.e., from white to brown and black and then back to white, after calcining in both air and Ar. Such a variation in color is essentially linked to the carbonization of the organic components at the calcination temperatures. After the organic components were completely eliminated, the samples turned white in color. At the calcination temperature of 400 °C, the molten organic components started to

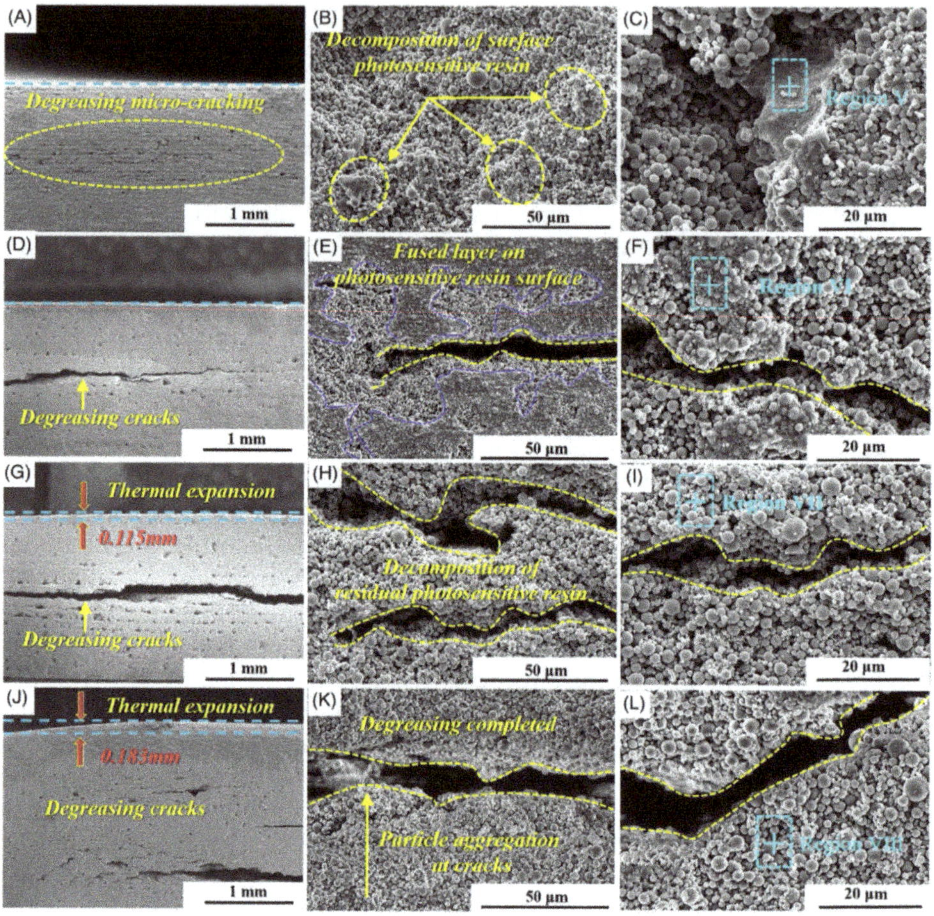

Figure 2.36. Microstructure and morphology of cracks in the green bodies after calcining at different temperatures in Ar: (A–C) 400 °C, (D–F) 500 °C, (G I) 600 °C and (J–L) 700 °C. [79] John Wiley & Sons.

decompose and diffused to the surface of the green bodies, due to the synergetic effect of thermal expansion pressure and the capillary action among the Al_2O_3 particles, leading to carbonization and hence color deepening, as presented in figure 2.37(C).

The temperature inside the green bodies was slightly lower than that at the surface, so that there was an uneven temperature distribution in the samples. The organic components inside the green bodies were unable to overflow to the surfaces due to an insufficiently high thermal driving force. Instead, they served to form channels in the interior of the green bodies. After calcining at 500 °C, the content of the organic components on the surface was increased, which wrapped the Al_2O_3 particles, as a membrane-like thin layer, as given in figure 2.37(D). Al_2O_3 particles near the cracks would gather at the inner edges, where the organic components experienced rapid decomposition. Through the calcination at 600 °C, the

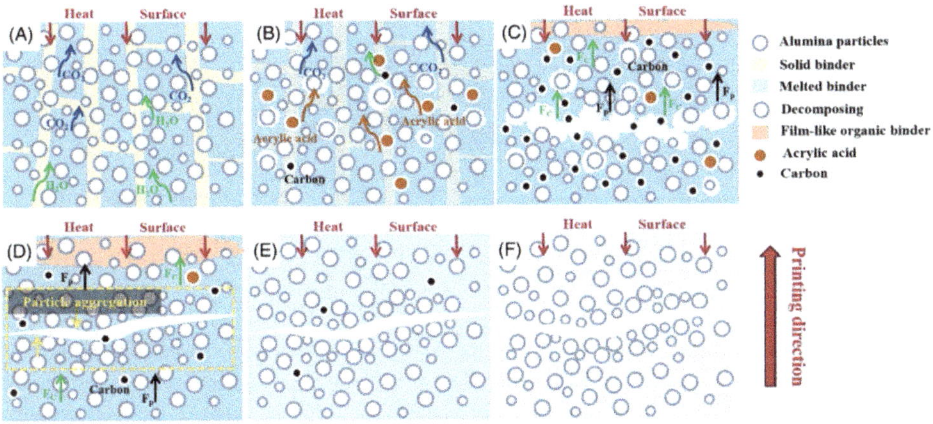

Figure 2.37. Schematic diagram of SLA in alumina ceramic green body thermal degreasing process: (A) 200 °C, (B) 300 °C, (C) 400 °C, (D) 500 °C, (E) 600 °C, and (F) 700 °C. [79] John Wiley & Sons. © 2023 The American Ceramic Society.

decomposition of the organic components was nearly finished (figure 2.37(E)), while complete debinding was achieved at 700 °C (figure 2.37(F)).

Chen *et al* fabricated Al_2O_3 molds by using the SLA printing process, with focus on the effects of particle size distribution and sintering temperature on properties of the final ceramic products [80]. Viscosity of the ceramic slurries was decreased with increasing content of the powder with smaller particles, while the particle size distribution had almost no effect on curing behaviors of the slurries. Densification rate was increased with increasing content of fine powder and sintering temperature. As the content of fine powder was increased, creep resistance of the final ceramics increased first and then decreased. Specifically, the sample derived from the slurry with 10% fine powder exhibited the highest creep resistance, with the droop distance of 4.44 mm. After sintering at 1550 °C, the sample exhibited linear shrinkages of 6.36% and 11.39% in the *x*-/*y*-direction and *z*-direction, respectively, with corresponding flexural strength of 78.2 MPa and porosity of 30%.

Al_2O_3 powders with two groups of particle size, with $D_{50} = 5.8$ and 1.14 μm, were used to prepare the slurries for SLA printing process. The weight ratios of the coarse powder and the fine powder were 10:0, 9:1, 8:2 and 7:3 to obtain powders with different particle size distributions, which were denoted as C10F0, C9F1, C8F2 and C7F3, respectively. Acrylate monomers 1,6-hexanediol diacrylate (HDDA) and polyethylene glycol diacrylate (PEGDA) were utilized as photocurable resins. Photoinitiator 184 was employed as initiator for photopolymerization. BYK-9076 (BYK) served as dispersant, while BYK-410 (BYK) acted as rheological agent. The printed samples were cleaned with PEG-200.

Figure 2.38 shows a schematic diagram describing the fabrication process of Al_2O_3 mold materials with different distribution profiles of particle size. Before preparation of the slurries, a low viscosity photocurable resin was first obtained, by mixing acrylate monomer mixture with weight ratio of HDDA:PEGDA = 6:4, along with 1 wt% photoinitiator. 2 wt% BYK-9076 dispersant was added to ensure

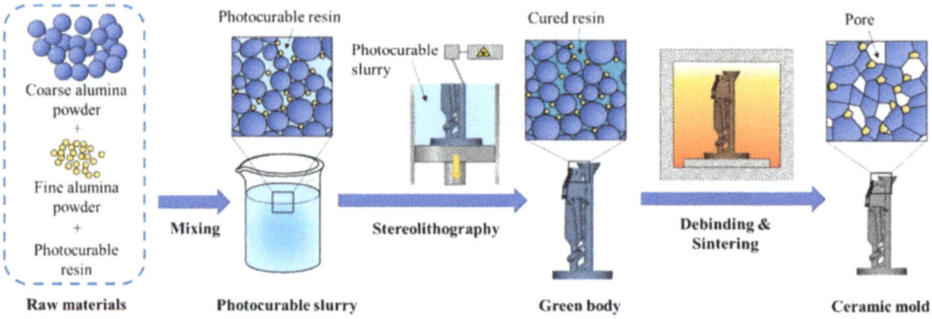

Figure 2.38. Schematic diagram of preparation of the Al_2O_3 mold materials from powders with different particle size distributions by using SLA 3D printing. Reprinted from [80], Copyright (2022), with permission from Elsevier.

low viscosity of the slurries, while 0.5 wt% more BYK-410 was introduced to adjust rheological behaviors of the slurries without sedimentation. Then, Al_2O_3 powders were slowly dispersed into the photocurable resin mixtures, at a final solid loading of 55 vol%. Workable slurries were achieved after ball milling for 2 h at 300 rpm to minimize particle agglomeration.

Green bodies were printed with a 355 nm diode-pumped laser, while the thickness of the single printing layer was 50 μm. The printed samples were debinded to burn off the organic components after drying. The temperature was raised to 600 °C at a heating rate of 0.5 $°C \cdot min^{-1}$ for 2 h. Then, the samples were reheated at 600 °C in air for 2 h, at a heating rate of 2 $°C \cdot min^{-1}$ to eliminate the residual carbon. The debinded samples were sintered in air for 2 h at 1350 °C, 1450 °C and 1550 °C, at a heating rate of 5 $°C \cdot min^{-1}$.

Viscosities of the slurries with different contents of fine powder were measured at the shear rate of 100 s^{-1}. It was found that the C10F0 slurry exhibited the highest viscosity of 6.90 Pa · s. With increasing fraction of fine powder, the viscosity was decreased. Correspondingly, the viscosities of the C9F1, C8F2 and C7F3 slurries were 5.58, 4.80 and 3.46 Pa · s, respectively. It has been reported that the viscosity of ceramic slurries at high concentrations was effectively reduced by using powders with bimodal particle size distributions [80].

The quality of the green bodies and hence their sintering characteristics are highly dependent on curing behaviors of the slurries. Curing depth and width are important parameters for evaluating slurry curing behaviors. The C10F0 slurry comparatively exhibited promising curing effects in both depth and width directions, with irradiation at all energy dose levels. The addition of fine powder had no significant influence on curing behaviors of the slurries. Experimentally, the printing parameters included 1.2 W laser power and 6000 $mm \cdot s^{-1}$ scanning speed, corresponding energy dose of 166.7 $mJ \cdot cm^{-2}$.

Figure 2.39 shows the microstructures of the natural surface of sintered specimens prepared with different particle size distributions. After sintering at 1550 °C, all the samples displayed porous microstructure, making it possible to be removed after casting. Almost no sintering necks were present in the samples sintered at 1350 °C,

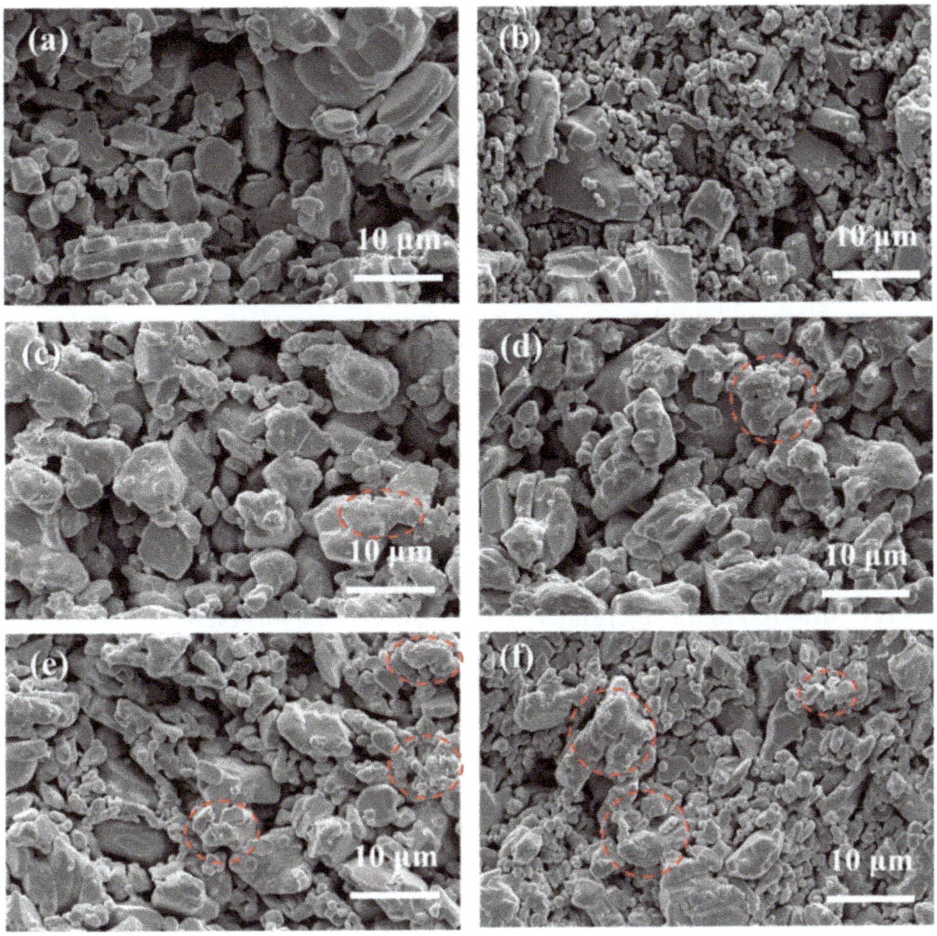

Figure 2.39. Surface SEM images of the Al_2O_3 ceramics: (a) C10F0 and (b) C7F3 sintered at 1350 °C, (c) C10F0, (d) C9F1, (e) C8F2 and (f) C7F3 sintered at 1550 °C. Reprinted from [80], Copyright (2022), with permission from Elsevier.

as observed in figures 2.39(a) and (b). The samples sintered at 1550 °C contained large grains made of small ones, as seen in figures 2.39(c)–(f). Generally, powders with small particle sizes have stronger densification behavior than those with large particle size. Mass transfer tends to occur in between fine particles, thus forming merged grains [81]. The higher the content of fine particles, the more the merged grains are formed, owing to the higher sinterability of fine sized powders.

Porosity of the final ceramic samples was decreased with increasing the content of fine powders, while the sintering shrinkage and porosity of a given sample were decreased with increasing sintering temperature. For casting applications, the ceramic molds should have a certain level of porosity, for the internal ceramic cores to be ventilated and removed after casting. Usually, silica ceramic molds produced by using the conventional ceramic process possess porosity of 20%–30%.

Because Al_2O_3-based ceramic molds are more resistant to NaOH and KOH, they should have higher porosity for etching out. In this regard, both the C10F0 and C9F1 samples sintered at 1550 °C are suitable for such purpose, due to their porosities of 33% and 30% porosity.

Li *et al* used SLA printing to form Al_2O_3 ceramic cores, with focusing on the effect of sintering temperature on microstructure and mechanical properties [82]. With an increasing sintering temperature, the grain size of the final ceramics gradually increased, while the interlayer spacing was first reduced and then enlarged. Correspondingly, open porosity of the ceramics was significantly decreased. Both flexural strength and hardness of Al_2O_3 ceramics increased as the sintering temperature increased. After sintering at 1150 °C, the sample had a flexural strength of 33.7 MPa, with shrinkages of 2.3%, 2.4% and 5.3% in the *x*-, *y*- and *z*-directions, respectively, while the open porosity was 37.9%.

To prepare printing slurries, 495 g Al_2O_3 powder dried at 200 °C for 5 h was slowly poured to 100 g photosensitive resin with strong stirring. After that, the mixtures were ball milled for 2 h with a planetary mill at 400 rpm. Then, the slurries were vacuum degassed for 10 min before printing. A cuboid model (50 mm × 4 mm × 3 mm) was drawn with UG software and imported into STL format. The model file was transferred to the 3D printer. The ceramic slurry was then transported to the 3D printer. The irradiation source was a 405 nm LED, with exposure energy of 10 mW · cm^{-2}, whereas the single layer was irradiated for 10 s. Green bodies were layer-by-layer printed with a layer thickness of 100 μm. After printing, the excessive slurry was cleaned with ethanol.

For debinding, the green bodies were heated to temperatures of 200 °C, 550 °C and 1000 °C for 2 h, at heating rates of 2, 1 and 5 °C · min^{-1}, respectively. The debound samples were cooled to 600 °C at a heating rate of 5 °C · min^{-1} and then to room temperature. For vacuum sintering, the samples were heated at 800 °C, at a heating rate of 10 °C · min^{-1} at vacuum of 10^{-3} Pa, while the sintering was conducted at different temperatures in the range of 1100 °C–1350 °C for 2 h, at a heating rate of 5 °C · min^{-1}.

Figure 2.40 shows SEM images of the ceramic samples vacuum sintered at different temperatures, revealing obvious delamination structure. The interlayer spacing was first decreased and then increased with increasing sintering temperature. The minimum interlayer spacing was 5.94 μm, observed in the sample after sintering at 1200 °C. No cracks were present in all samples, mainly because of the slow heating rate and vacuum atmosphere, so the densification of the green bodies was enhanced. Moreover, layers in the *x*–*y* plane were exposed to homogeneous planar for irradiation, while joining occurred through uniform polymerization in the *x*–*y* plane.

Interfacial joining between the adjacent layers was realized due to the irradiation in the *z*-direction, whereas the energy distribution could be different from that in the *x*–*y* plane. In other words, the efficiency of the interfacial polymerization was different from that in the *x*–*y* plane, since more photopolymer molecules are attracted. As a result, layer delamination of the printed samples was reflected in the sintered ceramics. In addition, because the sintering process was carried out in

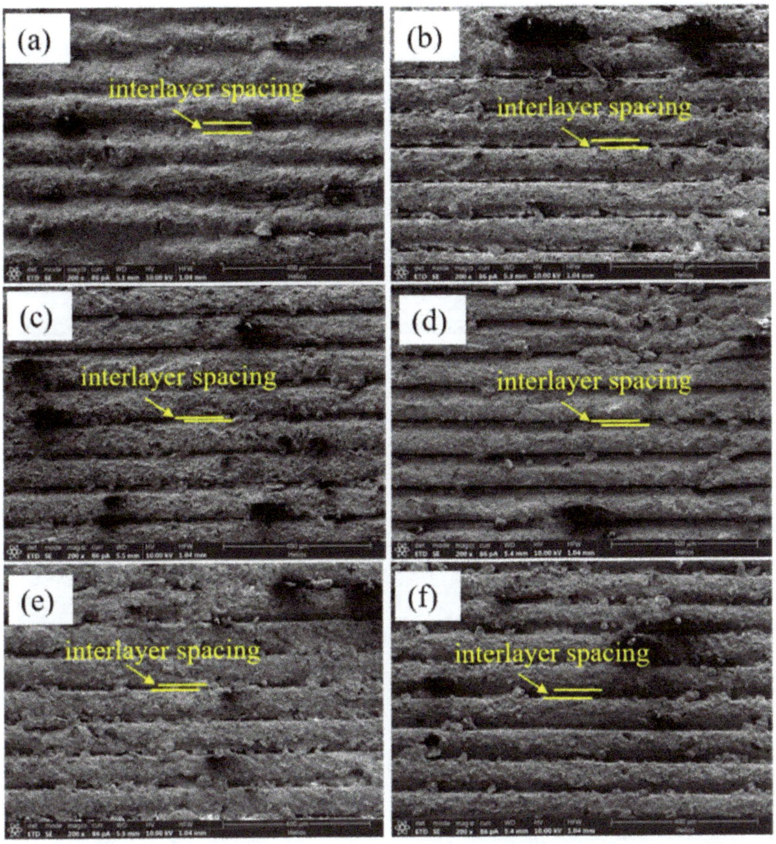

Figure 2.40. Low magnification SEM images of the samples with delamination microstructure after sintering at different temperatures: (a) 1100 °C, (b) 1150 °C, (c) 1200 °C, (d) 1250 °C, (e) 1300 °C and (f) 1350 °C. Reprinted from [82], Copyright (2020), with permission from Elsevier.

vacuum, relative pressures could be produced in the samples at high temperatures [83], thus leading enhancement in densification of the green bodies during the sintering process.

In the samples sintered at 1100 °C, the interlayer spacing was relatively large, owing to the volatilization of resin. As the sintering temperature was raised to 1150 °C, the driving force for densification was increased, so that the connection of the adjacent layers was enhanced, thus leading to reduction in the interlayer spacing. However, as the sintering temperature is further raised to 1300 °C, serious shrinkage took place in the samples. Consequently, the interlayer spacing increased.

With increasing sintering temperature, the increase in vapor pressure followed an exponential manner, while the grain growth was strongly accelerated. As a result, densification of the green bodies was promoted. Figure 2.41 shows SEM images of the ceramics after sintering at different temperatures. At lower sintering temperatures (e.g., 1100 °C–1150 °C), numerous pores and voids were observed between ceramic particles, indicating incomplete densification. In contrast, at higher

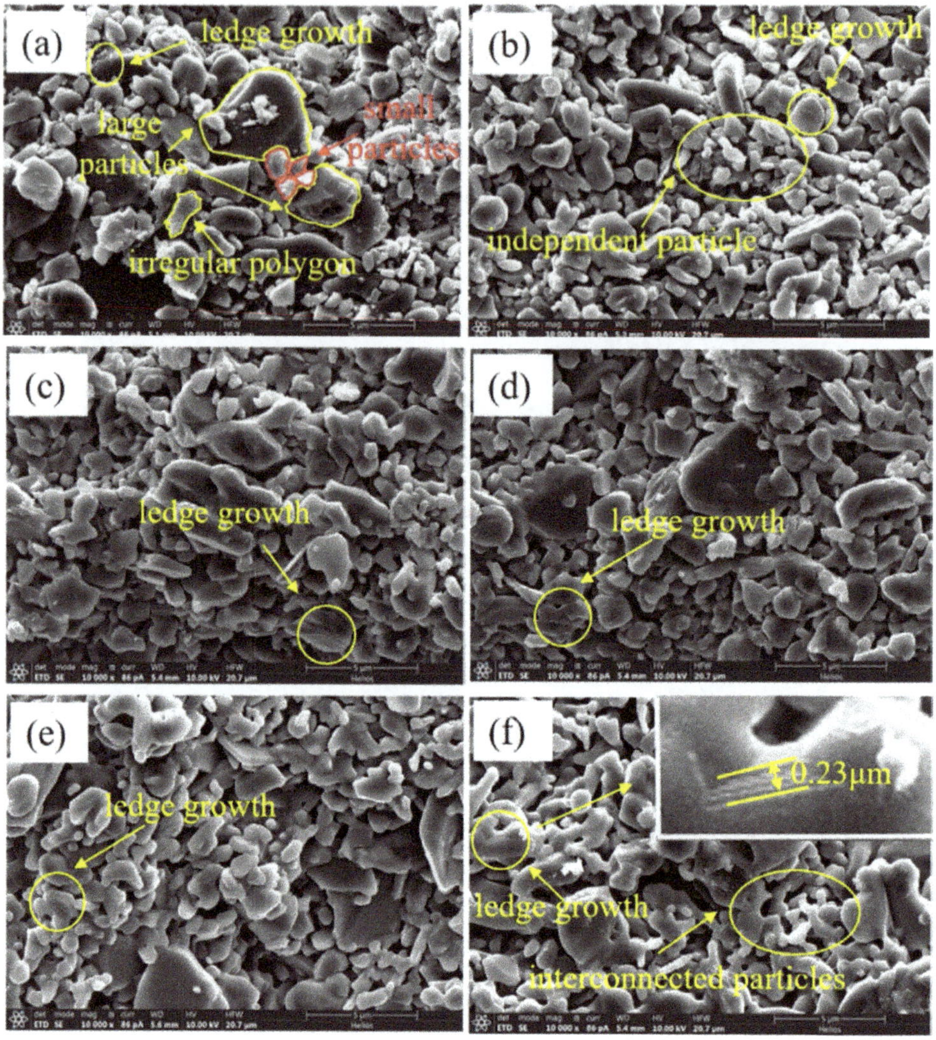

Figure 2.41. High magnification SEM images of the samples sintered at different temperatures: (a) 1100 °C, (b) 1150 °C, (c) 1200 °C, (d) 1250 °C, (e) 1300 °C and (f) 1350 °C. Reprinted from [82], Copyright (2020), with permission from Elsevier.

temperatures (e.g., 1300 °C–1350 °C), particle coalescence became evident, and the grain size increased, suggesting a tendency for Al_2O_3 particles to merge into larger grains. Additionally, ledges formed around larger particles, likely due to their curved edges extending into the surrounding matrix of finer grains.

Figure 2.42 shows representative TEM images of the ceramics sintered at different temperatures. Large particles with irregular shapes consisted of numerous small ones, which were formed due to the sintering effect at high temperatures. This is evident from the clear edges and lines visible in the samples sintered between 1150 °C and 1200 °C. It was reported that the boundary between ceramic particles was not

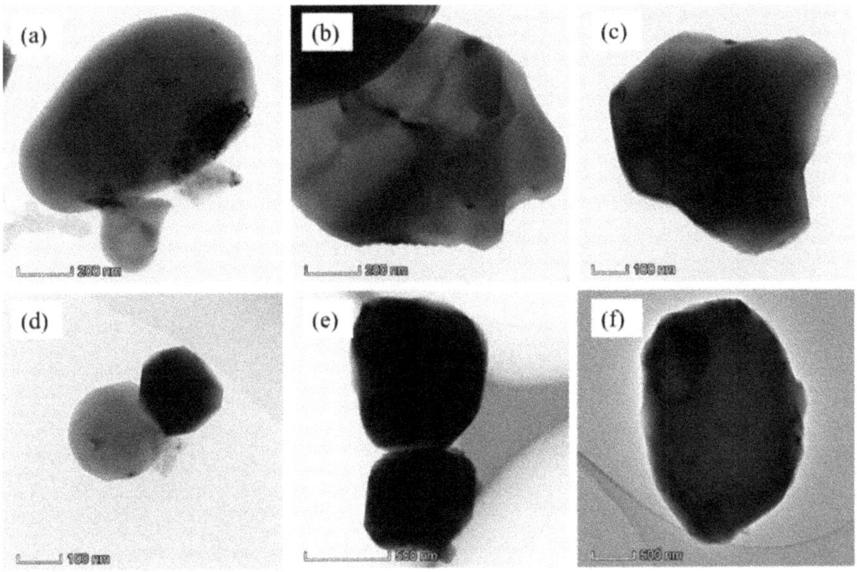

Figure 2.42. Representative TEM images of the samples after sintering at different temperatures: (a) 1100 °C, (b) 1150 °C, (c) 1200 °C, (d) 1250 °C, (e) 1300 °C and (f) 1350 °C. Reprinted from [82], Copyright (2020), with permission from Elsevier.

distinguishable, if an excessively thick α-Al_2O_3 support was present [84]. Since sintering drove the particles together, the interface between particles would be invisible.

It was observed that the shrinkage in the *z*-direction was more pronounced than that in the *x*- and *y*-directions, with a ratio in shrinkage of about 2, implying that the densification of the 3D-printed Al_2O_3 green bodies was anisotropic during the sintering process. Bulk density of the samples was increased from 2.4 to 2.9 $g \cdot cm^{-3}$, corresponding to a reduction in open porosity from 38.7% to 23.8%, as the sintered temperature was raised from 1100 °C to 1350 °C. As mentioned earlier, porosity of Al_2O_3 ceramic molds should be sufficiently high to ensure removal after casting, which could be readily controlled by adjusting the sintering temperature.

Zhang *et al* attempted to fabricate Al_2O_3 ceramics by using the SLA printing process, aiming to achieve high densification behaviors and hence high mechanical strengths, by utilizing fine powders and sintering aids. Meanwhile, defect-free Al_2O_3 ceramic lattice structures were prepared, with high precise and high promising mechanical strength [85]. Al_2O_3 powder, with coarse particles ($D_{50} = 10.3$ μm), was used as the main raw material to prepare Al_2O_3 ceramic slurries. Al_2O_3 powder, with fine particles ($D_{50} = 1.05$ μm), was incorporated as a minor constituent. The coarse and fine Al_2O_3 powders are denoted as c-Al_2O_3 and f-Al_2O_3, respectively. TiO_2 (anatase, 60 nm) and MgO (purity > 98%, 50 nm) were utilized as sintering aids.

HDDA and TMPTA were mixed with a volume ratio of 4:1, as photosensitive resin monomers. TPO and KOS110 were used as photoinitiator and dispersant, respectively. Firstly, Al_2O_3 powders, photosensitive resin monomers (HDDA and

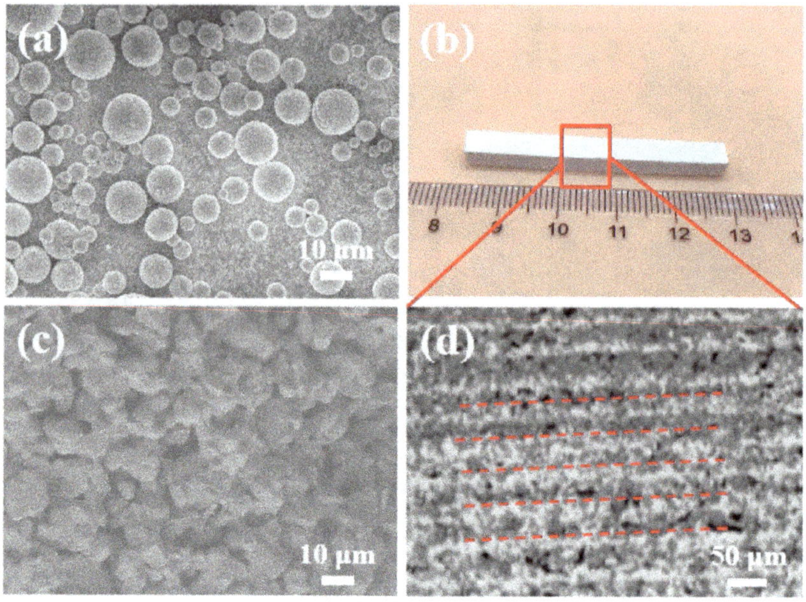

Figure 2.43. Microstructure of c-Al_2O_3 and f-Al_2O_3: (a) SEM image of the c-Al_2O_3 and f-Al_2O_3 particles. (b) Sintered body of A_c. (c, d) Fracture surface and cross-section SEM images of A_c. Reprinted from [85], Copyright (2021), with permission from Elsevier.

TMPTA), photoinitiators (TPO), dispersant (KOS110), and sintering additives (MgO and TiO_2) were mixed, through ball milling, with ZrO_2 balls, for 12 h at 400 rpm, leading to homogeneous stable photosensitive Al_2O_3 slurries, with different compositions. The Al_2O_3 slurries, with the compositions of c-Al_2O_3, c-Al_2O_3/TiO_2, c-Al_2O_3/f-Al_2O_3/TiO_2 and c-Al_2O_3/f-Al_2O_3/TiO_2/MgO, were denoted as A_c, A_cT, A_cA_fT and A_cA_fTM, respectively.

During printing, the slurries were irradiated with an ultraviolet light at a wavelength of 405 nm. The 3D models were constructed by using Solidworks software, which were then sliced using the AutoCera facility. The printed thickness of each slice was set at 50 μm, with an irradiation time of 4 s. Then, Al_2O_3 green bodies were obtained, and the excessive slurry was thoroughly cleaned with ethanol. After drying, the green bodies were debinded through calcination at high temperatures. The debinding process was conducted at 600 °C for 2 h and then pre-sintered at 1000 °C for 2 h in air. The debinded samples were sintered at 1600 °C for 2 h in air.

Both the c-Al_2O_3 and f-Al_2O_3 consisted of spherical particles, as shown in figure 2.43(a). Without involving the fine powder and sintering aids, pristine A_c was made first from c-Al_2O_3 slurry with a concentration of 50 vol%. A representative sintered sample of A_c is demonstrated in figure 2.43(b). It was obviously observed that the sample exhibited typical layer-by-layer characteristics, due to the printing process. Fracture surface and cross-sectional SEM images of the A_c ceramics are depicted in figures 2.43(c) and (d). After sintering, the thickness of each printed layer in the A_c ceramics was 45.5 μm, corresponding to a linear shrinkage of 9.1% in height direction. Defects were observed occasionally, with relatively low relative density of about 50%.

To reveal the effect of fine powder and sintering aids on densification behaviors of Al_2O_3 ceramics, A_cT, A_cA_fT and A_cA_fTM were prepared in a similar way. Figure 2.44 shows SEM images of the three Al_2O_3 ceramics, with low and high magnifications. As seen in figures 2.44(a) and (b), the addition of fine powder promoted densification and grain growth of the Al_2O_3 ceramics. With the introduction of TiO_2 as the sintering aid, a solid solution would be formed in the Al_2O_3 matrix, because they have a very close lattice constant. Once Al^{3+} in Al_2O_3 was replaced by Ti^{4+}, lattice distortion would be induced and cation vacancies were

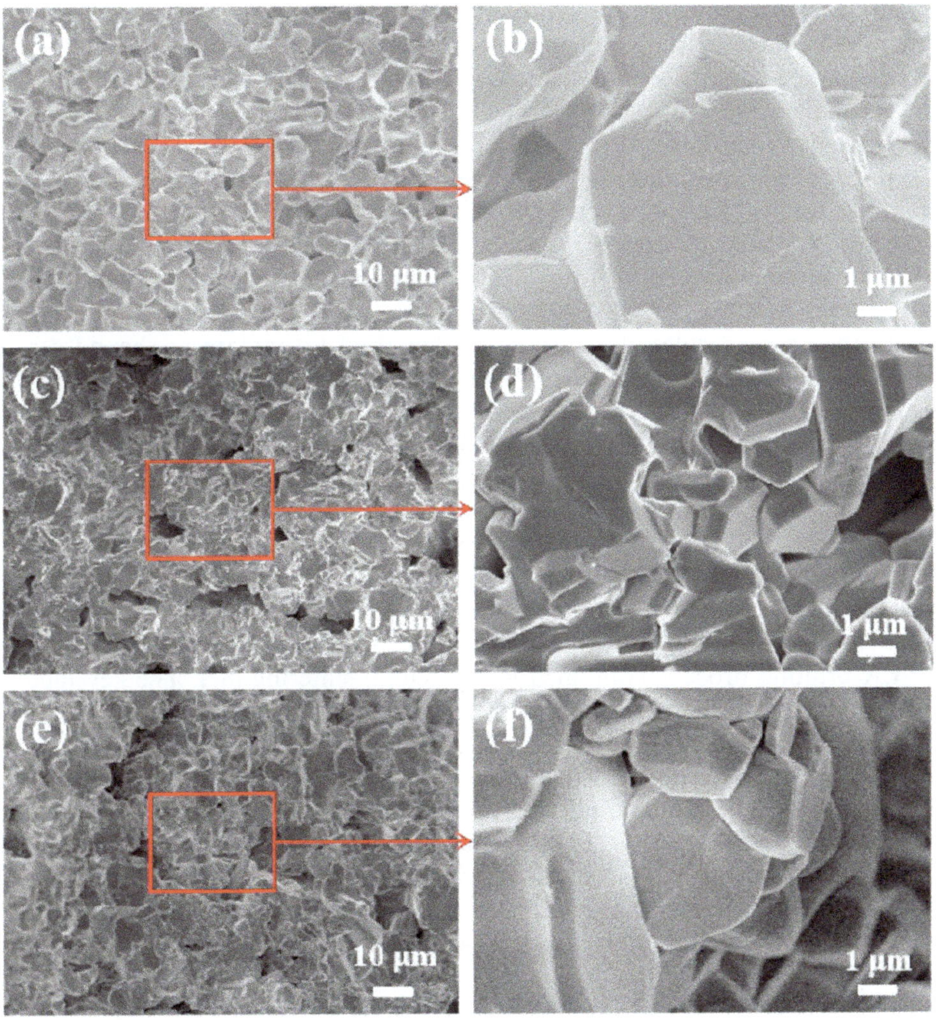

Figure 2.44. SEM images of the Al_2O_3 ceramics with different compositions: (a, b) A_cT, (c, d) A_cA_fT and (e, f) A_cA_fTM. Reprinted from [85], Copyright (2021), with permission from Elsevier.

generated, thus triggering the mechanism of solid phase sintering. Furthermore, the incorporation of Ti^{4+} increased the content of Al vacancies, thus enhancing the ion diffusion and hence densification of the Al_2O_3 ceramics. Nevertheless, the addition of TiO_2 was unable to eliminate the pores in the ceramics, as illustrated in figures 2.44(c) and (d).

When TiO_2 and MgO were introduced simultaneously, together with the addition of f-Al_2O_3, the pore size was reduced and the grain growth was suppressed, as presented in figures 2.44(e) and (f). With the presence of MgO, a liquid phase was formed between MgO and Al_2O_3 at the sintering temperature, due to its relatively low melting point. As a result, the fluidity of the Al_2O_3–TiO_2–MgO system enabled the occurrence of liquid phase sintering, in which the liquid phase promoted the rearrangement rate of ceramic particles. Normally, a high content of MgO liquid phase is beneficial to the accumulation of ceramic particles. As the fraction of liquid phase was sufficiently high, the sintering mechanism of dissolution-precipitation started to take action. Therefore, the A_cA_fTM sample exhibited no abnormal grain growth behavior [86].

Relative densities of the A_cT, A_cA_fT and A_cA_fTM ceramics are 80.8%, 84.7% and 93%, respectively, consistent with the observation in their microstructure. Flexural strength of the A_cT ceramics was 37.8 MPa, while that of the A_cA_fTM sample was raised to 178.8 MPa. Therefore, it is concluded that mechanical strength of Al_2O_3 ceramics can be enhanced by using the combination of fine powder addition and dual sintering aids.

With the above results, the authors further tried to prepare a -shaped ceramic lattice structure, by using the slurry made of A_cA_fTM. The lattice was designed to have a geometry with the considerations of the difference in shrinkage in different directions, as shown in figure 2.45(a), while the printing process and facility are illustrated in figure 2.45(b). The corresponding printed green body and sintered sample of the Al_2O_3 ceramic lattice structure are depicted in figures 2.45(c) and (d).

The 3D Al_2O_3 ceramic lattice structure exhibited a uniformly distributed load when compressed, indicating that they were free of delamination and defects. Specifically, the compressive strength was 34.9 MPa.

Sun *et al* reported translucent Al_2O_3 ceramics with complicated shapes by using the SLA printing process, combined with vacuum sintering [87]. The key factor to the development of Al_2O_3 ceramics with desired optical performance is the bonding characteristics of the interlayer. Commercially available high purity α-Al_2O_3 (99.99%), with an average particle size of $D_{50} = 0.4$ μm, specific surface area (SSA) of $3.7\ m^2 \cdot g^{-1}$ and theoretical density of $3.96\ g \cdot cm^{-3}$, was utilized as the raw material for slurries. $MgAl_2O_4$ (99.99%) was used as a sintering aid. Photosensitive slurries with a csolid loading level of 55 vol% were obtained by dispersing the powders in commercial photocurable resin. Diphenyl (2,4,6-trimethyl benzoyl) phosphine oxide (TPO, molar mass $348.4\ g \cdot mol^{-1}$) was employed as the photoinitiator.

Green bodies were printed with the slurries irradiated with UV light at a wavelength of 405 nm. The printed items were set to have a wall thickness of <2 mm to ensure integration of the green bodies. The pixel size was 50 μm, while the printed thickness of

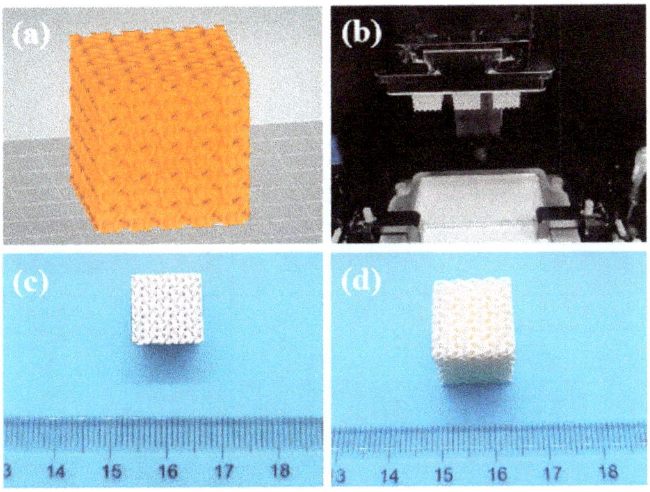

Figure 2.45. Complex-shaped ceramic lattice structure made of A_cA_fTM: (a) 3D model of the ceramic lattice. (b) Stereolithography additive manufacturing facility. (c) Photograph of the green body of the ceramic lattice. (d) Photograph of the sintered body of the ceramic lattice. Reprinted from [85], Copyright (2021), with permission from Elsevier.

the single layer was 50 μm. Intensity of the UV light was 24 mW · cm^{-2}. The green bodies were debounded at 1100 °C in air and sintered in a vacuum at temperatures of 1650 °C–1700 °C for 2 h, with vacuum levels to be 10^{-3}–10^{-4} Pa.

It was found that the curing depth (C_d) increased with increasing irradiation time, as demonstrated in figure 2.46(a). In the time range of 1–5 s, the curing depth increased almost linearly, reaching about 160μm for 5 s. After that, the increase in the curing depth was flattened. This nonlinear increment in the curing depth is associated with the dynamic change in the light transmission through the printed thin layers of the ceramic resin during the curing process.

The conversion extent of polymerization versus irradiation time is shown in figure 2.46(b), which is estimated from the Raman shift bands at 1720 and 1636 cm^{-1}. With increasing irradiation time, the extent of conversion at the upper side experienced a strong increase below 7 s, which was then stable at a level of about 59%. The conversion rate in the bottom layer increased from 55% to 70% until 8 s, owing to the repeated irradiation. However, over the irradiation time range of 3–15 s, the variation in the degree of polymerization was not very significant. As observed in the inset in figure 2.46(b), the variation trend in residual stress of the samples was not monotonic. The optimal irradiation time seemed to be 3 s, because the thin sheet was entirely flat, while the curing depth reached about 90 μm and the polymerization extent approached 40%–55%. Figure 2.46(c) shows photographs of the samples with different irradiation time.

The green bodies had an average density of 2.23 g · cm^{-3}, corresponding to a relative density of about 57% and a total porosity of 43%, according to the theoretical density of Al_2O_3. Thermogravimetric (TG) and differential scanning calorimetry (DSC) were used to analyze weight loss and thermal decomposition

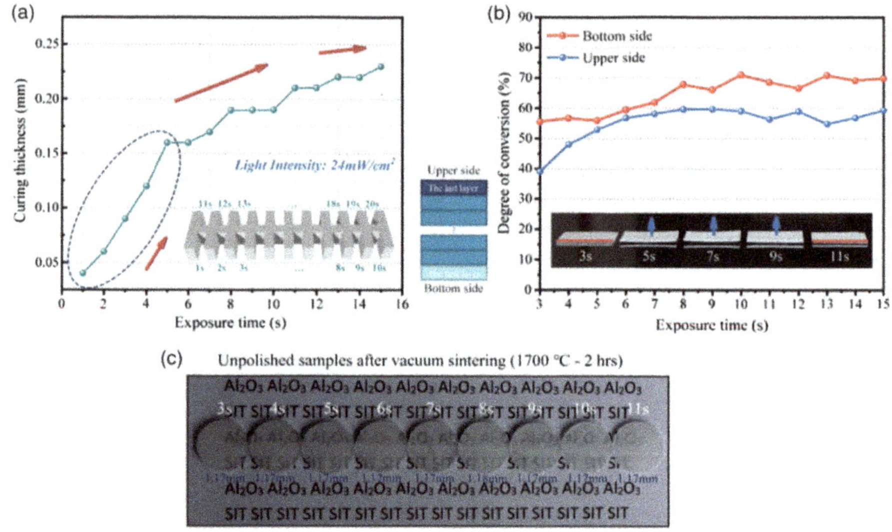

Figure 2.46. (a) Curing depth (C_d) versus irradiation time in the range of 1–20 s. (b) Degree of conversion on bottom and upper sides of the samples estimated from Raman spectra. (c) Photographs of the as-sintered Al_2O_3 ceramic samples with irradiation times of 3–11 s. [87] John Wiley & Sons. © 2021 Wiley-VCH GmbH.

behavior of the printed green bodies. Figure 2.47(a) shows representative TG and DSC curves of the green bodies. Based on these observations, the debinding process included heating stages at 200 °C and 500 °C, with heating rates of 1 and 0.5 °C · min^{-1}, respectively. The first weight loss stage occurred at about 20 °C and stopped at about 150 °C, corresponding to evaporation of the absorbed water molecules and residual organic solvents. Therefore, the green bodies could be dried at 150 °C for 1 h without the formation of cracks.

The second stage began at 150 °C until 600 °C, over which solvents and other organic additives were all burned out. The samples were heated to 1100 °C to ensure strength for sample treatment. SEM images of the Al_2O_3 ceramics sintered at 1650 °C and 1700 °C are depicted in figures 2.47(c)–(e). The samples all exhibited homogeneous microstructures, with grain sizes of 7–18 and 6–25 μm, after sintering at 1650 °C and 1700 °C, respectively.

Figure 2.48(a) shows inline transmission curves of the translucent Al_2O_3 ceramics. The sample derived from the green body printed with an irradiation time of 3 s exhibited higher optical performances. On top of that, the authors fabricated translucent Al_2O_3 ceramics with different shapes and sizes, as demonstrated in figures 2.48(b)–(g), indicating feasibility of the SLA 3D printing technology in producing ceramics with complicated structures.

Xu *et al* further optimized mechanical performances of SLA-printed Al_2O_3 ceramics by using powders with a wider size distribution [88]. Specifically, the slurry made of three Al_2O_3 powders, with average particle sizes of 5, 3 and 1 μm, at compositions of 40, 30 and 30 wt%, with solid loading of 50 vol%, led to ceramics

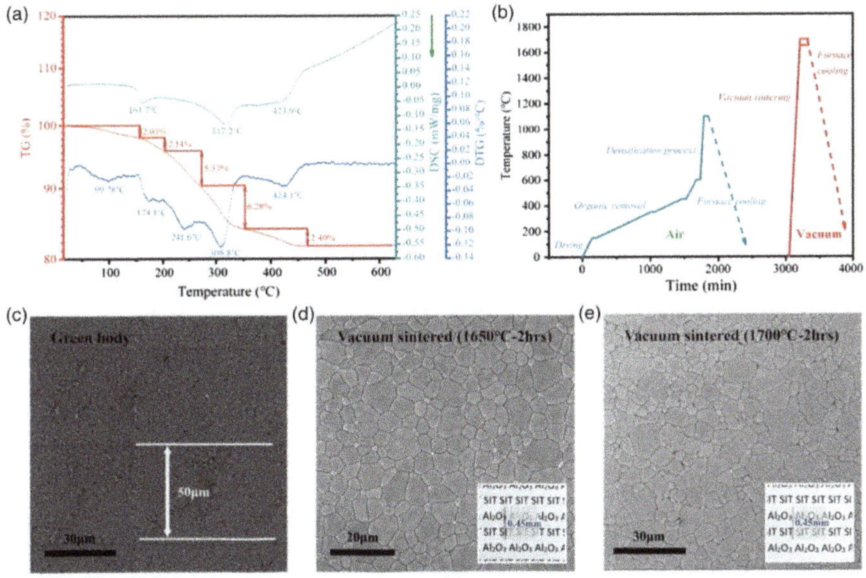

Figure 2.47. (a) Representative TG-DSC curves of the green bodies. (b) Debinding and vacuum sintering process curves in the experiment. SEM images of Al_2O_3 ceramics: (c) printing with UV light irradiation for 3 s, (d) vacuum sintered at 1650 °C for 2 h and (e) vacuum sintered at 1700 °C for 2 h. [87] John Wiley & Sons. © 2021 Wiley-VCH GmbH.

with porosity of 0.06% and flexural strength of 482 MPa, after sintering at 1650 °C for 4 h. Stearic acid (SA, $C_{18}H_{36}O_2$) and TW were used as dispersing agents. MgO (D_{50} = 0.3 μm) and Y_2O_3 powder were included as sintering aids, with a total concentration of 1 wt%.

A mixed resin (NIAM-Al-022) was utilized to prepare slurries, with a small curing shrinkage and high curing efficiency, owing to the low degree of functionality, high reactivity and high refractive index to minimize the lateral scattering. The ceramic powders and dispersants were mixed through ball milling for 12 h to ensure homogeneity of the slurries. Then, a photoinitiator was introduced, fully dissolved by ball milling for another 20 min. After that, the mixtures were stirred for 20 min without grinding balls for further homogenization with a planetary mixer.

Light irradiation was realized with a 355 nm quasi-continuous wave UV laser. The beam diameter was 30 nm, platform size was 300 mm × 100 mm, the output power was 80 mW, the layer thickness was 50 μm, the hatching space was 0.04 mm, and scan speed was 2400 mm · s^{-1}. For debinding, the green bodies were heated to 200 °C, 600 °C and 800 °C, at heating rates of 1, 0.5 and 1 °C · min^{-1}, respectively. The samples were held for 100 min at 300 °C, 400 °C, 500 °C, 600 °C and 800 °C. The debound bodies were sintered at 1500 °C–1650 °C for time durations of up to 6 h.

It was found that densification behaviors and flexural strength of the Al_2O_3 ceramics were strongly dependent on sintering temperature. For a given sintering time, the porosity and flexural strength were decreased and increased gradually with

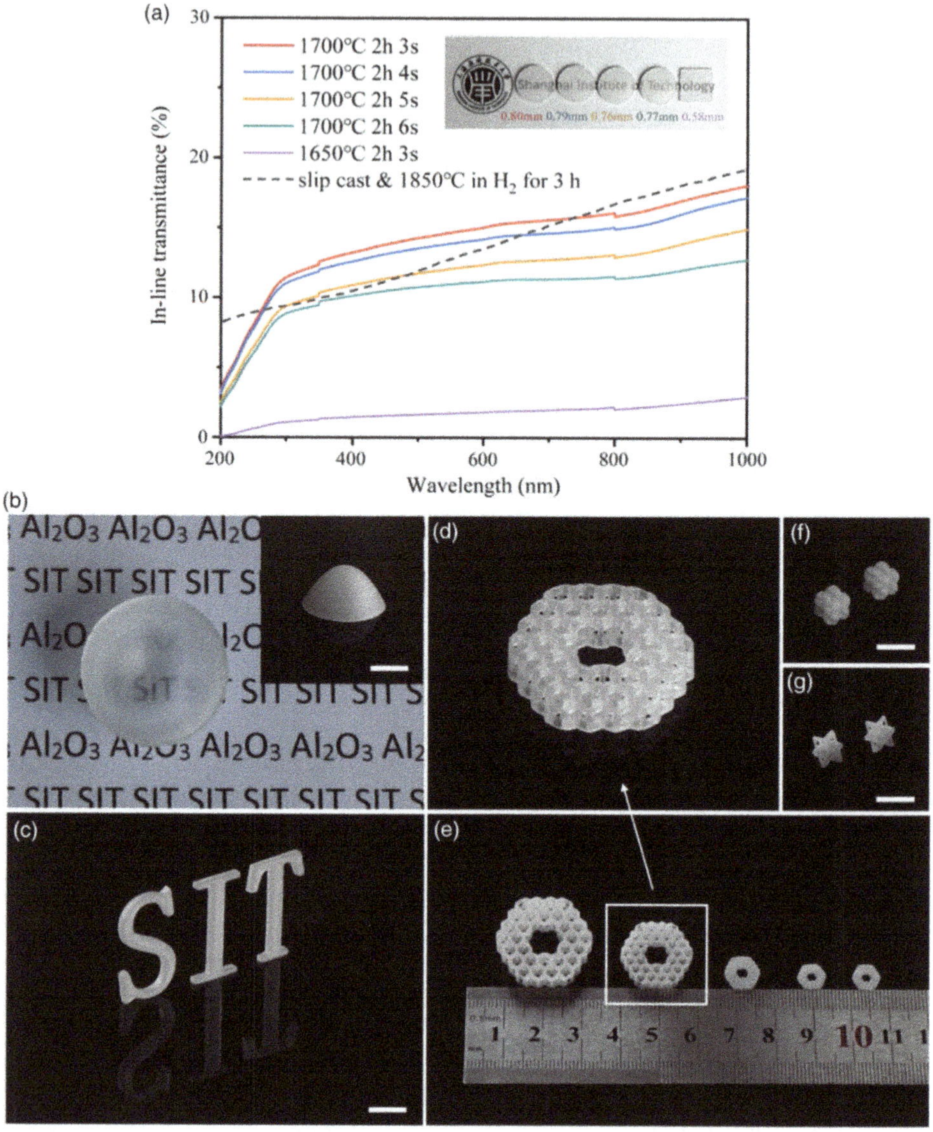

Figure 2.48. (a) Inline transmittance spectra of the samples with thickness normalized to 0.8 mm. Photographs of the translucent Al_2O_3 ceramics: (b) dome (φ20mm × 15 mm, scale bar = 10 mm), (c) 'SIT' letters (scale bar = 10 mm), (d, e) honeycomb structure, (f) P-cell TPMS structure (scale bar = 5 mm), (g) octet-truss structure (scale bar = 5 mm). [87] John Wiley & Sons. © 2021 Wiley-VCH GmbH.

increasing sintering temperature. Both the linear and volume shrinkages of the samples were increased with increasing sintering temperature. In the sample sintered at 1650 °C, the linear shrinkages in the directions of the *x*-axis, *y*-axis and *z*-axis were 20.97%, 19.7% and 21.5%, respectively. Relative densities of the Al_2O_3

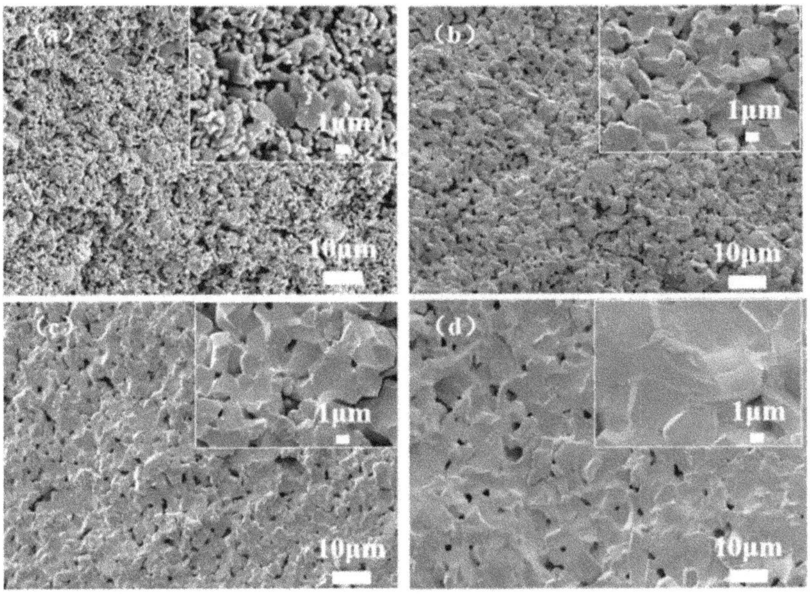

Figure 2.49. Fracture surface SEM images of the Al_2O_3 ceramic samples sintered at different temperatures: (a) 1500 °C, (b) 1550 °C, (c) 1600 °C and (d) 1650 °C. Reproduced from [88], with permission from Springer Nature.

ceramics sintered at 1500 °C, 1550 °C, 1600 °C and 1650 °C were 70.9%, 77.1%, 87% and 92.8%, respectively.

Figure 2.49 shows SEM images of Al_2O_3 ceramic samples after sintering at different temperatures. Obviously, the grain size of the samples was monotonically increased with increasing sintering temperature. With the progress of the sintering process, grain boundary diffusion of the ceramics was accelerated. Meanwhile, small gains grew into large ones.

Figure 2.50 depicts fracture surface SEM images of the Al_2O_3 ceramics sintered at 1650 °C for different time durations. As seen in figure 2.50(a), the Al_2O_3 particles were essentially loosely packed, with relatively small grain size and high porosity, thus having relatively low mechanical strength. As the time duration was prolonged, both the grain growth and densification were promoted, leading to a rapid increase in mechanical strength. In the sample sintered for 2 h, irregular strips were present, owing to the interconnection of the multiple irregular pores, as observed in figure 2.50(b). As the sintering time increased to 4 h, most of the pores were absent, while the sample was highly densified, as revealed in figure 2.50(c). However, a too-long sintering time would bring out abnormal grain growth, while the mechanical strength would not further increase.

Beyond the fundamental study, the authors further prepared combustion chamber carriers with the slurries through SLA printing, as presented in figure 2.51. Such devices could be used at working temperatures of above 1000 °C. A Kelvin cell lattice with tetrakaidecahedron cells, with dimensions of 18 mm × 52 mm × 18 mm,

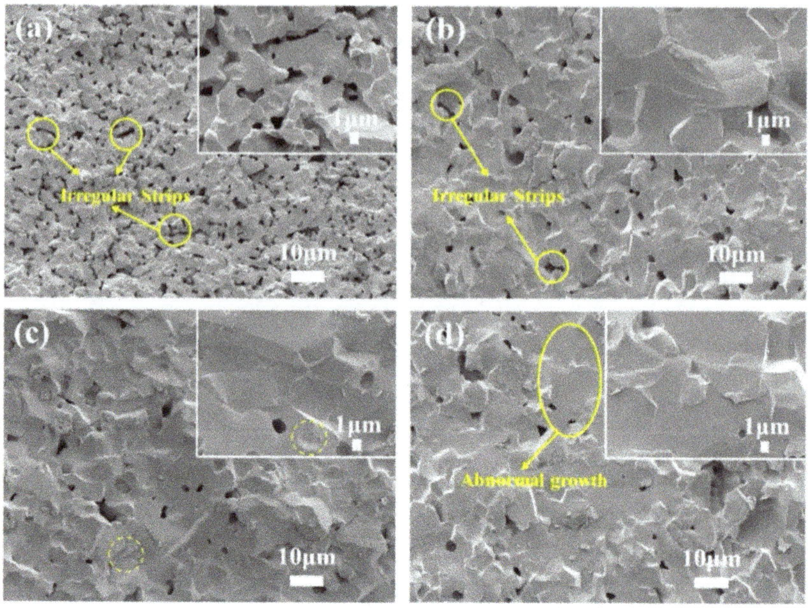

Figure 2.50. Fracture surface SEM images of the Al_2O_3 ceramic samples after sintering at 1650 °C for different time durations: (a) 0 h, (b) 2 h, (c) 4 h and (d) 6 h. Reproduced from [88], with permission from Springer Nature.

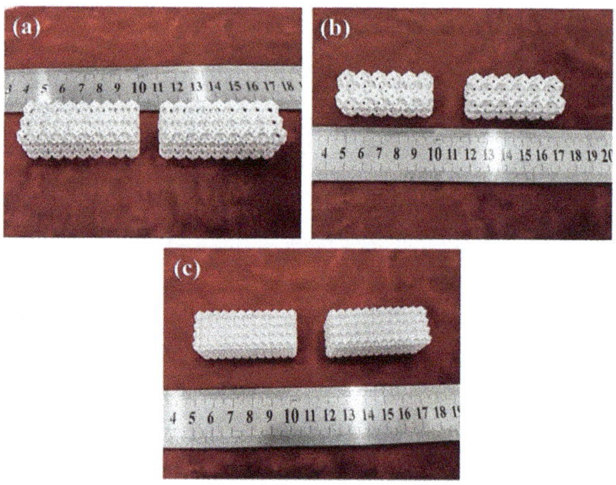

Figure 2.51. Combustion chamber carriers made of SLA-printed Al_2O_3 with different wire diameters: (a) 1.2 mm, (b) 1.8 mm and (c) 0.9 mm. Reproduced from [88], with permission from Springer Nature.

were printed, while sintered structures exhibited high accuracy to the models. The three templates had different wire diameters, with porosities in the range of 85%–95%.

Li *et al* developed ceramic cores based on Al_2O_3 enhanced with SiO_2, by using the SLA printing process [89]. The effects of the content of SiO_2 on mechanical properties of the ceramic cores were examined. As the content of SiO_2 increased from 0 to 30 wt%, flexural strength of the ceramics was increased from 13.3 MPa to 46.3 MPa, owing to formation of mullite ($Al_6Si_2O_{13}$). Too high a content of SiO_2 led to a reduction in flexural strength, since too much of the liquid phase was formed. By considering the requirement of real applications, 10 wt% SiO_2 was suitable to obtain ceramics with sufficiently high flexural strength of 35.6 MPa and open porosity of 47.5%.

Al_2O_3 and SiO_2 powders with different compositions were dispersed in photosensitive resin (E-Mode Light) to form ceramic slurries. For 3D printing, the laser had a wavelength of 385 nm, while the thickness of each printed layer was 100 μm and the model had dimensions of 40 mm × 5 mm × 4 mm. Microstructures of the Al_2O_3–SiO_2 ceramics are shown in figures 2.52 and 2.53. As the content of SiO_2 was

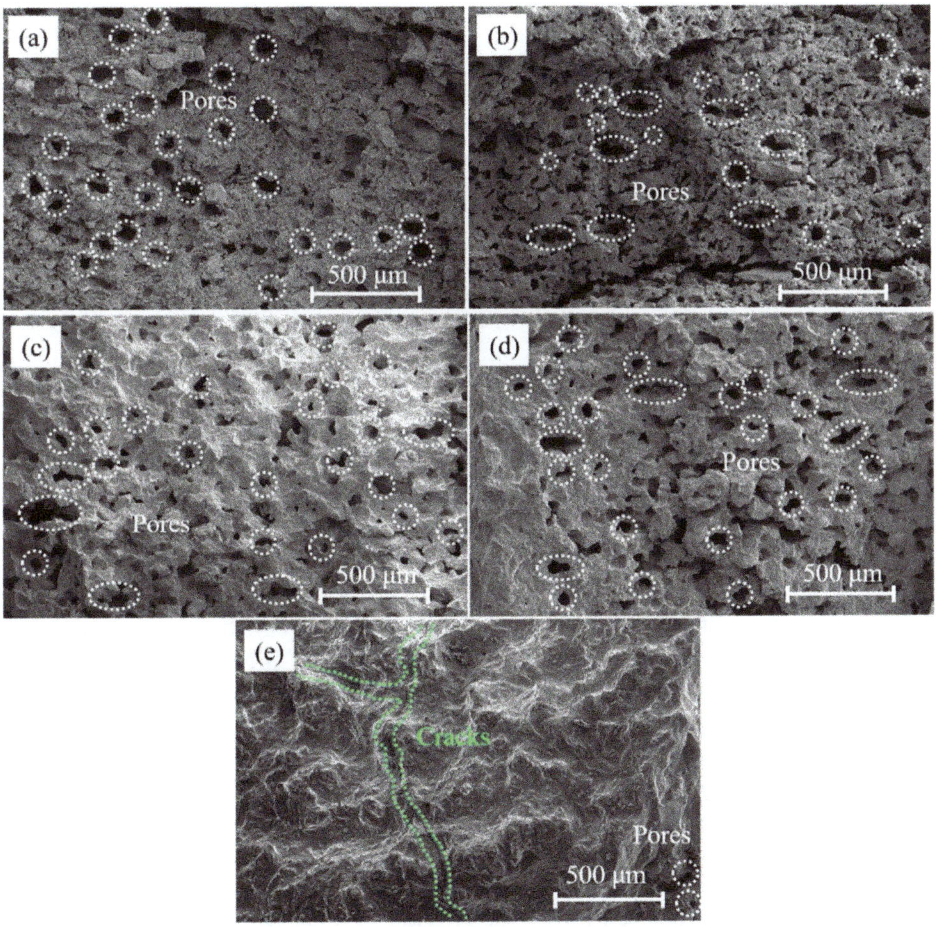

Figure 2.52. Low magnification SEM images of the sintered ceramics with different contents of SiO_2: (a) 0 wt %, (b) 10 wt%, (c) 30 wt%, (d) 50 wt% and (e) 70 wt%. Reprinted from [89], Copyright (2021), with permission from Elsevier.

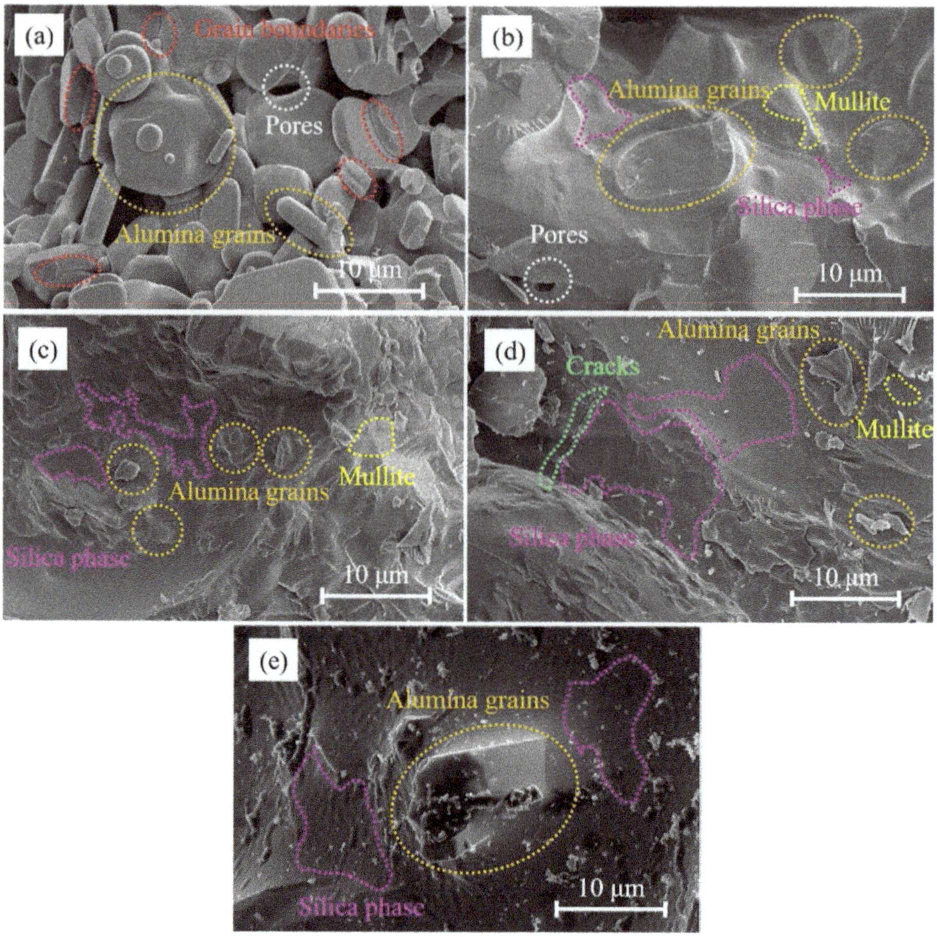

Figure 2.53. High magnification SEM images of the sintered ceramics with different contents of SiO_2: (a) 0 wt%, (b) 10 wt%, (c) 30 wt%, (d) 50 wt% and (e) 70 wt%. Reprinted from [89], Copyright (2021), with permission from Elsevier.

increased from 0 to 30 wt%, Al_2O_3 was the major phase in the final ceramics. The presence of SiO_2 led to formation of a liquid phase at high temperatures during the sintering process. The liquid phase would fill the open pores, thus promoting the densification. Meanwhile, the formation of mullite would increase density of the ceramic samples, because the density of mullite is higher than that of SiO_2. However, as the content of SiO_2 was higher than 50 wt%, the bulk density was decreased, because the density of SiO_2 is higher than those of SiO_2 and mullite. In addition, excessive SiO_2 resulted in a slight increase in open porosity, because of the increase in the number of cracks.

The sample with 70 wt% SiO_2 was nearly free of porosity, according to the SEM image, as observed in figure 2.52(e), while the measured open porosity was as high as 26%. This was probably because micropores were not visible in SEM images, owing to their relatively small sizes, while they could be measured based on capillary adsorption. Additionally, in the samples with excessively high content of SiO_2, mesopores might be formed, thus contributing to porosity. This assumption has been validated by experimental results.

High magnification SEM images of the sintered ceramics with various SiO_2 contents are depicted in figure 2.53. In the sample without SiO_2, Al_2O_3 grains, grain boundaries and pores are clearly visible in the SEM image, as seen in figure 2.53(a). The pores were mainly formed due to the stacking of the Al_2O_3 grains with irregular shapes, together with a relatively low extent of densification. With the addition of 10 wt% SiO_2, a liquid phase was produced in the sample. As revealed in figure 2.53 (b), the liquid phase was present in between the Al_2O_3 grains, filling into some pores. As the content of SiO_2 was $\geqslant$30 wt%, pores were entirely absent.

As observed in figure 2.53(c), the Al_2O_3 grains were covered by the SiO_2 phase, so that the Al_2O_3 grains were hardly visible. Comparatively, the Al_2O_3 grains were present in the samples with 50 and 70 wt% SiO_2, because of the significant increase in the fraction of the liquid phase. There were also cracks in the sample with 50 wt% SiO_2, as presented in figure 2.53(d). This observation was ascribed to the fact that SiO_2 has a higher Young's modulus and larger coefficient of thermal expansion than Al_2O_3. At the same time, internal stress was generated in the ceramic samples, owing to the shrinkage during the sintering process, thus triggering the formation of cracks in the SiO_2 phase.

The microstructures of the sample with 10 wt% SiO_2 were further characterized with TEM. It was confirmed that the ceramic grain boundaries in between the Al_2O_3 grains could be clearly observed, while the Al–Si boundaries could not be readily distinguished. In other words, the two phases were merged with each other, thus leading to high mechanical strength. Moreover, the binding force of Al–Si is stronger than that of Al–Al, owing to the difference in characteristics of the contact regimes. The boundaries of Al–Si could be identified by using the elemental distribution of Al and Si. The Al_2O_3 grains displayed boundaries with a higher content of Al, while the regions without clear boundaries possessed a higher content of Si.

More detailed TEM characterization results are shown in figure 2.54. Microstructural characteristics of the Al and Si regions in the Al_2O_3 ceramics with 10 wt% SiO_2 are almost the same. As demonstrated by the selected area electron diffraction (SAED) pattern, SiO_2 was present in an amorphous state to a certain degree, as illustrated in figure 2.54(b). A similar result was observed in the high-resolution transmission electron microscopy (HRTEM), as presented in figure 2.54(c). The presence of amorphous SiO_2 was also supported by XRD results.

As mentioned earlier, over the SiO_2 content range of 0–30 wt%, flexural strength of the ceramics was increased. Above 30 wt%, the flexural strength tended to decline. At the maximum point, the flexural strength was 46.3 MPa. In comparison, the sample with 10 wt% SiO_2 had a flexural strength of 35.6 MPa. As stated before, the application of ceramic cores requires flexural strengths of the materials to be just >20 MPa. Therefore, the ceramics with 10 and 30 wt% SiO_2 could be candidates in terms of mechanical performance. In contrast, flexural strength of the pure Al_2O_3

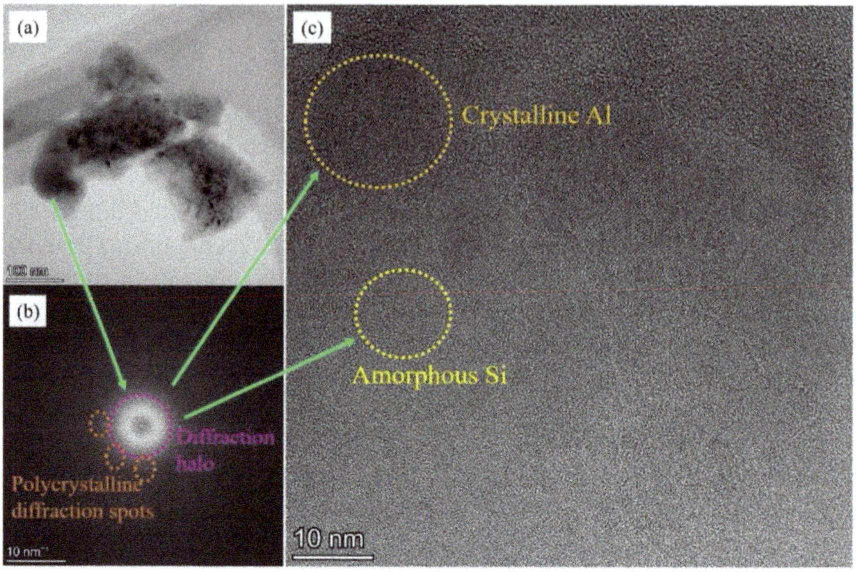

Figure 2.54. Representative TEM images of the sintered ceramics with 10 wt% SiO_2: (a) TEM image, (b) selected area electron diffraction (SAED) pattern and (c) high-resolution transmission electron microscopy (HRTEM) image. Reprinted from [89], Copyright (2021), with permission from Elsevier.

ceramics was just 13.3 MPa, while those of the samples with 50 and 70 wt% SiO_2 were decreased to 6.3 and 10.3 MPa, respectively. In this regard, excessive SiO_2 should be avoided.

Xing *et al* prepared ZrO_2 ceramic bars with three different dimensions by using the SLA 3D printing process [90]. Surface of the ZrO_2 ceramics exhibited anisotropic characteristics. Roughness of the horizontal surface was <0.41 μm, while that of the vertical surface was 1.07 μm. The warpage and flatness were below 40 and 27 μm, respectively, for the ceramic bars with dimensions of 3 mm × 4 mm × 80 mm. Meanwhile, relative density, flexural strength, fracture toughness and hardness of the ZrO_2 ceramics were 99.3%, 1154 ± 182 MPa, 6.37 ± 0.25 mPa · $m^{1/2}$ and 13.90 ± 0.62 GPa, respectively.

The printing slurries were prepared by using a photosensitive acrylic resin with a solid loading of 55 vol%. ZrO_2 powder with particle size of about 200 nm was used as the raw materials. An ultraviolet laser with a wavelength of 355 nm was employed to scan the ceramic slurry layers in a layer-by-layer way. A standard alternating x/y-raster scanning pattern was utilized to cure the printed layers. In this way, the concentration of stress could be minimized, so that the warpage was more effective than either an x- or y-scan. Two curing hatch spacings, with $H_S = 50$ and 160 μm, were applied in the individual scanning directions of the x- and y-axes, respectively.

Once the previous pattern cured, the solid platform was moved downwards by a distance of 25 μm with respect to the thickness of the printed layer, so that the ceramic slurry was re-coated on the surface of the cured layer. As a result, the next layer was cured analogously. ZrO_2 green bodies would be developed by repeating

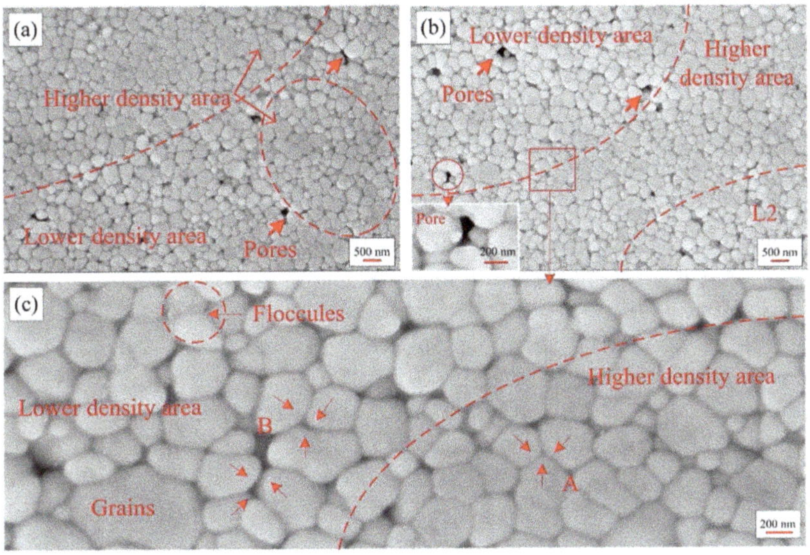

Figure 2.55. Polished surface SEM images of the ZrO_2 ceramics: (a) horizontal surface and (b, c) vertical surface. Reprinted from [90], Copyright (2017), with permission from Elsevier.

the steps. The laser power for irradiation, the light spot diameter, the laser beam scanning speed, the curing depth and curing width were 300 mW, 35 μm, 5 m s^{-1}, 50 and 80 μm, respectively.

The printed green bodies after drying were subject to debinding and sintering processes. The organic components were removed by calcining at 500 °C in N_2 for 18 h, at the heating rate of 0.2 °C · min^{-1}. The samples were then sintered at 1450 °C for 2 h, at the heating rate of 7 °C · min^{-1} in N_2. The final ZrO_2 ceramics bars had *x*-, *y*- and *z*-dimensions of 3 ± 0.032 mm, 4 ± 0.026 mm and 80 ± 0.023 mm, respectively. Meanwhile, ZrO_2 ceramic gears and fiber ferrules could also be fabricated with this process.

Figure 2.55 shows SEM images of horizontal (*XOY*) and vertical (*XOZ*) surfaces of the sample, revealing grain boundary structures. The two surfaces exhibited similar profiles. The ZrO_2 ceramics had grains with an average size of 0.3 μm, without significant grain growth. As observed in figure 2.55(a), the sample possessed a microstructure with different regions that were different in degree of densification. In the dense area, the grains were closely compacted, corresponding to flat and compact characteristics. In the loose area, pores were present with randomly distributed oval features. A similar profile was observed on the *XOZ* surface, as illustrated in figure 2.55(b).

As labeled as A in figure 2.55(c), the three grains are tightly connected, forming a triangle grain boundary with strong cohesion. According to mechanics, the samples subjected to higher strain rates might be fractured when the tension stress is higher than the bonding strength of the grain boundaries. By contrast, in the low-density area, labeled as B, there is an inter-grain space between two grains. In this case, the bonding strength is relatively weak. Therefore, an intergranular fracture could occur predominantly in the low-density area. This fracture mode is unable to block the propagation of cracks.

In addition, a certain number of floccules composed of superfine particles were present on the ZrO_2 grains. Wrinkles are observed on the fractured surface of the ceramics, as a toughening mechanism, to deflect cracks. At overloadings, this fracture mode is capable of inhibiting the propagation if cracks form. As a result, ceramics with such a surface morphology would have strong fracture toughness. Additionally, the small grain size and narrow grain size distribution would be favorable for improving the flexural strength of the ZrO_2 ceramics.

Sun *et al* compared five dispersants, including stearic acid (SA), oleic acid (OA), Disperbyk (BYK), coupling agent KH560 and variquat CC 42 NS (CC), when they were used to prepare slurries for SLA 3D-printed ZrO_2 ceramics [91]. Then, the authors claimed that BYK, KH560 and OA were promising dispersants for producing slurries with sufficiently low viscosity, for nanosized ZrO_2 powder. Among the three dispersants, BYK was the best one. Slurries with 3 wt% BYK exhibited shear thinning behavior, with a viscosity of 1680 mPa · s at the shear rate of 18.6 s^{-1}.

The nanosized powder was ZrO_2 doped with 3 mol% Y_2O_3 (3YSZ). The UV curable resin was composed of ACMO and PEGDA, which had a relatively low viscosity of <40 mPa · s, thus guaranteeing the effect of recoating during the 3D printing process. Before making slurries, the 3YSZ powder was heated in vacuum at 100 °C for one day for drying and removal of moisture, to prevent particle agglomeration. Dispersants, with concentrations of 0.1–8 wt%, were mixed with an ethanol solution by stirring for 1 h at room temperature. After the addition of the ceramic powder, the suspension was stirred at 70 °C for 4 h to promote the attachment of the dispersants on the particle surface of ceramic powder. The excess dispersant was filtrated with the aid of centrifugation and rinsing with ethanol. The precipitates were dried at 80 °C for 12 h. After drying, the powder was screened through an 80-mesh sieve.

The UV curable resin was a mixture of 48 ml ACMO and 32 ml PEGDA, which had a viscosity of 23.4 mPa · s. The dispersant treated 3YSZ powder, with a solid loading of about 42 vol%, was dispersed in the photosensitive resin. The mixture was stirred for 24 h and then homogenized with homogenizer, forming slurries for printing. UV light with a wavelength of 405 nm was used to cure the printed layers, at an irradiation energy density of 8.5 mJ · cm^{-2}. The green bodies were sintered at 1450 °C for 2 h, at heating and cooling rates of 3 and 5 °C · min^{-1}, respectively.

Figure 2.56 shows photographs of representative green body and sintered zirconia towers, which were derived from the slurry with 3 wt% BYK, while the thickness of the printed layer was 10μm and the irradiation energy was 20 mJ · cm^{-2}. The shrinkages in the x–y- and z-directions were 21.9% and 28.9%, respectively. Figure 2.57 shows representative SEM images of the sintered samples, revealing that the 3YSZ ceramics were fully densified with almost no pores and abnormal grain growth.

Song *et al* used SLA 3D printing technology to develop ZrO_2 ceramics, with nanosized and microsized ZrO_2 powders, with average particle sizes of D_{50} = 20 nm

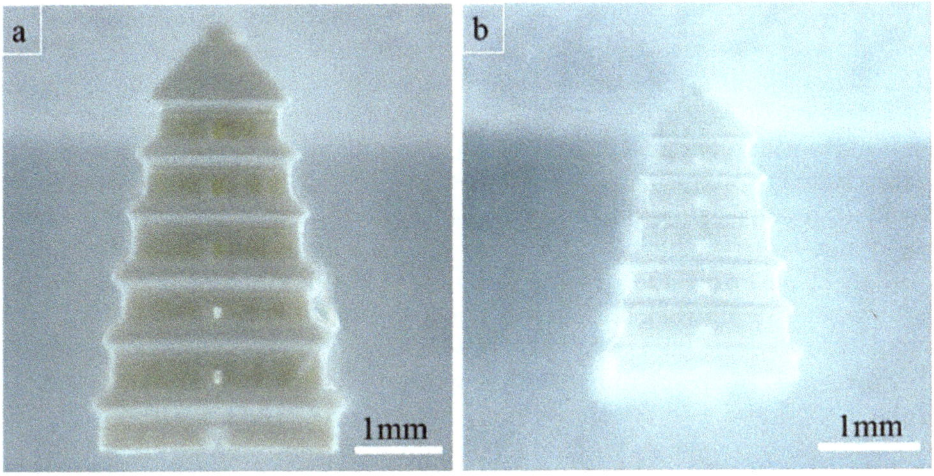

Figure 2.56. Photographs of the zirconia towers: (a) green body and (b) sintered item. Reprinted from [90], Copyright (2017), with permission from Elsevier.

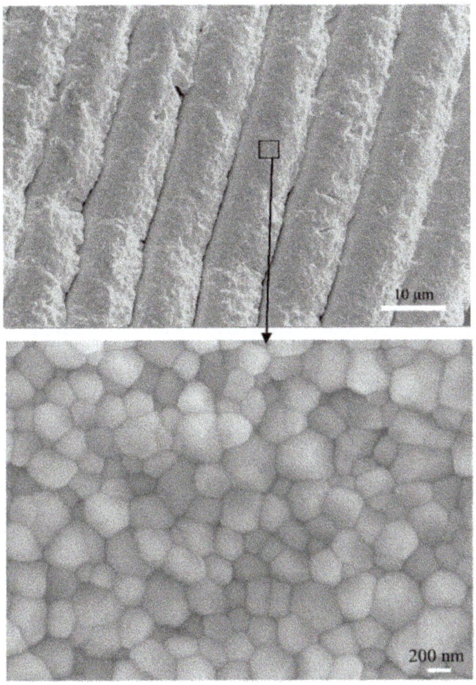

Figure 2.57. Representative SEM images for the sintered 3YSZ ceramics. Reprinted from [90], Copyright (2017), with permission from Elsevier.

and $D_{50} = 13$ μm [92]. To ensure efficient dispersion, the ZrO_2 powders were modified with acrylate groups through hydrolysis and condensation of a silane coupling agent (APTMS, 3-acryloxypropyl trimethoxysilane). The APTMS-modified ZrO_2 ceramic powders were dispersed in mixtures of di-functional and tri-functional acrylate monomers with non-reactive diluents via interpenetrating networks.

The nanosized and microsized ZrO_2 powders were mixed at volume ratios of 70:30, 50:50, 30:70 and 0:100. After surface functionalization, the ceramic powders were blended with monomers, with 2 wt% photoinitiator, 2 wt% dispersant and non-reactive diluent. Volume content of the APTMS-coated ZrO_2 was 50 vol%. Slurries were obtained after ball milling for 4 h at a rotation speed of 400 rpm. The optimal ratios of non-reactive diluents (IPA) were studied by adjusting the IPA contents to be 10, 15, 20, 30 and 40 vol%. In addition, the effect of the polymer network structures was examined by adopting different weight ratios of HDDA and TMPTA monomers, including 100:0, 70:30, 50:50, 30:70 and 0:100. In the case of controlling the content of IPA, the ceramic nanocomposite slurries were denoted as HTD-Ix, with H, T, D, I and x to be referred to the contents of HDDA, TMPTA, DMPA, IPA and IPA, respectively. When the photoinitiator was replaced from DMPA to BASF, 'D' in the sample name was changed to 'B'. When the control factor was the content of TMPTA, the ceramic nanocomposite slurries were named HIB-Ty, where y is the content of TMPTA.

The SLA printing process was conducted at the single layer thickness of 50 μm, with a UV laser with a wavelength of 355 nm, at intensities of 0.2–0.4 W, as depicted in figure 2.58(a). The as-printed green bodies were rinsed with IPA, followed by drying at 100 °C for 24 h in vacuum. The dried green bodies were debinded by calcining at temperatures of 150 °C, 465 °C and 650 °C, for 2 h, 4 h and 2 h, respectively. They were then sintered at 1000 °C for 1 h and 1450 °C for 2 h.

To print structures with different angles and precise dimensions using low viscosity resins, supports are required for the printing process, otherwise, the printed items would be sagged and distorted. However, if the slurries have sufficiently high viscosity, any structures could be printed, as demonstrated, as shown in figure 2.58 (b). Interestingly, it was found that the high viscosity slurries themselves acted as supports, due to their low flowability, if the fraction of the nanosized powder was sufficiently high. Therefore, high viscosity ceramic nanocomposite slurries could be used to realize supportless SLA 3D printing, thus avoiding the washing process of supports and hence minimizing the materials consumption.

For the APTMS-coated ZrO_2 ceramic nanocomposite slurries with a solid loading level of 50 vol%, the 3D-printed green bodies possessed a dimension of 10 mm × 10 mm × 1 mm. With increasing content of TMPTA, both the rate of curing and shrinkage of the green bodies increased, owing to the increase in cross-linking density of the systems. As a result, 3D printing could not be realized, for thickness of the printed layers to be over the critical value. To modify the shrinkage rate of TMPTA, UV absorber, 2-(2-hydroxy-5-methylphenyl) benzotriazole, Tinuvin P, was incorporated in the ceramic nanocomposite slurries, which acted through influencing the photo-curing reaction by reducing the depth of curing. The content of the UV absorber was 0.5 wt% of monomers.

In the ceramic nanocomposite resins without the addition of the UV absorber, the curing depth with nearly the same, with a value of 141 μm, as the laser power was 200 mW. Even as the laser intensity was raised to 400 mW, the curing depth was

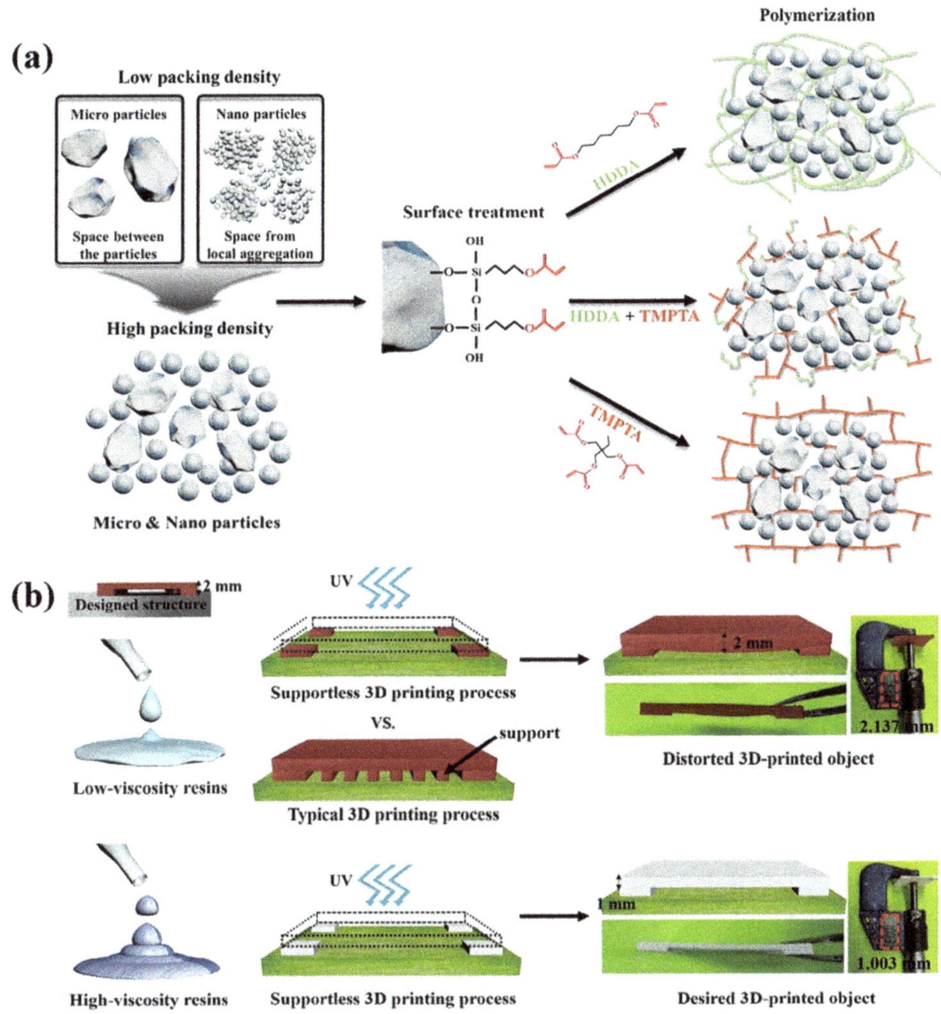

Figure 2.58. High-viscosity ZrO_2 ceramic nanocomposite resins: (a) Schematic diagram of preparation processes of the high-viscosity ZrO_2 ceramic nanocomposite resins with different polymer network structures for supportless SLA printing. (b) Comparison of the 3D-printed items with low-viscosity and high-viscosity resins produced by using supportless SLA printing. Reprinted from [92], Copyright (2019), with permission from Elsevier.

decreased, for the ceramic nanocomposite resins with 0.5 wt% of UV absorber. This variation trend was similar for ceramic nanocomposite resins with a controlled UV absorber. Comparatively, the curing depth of slurries with the controlled UV absorber was larger than those of the ones with 0.5 wt% UV absorber.

Although the curing depth decreased when introducing the UV absorber, the printed green bodies with a thickness of 1 mm could be readily fabricated, with slurries having different contents of TMPTA content. For the green bodies from the ceramic nanocomposite slurries with 0.5 wt% UV absorber, the volume shrinkages during sintering were 50.4%, 46.8%, 43.6%, 44.0% and 49.5%, for HIB-T0, HIB-T30, HIB-T50, HIB-T70 and HIB-T100, respectively. As the content of the UV absorber was too high, the samples exhibited cracks and pores. In this respect, the content of the UV absorber should be adjusted according to the content of TMPTA.

Photographs of the 3D-printed green bodies and sintered ceramics are depicted in figure 2.59(a), while cross-sectional SEM images of the sintered ceramics are presented in figure 2.59(b), which were derived from the ceramic nanocomposite

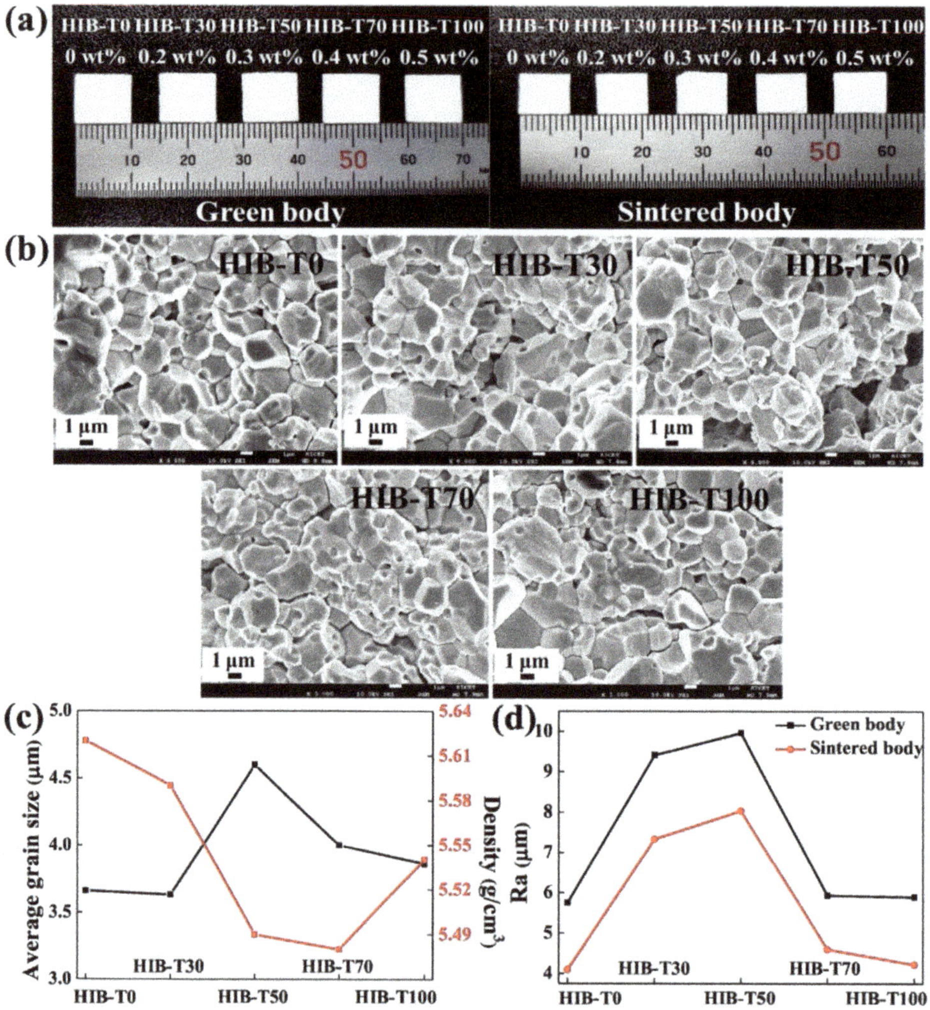

Figure 2.59. Properties of the 3D-printed samples from the APTMS-coated ZrO_2 ceramic nanocomposite resins with various contents of TMPTA and UV absorber: (a) Photographs of the green bodies and sintered bodies, (b) cross-sectional SEM images of the sintered bodies, (c) average grain size and density of the sintered bodies and (d) surface roughness of the green bodies and sintered bodies. Reprinted from [92], Copyright (2019), with permission from Elsevier.

slurries with controlled UV absorber. Even though the content of the UV absorber was limited, there were cracks and pores in both the green bodies and sintered ceramics, for HIB-T30, HIB-T50 and HIB-T70. The number of the cracks and pores was reduced, as the UV absorber was incorporated. Such a problem still cannot be tackled, because it is linked to the phase transition of ZrO_2 during cooling process from high sintering temperatures to room temperature.

Figure 2.59(c) shows average grain size and density of the sintering ceramics, estimated from the SEM images in figure 2.59(b), as functions of the content of TMPTA. The average grain sizes of the HIB-T0 and HIB-T100 derived samples were 3.66 and 3.86 μm, respectively, while that of the HIB-T50 one was the largest, with a value of 4.60 μm. The samples related to HIB-T0 and HIB-T100 exhibiting less cracks and pores, displayed small average grain sizes and high densities. Generally, the density of ZrO_2 ceramics is inversely proportional to the average grain size. Moreover, ceramics with large average grain sizes are usually of more cracks and high porosity.

Figure 2.59(d) depicts surface roughness of the 3D-printed green bodies and sintered ceramics. After sintering, the surface roughness was reduced by 1–2 μm, which means that the surface of the green bodies became smooth during the sintering process. The sintered ceramics derived from the HIB-T30 and HIB-T50 slurries possessed rough surfaces, owing to their low curing effect. In comparison, the surfaces of the samples from the HIB-T0, HIB-T70 and HIB-T100 based slurries displayed high smoothness.

Fully dense 8 mol% yttria-stabilized zirconia (8YSZ) ceramics have been fabricated by using the SLA printing technique for electrolyte self-supported fuel cell applications [93]. The printed planar green bodies could be fully densified after sintering at 1450 °C. The sintered 8YSZ ceramics exhibited a total electrical conductivity of 2.18×10^{-2} S · cm^{-1} at 800 °C, showing strong potential for practical applications. The electrolyte self-supported fuel cell displayed a power density of 114.3 mW · cm^{-2}, when using Ni-8YSZ cermet as the anode and $La_{0.8}Sr_{0.2}MnO_3$ (LSM) as the cathode.

The 8YSZ UV resin slurry (CPM-S-8YSZ) had a solid loading of 50 vol%. The preparation of the 8YSZ planar electrolytes involved five stages, including model design, 3D printing, cleaning/washing, debinding and sintering. A top-down DLP printer with *x*–y-plane resolution of 50 μm was used to print the slurry with designed structures. The UV light has a wavelength of 405 nm, while the thickness of each printed layer was 25 μm. After printing, the excessive slurry was washed out with ethanol, followed by an additional UV curing for 1 min. The green bodies were debinded at 600 °C for proper time durations in air, at a heating rate of 1 °C · min^{-1}. After the debinding process, the samples were sintered at different temperatures, in the range of 1300 °C–1500 °C, at a heating rate of 5 °C · min^{-1}.

Two different sizes of self-supported planar 8YSZ ceramics were fabricated for different purposes. The samples used to examine densification behavior and electrical conductivity had a dimension of 10 × 10 × 0.5 mm, while those for fuel cell assembly possessed a dimension of 15 × 15 × 0.2 mm. To construct the fuel cells, NiO-8YSZ and LSM slurries were brushed onto the surface of the sintered 8YSZ electrolyte plates. The NiO-8YSZ anode was sintered at 1400 °C, whereas the LSM cathode was heated at 1000 °C. Then, Au paste was coated on both sides of the plate and baked at 800 °C for 30 min to serve as electrodes for electrochemical testing.

The slurry had a low viscosity of 3.5 Pa · s at a relatively high solid loading level of 50 vol%. XRD results indicated that the ceramic powder was phase pure 8YSZ. Relative density of the sintered samples was increased with increasing temperature, reaching 97.5% after sintering at temperatures of ⩾1350 °C. The sample sintered at 1450 °C had a relative density of as high as 99.5%, which was kept unchanged with further increase in the sintering temperature.

Electrical conductivity of ceramic electrolytes is strongly dependent on grain size and density. Figure 2.60 shows fracture surface SEM images of the 8YSZ ceramics after sintering at different temperatures. Small-sized pores were present in the

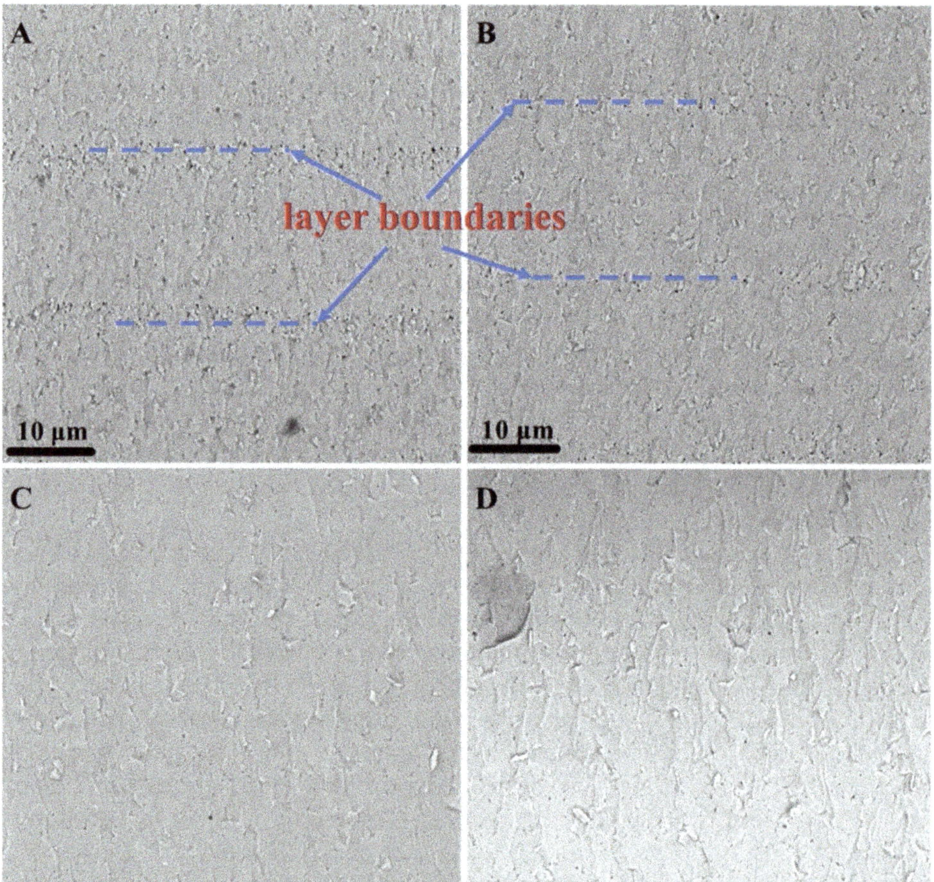

Figure 2.60. Fracture surface SEM images of the 8YSZ planar ceramics after sintering at different temperatures: (A) 1350 °C, (B) 1400 °C, (C) 1450 °C and (D) 1500 °C. Reprinted from [93], Copyright (2020), with permission from Elsevier.

samples sintered at 1350 °C and 1400 °C, corresponding to densification of >97.5%. As the sintering temperature was raised to ⩾1450 °C, almost no pores were observed in the samples. The samples sintered at 1350 °C and 1400 °C exhibited boundaries between adjacent printed layers, with more pores to be near the boundaries. Therefore, to achieve full densification, the sintering temperature was at least 1450 °C.

Total electrical conductivities of the samples sintered at 1350 °C, 1400 °C, 1450 °C and 1500 °C were estimated, with values of 1.96×10^{-2} S · cm^{-1}, 2.08×10^{-2} S · cm^{-1}, 2.18×10^{-2} S · cm^{-1} and 1.91×10^{-2} S · cm^{-1}, respectively. The drop in electrical conductivity of the sample sintered at 1500 °C was caused by the decrease in the conductivity of the grain boundary, owing to the increase in grain size. Therefore, sintering temperatures in the range of 1400 °C–1450 °C are recommended to ensure adequate electrical performance for solid oxide fuel cell (SOFC) electrolytes, which require electrical conductivities between 2 and 4×10^{-2} S · cm^{-1}.

The 8YSZ ceramic sample sintered at 1450 °C was used to assemble SOFC cells, in the form of a membrane with a thickness of 200 μm. Figure 2.61 shows cross-sectional SEM images of the NiO-8YSZ/8YSZ/LSM cell, with the SLA-printed 8YSZ as the electrolyte. Due to the simple nature in coating of the samples, an exfoliating phenomenon was observed at the interface between the electrolyte and the cathode. It is believed that this issue can be overcome by using more sophisticated processing techniques.

Li *et al* studied light absorption characteristics of ZrO_2 and three colorful ZrO_2 mixtures when developing ZrO_2 ceramics by using SLA printing [94]. The absorbance of the colored ZrO_2 powder increased with increasing content of colorant, whereas the value of the yellow-colored powder was the strongest at the wavelength of 405 nm. Meanwhile, with increasing content of colorant, both the cure depth and excess cure width were decreased, owing to the light absorption of the ceramic components. The cure depth of the colored slurries followed a linear relationship with the logarithm of irradiation energy, while the excess cure width was governed by a nonlinear increasing behavior.

3 mol% Y_2O_3 doped ZrO_2 (3Y-TZP) powders, together with three ceramic pigments ($CoAl_2O_4$, $ZrSiO_4$ (Fe_2O_3) and $ZrSiO_4$ (Pr_2O_3), were used as the raw materials. The uniform photocurable resin was a mixture of PPTTA (ethoxylated pentaerythritol tetraacrylate), HDDA (1,6-hexanediol diacrylate), U600 (di-functional aliphatic urethane acrylate), 1-Octanol, PEG400 (polyethylene glycol) and hyper-dispersant. Ceramic slurries were obtained by blending the ceramic powders and the photocurable resin with ball milling for 2 h.

The UV light for curing had a wavelength of 405 nm. The cure behaviors of colorful slurries were measured as using cure depth and excess cure width, with single printed layers. The printed green bodies were subject to debinding with two steps, consisting of a vacuum debinding for quick pyrolysis of the organic components to carbon and an air debinding to eliminate the carbon. Then, the samples were sintered in air at 1450 °C for 2 h to obtain colorful 3Y-TZP ceramics.

Figure 2.62 shows SEM images and EDS spectra of all the ceramic powders. The 3Y-TZP powder had an average particle size of 150 nm (figure 2.62(a)), being finer

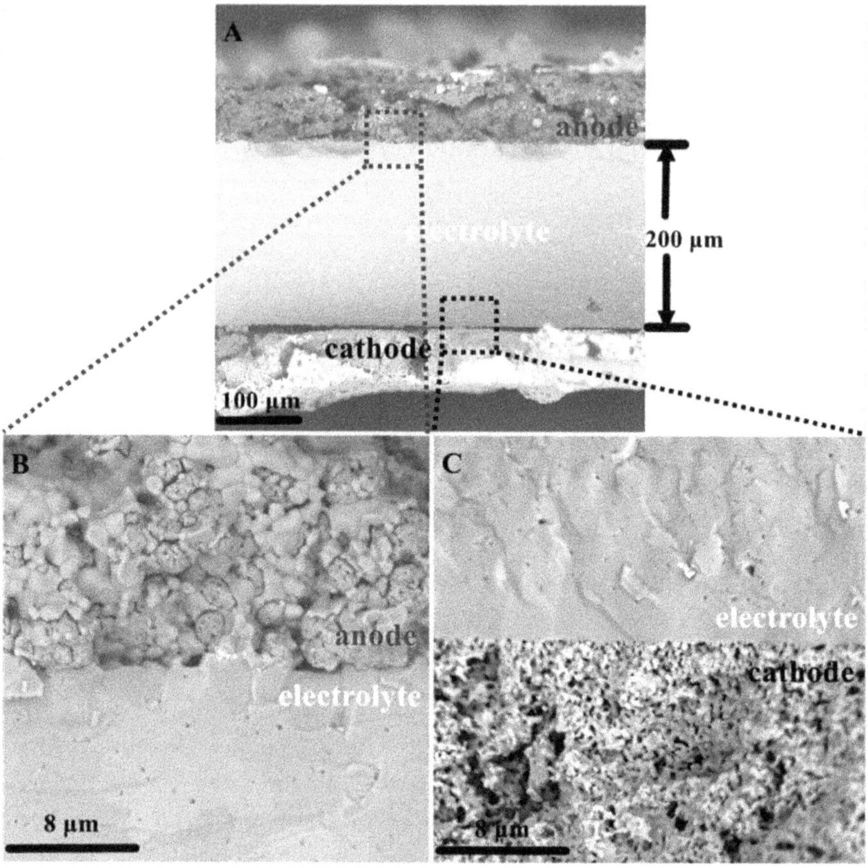

Figure 2.61. Cross-sectional SEM images of the flat full cell after being subject to electrochemical test. Reprinted from [93], Copyright (2020), with permission from Elsevier.

than the three ceramic pigments, $CoAl_2O_4$ (figure 2.62(b)), $ZrSiO_4$ (Fe_2O_3) (figure 2.62(c)), and $ZrSiO_4$ (Pr_2O_3) (figure 2.62(d)), which particle sizes of 2–4 µm. The 3Y-TZP exhibited a nearly zero absorption rate in the wavelength range of 350–800 nm. $CoAl_2O_4$ displayed a broad absorption peak over 500–700 nm, centered at 600 nm, while the other two pigments possessed a weak peak, at 410 and 480 nm, respectively. The $ZrSiO_4$ (Fe_2O_3) presented strong absorption over 350–550 nm, while the $ZrSiO_4$ (Pr_2O_3) showed an almost complete absorption at 350 nm.

The depth sensitivity followed a linear reduction with the content of colorant, suggesting that the depth attenuation was increased with the content colorant, since the light absorption rate of the colorful ceramic powders was increased with the content of colorant. At a given content, moreover, the blue slurry ($ZrO_2 + CoAl_2O_4$) had the highest depth sensitivity, with a range of 15.7–16.9 µm, while those of the red slurry ($ZrO_2 + ZrSiO_4$ (Fe_2O_3)) and the yellow slurry ($ZrO_2 + ZrSiO_4$ (Pr_2O_3)) were 10.6–12.3 and 11.7–13.3 µm, respectively.

Figure 2.62. SEM images and energy dispersive spectrometer (EDS) results of pure ZrO_2 and colorful powder mixtures: (a) pure ZrO_2 powder, (b) blue powder mixture (ZrO_2 + $CoAl_2O_4$), (c) red powder mixture (ZrO_2 + $ZrSiO_4$ (Fe_2O_3)) and (d) yellow powder mixture (ZrO_2 + $ZrSiO_4$ (Pr_2O_3)). Reprinted from [94], Copyright (2019), with permission from Elsevier.

The critical energy was also increased with increasing content of colorant. Specifically, as the content of the colorant was increased from 3 to 7 wt%, the increments in the critical energy, for slurries of blue slurry (ZrO_2 + $CoAl_2O_4$), red slurry (ZrO_2 + $ZrSiO_4$ (Fe_2O_3)) and yellow slurry (ZrO_2 + $ZrSiO_4$ (Pr_2O_3)), were increased from 4.0, 5.5 and 9.9 $mJ \cdot cm^{-2}$ to 7.1, 9.8, and 14.9 $mJ \cdot cm^{-2}$, respectively.

The excess cure width of all slurries was increased with the logarithm of irradiation energy in a nonlinear manner. The presence of the colorants was found to reduce the cure excess width, which became more pronounced at high irradiation energy levels. Such an effect was the most significant for the yellow colorant ($ZrSiO_4$ (Pr_2O_3)), followed by the red ($ZrSiO_4$ (Fe_2O_3)) and blue ($CoAl_2O_4$). The colorants would consume additional irradiation energy, resulting in an increase in the critical energy of width and hence a decrease in the excess width.

With 5 wt% addition level, the printing energy densities were 41, 60 and 60 $mJ \cdot cm^{-2}$, for $CoAl_2O_4$ (blue), $ZrSiO_4$ (Fe_2O_3) (red) and $ZrSiO_4$ (Pr_2O_3) (yellow), respectively. ZrO_2 based colorful ceramics with high densification rate and strong mechanical strength were obtained after sintering, corresponding to relative densities of 95.8%, 96.98%, 94.8% and 94.1%, for 3Y-TZP, blue ZrO_2 (ZrO_2 + 5 wt% $CoAl_2O_4$), red ZrO_2 (ZrO_2 + 5 wt% $ZrSiO_4$ (Fe_2O_3)) and yellow ZrO_2 (ZrO_2 + 5 wt% $ZrSiO_4$ (Pr_2O_3)), respectively.

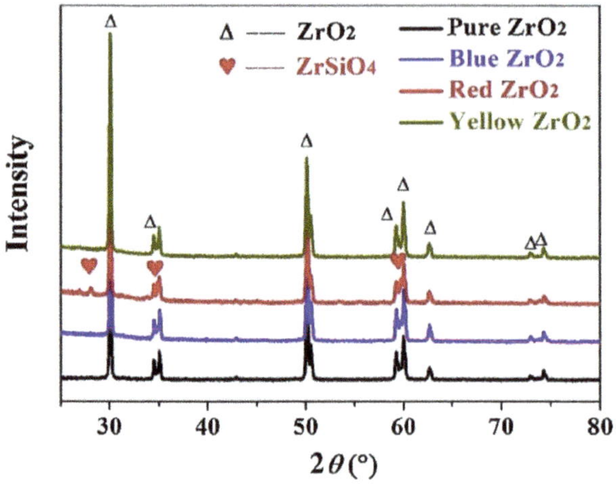

Figure 2.63. XRD patterns of the colorful ZrO_2 ceramics with 5 wt% colorant after sintering at 1450 °C for 2 h. Reprinted from [94], Copyright (2019), with permission from Elsevier.

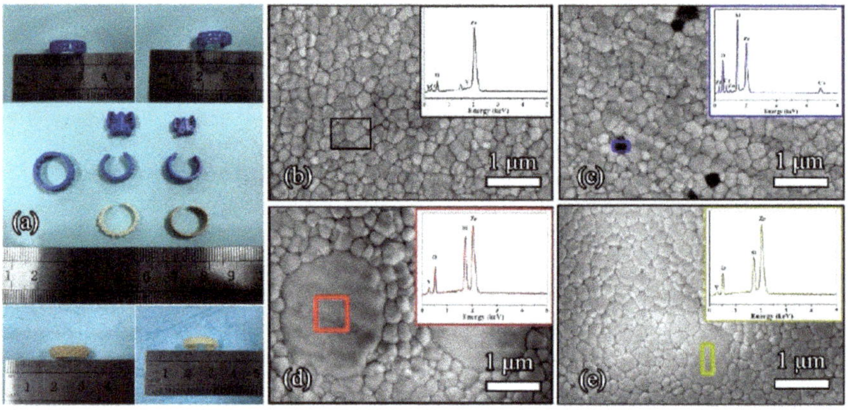

Figure 2.64. Photographs (a), SEM images and colorant EDS analyzed results (b–e) of the colorful ZrO_2 ceramics after sintering at 1450 °C for 2 h: (b) pure ZrO_2 (3Y-TZP), (c) blue ZrO_2 (ZrO_2 + $CoAl_2O_4$), (d) red ZrO_2 (ZrO_2 + $ZrSiO_4$ (Fe_2O_3)) and (e) yellow ZrO_2 (ZrO_2 + $ZrSiO_4$ (Pr_2O_3)). Reprinted from [94], Copyright (2019), with permission from Elsevier.

Figure 2.63 depicts XRD patterns of the ZrO_2 ceramic samples with 5 wt% colorant after sintering at 1450 °C for 2 h. Zirconia with a tetragonal crystal structure was the main phase for the ZrO_2 (3Y-TZP) and colored ZrO_2. Other phases, such as monoclinic and cubic, were not present in the XRD patterns. However, a slight quantity of $ZrSiO_4$ was observed, with very weak diffraction peaks.

Figure 2.64 shows photographs, SEM images and colorant EDS spectra of the colorful ZrO_2 ceramics. Both pure ZrO_2 (3Y-TZP) and colorful ZrO_2 ceramics had submicron grain size of ZrO_2 phase, in the range of 400–500 nm. The grain size of blue $CoAl_2O_4$ was very close to that of ZrO_2, while the grain sizes of the colorants

$ZrSiO_4$ (Fe_2O_3) (red) and $ZrSiO_4$ (Pr_2O_3) (yellow) were much larger than that of ZrO_2 grains, with a range of 1–4 μm.

Li *et al* reported 5 mol% yttria-partially stabilized zirconia (5Y-PSZ) ceramics, with promising mechanical strengths and optical translucency, by using SLA 3D printing [95]. The 5Y-PSZ ceramic slurries at solid loading of 52 vol% is suitable for the printing process. After sintering at 1550 °C, the ceramics reached maximum density and flexural strength values of 5.95 g · cm^{-3} and 685.6 MPa, respectively. The contrast ratio (CR) arrived at the minimum value of 0.43 at the sintering temperature of 1500 °C. Meanwhile, the final ceramics displayed hardness and fracture toughness of 12.7 GPa and 3.2 mPa · $m^{1/2}$, respectively. The content of *Y* in large grains was higher than that in small ones, so that the former was of a cubic crystal structure. In addition, the grain boundaries were strengthened due to the segregation of Y, making the fracture mode to be a mixture of transgranular and intergranular fracture for the sample sintered at 1550 °C.

Commercial chemicals, including HDDA (1,6-hexanediol diacrylate), photo-initiator (diphenyl (2,4,6-trimethyl benzoyl phosphine oxide, TPO), and dispersant (copolymer alkylammonium salt with acidic groups), were used to prepare the ceramic slurries, with five solid loadings (46–54 vol%). UV light with the wavelength of 405 nm was utilized for the printing machine. The thickness of the single printed layer was set to be 25 μm, at a scanning speed of 7000 mm · s^{-1} and a laser power of 100 mW. Green bodies to be printed included strips (30 × 5 × 1.65 mm, 37.5 × 5 × 3.75 mm) and pellets (Φ30 × 1 mm).

The as-printed samples were washed with isopropyl alcohol for cleaning. After that, the green bodies were calcined for debinding at 1000 °C in air, at a heating rate of 10 °C · min^{-1}. Then, the debound samples were sintered at temperatures of 1450 °C–1550 °C for 2 h. The maximum linear shrinkages of the samples sintered at 1500 °C in the *x*-, *y*- and *z*-directions were 19.2%, 19.6% and 22.6%, respectively. Figure 2.65 shows SEM images of the 5Y-PSZ ceramic samples.

Figure 2.66 shows SEM images and grain size statistic data of the ceramics sintered at different temperatures. Some extremely large and small grains are labeled in the images. EDS data indicated that the content of *Y* in large grains was higher than that in small ones, whereas the contents of Zr and *O* were nearly constant for either large or small grains. It was hence suggested that the large grains were of cubic phase [96]. The presence of the cubic phase was confirmed by XRD results.

As compared with the sample sintered at 1450 °C, the two sintered at 1500 °C and 1550 °C exhibited increases in the content of cubic phase by 7.5% and 8.7%, respectively. It has been accepted that the flexural strength of YSZ based ceramics is related to the content of the tetragonal phase, while the optical translucency is associated with the level of the cubic phase [97]. The transparency of the 5Y-PSZ ceramics after sintering at 1500 °C was higher than that of one sintered at 1550 °C, owing mainly to the increase in tetragonality and anisotropy.

Three-point bending test results indicated that predominantly the intergranular fracture mode, instead of the transgranular fracture mode, was present in the samples sintered at 1450 °C and 1500 °C. The tetragonal and cubic phases are different in volumetric variation in all directions. When stress is concentrated at the

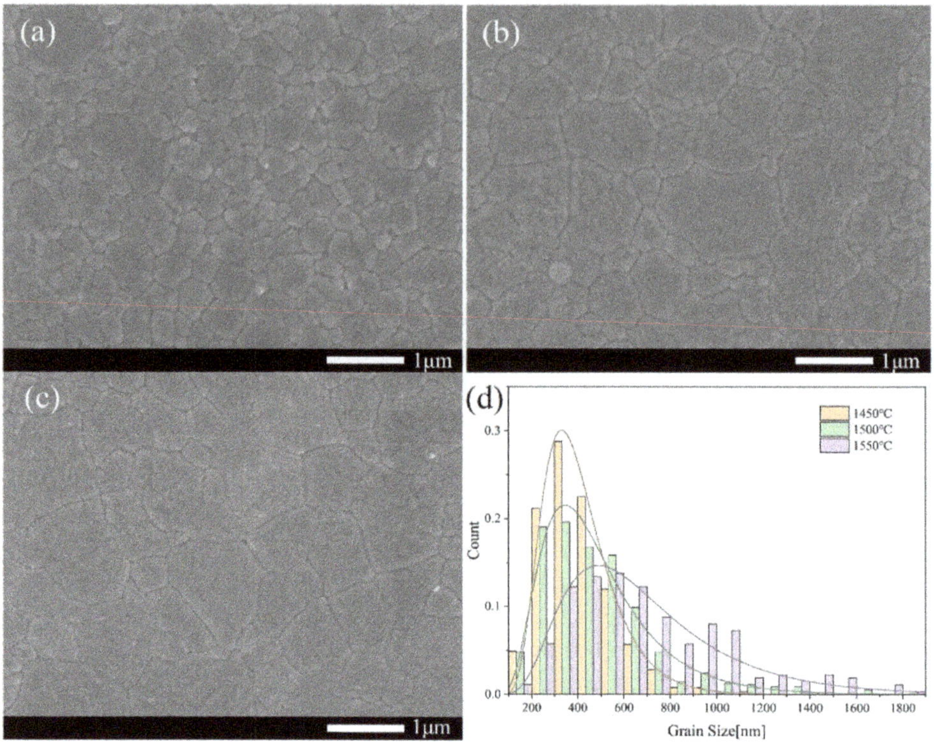

Figure 2.65. Surface SEM images of the ceramic samples after sintering at different temperatures: (a) 1450 °C, (b) 1500 °C and (c) 1550 °C. (d) Grain size distribution of the three samples. Reprinted from [95], Copyright (2023), with permission from Elsevier.

grain boundaries, the bonding strength between grains decreased, thus leading to weak grain boundaries. In other words, the resistance to the propagation of cracks along the grain boundaries was reduced. As a result, the grains tended to be separated entirely from the matrix of the ceramics. Meanwhile, the strength of the grains was inevitably influenced, so that intergranular fracture took place. As the sintering temperature was raised to 1550 °C, coarse grains with large sizes were formed, which were relatively weak, thus corresponding to the transgranular fracture mode [98].

The fracture features were represented by the compression curls, which are usually measures for the tensile and compressive sides of the fracture surface [99]. The fracture surfaces of the long strips in the 5Y-PSZ ceramic samples displayed similar features in terms of morphology, where the bottom surface was much smoother and closer to the origin of failure. Furthermore, the top surface at the end was far away from the origin of failure and quite rough in nature. Additionally, as depicted in figure 2.67, high-resolution SEM images further confirmed that the type of defects were similar to the conventional isostatic pressing. Various defects, such as pores, agglomerates and edge damages, were present in the SLA-printed 5Y-PSZ ceramics.

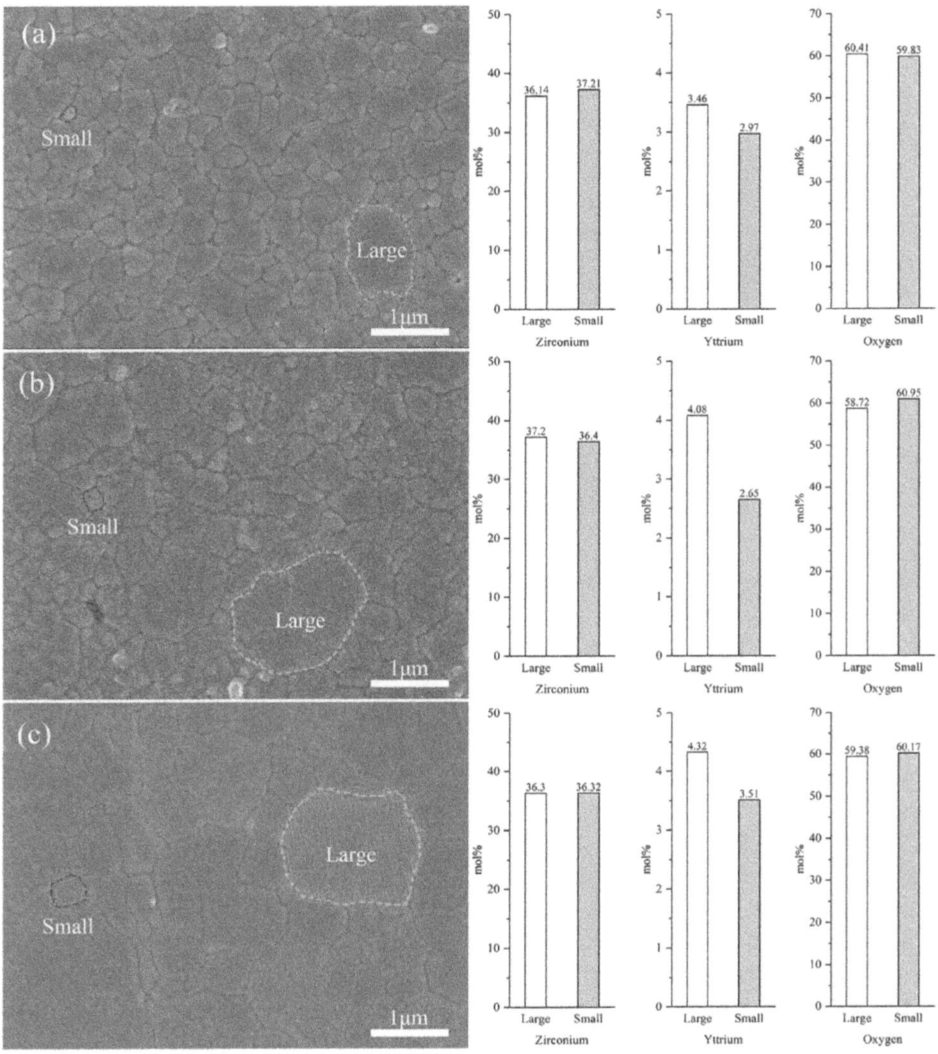

Figure 2.66. SEM images and fractions of large and small grains of the ceramic samples sintered at different temperatures: (a) 1450 °C, (b) 1500 °C and (c) 1550 °C. Reprinted from [95], Copyright (2023), with permission from Elsevier.

Ma *et al* used the SLA 3D printing technique to prepare porous honeycomb mullite ceramics [100]. The authors found that low-viscosity and highly stable slurries could be readily obtained by using oleic acid (OA), 3-glycidoxypropylthrimethoxysilane (KH560) and disperbyk (BYK111), with KH560 proving to be the most effective one. The mullite precursor slurry with 4 wt% KH560 followed a shear thinning behavior, with a viscosity of 0.26 Pa · s, at a shear rate of 30 s^{-1}. At irradiation intensity of 5.47 mJ · cm^{-2}, the mullite green bodies had sufficiently high resolution. The linear shrinkages, in the directions of x, y and z were 5.81%, 6.33%

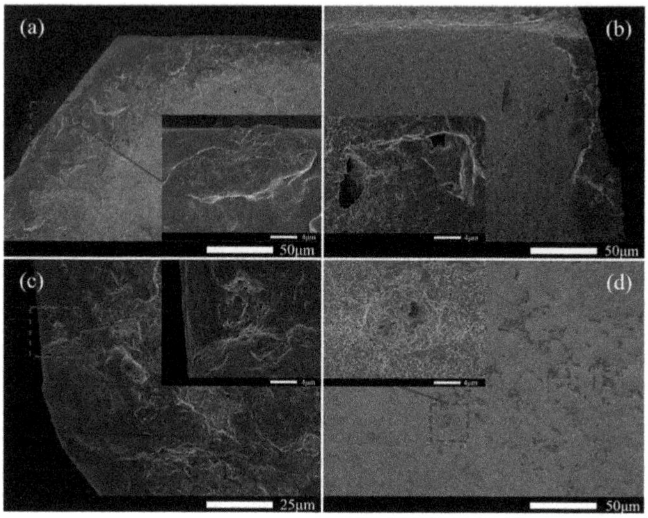

Figure 2.67. Common types of fracture defects. (a) Agglomerates below the surface, (b) pores below the surface, (c) edge damages and (d) internal pores. Reprinted from [95], Copyright (2023), with permission from Elsevier.

and 10.26%, respectively. After sintering at 1600 °C, mullitization process was completed.

Commercial powders of α-Al_2O_3 and SiO_2, with D_{50} = 1.33 μm and 1.36 μm, were used as the raw materials. 1,6-hexanediol diacrylate (HDDA) and trimethylolpropane triacrylate (TMPTA) were employed to prepare the UV-curing resins, with diphenyl (2,4,6-trimethyl benzoyl) phosphine oxide (TPO) and disperbyk (BYK111), oleic acid (OA) and 3-glycidoxypropylthrimethoxysilane (KH560) as the photoinitiator and dispersants. UV curable resins were made of HDDA and TMPTA at volume ratio of 4:1, with stirring for 2 h at 300 rpm. Ceramic slurries with desired compositions were prepared by mixing all components through ball milling for 8 h at a rotation speed of 300 rpm. Before making slurries, the ceramic powders were modified, while the solid loading level was 45 vol%.

The UV light used for the 3D printing had a wavelength of 355 nm, at the scanning speed of 3000 mm $\cdot$ s^{-1}, with printed single layer thickness of 50 μm. The as-printed items were cleaned with HDDA by using an air gun. The green bodies were then debound in Ar, followed by sintering at 1500 °C–1600 °C for 2 h, at a heating rate of 5 °C $\cdot$ min^{-1}.

The penetration depth and critical irradiation energy intensity were examined through data fitting, with values of 51.02 μm and 5.47 mJ $\cdot$ cm^{-2}, respectively. Therefore, the energy irradiation intensity should be $\geqslant$5.47 mJ $\cdot$ cm^{-2}, to ensure the strength of connection between adjacent layers and prevent delamination of the printed items during the debinding process. Figure 2.68 shows dried green bodies with various dimensions and shapes.

Photographs of the representative green bodies after debinding in air and Ar are depicted in figure 2.69. The sample debound in air had numerous cracks on surface

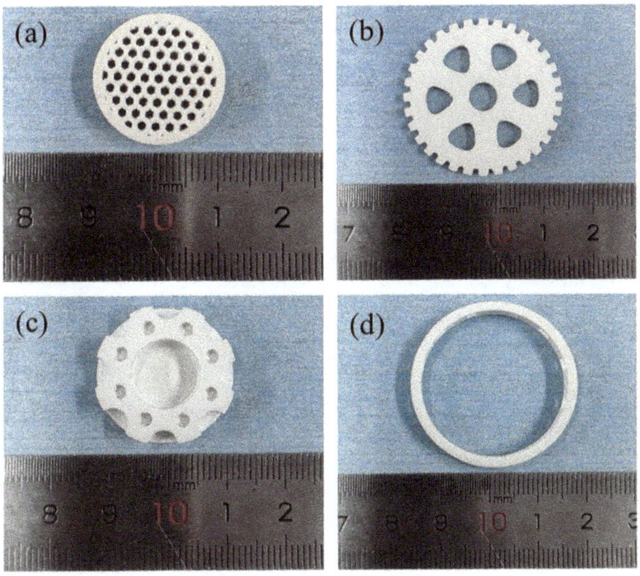

Figure 2.68. Photographs of the printed green bodies with different dimensions and shapes: (a) porous honeycomb ceramics, (b) cutting blades, (c) wheels and (d) sealing rings. Reprinted from [100], Copyright (2024), with permission from Elsevier.

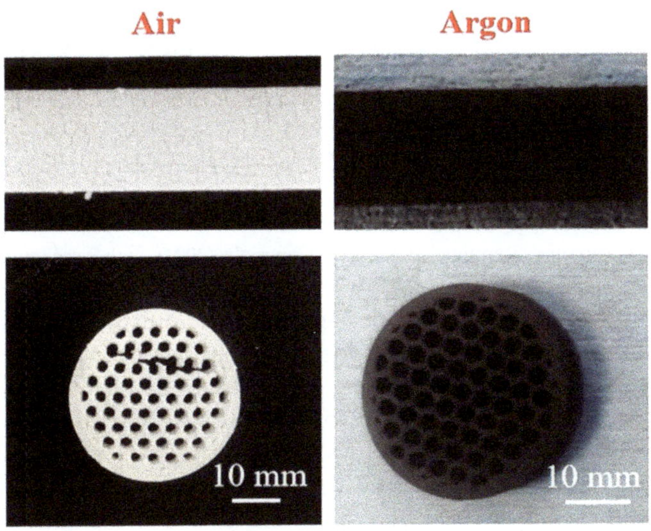

Figure 2.69. Photographs of representative samples after debinding in air and Ar. Reprinted from [100], Copyright (2024), with permission from Elsevier.

and deformation in shape. The formation of the cracks is related to two factors. On one hand, the decomposition and burning of the organic components in the green bodies occurred too rapidly, resulting in huge volume of gases in a short time. On the

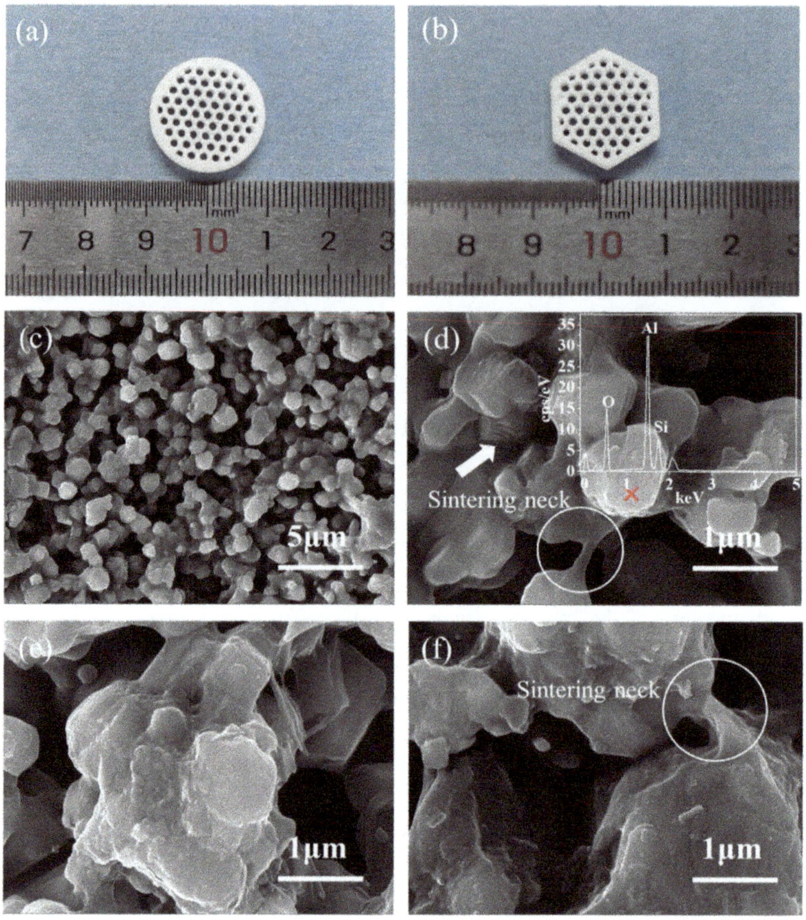

Figure 2.70. (a, b) Photographs of the sintered ceramics. (b–f) Cross-sectional SEM images of the representative sintered ceramics. Reprinted from [100], Copyright (2024), with permission from Elsevier.

other hand, the ceramic particles tended to aggregate during the debinding process, leading to the formation of cracks and pores. In comparison, when the samples were heated in Ar, the organic components were carbonized instead of oxidized. As a result, the issue of rapid generation of gases was avoided. In this case, no deformation and cracks were induced. After sintering at 1600 °C, the contents of mullite and Al_2O_3 were 96.56 and 3.44 wt%, respectively.

Figure 2.70 shows photographs and SEM images of representative printed porous honeycomb ceramics after sintering at 1600 °C. The porous honeycomb ceramic items were free of defects, as seen in figures 2.70(a) and (b). As observed in figures 2.70(c) and (d), the ceramic grains were connected to one another to form a skeleton structure, uniformly distributed pores with different sizes. EDS spectrum confirmed the presence of the elements of mullite, as illustrated as the inset in figure 2.70(d). Meanwhile, grain growth steps were present on sidewalls of the structure, with stacking faults on their tips, suggesting two-dimensional nucleation mechanism of the mullite grain growth mode [101].

Generally, the morphology of mullite is dependent on both the nucleation and growth. For a Al_2O_3–SiO_2 binary system, there is a eutectic point at about 1587 °C [102–104]. At the sintering temperature of 1600 °C, a liquid phase was formed, coating the cristobalite silica grains. The α-Al_2O_3 particles were dissolved in the silica-rich liquid phase, thus facilitating the nucleation of the mullite phase. Owing to the gradient in concentration, mass migration toward the nucleus occurred. Because the nuclei grew in certain crystallographic directions, this led to the presence of steps. In addition, there were sintering necks in the microstructures of ceramics, as demonstrated in figures 2.70(d) and (f).

The same authors further developed honeycomb mullite ceramics with more complicated structures [105]. In this study, anisotropic grain growth of mullite was observed, while the relationship between geometry of the honeycomb structures and compressive strength was established. It was found that the honeycomb mullite ceramics with square holes exhibited the highest compressive strength, with a level of 206.7 MPa.

Figure 2.71 shows SEM images of the samples sintered at temperatures of 1400 °C–1600 °C. In the sample sintered at 1400 °C, alumina and silica particles with spherical morphology were still present, due to incomplete mullitization. Consequently, mullite grains were visible but relatively small, owing to the slow grain growth rate at this temperature. Some spherical grains were connected by the liquid phase, in the form of a sintering neck. Meanwhile, the sample was of quite high porosity, as observed in figure 2.71(a). At 1450 °C, the driving force for mullite phase formation and grain growth was increased, so that the number of spherical grains was reduced, while irregular grains could be observed. Therefore, porosity of the sample was decreased, as illustrated in figure 2.71(b). As the sintering temperature further increased to 1500 °C,

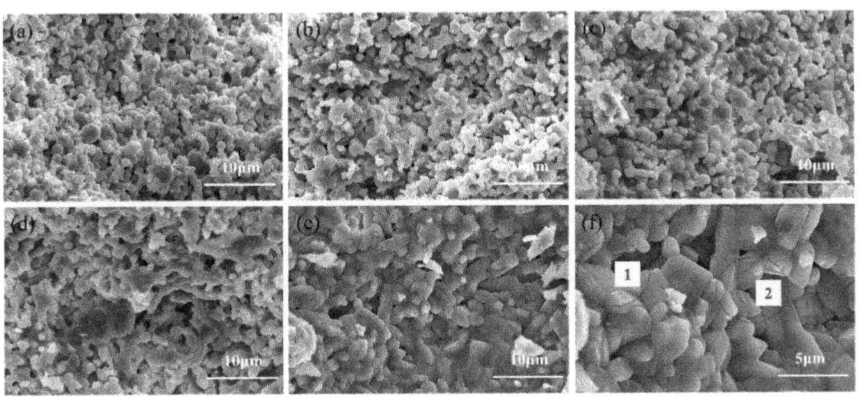

Figure 2.71. SEM images of the ceramic samples sintered at different temperatures: (a) 1400 °C, (b) 1450 °C, (c) 1500 °C, (d) 1550 °C, (e) 1600 °C and (f) higher magnification image of (e). Reprinted from [105], Copyright (2024), with permission from Elsevier.

a liquid phase was formed in the MgF_2–Al_2O_3–SiO_2 ternary system. The grains of the sample had sizes of 0.5–1 μm, as seen in figure 2.71(c).

In the sample sintered at 1550 °C, mullite grains with short rod-like shape were developed, as observed in figure 2.71(d). During sintering at this temperature, mullite phase was formed before the occurring of densification, so that the mullite grains would be able to grow freely without restriction. Owing to its crystal structure, the mullite grains tended to grow anisotropically. As a result, rod-like grains were derived. To reduce the nucleation energy, the seeds of mullite were first nucleated on the surface of the initial mullite grains. Then, more mullite seeds would be nucleated along the boundaries of these grains, and the surfaces of the grains were entirely covered. In this case, the mullite grains followed a gradual axial growth mode, which was responsible for the presence of the layered steps.

At the highest sintering temperature of 1600 °C, the driving force was further increased, so that the grain growth of mullite was further enhanced. As a consequence, rod-like grains of mullite were enlarged, while the binding state of the grains was strengthened. Accordingly, the sample exhibited a much denser microstructure, with just closed pores, as presented in figures 2.71(e) and (f). At the same time, layered steps were still visible on the side of the mullite grains, reflecting the two-dimensional mechanism of nucleation of mullite.

Density of the ceramics was gradually increased, from 1.73 to 2.74 $g \cdot cm^{-3}$, while the porosity was decreased from 45.6% to 8.5%, when the sintering temperature was raised from 1400 °C to 1600 °C. It has been recognized that the densification mechanism of mullite ceramics derived from oxide precursors is primarily liquid phase sintering [106]. The presence of a liquid phase promoted rearrangement of ceramic particles, due to the high capillary attraction effect of the wetting liquids, thus accelerating the densification process [107].

The stress–displacement curves of all samples sintered 1400 °C–1600 °C exhibited three stages, i.e., elastic, plateau and densification. The stress–displacement relationship at the elastic stage ($d < 0.1$ mm) was characterized by a linear trend, while cracks resulted in almost no damage. Then, a quasi-plastic regime ($0.1 \text{ mm} < d < 0.5 \text{ mm}$) was reached, where the stress was nearly constant and the displacement started to increase. In this case, whether the stress–displacement curves were smooth or not was dependent on size and distribution of the pores in the ceramics. At the final failure stage ($d > 0.5$ mm), the stress was significantly increased with progressing in displacement. Specifically, the samples with circular pores displayed the lowest engineering stress of 77.2 MPa, while those with square pores had a maximum engineering stress of 206.7 MPa.

Fracture morphologies of representative samples with different pore geometries are depicted in figure 2.72. The samples with circular pores were damaged most seriously, in good agreement with the engineering stress characterization results. The fractures occurred on both sides of the samples, as seen in figures 2.72(a) and (a1). In comparison, the samples with triangular pores were less damaged, with fractures at a 45° angle in the diagonal direction (figures 2.72(b) and (b1)). The samples with square pores were the strongest against the failure testing, with the lowest extent of

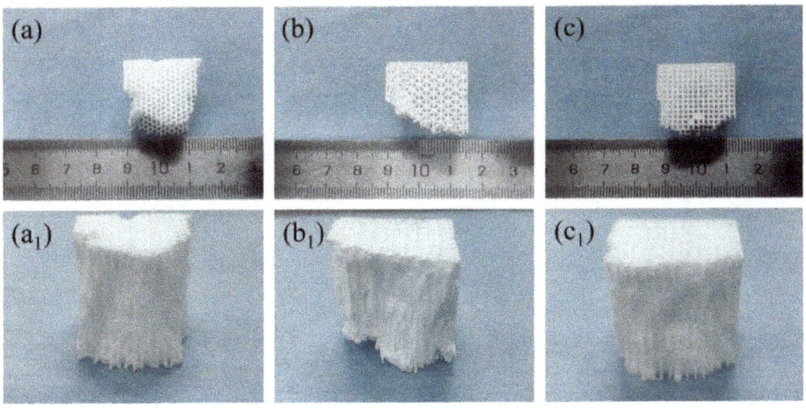

Figure 2.72. Photographs of fractured samples with different pore structures: (a, a1) circular pores, (b, b1) triangular pores and (c, c1) square pores. Reprinted from [105], Copyright (2024), with permission from Elsevier.

fracture. The fracture model was similar to that of the samples with triangular pores, as demonstrated in figures 2.72(c) and (c1).

Si_3N_4 ceramics have a wide range of applications with unique physical and chemical properties, which have been extensively studied with SLA printing technologies [108–113]. Owing to the large difference in refraction index between Si_3N_4 and resin, it is difficult to fabricate Si_3N_4 ceramics through SLA 3D printing. Huang *et al* addressed this problem by modifying Si_3N_4 powder through surface oxidation in increasing the curing depth, thus realizing 3D printing of Si_3N_4 [114]. After surface oxidation, the Si_3N_4 ceramic particles were coated with a layer of amorphous SiO_2, with which optical absorbance of the Si_3N_4 powder was effectively reduced and the refractive index difference was minimized. At the irradiation energy intensity of 500 $mJ \cdot cm^{-2}$, the cure depth of the Si_3N_4 resin slurries reached 34 μm.

After oxidizing at 1150 °C and 1200 °C for 1 h, the cure depth was deepened to 42 and 51 μm, respectively.

The raw material for experiment was α-Si_3N_4 powder with $D_{50} = 0.2$ μm, with XRD patterns to be depicted in figure 2.73(a). The powder was oxidized in air at temperatures of 1150 °C–1200 °C for 1 h or 3 h, resulting in four groups of samples, denoted as 1150-1, 1150-2, 1200-1 and 1200-3. After oxidation, the ceramic particles became core–shell structures, where the core and shell were crystalline Si_3N_4 and amorphous silica, respectively, as observed in figures 2.73(b) and (c). The core–shell structure was also confirmed by XPS characterization results.

The resin for slurry preparation included 1,6-hexanediol diacrylate (HDDA) and trimethylolpropane triacrylate (TMPTA). The powder was dispersed in the resin, followed by the addition of dispersant, through ball milling. After that, a photo-initiator (phenylbis (2,4,6-trimethyl benzoyl)-phosphine oxide) was added into the mixtures. After ball milling for 1 h at 350 rpm, printing slurries were obtained. The printed items were constructed through curing in a layer-by-layer way, with irradiation energy intensities of 125–1500 $mJ \cdot cm^{-2}$. The green bodies were debinded and sintered at 1820 °C for 90 min in N_2 at 1.0 atm.

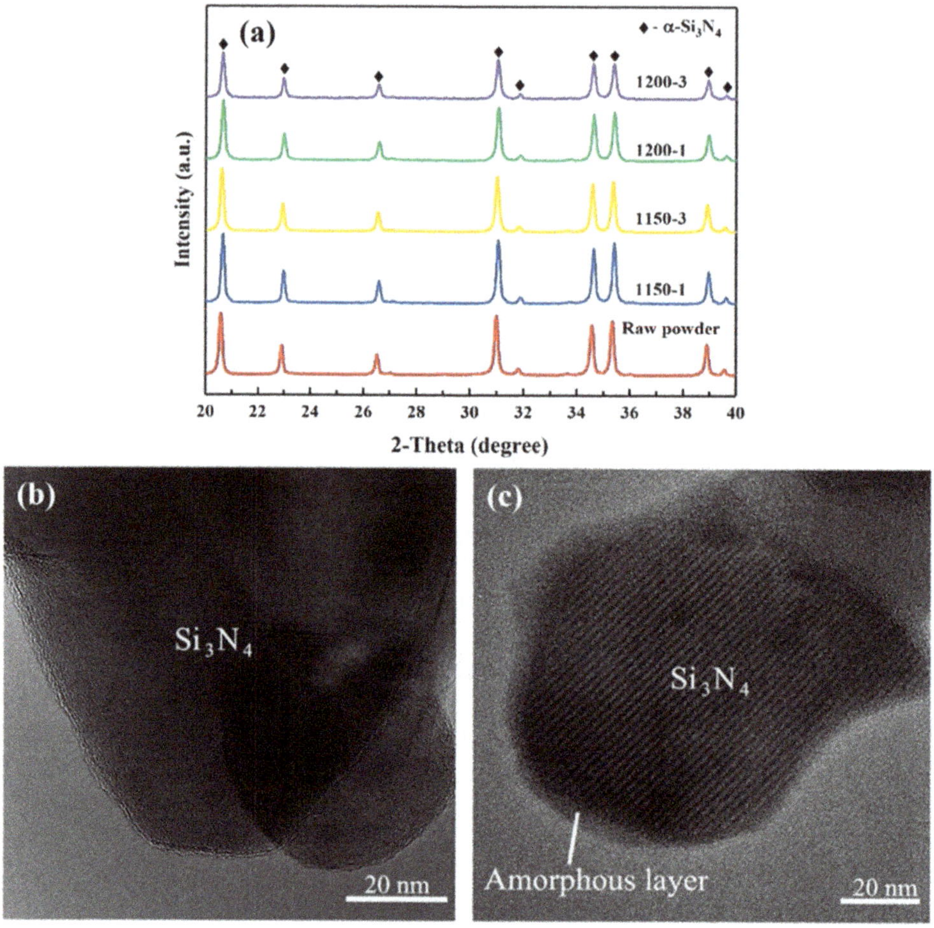

Figure 2.73. Characterizations of the Si_3N_4 powders: (a) XRD patterns of the Si_3N_4 powders before and after oxidation in various conditions. Representative TEM images of the Si_3N_4 powders: (b) pristine powder and (c) sample 1150-1. Reprinted from [114], Copyright (2019), with permission from Elsevier.

The irradiation light had a wavelength of 405 nm. The curing depth was increased with increasing irradiation energy intensity. The increase in curing depth was quite rapid below 500 mJ · cm^{-2}, while it was slowed down at $\geqslant$500 mJ · cm^{-2}. At the irradiation energy intensity of 500 mJ · cm^{-2}, the curing depth was increased from 34 μm for the pristine Si_3N_4 to 42, 47, 51 and 68 μm, for 1150-1 and 1150-3, 1200-1 and 1200-3, respectively.

Figure 2.74 representatively shows photographs of the green bodies and sintered ceramics from the sample 1150-1. It was demonstrated that Si_3N_4 ceramics structures with complicated shapes could be readily fabricated by using the SLA printing technique. Green bodies of blades and vertebrae are shown in figures 2.74(a) and (b). A densified Si_3N_4 gear is demonstrated in figure 2.74(c). All the sintered ceramics had relative densities of >90%. Figure 2.74 shows a typical SEM image of

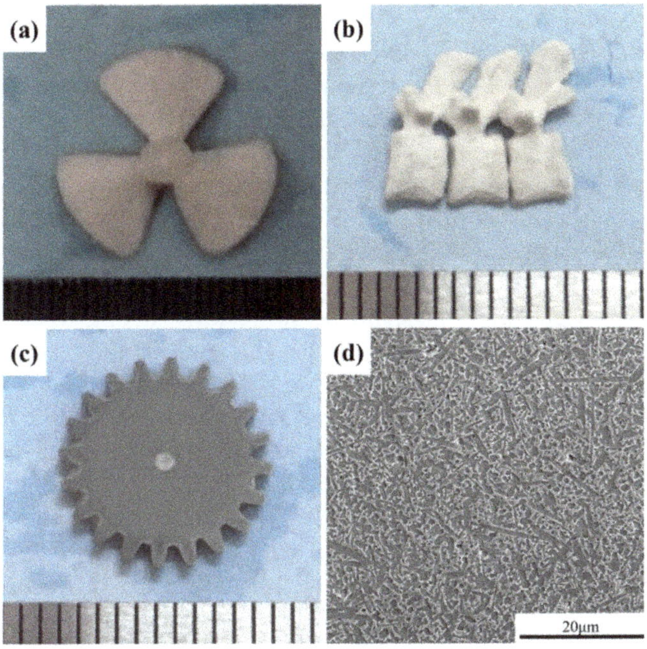

Figure 2.74. Photographs and SEM image of Si_3N_4 green bodies and sintered ceramics: (a) blade green body, (b) vertebrae green body, (c) sintered gear and (d) SEM micrograph of the Si_3N_4 ceramics. Reprinted from [114], Copyright (2019), with permission from Elsevier.

the sintered Si_3N_4 ceramics. The elongated grains in the microstructure were β-Si_3N_4. The rod-like grains were uniformly embedded in the matrix of fine grains, forming an interlocking network structure, which had a positive effect in terms of mechanical strength.

Liu *et al* examined the effects of particle size and color of Si_3N_4 powders on photocuring behaviors of the ceramic slurries for SLA 3D printing [115]. The UV light transmission of ultraviolet light and curing performance of the Si_3N_4 slurries were gradually increased, as the color of the Si_3N_4 powders was varied from black to light gray. At a given particle size, the Si_3N_4 ceramic slurries from lighter powders exhibited a lower scattering effect. The gray powder with an average particle size of 0.8 μm had the lowest scattering coefficient. For a given color, the larger the particle size, the smaller the scattering coefficient of the slurries would be.

SiC and SiOC ceramics have also been fabricated by using the SLA 3D printing process [116–129]. For instance, Li *et al* incorporated SLA and reactive melt infiltration (RMI) methods to obtain SiC ceramics with complicated shapes and highly precise structures [121]. At optimized conditions, the SLA/RMI processed SiC ceramics had a density, flexural strength and elastic modulus of 2.89 $g \cdot cm^{-3}$, 244.17 MPa and 402.39 GPa, respectively.

Raw materials for the experiment included phenolic (PF) resin, ethyl alcohol and commercially available micron-sized α-SiC with purity ⩾99.5% and $D_{50} = 54.8$ μm. Photosensitive resin (UV-3050) and photoinitiator (TPO, purity 97%) with high sensitivity to 405 nm UV light were used to prepare the ceramic slurries. Commercial silicon powder (purity = 99%, D_{50} = 5 μm) was utilized for infiltration.

SiC powder was dispersed in the resin at a volume ratio of 47 vol%, together with photosensitive resin and a photoinitiator, through ball milling for 5 h, forming printing slurries. The wavelength of the UV light, light irradiation intensity, exposure time duration and the curing thickness were 405 nm, 90 $\mu w \cdot cm^{-2}$, 1.9 s and 50 μm, respectively. The printed SiC green bodies were debounded at 450 °C and 650 °C for 0.5 h, at a heating rate of 0.5 $°C \cdot min^{-1}$. The samples were pre-sintered at 1650 °C for 0.5 h.

RB-SiC based space mirrors should be sufficiently strong to prevent deflection due to the self-loading. The samples from the slurries with 40 vol% solid loading had the highest specific stiffness, along with high flexural strength, low coefficient of thermal expansion (CTE) and high thermal conductivity. Figure 2.75 shows photographs of the lightweight SiC spatial optical mirrors. The green bodies exhibited smooth surfaces, as seen in figures 2.75(a) and (b). After thermal debinding, the samples were free of cracks, pores and other defects, as observed in figures 2.75(c) and (d). The green bodies were subject to impregnation with 40 wt% PF resin solutions, so as to enhance mechanical performances of the final ceramics. The debounded SiC spatial optical mirrors were fully sintered through RMI, as demonstrated in figures 2.75(e) and (f). The polished SiC spatial optical mirrors had high structural integrety and optical properties, as illustrated in figure 2.75(g).

Wang *et al* elaborated on the method and mechanism to increase the curing thickness of SiC green bodies when using the SLA 3D printing technique [127]. Rheological properties of the SiC slurries were optimized, while the cracking problem of the green bodies during the debinding process was effectively tackled. The optimal conditions for the surface modification of SiC powder included oxidation at 1180 °C for 1 h and hydroxylation in 0.6 M H_2O_2 solution. SiC slurries with promising rheological behaviors and large curing thickness were obtained with a resin mixture consisting of BPA, HDDA, TPGDA, TMPTA,

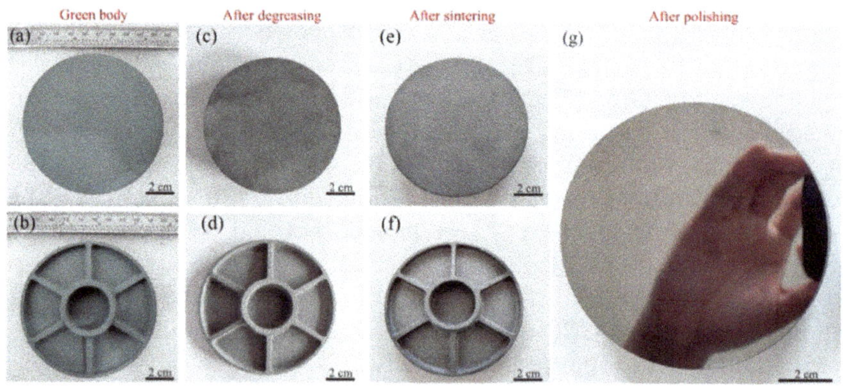

Figure 2.75. Photographs of the lightweight SiC spatial optical mirrors: (a, b) green bodies made with SLA printing process, (c, d) after debinding, (e, f) after RMI and (g, h) after polishing. Reprinted from [121], Copyright (2021), with permission from Elsevier.

DPHA and A-BPEF, with mass ratio = 9:7:5:20:15:6. The best debinding process for SLA SiC was segmented debinding. Specifically, the SiC ceramics developed at the optimal conditions possessed bulk density, total porosity and room temperature flexural strength of 2.13 $g \cdot cm^{-3}$, 10.2% and 229 MPa, respectively. The corresponding nanoindentation elastic modulus and hardness of the SiC ceramics were 18.7 and 1.66 GPa.

Commercial SiC powder with an average particle of 1.6 μm was used as the main raw material, while Y_2O_3 and Al_2O_3 with particle sizes of 0.5 μm were employed as sintering aids. The photoinitiators were diphenyl (2,4,6-trimethyl benzoyl) phosphine oxide (TPO) and 2-isopropylthioxanthone (ITX). The addition of 6-functional DPHA for the SiC slurries could enhance the curing efficiency, thus leading to increase in curing depth. With BPA and A-BPEF, the SiC green bodies after curing exhibited sufficiently high mechanical strength. Unfortunately, slurries made with these two resins had high viscosity and low fluidity. In comparison, HDDA, TMPTA and TPGDA were advantageous in this regard.

For oxidation, the SiC powder was heated at 1180 °C in air for time durations of 0, 0.5, 1 and 1.5 h, at the heating rate of 2 $°C \cdot min^{-1}$. Then, the powders were merged in 0.5 M NaOH solution for 24 h. After washing and drying, the SiC powders were sieved through 100 mesh and dispersed in H_2O_2 at a volume ratio of 1:2, followed by stirring for 0.5 h. Hydroxylated SiC powders were achieved. After drying, the hydroxylated SiC powders, together with Y_2O_3, Al_2O_3, silane coupling agent (KH560) and ethanol, were mixed through ball milling at 240 rpm for 28 h, leading to modified SiC powders.

BPA, HDDA, TPGDA, TMPTA, DPHA and A-BPEF were mixed at the weight ratio of 9:7:5:20:15: (0, 2, 4, 6, 8, 10). The contents of photoinitiators TPO and ITX were 4.6% and 6.2%, with respect to the mixed resin. Eventually, SiC slurries were made of the modified SiC powder, solvent and dispersant, at a weight ratio of 4.5:15:0.4 through thorough blending. The weight content of SiC was 70.3% in the slurries. The UV laser used for printing had a wavelength of 355 nm.

After heating at 1180 °C in air, a thin layer of SiO_2 was produced on the surface of the SiC particles. With the presence of the SiO_2 thin layer, the difference in refractive index between the SiC powder and resin was reduced, so that the curing depth of the ceramic slurries was increased. Meanwhile, the SiO_2 would promote the formation of hydroxyl groups on the surface of the SiC particles through H_2O_2. The higher the content of hydroxyl groups, the tighter the connection of the silane coupling agent (KH560) and the SiC particles would be.

Figure 2.76 shows representative TEM images of the SiC powder before and after oxidation at 1180 °C. The pristine SiC particles have a smooth surface, as revealed in figure 2.76(a). The crystal lattice fringe spacing of 0.25 nm corresponds to the (111) of β-SiC. The layer of amorphous SiO_2 on the surface of SiC particles was confirmed by the HRTEM image, as seen in figure 2.76(d). The SiO_2 layer had a thickness of 12 nm, without visible ordering structure, confirming its amorphous nature.

Figure 2.77 depicts photographs of the SiC slurries attached on a rod, which were derived from the SiC powders treated with H_2O_2 at different concentrations. Solid loading of the SiC powders was 70.3 wt%. As illustrated in figure 2.77(a), without

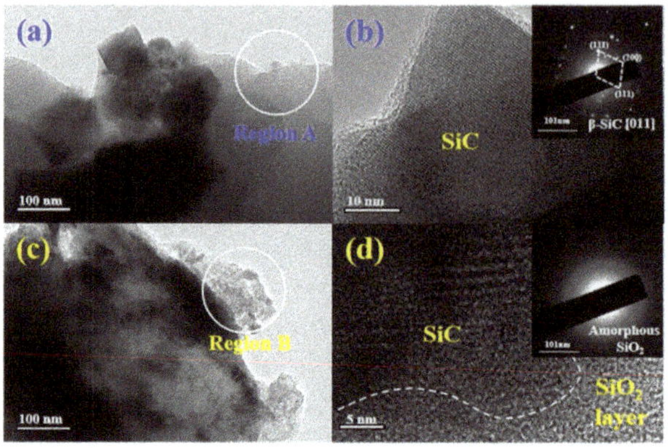

Figure 2.76. Representative TEM images of the SiC powders: (a) BF TEM image of the unoxidized SiC powder, (b) electron diffraction pattern of region A in panel (a), (c) BF TEM image of the powder after oxidation at 1180 °C for 1 h and (d) electron diffraction pattern of region B in panel (c). Reprinted from [127], Copyright (2022), with permission from Elsevier.

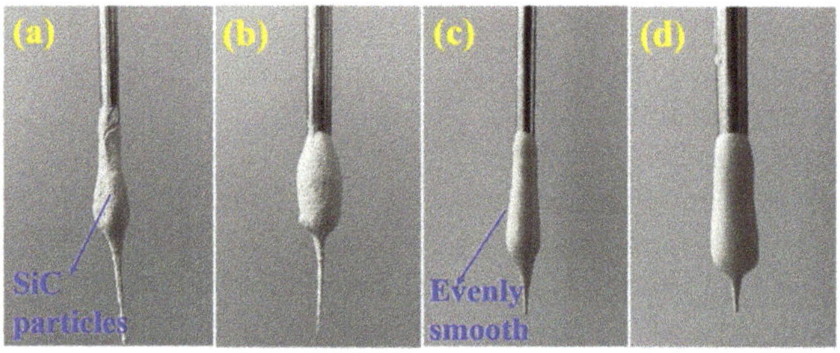

Figure 2.77. Appearances of the SiC slurries from the powders treated in H_2O_2 with different concentrations: (a) 0 M, (b) 0.3 M, (c) 0.6 M and (d) 0.9 M. Reprinted from [127], Copyright (2022), with permission from Elsevier.

the treatment of hydroxylation, the SiC powder led to slurry with the ceramic particles to be visible, reflecting inhomogeneity of the slurry. The homogeneity was improved when the SiC powders were treated in H_2O_2, with the homogeneity increased with increasing concentration of H_2O_2. For example, the slurry was fine and uniform, as the SiC powder that was treated with 0.6 and 0.9 M H_2O_2, as demonstrated in figures 2.77(c) and (d).

2.3.2 Microstereolithographic approach

Microstereolithography (μSLA) is a 3D printing technology that enables the fabrication of ceramic components with relatively small sizes and complicated shapes, which cannot be achieved by using the conventional processing techniques. Bertsch *et al* reported a new type of photosensitive resins based on polymer/composite, allowing the fabrication of 3D structures via the μSLA process [130]. A high solid loading of up to 80 wt% Al_2O_3 nanoparticles in photosensitive polymer matrix could be achieved. Al_2O_3 ceramic components could be obtained from the green bodies after debinding and sintering, while retaining the printed shapes.

Monri and Maruo developed a 3D approach to mold piezoelectric elements made of ceramics, with master polymeric molds that were generated by using μSLA printing [131]. Ceramic slurries were filled into the 3D polymer molds through centrifugal casting. 3D piezoelectric ceramic elements were obtained, after the polymer master molds were burned out. This process yielded 3D piezoelectric ceramic elements with helical geometries, capable of converting multidirectional loads into electrical voltages. The piezoelectric effect was present in both the parallel and lateral directions along the helical axis. Specifically, a power output of 123 pW was achieved, as the elements were subject to a load of 2.8 N at 2 Hz in the direction along the helical axis.

A He-Cd laser with a wavelength of 325 nm and a laser power of 15 mW was used as the light source for curing. The laser beam was focused using a lens with a focal length of 150 mm. A commercial epoxy-type photopolymer was used for printing. The lateral resolution and accumulated layer thickness were 25 μm and 50 μm, respectively. Ceramic slurries were made of $BaTiO_3$ (BT) nanoparticles with two average particle sizes of 400 and 150 nm. The density of the green bodies could be maximized by using the two BT powders. The BT nanoparticles were dispersed in deionized water and dispersant. The dispersant also served as a binder, due to its relatively high heat resistance. The slurry consisted of 12.6 wt% deionized water, 42 wt% 400 nm diameter BT powder, 42 wt% 150 nm diameter BT powder and 3.4 wt% dispersant. The green bodies were dried and then calcined at 600 °C, after which the samples were sintered at 1300 °C.

The 3D CAD model of the master molds with a helically curved channel is shown in figure 2.78(a), while photographs of the μSLA-printed 3D polymer molds and

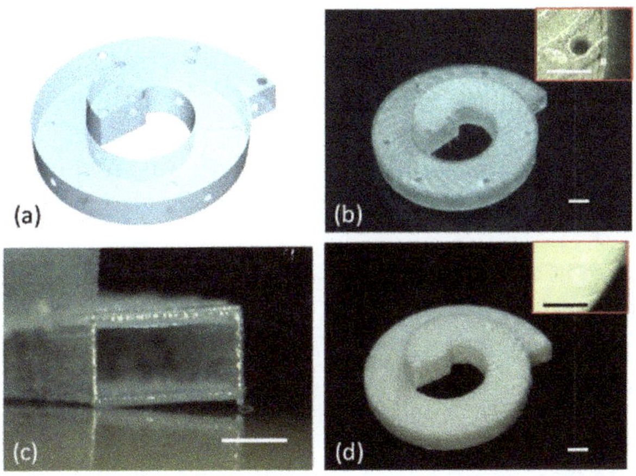

Figure 2.78. Fabrication and appearances of the helically curved piezoelectric element: (a) 3D CAD model of polymer mold, (b) polymer mold with inlet diameter of 400 μm, (c) cross-sectional view of the polymer mold with a rectangular shape and (d) polymer mold filled with BT slurry after centrifugal casting and drying (scale bar = 1 mm). Reprinted from [131], Copyright (2013), with permission from Elsevier.

cross-sectional view are depicted in figures 2.78(b) and (c). The mold had an inner diameter of 400 μm to guarantee filling efficiency of the ceramic slurries, while the height and maximum diameter were 3 and 10 mm, respectively. The helical tube had a rectangular cross-section, with thicknesses of 100–200 μm, which were optimized for easy removal and sufficiently high mechanical strength. The thin outer wall can be easily decomposed in a pyrolysis process.

To ensure high filling efficiency during centrifugal casting, the process was carried out for 5 min at a rotation speed of 4200 rpm. Figure 2.78(d) shows photographs of the dried green bodies. SEM results indicated that the green bodies exhibited a dense microstructure, due to the use of the BT powders with two groups of particle sizes. After sintering at 1300 °C for 2 h, the total shrinkage was 17%. The helical shape was well retained, as illustrated in figure 2.79(a). Figure 2.79(b) shows a representative SEM image of the sintered samples, showing grain sizes in the range of 3–15 μm. After poling at DC electric field of 1.5 $kV \cdot mm^{-1}$, the helically curved piezoelectric element had piezoelectric coefficients, d_{33} and d_{31}, at 65.4 $pC \cdot N^{-1}$ and −43.8 $pC \cdot N^{-1}$, respectively. At the load resistance of 89 MΩ, the maximum electrical power reached 123 pW.

Micron-sized dimensional Al_2O_3 ceramics have been prepared by using μSLA 3D printing, with non-aqueous colloidal Al_2O_3 slurries [132]. Two types of Al_2O_3 powders, with $D_{50} = 0.5$ and 2.9 μm, specific surface areas of 8.9 $m^2 \cdot g^{-1}$ (BET), were used to prepare the ceramic slurries. HDDA, TMPTA monomers, TOPO and decalin were employed as additives. BP was recrystallized in ethanol (EtOH) two times before usage. TMPTA and HDDA were mixed at a ratio of 20:80 vol% at room temperature to obtain a TH2080 monomer solution.

Before the Al_2O_3 monomer slurries were prepared, the Al_2O_3 powders were modified with TOPO at different concentrations. The concentrations of TOPO were in the range of 0 5 wt%, with respect to Al_2O_3 in EtOH. Initially, the TOPO was dissolved in EtOH, followed by the addition of Al_2O_3 powder. In order for the Al_2O_3 particles to be entirely covered by TOPO, the suspensions were subject to strong sonication. After that, the suspensions were dried at 45 °C to vaporize the EtOH. The TOPO modified Al_2O_3 powders were passed through a 120 mesh sieve to minimize the agglomeration.

To optimize the concentration of the dispersant, Al_2O_3 slurry with solid loading of 25 vol% with respect to the TH2080 monomer and decalin (diluent) was prepared.

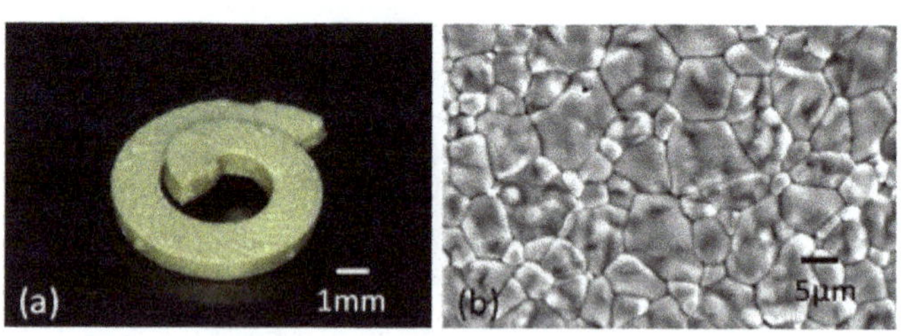

Figure 2.79. (a) Photograph of the sintered ceramics of the helically curved piezoelectric element. (b) Representative surface SEM image of the the sintered ceramic body. Reprinted from [131], Copyright (2013), with permission from Elsevier.

The ratio of monomer (M) to diluent (D) was 4:1, which was denoted as TD41. The monomer–diluent mixtures were thoroughly mixed through magnetic stirring. The TOPO modified Al_2O_3 powders were dispersed into the monomer–diluent mixtures, with stirring for 5 h, resulting in homogeneous slurries. Once the optimal dispersant concentration was determined, slurries with varying Al_2O_3 solid loading levels ranging from 10 to 40 vol% were prepared to further investigate and optimize the solid loading.

Al_2O_3 green bodies were derived from the slurries through μSLA of the ceramic slurries with an Ar ion laser at a wavelength of 364 nm. The as-printed green bodies were cleaned and dried, followed by debinding at 500 °C for 0.5 h at a heating rate of 0.2 °C · min^{-1} in N_2. After the debinding process, the samples were sintered at 1550 °C for 5 h, at a heating rate of 15 °C · min^{-1}, leading to Al_2O_3 micron-sized ceramic structures.

Samples with complicated structures were fabricated with slurries having 40 vol% of Al_2O_3. Figure 2.80 shows SEM images of the Al_2O_3 green bodies with 2D net shapes. The structures included a square mesh, impeller (with a hole at center), crosshatch mesh and hollow gear. The green bodies were characterized by high porosity, as seen as the inset in figure 2.80(a). A confocal microscope image of the impeller is shown in the inset in figure 2.80(b), with a 2D structure. Representative SEM images of the sintered samples with gear structures with and without a central hole are depicted in figure 2.81. Microstructures of the sintered gear and mesh are illustrated as insets in figures 2.81(a)–(c). All the samples experienced shrinkages of 25%–28%.

Troksa *et al* developed ceramic inks that could be used for projection micro-stereolithography (PμSLA) and direct ink write (DIW), to make ceramic structures with different sizes and porosity [133]. 3YSZ ceramic nanopowders were blended with diacrylate polymer to obtain printing slurries with different solid loadings. The 3YSZ ceramic structures exhibited micron-sized cavities with span lengths at millimeter scales, wall thicknesses of 200–540 μm and porosity within the wall structure of 0.1 μm.

To prepare ceramic slurries, 3YZ nanopowders, polyethylene glycol diacrylate and a thermal initiator were thoroughly mixed and blended. Two 3YSZ nano-powders were denoted as 3Y-55 (D_{50} = 0.055 μm) and 3YS-120 (D_{50} = 0.12 μm). The solids loading levels in PEGDA were in the range of 55–70 wt% or 18–30 vol%. In addition, photoinitiator of 4-methoxyphenol (MEHQ, 0.1 wt% with respect to PEGDA), isopropylthioxanone (ITX, 0.1–0.3 wt% with respect to PEGDA) and 2-ethylhexyl 4-(dimethylamino)benzoate (EHDA, 0.2–0.6 wt%), which were dissolved in 0.1 ml tetrahydrofuran (THF), were included.

The UV laser for irradiation had a wavelength of 405 nm, with a light intensity of 258 mW · cm^{-2}. After printing, the green bodies were dried at 100 °C for 12 h and then calcined at 200 °C, 300 °C, 400 °C and 800 °C for 2 h, 4 h, 2 h and 4 h, respectively, at a heating rate of 1 °C · min^{-1}. The samples were sintered at 1090 °C for 15 h, at heating and cooling rates of 2 °C · min^{-1}.

All slurries exhibited shear thinning behaviors, with an increase in viscosity with increasing solid loading levels of the ceramic powders, especially as the shear rate

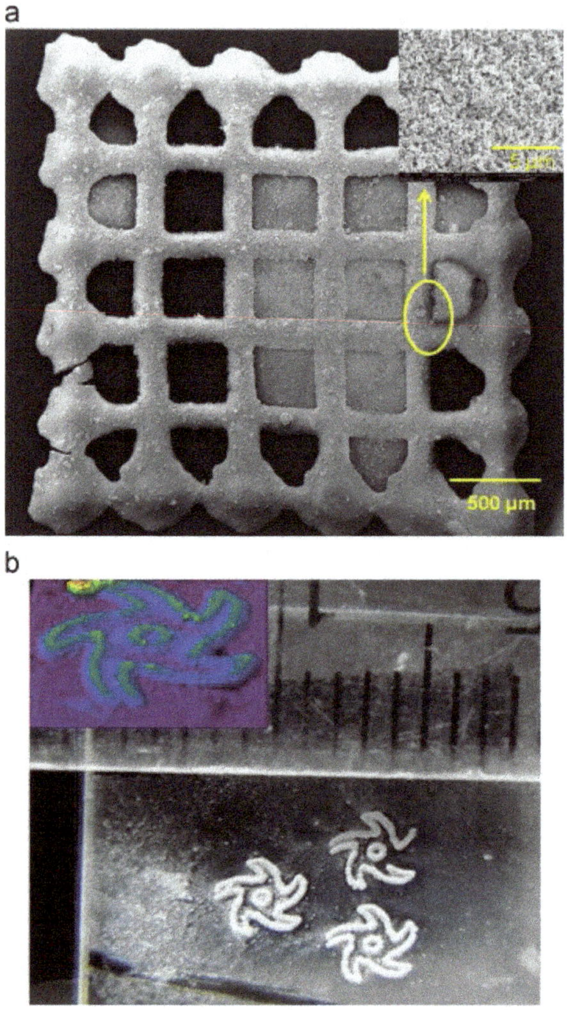

Figure 2.80. SEM images of the green bodies 2D structures: (a) square mesh and (b) impeller with a hole at center. Inset of panel (a) shows the porous microstructure of the ceramic green body. Inset of panel (b) shows the confocal image of impeller. Reprinted from [132], Copyright (2014), with permission from Elsevier.

was relatively low. For instance, at the shear rate of 1 s^{-1}, the difference in viscosity between the slurries with 55 and 70 wt% ceramic powders was at one order of magnitude. At same solid loading level, the slurries made of the 3Y-55 ceramic powder had higher viscosities than those from the 3YS-120 powder, mainly because the former had higher specific surface area than the latter. The slurries with 70 wt% (30 vol%) 3Y-55 powder for the DIW printing were extruded through nozzles with diameters in the range of 0.25–1.19 mm.

Figure 2.82 shows representative samples of partially sintered ceramics made with the DIW process, with different structures and dimensions. The tube had a height of 10 mm and a diameter of 4 mm after partial densification, as depicted in figure 2.82(a).

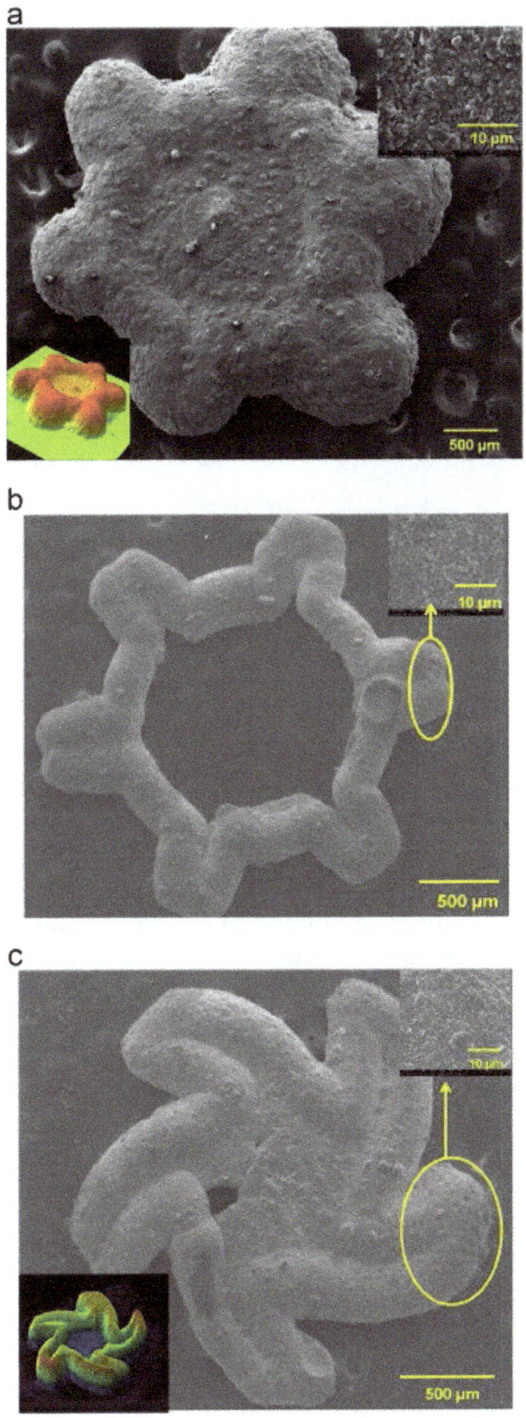

Figure 2.81. SEM images of the Al_2O_3 structures after sintering at 1550 °C for 5 h: (a) microgear without hole, (b) hollow gear with six teeth and (c) solid impeller. All are sintered at. Inset (right top corner) of panels (a, b, c) show SEM images of the sintered ceramics of all structures. Insets of panels (a, c) at left bottom show confocal images of the gear and impeller. Reprinted from [132], Copyright (2014), with permission from Elsevier.

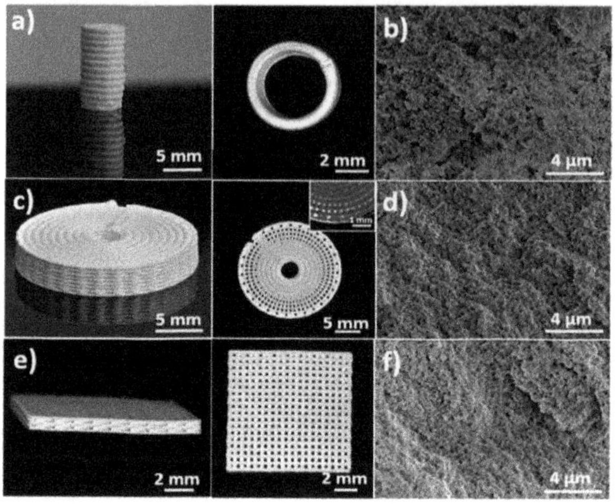

Figure 2.82. DIW printed items derived from the slurry of 3Y-55 ceramic powder with a solid loading of 70 wt% (30 vol%) with nozzle size of 406 μm and SEM images of the corresponding samples: (a, b) partially sintered tubes made with 1190 μm nozzle, (c, d) partially sintered 'wagon wheel' and (e, f) partially sintered sample with cubic lattice structure. Reproduced from [133]. CC BY 4.0.

The nozzle diameter was 1.19 mm, leading to a wall thickness of 1.05 mm, after the sample was sintered at 1090 °C. The total shrinkage of the green bodies after sintering was about 15%, while there were nanosized pores in the sintered ceramics, as presented in figure 2.82(b).

Figure 2.82(c) shows a photograph of a wagon wheel, where the structure was well retained after the curing and partial sintering processes. The item was constructed with alternating layers of concentric circles and angled lines from the center of the circle that spanned the radius of the circle. It was 20 mm in diameter and 5 mm in height. When the nozzle diameter was 4.06 mm, the resultant outer diameter, inner diameter, angle and layer height were 20 mm, 4 mm, 6° and 0.4 mm, respectively. The corresponding microstructure of the ceramics was homogeneous, distributed with nanosized pores, as illustrated in figure 2.82(d).

A cubic lattice structure is shown in figure 2.82(e), which was made of orthogonal layers consisting of parallel cylindrical rods, where each layer was rotated by 90°. The green body had a dimension of 15 × 15 mm^2, with strut width of about 0.45 mm. After sintering, the dimensions were shrunk to 12.75 × 12.75 mm^2, while the struts width became 0.38 mm. The diameter of the cylindrical rod and the spacing between the parallel cylindrical rods were the diameter of the nozzle (0.406 mm). The microstructure of this structure was similar to those of the other two, as observed in figure 2.82(f).

Figure 2.83 shows green bodies and sintered gyroids, from slurries of 3Y-55 (figure 2.83(a)) and 3YS-120 (figure 2.83(b)) powders, with solid loading level of 55 wt% or 18 vol%. Obviously, the green bodies made with larger particle sized powder had higher layer adhesion. As a result, items with a larger height could be

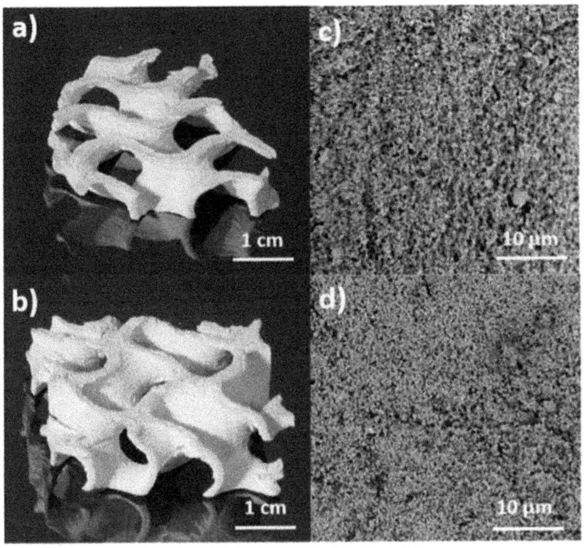

Figure 2.83. Partially sintered porous gyroids derived from 3Y-55 (a, c) and 3YS-120 (b, d) powders, along with the corresponding SEM images. Reproduced from [133]. CC BY 4.0.

printed. With 55 wt% 3Y-55 slurry, the projected wall thickness of the gyroids was 750 μm. As for the slurry with 55 wt% 3YS-120 powder, the projected wall thickness was 666 μm, which became 540 μm after sintering. The sintered items had a homogenous microstructure, as demonstrated in figures 2.83(c) and (d).

Another example was reported by Kirihara and Niki [134], who developed diamond-like structures of alumina microlattices, with electromagnetic band-gap characteristics at terahertz frequencies, by using the μSLA printing process. The acrylic resins dispersed with Al_2O_3 particles were printed with an accuracy at the micrometer order. After debinding and sintering processes, Al_2O_3 lattice structures were achieved with nearly full densification, which exhibited reflection at terahertz frequences, in almost all directions. The structures were designed to have twinned crystals, with mirror symmetric diamond lattices, to introduce defect interfaces. As a consequence, coupled resonation modes could be realized, owing to the double-cavity defects with unit cells hollowed from the diamond lattices.

The photonic crystals have diamond lattices with structural defects, as observed in figure 2.84. The twinned defect interfaces between mirror symmetric lattices were parallel to the (100) and (111) planes, as shown in figures 2.84(a) and (b). Double-cavity defects, consisting of hollowed unit cells, had center intervals of 1.5 and 2.0. Dielectric rods, with an aspect ratio of 1.5, were linked with a coordination number of four, resulting in a diamond structure with a lattice constant of 500 μm. The models were converted to μSLA files and then sliced as 2D layers. Figure 2.85 shows a schematic diagram of the μSLA printing process.

Al_2O_3 powder, with an average particle size of 0.2 μm, was dispersed in photosensitive acrylic resin, with a solid loading of 40 vol%. After mixing with

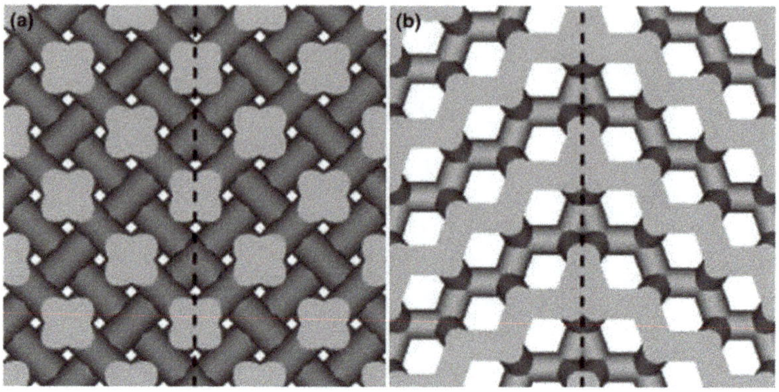

Figure 2.84. Graphic models of twinned photonic crystals with diamond structures designed with computer. Defect interfaces parallel to the planes indicated by dotted lines that were sandwiched between mirror symmetric lattice domains: (a) (100) and (b) (111). [134] John Wiley & Sons. © 2014 The American Ceramic Society.

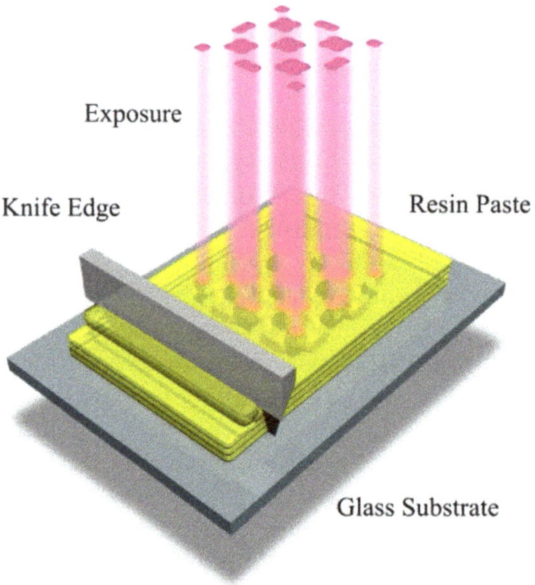

Figure 2.85. Schematic illustration of the μSLA 3D printer using photosensitive resin slurries with Al_2O_3 ceramic nanoparticles. Composite precursors were prepared by laminating 2D cross-sections exposed by a digital micromirror device (DMD). [134] John Wiley & Sons. © 2014 The American Ceramic Society.

rotation and revolution mixers, slurries with a viscosity of 2000 MPa at room temperature were obtained. The slurries were pressed onto the working stage with a dispenser nozzle and then spread uniformly by moving the knife. The speed of squeezing was controlled to be 5 mm $\cdot$ s^{-1}, while thickness of the layer was set to be 10 μm. Visible light, with a wavelength of 405 nm, was used to irradiate the surface of the resin layers, which were solidified owing to the photopolymerization.

A digital micromirror device (DMD) was utilized to ensure sufficiently high resolution. An array of $768 \times 1024 = 786,432$, with 14 μm edge length square aluminum mirrors, was achieved by using this optical device. The mirrors could be independently tilted, while 2D patterns could be dynamically exposed through an objective lens at the scale of 1/10, in the form of bitmap images with an edge size of 1.4 μm as square pixels. Large scale images could be obtained, as the micro patterns, with a dimension of 1.07×1.43 mm^2, were exposed in a side-by-side manner. Finally, 3D microstructures were formed after the layer-by-layer stacking of the 2D patterns. After printing, the uncured resin was cleaned out with ultrasonication. The printed samples were debinded at 600 °C for 2 h, followed by sintering at 1500 °C for 2 h. The heating rate for the debinding was 1.0 °C · min^{-1}, while that for the sintering was 8.0 °C · min^{-1}.

Acrylic photonic crystals distributed with Al_2O_3 ceramic particles were prepared by using μSLA process. The printed items had tolerances of $\leqslant \pm 5$ μm. SEM results indicated that the Al_2O_3 ceramic particles were homogeneously dispersed in the acrylic resin matrix. After debinding and sintering, the lattice constant and the linear shrinkage were 375 μm and 25%, respectively. The corresponding relative density of the Al_2O_3 ceramic was as high as 99%. The dielectric constant of the Al_2O_3 lattice was 9.8 by at THz frequencies. The photonic band gap was widened from 0.4 to 0.47 THz. The dense Al_2O_3 lattices exhibited isotropic propagation with a coordination number of four. Therefore, the lattices were shrunk equivalently in all crystalline directions, without deviation in dimension during the debinding and sintering processes.

Al_2O_3 ceramic-based photonic crystals with twinned diamond lattices were finally obtained. The defect interfaces of the (100) and (111) planes were sandwiched between the mirror symmetric domains, respectively with four and three periods, as revealed in figure 2.86. By optimizing the period number, precise localized modes of sharp transmission peaks in the bandgaps were achieved. Peaks of localized mode

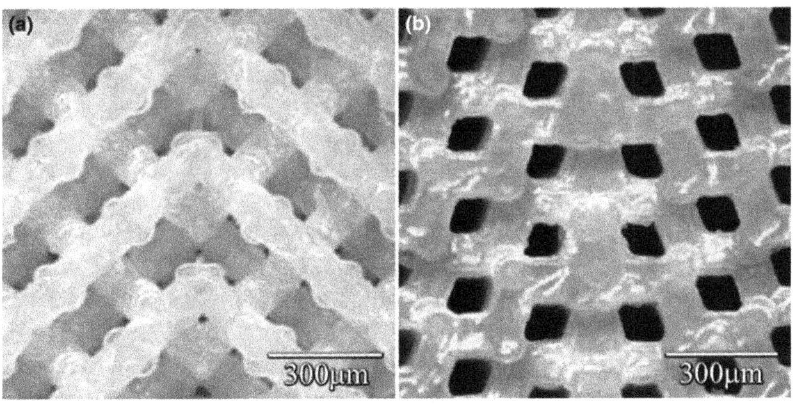

Figure 2.86. Twinned photonic crystals consisting of the sintered Al_2O_3 lattices. The defect interfaces were produced between the diamond lattice domains to be parallel to the planes: (a) (100) and (b) (111). [134] John Wiley & Sons. © 2014 The American Ceramic Society.

with transmission intensities of 22%–38% were developed at 0.414 and 0.409 THz, through the defect interfaces of (100) and (111), respectively.

2.4 Concluding remarks

Stereolithography (SLA) has emerged as one of the fastest-evolving 3D printing technologies for the fabrication of ceramic materials in recent years. From the process principles point of view, this technique aims to utilize light-induced polymerization to build ceramic green bodies, in a layer-by-layer manner. With ceramic slurries containing photosensitive agents or resins containing ceramic powders, combined with precise control over the curing process, structures with complicated geometries could be created, which is challenging through conventional ceramic processing strategies.

Currently, a wide range of ceramics have been processed by using the stereolithography technology, such as Al_2O_3, ZrO_2, and SiO_2. Focus has mainly been on the optimization of the formulation, composition and processing parameters to improve the rheological properties of slurries, ensure uniform dispersion of ceramic particles, and enhance mechanical strength of the green bodies and performances of the final ceramic products. In future, continued efforts should be made to explore the potential of stereolithography for fabricating ceramic materials with more advanced properties and functionalities.

References

[1] Chu T M G, Orton D G, Hollister S J, Feinberg S E and Halloran J W 2002 Mechanical and *in vivo* performance of hydroxyapatite implants with controlled architectures *Biomaterials* **23** 1283–93

[2] Chartier T, Duterte C, Delhote N, Baillargeat D, Verdeyme S, Delage C *et al* 2008 Fabrication of millimeter wave components via ceramic stereo- and microstereolithography processes *J. Am. Ceram. Soc.* **91** 2469–74

[3] Choi J W, Wicker R, Lee S H, Choi K H, Ha C S and Chung I 2009 Fabrication of 3D biocompatible/biodegradable micro-scaffolds using dynamic mask projection microstereolithography *J. Mater. Process. Technol.* **209** 5494–503

[4] Choi J W, Wicker R B, Cho S H, Ha C S and Lee S H 2009 Cure depth control for complex 3D microstructure fabrication in dynamic mask projection microstereolithography *Rapid Prototyp. J.* **15** 59–70

[5] Gao F, Yang S F, Hao P W and Evans J R G 2011 Suspension stability and fractal patterns: a comparison using hydroxyapatite *J. Am. Ceram. Soc.* **94** 704–12

[6] Bian W G, Li D C, Lian Q, Li X, Zhang W J, Wang K Z *et al* 2012 Fabrication of a bio-inspired β-tricalcium phosphate/collagen scaffold based on ceramic stereolithography and gel casting for osteochondral tissue engineering *Rapid Prototyp. J.* **18** 68–80

[7] Travitzky N, Bonet A, Dermeik B, Fey T, Filbert-Demut I, Schlier L *et al* 2014 Additive manufacturing of ceramic-based materials *Adv. Eng. Mater.* **16** 729–54

[8] Dufaud O and Corbel S 2003 Oxygen diffusion in ceramic suspensions for stereolithography *Chem. Eng. J.* **92** 55–62

[9] Wozniak M, Graule T, de Hazan Y, Kata D and Lis J 2009 Highly loaded UV curable nanosilica dispersions for rapid prototyping applications *J. Eur. Ceram. Soc.* **29** 2259–65

[10] Wozniak M, de Hazan Y, Graule T and Kata D 2011 Rheology of UV curable colloidal silica dispersions for rapid prototyping applications *J. Eur. Ceram. Soc.* **31** 2221–9
[11] Zhou W Z, Li D C and Wang H 2010 A novel aqueous ceramic suspension for ceramic stereolithography *Rapid Prototyp. J.* **16** 29–35
[12] Lingois P, Berglund L, Greco A and Maffezoli A 2003 Chemically induced residual stresses in dental composites *J. Mater. Sci.* **38** 1321–31
[13] Abouliatim Y, Chartier T, Abelard P, Chaput C and Delage C 2009 Optical characterization of stereolithography alumina suspensions using the Kubelka–Munk model *J. Eur. Ceram. Soc.* **29** 919–24
[14] Griffith M L and Halloran J W 1996 Freeform fabrication of ceramics via stereolithography *J. Am. Ceram. Soc.* **79** 2601–8
[15] Griffith M L and Halloran J W 1997 Scattering of ultraviolet radiation in turbid suspensions *J. Appl. Phys.* **81** 2538–46
[16] Jiang X N, Sun C, Zhang X, Xu B and Ye Y H 2000 Microstereolithography of lead zirconate titanate thick film on silicon substrate *Sens. Actuators A: Phys.* **87** 72–7
[17] Zhang X, Jiang X N and Sun C 1999 Micro-stereolithography of polymeric and ceramic microstructures *Sens. Actuators A: Phys.* **77** 149–56
[18] Gentry S P and Halloran J W 2013 Depth and width of cured lines in photopolymerizable ceramic suspensions *J. Eur. Ceram. Soc.* **33** 1981–8
[19] Gentry S P and Halloran J W 2013 Absorption effects in photopolymerized ceramic suspensions *J. Eur. Ceram. Soc.* **33** 1989–94
[20] Gentry S P and Halloran J W 2015 Light scattering in absorbing ceramic suspensions: effect on the width and depth of photopolymerized features *J. Eur. Ceram. Soc.* **35** 1895–904
[21] Badev A, Abouliatim Y, Chartier T, Lecamp L, Lebaudy P, Chaput C *et al* 2011 Photopolymerization kinetics of a polyether acrylate in the presence of ceramic fillers used in stereolithography *J. Photochem. Photobiol.* A **222** 117–22
[22] Lecamp L, Youssef B, Bunel C and Lebaudy P 1997 Photoinitiated polymerization of a dimethacrylate oligomer: 1. Influence of photoinitiator concentration, temperature and light intensity *Polymer* **38** 6089–96
[23] Lecamp L, Youssef B, Bunel C and Lebaudy P 1999 Photoinitiated polymerization of a dimethacrylate oligomer: 2. Kinetic studies *Polymer* **40** 1403–9
[24] Lecamp L, Youssef B, Bunel C and Lebaudy P 1999 Photoinitiated polymerization of a dimethacrylate oligomer—part 3. Postpolymerization study *Polymer* **40** 6313–20
[25] Wang D K, Carrera L and Abadie M J M 1993 Photopolymerization of glycidyl acrylate and glycidyl methacrylate investigated by differential photocalorimetry and FT-IR *Eur. Polym. J.* **29** 1379–86
[26] Scherzer T and Decker U 2000 The effect of temperature on the kinetics of diacrylate photopolymerizations studied by real-time FTIR spectroscopy *Polymer* **41** 7681–90
[27] Wu K C and Halloran J W 2005 Photopolymerization monitoring of ceramic stereolithography resins by FTIR methods *J. Mater. Sci.* **40** 71–6
[28] Oh S J, Lee S C and Park S Y 2006 Photopolymerization and photobleaching of n-butyl acrylate/fumed silica composites monitored by real time FTIR-ATR spectroscopy *Vib. Spectrosc.* **42** 273–7
[29] Lecamp L, Youssef B, Bunel C and Lebaudy P 1997 Photoinitiated polymerization of a dimethacrylate oligomer. 1. Influence of photoinitiator concentration, temperature and light intensity *Polymer* **38** 6089–96

[30] Chartier T, Penarroya R, Pagnoux C and Baumard J F 1997 Tape casting using UV curable binders *J. Eur. Ceram. Soc.* **17** 765–71

[31] Tanimoto Y, Hayakawa T and Nemoto K 2005 Analysis of photopolymerization behavior of UDMA/TEGDMA resin mixture and its composite by differential scanning calorimetry *J. Biomed. Mater. Res. B Appl. Biomater.* **72B** 310–5

[32] Azan V, Lecamp L, Lebaudy P and Bunel C 2007 Simulation of the photopolymerization gradient inside a pigmented coating: influence of TiO_2 concentration on the gradient *Prog. Org. Coat.* **58** 70–5

[33] Licciulli A, Corcione C E, Greco A, Amicarelli V and Maffezzoli A 2005 Laser stereolithography of ZrO_2 toughened Al_2O_3 *J. Eur. Ceram. Soc.* **25** 1581–9

[34] Hinczewski C, Corbel S and Chartier T 1998 Ceramic suspensions suitable for stereolithography *J. Eur. Ceram. Soc.* **18** 583–90

[35] Wang Y Y, Wang Z Y, Liu S H, Qu Z B, Han Z Q, Liu F T *et al* 2019 Additive manufacturing of silica ceramics from aqueous acrylamide based suspension *Ceram. Int.* **45** 21328–32

[36] Zocca A, Colombo P, Gomes C M and Günster J 2015 Additive manufacturing of ceramics: issues, potentialities, and opportunities *J. Am. Ceram. Soc.* **98** 1983–2001

[37] Manière C, Kerbart G, Harnois C and Marinel S 2020 Modeling sintering anisotropy in ceramic stereolithography of silica *Acta Mater.* **182** 163–71

[38] Zhou W Z, Li D C and Chen Z W 2011 The influence of ingredients of silica suspensions and laser exposure on UV curing behavior of aqueous ceramic suspensions in stereolithography *Int. J. Adv. Manuf. Technol.* **52** 575–82

[39] Chartier T, Chaput C, Doreau F and Loiseau M 2002 Stereolithography of structural complex ceramic parts *J. Mater. Sci.* **37** 3141–7

[40] Fraden S and Maret G 1990 Multiple light scattering from concentrated interacting suspensions *Phys. Rev. Lett.* **65** 512–5

[41] Jones A R 1999 Light scattering for particle characterization *Prog. Energy Combust. Sci.* **25** 1–53

[42] Wriedt T 2009 Light scattering theories and computer codes *J. Quant. Spectrosc. Radiat. Transf* **110** 833–43

[43] Zheng W, Wu J M, Chen S, Wang C S, Liu C L, Hua S B *et al* 2021 Influence of Al_2O_3 content on mechanical properties of silica-based ceramic cores prepared by stereolithography *J. Adv. Ceram* **10** 1381–8

[44] Huang L P, Duffrène L and Kieffer J 2004 Structural transitions in silica glass: thermo-mechanical anomalies and polyamorphism *J. Non-Cryst. Solids* **349** 1–9

[45] Ojovan M I and Tournier R F 2021 On structural rearrangements near the glass transition temperature in amorphous silica *Materials* **14** 5235

[46] Xiao Z H, Sun X Y, Li X Y, Wang Y Q, Wang Z Q, Zhang B W *et al* 2018 Phase transformation of GeO_2 glass to nanocrystals under ambient conditions *Nano Lett.* **18** 3290–6

[47] Xiao Z H, Li X L, Yu S J, Sun X Y, Li X Y, Wu M *et al* 2018 Effect of Bi_2O_3 on phase formation and microstructure evolution of mullite ceramics from mechanochemically activated oxide mixtures *Ceram. Int.* **44** 13841–7

[48] Xiao Z H, Li X L, Sun X Y, Jiang Y, Li X Y, Jiang H *et al* 2018 Phase formation and microstructure of Cr_2O_3 doped mullite ceramics through mechanochemical activation *J. Phys. Chem. Solids* **123** 198–205

[49] Yaroshenko V and Wilkinson D S 2001 Sintering and microstructure modification of mullite/zirconia composites derived from silica-coated alumina powders *J. Am. Ceram. Soc.* **84** 850–8

[50] Duan W J, Yang Z H, Cai D L, Zhang J W, Niu B, Jia D C *et al* 2020 Effect of sintering temperature on microstructure and mechanical properties of boron nitride whisker reinforced fused silica composites *Ceram. Int.* **46** 5132–40

[51] Katsura K, Shinoda Y, Akatsu T and Wakai F 2015 Sintering force behind shape evolution by viscous flow *J. Eur. Ceram. Soc.* **35** 1119–22

[52] Kazemi A, Faghihi-Sani M A, Nayyeri M J, Mohammadi M and Hajfathalian M 2014 Effect of zircon content on chemical and mechanical behavior of silica-based ceramic cores *Ceram. Int.* **40** 1093–8

[53] Breneman R C and Halloran J W 2015 Effect of cristobalite on the strength of sintered fused silica above and below the cristobalite transformation *J. Am. Ceram. Soc.* **98** 1611–7

[54] Dehghani P and Soleimani F 2022 Effects of sintering temperature and cristobalite content on the bending strength of spark plasma sintered fused silica ceramics *Ceram. Int.* **48** 16800–7

[55] Xia G B, He L T and Yang D A 2012 Preparation and characterization of $CaO–Al_2O_3–SiO_2$ glass/fused silica composites for LTCC application *J. Alloys Compd.* **531** 70–6

[56] Yang Z G, Zhao Z J, Yu J B and Ren Z M 2019 Preparation of silica ceramic cores by the preceramic pyrolysis technology using silicone resin as precursor and binder *Mater. Chem. Phys.* **223** 676–82

[57] Liang J J, Lin Q H, Zhang X, Jin T, Zhou Y Z, Sun X F *et al* 2017 Effects of alumina on cristobalite crystallization and properties of silica-based ceramic cores *J. Mater. Sci. Technol* **33** 204–9

[58] Yin Y H, Wang J, Huang Q Q, Xu S Z, Shuai S S, Hu T *et al* 2023 Influence of debinding parameter and nano-ZrO_2 particles on the silica-based ceramic cores fabricated by stereolithography-based additive manufacturing *Ceram. Int.* **49** 20878–89

[59] Hua S B, Su J, Deng Z L, Wu J M, Cheng L J, Yuan X *et al* 2021 Microstructures and properties of 45S5 bioglass® & BCP bioceramic scaffolds fabricated by digital light processing *Addit Manuf* **45** 102074

[60] Pan Z P, Guo J Z, Li S M, Xiong J Y and Long A P 2022 Experimental study on high temperature performances of silica-based ceramic core for single crystal turbine blades *Ceram. Int.* **48** 548–55

[61] Zhang J, Yu K B, Wu J M, Ye C S, Zheng W, Liu H *et al* 2023 Effects of $ZrSiO_4$ content on properties of SiO_2-based ceramics prepared by digital light processing *Ceram. Int.* **49** 9584–91

[62] Kazemi A, Faghihi-Sani M A and Alizadeh H R 2013 Investigation on cristobalite crystallization in silica-based ceramic cores for investment casting *J. Eur. Ceram. Soc.* **33** 3397–402

[63] Zhao G, Hu K H, Feng Q and Lu Z G 2021 Creep mechanism of zircon-added silica ceramic cores formed by stereolithography *Ceram. Int.* **47** 17719–25

[64] Wu H H, Li D C, Tang Y P, Sun B and Xu D Y 2009 Rapid fabrication of alumina-based ceramic cores for gas turbine blades by stereolithography and gelcasting *J. Mater. Process. Technol.* **209** 5886–91

[65] Lindqvist K M and Carlström E 2005 Indirect solid freeform fabrication by binder assisted slip casting *J. Eur. Ceram. Soc.* **25** 3539–45

[66] Ma J T, Xie Z P, Miao H Z, Huang Y, Cheng Y B and Yang W Y 2003 Gelcasting of alumina ceramics in the mixed acrylamide and polyacrylamide systems *J. Eur. Ceram. Soc.* **23** 2273–9

[67] Santacruz I, Nieto M I and Moreno R 2005 Alumina bodies with near-to-theoretical density by aqueous gelcasting using concentrated agarose solutions *Ceram. Int.* **31** 439–45

[68] Tong J F and Chen D M 2004 Preparation of alumina by aqueous gelcasting *Ceram. Int.* **30** 2061–6

[69] Jiang S W, Matsukawa T, Tanaka S and Uematsu K 2007 Effects of powder characteristics, solid loading and dispersant on bubble content in aqueous alumina slurries *J. Eur. Ceram. Soc.* **27** 879–85

[70] Li X D, Wang L, Iwasa M and Hayakawa M 2003 Effect of powder characteristics on centrifugal slip casting of alumina powders *J. Ceram. Soc. Jpn.* **111** 594–9

[71] Mohapatra D and Sarkar D 2007 Preparation of MgO–$MgAl_2O_4$ composite for refractory application *J. Mater. Process. Technol.* **189** 279–83

[72] Hu Y W, Li D C, Dai W, Wang M J, Wang H and Sun K 2012 Fabrication of three-dimensional electromagnetic band-gap structure with alumina based on stereolithography and gelcasting *J. Manuf. Syst.* **31** 22–5

[73] Zhou M P, Liu W, Wu H D, Song X, Chen Y, Cheng L X *et al* 2016 Preparation of a defect-free alumina cutting tool via additive manufacturing based on stereolithography: optimization of the drying and debinding processes *Ceram. Int.* **42** 11598–602

[74] An D, Li H Z, Xie Z P, Zhu T B, Luo X D, Shen Z J *et al* 2017 Additive manufacturing and characterization of complex Al_2O_3 parts based on a novel stereolithography method *Int. J. Appl. Ceram. Technol.* **14** 836–44

[75] Cao M Q, Yan Q Z, Li X H and Mi Y Y 2014 Effect of plate-like alumina on the properties of alumina ceramics prepared by gel-casting *Mater. Sci. Eng., A-Struct.* **589** 97–100

[76] Liu W, Xie Z and Cheng L 2015 Sintering kinetics window: an approach to the densification process during the preparation of transparent alumina *Adv. Appl. Ceram* **114** 33–8

[77] Liu W W, Li M S, Nie J B, Wang C Y, Li W L and Xing Z W 2020 Synergy of solid loading and printability of ceramic paste for optimized properties of alumina via stereolithography-based 3D printing *J. Mater. Res. Technol* **9** 11476–83

[78] Qian C C, Hu K H, Wang H Y, Nie L, Feng Q, Lu Z G *et al* 2022 The effect of particle size distribution on the microstructure and properties of Al_2O_3 ceramics formed by stereolithography *Ceram. Int.* **48** 21600–9

[79] Cui M M, Wang T, Zhao Y, Zhang Z, Wang X, Hou X *et al* 2023 Research on crack mechanism and kinetic model of alumina ceramic in the degreasing stage based on stereolithography *Int. J. Appl. Ceram. Technol.* **20** 3419–35

[80] Chen S, Wang C S, Zheng W, Wu J M, Yan C Z and Shi Y S 2022 Effects of particle size distribution and sintering temperature on properties of alumina mold material prepared by stereolithography *Ceram. Int.* **48** 6069–77

[81] Kaczmarski K and Bellot J C 2003 Effect of particle-size distribution and particle porosity changes on mass-transfer kinetics *Acta Chromatogr.* **13** 22–37

[82] Li H, Liu Y S, Liu Y S, Zeng Q F, Wang J, Hu K H *et al* 2020 Evolution of the microstructure and mechanical properties of stereolithography formed alumina cores sintered in vacuum *J. Eur. Ceram. Soc.* **40** 4825–36

[83] He Z M and Ma J 2000 Grain-growth rate constant of hot-pressed alumina ceramics *Mater. Lett.* **44** 14–8

[84] Montero C, Ochoa A, Castaño P, Bilbao J and Gayubo A G 2015 Monitoring Ni^0 and coke evolution during the deactivation of a Ni/La_2O_3–α-Al_2O_3 catalyst in ethanol steam reforming in a fluidized bed *J. Catal.* **331** 181–92

[85] Zhang K Q, He R J, Ding G J, Bai X J and Fang D N 2021 Effects of fine grains and sintering additives on stereolithography additive manufactured Al_2O_3 ceramic *Ceram. Int.* **47** 2303–10

[86] Pal S, Bandyopadhyay A K, Mukherjee S, Samaddar B N and Pal P G 2010 Function of magnesium aluminate hydrate and magnesium nitrate as MgO addition in crystal structure and grain size control of α-Al_2O_3 during sintering *Bull. Mater. Sci.* **33** 55–63

[87] Sun Y, Li M, Jiang Y L, Xing B H, Shen M H, Cao C R *et al* 2021 High-quality translucent alumina ceramic through digital light processing stereolithography method *Adv. Eng. Mater.* **23** 2001475

[88] Xu H M, Li S J, Liu R Z, Bao C G, Mu M Q and Wang K J 2023 Fabrication of alumina ceramics with high flexural strength using stereolithography *Int. J. Adv. Manuf. Technol.* **128** 2983–94

[89] Li H, Liu Y S, Liu Y S, Zeng Q F and Liang J J 2021 Silica strengthened alumina ceramic cores prepared by 3D printing *J. Eur. Ceram. Soc.* **41** 2938–47

[90] Xing H Y, Zou B, Li S S and Fu X S 2017 Study on surface quality, precision and mechanical properties of 3D printed ZrO_2 ceramic components by laser scanning stereolithography *Ceram. Int.* **43** 16340–7

[91] Sun J X, Binner J and Bai J M 2019 Effect of surface treatment on the dispersion of nano zirconia particles in non-aqueous suspensions for stereolithography *J. Eur. Ceram. Soc.* **39** 1660–7

[92] Song S Y, Park M S, Lee D, Lee J W and Yun J S 2019 Optimization and characterization of high-viscosity ZrO_2 ceramic nanocomposite resins for supportless stereolithography *Mater. Des.* **180** 107960

[93] Xing B H, Cao C R, Zhao W M, Shen M H, Wang C and Zhao Z 2020 Dense 8 mol% yttria-stabilized zirconia electrolyte by DLP stereolithography *J. Eur. Ceram. Soc.* **40** 1418–23

[94] Li Y H, Wang M L, Wu H D, He F P, Chen Y and Wu S H 2019 Cure behavior of colorful ZrO_2 suspensions during digital light processing (DLP) based stereolithography process *J. Eur. Ceram. Soc.* **39** 4921–7

[95] Wang L, Yu H, Hao Z D, Tang W Z and Dou R 2023 Fabrication of highly translucent yttria-stabilized zirconia ceramics using stereolithography-based additive manufacturing *Ceram. Int.* **49** 17174–84

[96] Chevalier J, Deville S, Münch E, Jullian R and Lair F 2004 Critical effect of cubic phase on aging in 3 mol% yttria-stabilized zirconia ceramics for hip replacement prosthesis *Biomaterials* **25** 5539–45

[97] Zhang F, Van Meerbeek B and Vleugels J 2020 Importance of tetragonal phase in high-translucent partially stabilized zirconia for dental restorations *Dent. Mater.* **36** 491–500

[98] Cui J P, Gong Z Y, Lv M and Rao P G 2018 Effect of notch depth on fracture toughness of zirconia ceramics tested by SEVNB method *Ceram. Int.* **44** 17218–23

[99] Boccaccini D N, Frandsen H L, Soprani S, Cannio M, Klemenso T, Gil V *et al* 2018 Influence of porosity on mechanical properties of tetragonal stabilized zirconia *J. Eur. Ceram. Soc.* **38** 1720–35

[100] Ma H Q, Fang X, Yin S, Li T Y, Zhou C, Jiang X W *et al* 2024 Preparation and characteristics of honeycomb mullite ceramics with controllable structure by stereolithography 3D printing and in-situ synthesis *Ceram. Int.* **50** 3176–86

[101] Zhang J H, Ke C M, Wu H D, Zhang S X and Yu J S 2015 Anisotropic grain growth mechanism of mullite derived from cobalt oxide doped diphasic-gels *Rare Met. Mater. Eng.* **44** 323–6

[102] Kong L B, Ma J, Huang H, Zhang T S and Boey F 2003 Anisotropic mullitization in CuO-doped oxide mixture activated by high-energy ball milling *Mater. Lett.* **57** 3660–6

[103] Kong L B, Zhang T S, Ma J and Boey F Y C 2009 Mullitization behavior and microstructural development of B_2O_3–Al_2O_3–SiO_2 mixtures activated by high-energy ball milling *Solid State Sci.* **11** 1333–42

[104] Zhang J H, Wu H D, Zhang S X, Yu J S and Xiao H Y 2013 Anisotropic grain growth in diphasic-gel-derived vanadium pentoxide doped mullite *J. Cryst. Growth* **364** 11–5

[105] Ma H Q, Meng T Y, Yin J W, Yin S, Fang X, Li T Y *et al* 2024 Mechanical properties and fracture mechanism of 3D-printed honeycomb mullite ceramics fabricated by stereolithography *Ceram. Int.* **50** 41499–508

[106] Zhao P L, Ma S H, Wang X H, Wu W and Ou Y J 2023 Properties and mechanism of mullite whisker toughened ceramics *Ceram. Int.* **49** 10238–48

[107] German R, Suri P and Park S 2009 Review: liquid phase sintering *J. Mater. Sci.* **44** 1–39

[108] Huang Z Y, Liu L Y, Yuan J M, Guo H L, Wang H M, Ye P C *et al* 2023 Stereolithography 3D printing of Si_3N_4 cellular ceramics with ultrahigh strength by using highly viscous paste *Ceram. Int.* **49** 6984–95

[109] Liu Y, Cheng L J, Li H, Li Q, Shi Y, Liu F *et al* 2020 Formation mechanism of stereolithography of Si_3N_4 slurry using silane coupling agent as modifier and dispersant *Ceram. Int.* **46** 14583–90

[110] Sun N, Wang T P, Du Y H, Ma X J, Xin W K, Dang H C *et al* 2024 Effect of TMAH as a modifier on the performance of Si_3N_4 stereolithography pastes *Ceram. Int.* **50** 15502–12

[111] Wu X Q, Xu C J and Zhang Z M 2021 Preparation and optimization of Si_3N_4 ceramic slurry for low-cost LCD mask stereolithography *Ceram. Int.* **47** 9400–8

[112] Wu X Q, Xu C J and Zhang Z M 2022 Development and analysis of a high refractive index liquid phase Si_3N_4 slurry for mask stereolithography *Ceram. Int.* **48** 120–9

[113] Zou W J, Yang P, Lin L F, Li Y H and Wu S H 2022 Improving cure performance of Si_3N_4 suspension with a high refractive index resin for stereolithography-based additive manufacturing *Ceram. Int.* **48** 12569–77

[114] Huang R J, Jiang Q G, Wu H D, Li Y H, Liu W Y, Lu X X *et al* 2019 Fabrication of complex shaped ceramic parts with surface-oxidized Si_3N_4 powder via digital light processing based stereolithography method *Ceram. Int.* **45** 5158–62

[115] Liu Y, Zhan L N, Wen L, Cheng L, He Y, Xu B *et al* 2021 Effects of particle size and color on photocuring performance of Si_3N_4 ceramic slurry by stereolithography *J. Eur. Ceram. Soc.* **41** 2386–94

[116] Bai X J, Ding G J, Zhang K Q, Wang W Q, Zhou N P, Fang D N *et al* 2021 Stereolithography additive manufacturing and sintering approaches of SiC ceramics *Open Ceramics* **5** 100046

[117] Chen J S, Wang Y J, Pei X L, Bao C G, Huang Z R, He L *et al* 2020 Preparation and stereolithography of SiC ceramic precursor with high photosensitivity and ceramic yield *Ceram. Int.* **46** 13066–72

[118] de Hazan Y and Penner D 2017 SiC and SiOC ceramic articles produced by stereolithography of acrylate modified polycarbosilane systems *J. Eur. Ceram. Soc.* **37** 5205–12
[119] Ding G J, He R J, Zhang K Q, Xia M, Feng C W and Fang D N 2020 Dispersion and stability of SiC ceramic slurry for stereolithography *Ceram. Int.* **46** 4720–9
[120] Ding G J, He R J, Zhang K Q, Xie C, Wang M, Yang Y Z *et al* 2019 Stereolithography-based additive manufacturing of gray-colored SiC ceramic green body *J. Am. Ceram. Soc.* **102** 7198–209
[121] Li W, Cui C C, Bao J X, Zhang G, Li S and Wang G 2021 Properties regulation of SiC ceramics prepared via stereolithography combined with reactive melt infiltration techniques *Ceram. Int.* **47** 33997–4004
[122] Essmeister J, Altun A A, Staudacher M, Lube T, Schwentenwein M and Konegger T 2022 Stereolithography-based additive manufacturing of polymer-derived SiOC/SiC ceramic composites *J. Eur. Ceram. Soc.* **42** 5343–54
[123] He R J, Ding G J, Zhang K Q, Li Y and Fang D N 2019 Fabrication of SiC ceramic architectures using stereolithography combined with precursor infiltration and pyrolysis *Ceram. Int.* **45** 14006–14
[124] Hu C Q, Chen Y F, Yang T S, Liu H L, Huang X T, Huo Y L *et al* 2021 Effect of SiC powder on the properties of SiC slurry for stereolithography *Ceram. Int.* **47** 12442
[125] Tang J, Guo X T, Chang H T, Hu K H, Shen Z, Wang W X *et al* 2021 The preparation of SiC ceramic photosensitive slurry for rapid stereolithography *J. Eur. Ceram. Soc.* **41** 7516–24
[126] Tian X Y, Zhang W G, Li D C and Heinrich J G 2012 Reaction-bonded SiC derived from resin precursors by Stereolithography *Ceram. Int.* **38** 589–97
[127] Wang K, Liu R Z and Bao C G 2022 SiC paste with high curing thickness for stereolithography *Ceram. Int.* **48** 28692–703
[128] Liu T L, Yang L X, Chen Z F, Yang M M and Lu L 2023 Effects of SiC content on the microstructure and mechanical performance of stereolithography-based SiC ceramics *J. Mater. Res. Technol.* **25** 5184–95
[129] Qu P, Liang G Z, Hamza M, Mo Y, Jiang L, Luo X *et al* 2024 3D printing of high-purity complex SiC structures based on stereolithography *Ceram. Int.* **50** 23763–74
[130] Bertsch A, Jiguet S and Renaud P 2004 Microfabrication of ceramic components by microstereolithography *J. Micromech. Microeng.* **14** 197–203
[131] Monri K and Maruo S 2013 Three-dimensional ceramic molding based on microstereolithography for the production of piezoelectric energy harvesters *Sens. Actuators A: Phys.* **200** 31–6
[132] Goswami A, Ankit K, Balashanmugam N, Umarji A M and Madras G 2014 Optimization of rheological properties of photopolymerizable alumina suspensions for ceramic microstereolithography *Ceram. Int.* **40** 3655–65
[133] Troksa A L, Eshelman H V, Chandrasekaran S, Rodriguez N, Ruelas S, Duoss E B *et al* 2021 3D-printed nanoporous ceramics: tunable feedstock for direct ink write and projection microstereolithography *Mater. Des.* **198** 109337
[134] Kirihara S and Niki T 2015 Three-dimensional stereolithography of alumina photonic crystals for terahertz wave localization *Int. J. Appl. Ceram. Technol.* **12** 32–7

Ling Bing Kong, Zhuohao Xiao, Bin He and Yin Liu

Chapter 3

Inkjet printing

Inkjet printing is a special type of 3D printing technology, which has been employed to fabricate a wide range of materials. Inkjet printing has been used to develop both structural and functional ceramics and ceramic materials. In this chapter, ceramics that have been prepared by using the inkjet printing process will be elaborated.

3.1 Brief description

Inkjet printing has been widely used in our daily life through inkjet printers for newspapers, magazines, documents, and so on. For industrial applications, inkjet techniques have been employed to fabricate various structural and functional materials, including ceramic materials [1, 2]. Inkjet printing was developed as a non-contact, direct-write technology for the decoration of ceramic tiles to replace conventional flat screen-printing techniques [3–5].

As a promising technique, inkjet printing is suitable for producing complicated structures and patterns. Inkjet printing is realized through digitally controlled ejection of droplets of specific inks from the print head onto substrates. There are essentially two types of inkjet printers, i.e., continuous inkjet (CIJ) and droplet-on-demand (DOD) [6–12]. The DOD mode has been widely used for industrial applications, where functional inks are deposited on given substrates and the quantity of ink droplets can be precisely controlled.

The functional inks are made of specific components through dissolving or dispersing in solvents. During the inkjet printing process, a given quantity of ink in a chamber is ejected from a nozzle through a rapid quasi-adiabatic reduction in the chamber volume by using a piezoelectric actuator. The chamber containing the ink contracts in response to the piezoelectric actuator when an external voltage is applied. The ink is then subjected to a shockwave, leading to the ejection of liquid droplets from the nozzle. Owing to gravity, the ejected droplets fall and impinge onto the substrate. The droplets spread due to the momentum of motion and flow

doi:10.1088/978-0-7503-4831-7ch3

along the surface of the substrate depending on surface tension. The droplets then gradually dry as the solvent evaporates.

Inkjet printing for developing ceramic materials includes two aspects. On one hand, it enables us to prepare ceramics with fine and complicated structures. On the other hand, it has strong potential in the decoration of ceramic tiles, as the patterns that can be coated on ceramic tiles nearly have no limitation compared to those achievable by conventional printing techniques. Besides, inkjet printing offers various advantages. For instance, inkjet printing is a non-contact process with much lower noise, as compared with conventional printing machines. The tiny ink droplets pose minimal force on the substrate during the printing process. As a result, substrates with various mechanical properties and even non-flat surfaces can be printed.

Moreover, inkjet printing is a fully digitized process, which is simple from design to production, thus ensuring a short product cycle and cost-effectiveness. The products made with inkjet printing have high image resolution and accuracy. The droplet deposition sites on the substrate can be varied according to the instantaneous requirements of the real scenario. In practice, patterns with different configurations on ceramic tiles can be printed sequentially or even simultaneously.

This chapter aims to provide an overview of the applications of inkjet printing in the fabrication of various ceramic materials, including ceramic bulks, thin films, pattern structures, pigments, and tile decoration. First of all, the principles of inkjet printing will be briefly described, followed by a discussion on the manufacturing of various ceramics according to material types. Finally, ceramic inks for tile decoration applications will be discussed. The chapter will end with a conclusion.

3.2 Principles of inkjet printing

In the continuous inkjet process, the first step is to form droplets from a continuously flowing jet of ink, which is forced out of a nozzle under pressure [13]. Figure 3.1 shows a schematic diagram of a single-jet printing system. When the disturbance at a particular wavelength along the jet gradually increases to a certain level, the jet breaks into droplets. By regulating the disturbance at a suitable frequency, a uniform stream of droplets can be produced.

In most cases, a piezoelectric transducer is used to control the break-up of the jet. Selected droplets from the stream are used in inkjet printing. For effective selection of the droplets, electrically conductive inks are usually used, while the droplets are inductively charged at certain applied potentials. Once the droplets break off from the stream, the induced charges are unable to flow along the liquid column, so they are retained on the droplets. Then, the applied voltages are varied to charge the next droplets being formed. Depending on the applied electric fields, the droplets pass through and are deflected.

The charged droplets are deposited onto the substrate, while the uncharged ones are recycled. The position where the droplets strike is related to the level of charges. Therefore, lines can be printed with the droplets by controlling the charge levels. Meanwhile, more complicated structures can be obtained by either moving the

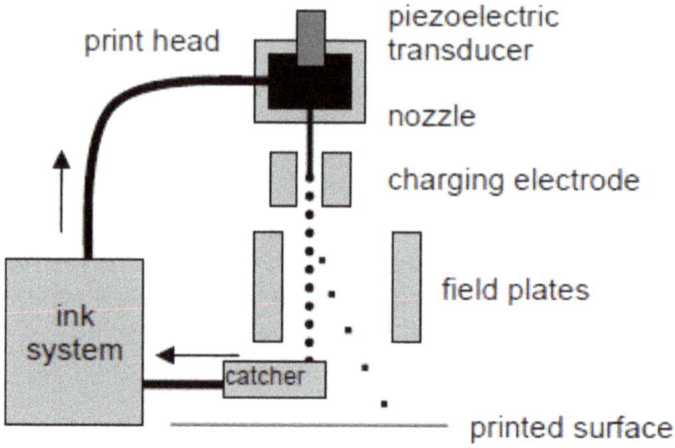

Figure 3.1. Schematic diagram of a continuous inkjet printer. Reproduced from [13]. © IOP Publishing Ltd. CC BY 3.0.

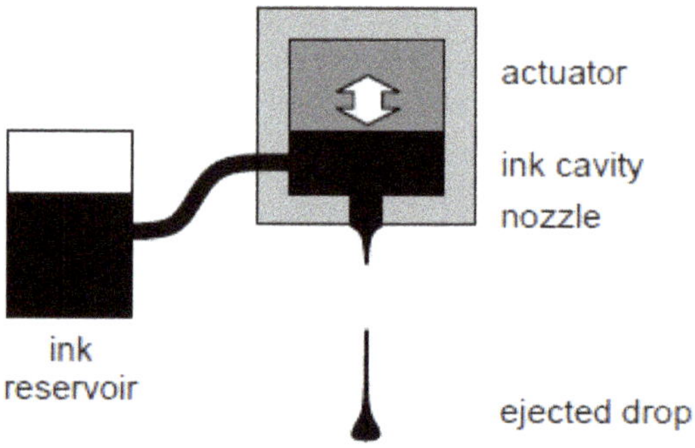

Figure 3.2. Schematic diagram of a DOD print head. Reproduced from [13]. © IOP Publishing Ltd. CC BY 3.0.

substrate or printing successive lines. When a sufficiently large number of arrays are continuously printed, full-color images can be developed in a much faster way.

In DOD printing, the heads generally have nozzles arranged in an array, each of which ejects ink droplets only when required to form the image. Figure 3.2 shows a schematic diagram of the nozzles in operation. The actuator triggers expansion/contraction of the chamber, which generates momentum to eject the ink droplets. In this dynamic process, the ink is subject to wave propagation, while the volume of the chamber varies periodically. There are mainly two ways to trigger the ejection of inks. One way is to generate vapor bubbles in the ink, using a heater pad, known as a bubble jet. The other one is through the distortion of ceramic piezoelectric actuators. The droplets ejected from the ink first appear as a jet, which then forms a ligament or tail, but remains connected to the ink inside the nozzle, as illustrated in figure 3.3.

At the final stage of the ligament, the ink is partly sucked back into the nozzle, while the rest of the tail is linked to the droplets, which may also be broken into smaller satellite droplets. Before the next droplet is ejected from the nozzle, the

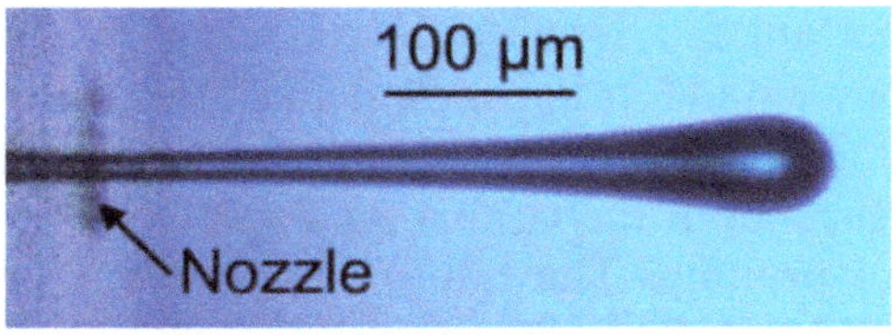

Figure 3.3. High-speed photograph showing a jet of ink emerging from a DOD nozzle. The jet is typically travelling at rates of 5–10 $m \cdot s^{-1}$. Reproduced from [13].

chamber should be refilled and acoustic disturbances should be regulated to avoid any negative effects on the formation of the next droplet.

Because external droplet selection and recovery systems are not required, droplet-on-demand printing is simpler than continuous printing. Nevertheless, a relatively large number of nozzles are necessary for this technique. In DOD printing, the ink after printing should be dried or densified on the substrate to avoid clogging the nozzles. In this case, the print head should be properly cleaned and capped or low volatility inks and absorbing substrates should be utilized.

Electrostatic inkjet printing is a technique by using electric fields to generate liquid streams and droplets. Conical surfaces are present and then the liquid is jetted from the tip, if the electric field applied between the liquid in the nozzles and the substrate is sufficiently high. This process has been employed to fabricate various ceramic materials. The jet of liquid extending from the nozzle is unstable, and strongly tends to form droplets, even with very weak disturbances.

For inkjet printing via continuous jets, the formation of satellites should be under control, otherwise, the printing process could be disrupted. Once the satellites tend to merge before deflection in the electrostatic field, they can enter the field and will be deflected at a large magnitude, owing to their relatively larger charge-to-mass ratio. In the worst case, the satellites could attack the electrodes, and an electrical breakdown could be triggered, causing the printing process to fail. Generally, the formation of satellites can be controlled by varying the magnitude of disturbance.

In the printing process, the jets are generated as the inks are ejected from the nozzles. According to the structure of the head of the printers and the characteristics of the inks, the waveform of the electric fields should be predetermined, so as to effectively drive the actuators. The electric fields are applied to either trigger the movement of piezoelectric actuators to expand/contract the ink chambers or the heat resistive pad to vaporize the inks. As a result, inks are ejected from the nozzles.

Because the drive electric waveforms are applied in a very short time, while the formation of the jet needs a relatively longer time, droplets can be effectively formed. In the beginning, the emergence rate of the droplets is quite high. As the drive impulse is reduced, the droplets continuously move forward, which are connected to the ink in the nozzle due to the stretching ligament. Because the ligament stretches, the droplets would be decelerated, due to the dissipative energy loss from viscous forces, the energy needed to generate new liquid surfaces, and air dragging effects. Once the ligaments break, the droplets tend to be spherical in shape, due to the surface tension. If the ligaments were too long, satellites would be formed.

Velocity is an important factor to characterize the performance of the printer head for a droplet-on-demand printing system. This measure is very useful in controlling the uniformity of the arrays to be printed and printing parameters of the droplets. In an ideal case, both the volume and velocity of droplets should not be affected by the rate of printing. The rate of droplets has an upper limit, beyond which the printer would fail. This occurrence can be avoided by monitoring the volume and velocity of the droplets.

3.3 Methods for preparation of printing inks

The effectiveness of the inkjet printing process is essentially dependent on various parameters, such as the properties of inks, the functions of printers, the characteristics of substrates and so on. Among them, physical and chemical properties of the inks are the most critical to the qualities of inkjet-printed products. Specifically, the properties of the inks affect the efficiency of ink droplets, the ink droplet-substrate interactions, and the drying behaviors of the ink droplets during the formation of patterns to be printed. In practice, it is necessary to define the hardware and implement techniques, according to the physical and chemical properties of the inks.

Currently, inks for inkjet printing include phase-change, solvent-based, water-based and UV curable. Phase-change inks are usually made of inorganic materials, such as alumina (Al_2O_3), titania (TiO_2), zirconia (ZrO_2), perovskite ferroelectric lead zirconate titanate ($Pb(Zr_{0.52}Ti_{0.48})O_3$, PZT), etc. Solvent-based inks are composed of Au-Cu metals/alloys, barium titanate ($BaTiO_3$), ZrO_2, nickel oxide (NiO) and various inorganic pigments. Generally, ethanol or its mixtures with other alcohols are used as solvents to prepare solvent-based printing inks. For UV curable printing inks, solidification of the printed items is carried out by the irradiation of UV light.

The development of inkjet printing inks is always a challenge, owing to the special application scenarios. In addition to normal requirements, such as long-life stability, color and so on, special physical and chemical properties, such as storage behavior in cartridges, jetting efficiency, interaction characteristics with substrates, human health and environmental friendliness, should also be considered. Specific physical and chemical properties include suspension stability, viscosities, surface tensions, pH values, type of electrolyte, chemical compositions, solid loading levels and so on.

More recently, ceramic pigment powders with submicron and nanosized particle sizes have attracted huge attention, because of their large surface coverage and strong light scattering. Submicron-sized powders are those with particle sizes in the range of 0.2–0.5 μm, while nanosized powders have average particle sizes of 1–100 nm. The key challenge in preparing the inkjet printing inks with such ultrafine particles is their dispersion in the given solvents, thus ensuring the desired physical and chemical properties. One of the most important requirements is that ceramic ink should have sufficiently high solid loading levels. Due to the high surface energy, ultrafine particles tend to agglomerate. Therefore, strong attempts have been made in developing strategies to disperse ultrafine ceramic powders with high solid loadings.

3.3.1 Physical processes

In physical processes, ceramic powders with large sizes are ground by using mechanical milling, without involving chemical reactions. The efficiency of mechanical processes can be ensured by adjusting various processing parameters, such as milling media, materials of balls, ball-to-powder ratio, milling speed, milling agent and so on. Ceramic inks have been prepared by using mechanical milling process, including Al_2O_3 [14, 15], ZrO_2 [16–19], TiO_2 [20, 21], Si_3N_4/MoS_2 [22], $Pb(Zr_{0.53}Ti_{0.47})O_3$ (PZT) [23], $BaTiO_3$ [24, 25] and so on.

3.3.2 Chemical synthesis processes

Chemical processes have been widely used to synthesize various materials with nanosized scales, mainly involving precipitation or crystallization in solutions. In most cases, the products should be thermally calcined to obtain final powders. In this section, various chemical processes that have been employed to prepare ceramic powders for inkjet printing will be briefly summarized and discussed.

3.3.2.1 Sol–gel reaction process

Sol–gel is a multi-step process, involving precursor preparation, hydrolyzation, condensation, formation of sols, gelation to gels, drying of wet gels, calcination, and final powders [26–36]. In ceramic industries, ceramic inks are usually prepared by dispersing ceramic pigments in solvents, through sol–gel reaction [37, 38]. Meanwhile, the sols obtained in the sol–gel processes can be directly applied as ceramic inks, forming color coatings on ceramic tiles after sintering.

Microporous membranes on porous ceramic or stainless steel supports are promising candidates for the separation of H_2 from its mixtures with CO_2, CH_4 or N_2 [39, 40]. Inkjet printing was combined with rapid thermal processing (RTP) to fabricate metal-supported microporous SiO_2 membranes from SiO_2 sols [41]. The SiO_2 membranes displayed high selectivity for He and H_2 to CO_2 and N_2.

The SiO_2 sols were prepared through the acid-catalyzed hydrolyzation in ethanol as the solvent [42–46]. Tetraethyl orthosilicate (TEOS) and ethanol with given volume ratios were mixed at room temperature. After that, a mixed solution consisting of ethanol, DI water and nitric acid was added into the TEOS solution with a metering pump at a rate of $1.5\ ml \cdot min^{-1}$ with the aid of stirring. The mixture was then refluxed at 60 °C for 3 h to promote the hydrolyzation and condensation reactions, resulting in transparent SiO_2 sols.

Inorganic inks based on sol–gel derived SiO_2 sols have also been inkjet printed on glasses for architecture building applications [47]. In detail, 29.27 g of TEOS was mixed with 52.07 g of ethanol absolute at room temperature, aided with magnetic stirring. Hydrochloric acid (HCl) was diluted with DI water to have a concentration of 0.1 M. After that, 5 ml of 0.1 M HCl was added to the mixture. After stirring overnight, SiO_2 sol was formed, with a solid loading level of 14 wt% SiO_2. The sol was then heated to concentrate to solid loading level of 50 wt% SiO_2 by evaporating solvents. In 2 g transparent viscous sol, 1.5 g cyclohexanol, 1.5 g 2,6-dimethyl-4-heptanol and 4 g hexylene glycol were added, resulting in inkjet printing ink.

The sol–gel process is advantageous for synthesizing nanosized powders of various materials with desired properties, owing to the atomic level homogeneity when mixing the precursors. Most likely, in the sol–gel processes, metal alkoxides are used for hydrolyzation reactions, which are expensive and very sensitive to moisture. Therefore, sol–gel processes are not superior for large-scale industrial applications, due to the high cost of raw materials, large volume shrinkage, health hazards of organic components, complicated processing steps, strict environmental requirements and so on. Therefore, aqueous sol–gel processes have attracted increasing interest from both researchers and industries [48–50].

In aqueous sol–gel processes, cheap precursors, such as inorganic salts, can be used. Meanwhile, the by-products require much simpler treatment. There are four types of aqueous sol–gel processes, including (i) colloidal dispersions of oxide particles synthesized with other processes, e.g., vapor phase reaction, (ii) colloidal dispersions of hydroxides or hydrated oxides made of peptized precipitate, (iii) polymerization of hydrolyzable cations with the polynuclear cations >1 nm in size and (iv) precipitation of hydrated oxide or hydroxide in the presence of an organic gelling agent. The dispersions are stable owing to electrostatic interactions, while gels can be formed when they are concentrated by evaporating water or adjusting the pH level [51–53].

Sols derived from aqueous sol–gel processes have various advantages, when used as ceramic inks for inkjet printing. First of all, the sols consist of colloidal particles generated by chemical reactions. Meanwhile, the viscosity of the dispersions can be adjusted to meet the requirement of inkjet printing. Also, the droplets formed from sols transition into gels upon deposition on the substrate, as water partly evaporates during the drying. Due to gel formation, the components in the ceramic inks remain uniformly distributed, avoiding segregation. Compared to inks containing conventional pigments, inks prepared via sol–gel processes contain only precursors, with no pigments. As a result, colors are not displayed during the printing process. In this case, the reaction of the precursors occurs during the calcination process, thus forming desired pigment products [54–56]. As an example, the sol–gel process was developed to prepare a pink ceramic stain, with a weight composition of 40.5% Al_2O_3, 3.5% B_2O_3, 37.4% ZnO and 18.6% Cr_2O_3 [54]. In 13 ml alumina sol of aluminum chlorohydrate, 13.7 g zinc nitrate and 4.87 chromic nitrate were added, followed by the addition of a boric acid solution, which was formed by dissolving 0.35 g boric acid in 30 ml water. The final sol contained oxides with a total concentration of 200 $g \cdot ml^{-1}$. The sol was calcined at 700 °C to decompose into oxides and at 1020 °C to form the pink pigment, $ZnAl_2O_4$ with a spinel crystalline structure.

Functional CeO_2 thin films have been deposited by using inkjet printing with aqueous inks [57]. The CeO_2 sol was synthesized by employing an aqueous sol–gel process, with water-soluble materials, including cerium (III) nitrate hexahydrate ($Ce(NO_3)_3 \cdot 6H_2O$) and cerium (III) acetate monohydrate ($Ce(AC)_3 \cdot H_2O$). For instance, cerium acetate was dissolved in a mixture of water and acetic acid, with a Ce^{3+} to acetic acid ratio of 1:10. EDTA in acid form (H_4EDTA) was dissolved in EDA solution with a ratio Ce^{3+}:EDTA to be 1.0:0.8. The final pH value of the solution was controlled with EDA in the range of 5.5–6.0, resulting in transparent sols that could be stored for up to one year at room temperature. The inkjet-printed CeO_2 thin films were printed on substrates with or without post-annealing [58].

Ferroelectric $BaTiO_3$ ceramics have a wide range of applications in various fields [59–62]. $BaTiO_3$ ceramic inks were prepared via a sol–gel process using barium acetate and titanium isopropoxide as the raw materials, which were dissolved in DI water at room temperature under continuous stirring [24]. Acetic acid, acetylacetone and KOH were added to control the hydrolysis rate of the titanium alkoxide, thus adjusting the phase formation rate of the perovskite $BaTiO_3$. $BaTiO_3$ powder derived in this way had an average particle size of 50 nm and a specific surface area of 68 $m^2 \cdot g^{-1}$. To prevent sedimentation and ensure compatibility with inkjet printing, small amounts of polyacrylic acid, ammonium nitrate and polyvinylbutyral (PVB) were incorporated.

TiO_2 is a semiconductor with a relatively large bandgap and can be used to form functional films on different substrates for applications such as self-cleaning, anti-fogging, and hydroplethilic surfaces [63–65]. More recently, TiO_2 films have been deposited via inkjet printing, using sols prepared by sol–gel processes. Because Ti-alkoxide is highly reactive with H_2O, it is necessary to use complexing ligands for stabilization to ensure controllable hydrolysis [66]. The aqueous TiO_2 sol was prepared by directly blending titanium tetraisopropoxide with stabilizers, such as tetramethylammonium hydroxide, triethylamine, diethylamine, lactic acid, citric acid and triethanolamine, at room temperature in ambient conditions, followed by the addition of an appropriate amount of water.

TiO_2 thin films derived from sols prepared from titanium tetraisopropoxide and tetramethylammonium hydroxide exhibited a relatively low crystallization temperature, homogeneous microstructure and high refractive index [67]. The TiO_2 films had higher performances than those made from commercial anatase colloidal suspensions. Aqueous TiO_2 precursor sols were synthesized with tetrabutyl orthotitanate as the Ti source and citric acid or triethanolamine as a stabilizer. The films were inkjet-printed onto glass substrates, demonstrating promising photocatalytic activity.

3.3.2.2 Reverse microemulsion processes

Microemulsions are clear mixtures of liquids with high thermodynamic stability and isotropic properties, usually consisting of oil, water and surfactants [68–71]. The aqueous phase may be composed of various salts, while the oil phase may comprise mixtures of hydrocarbons and olefins. There are mainly three categories of microemulsions, i.e., (i) direct (oil dispersed in water, o/w), (ii) reversed (water dispersed in oil, w/o) and (iii) bi-continuous. Due to their high thermodynamic stabilities, ultra-low interfacial tensions and large interfacial areas, microemulsions are potential candidates as inks for inkjet printing.

With suitable surfactants and solvents, microemulsions can form spontaneously. Thermodynamically stable microemulsions have droplets with sizes in the range of 10–20 nm, making them outstanding in terms of long-time stability. Meanwhile, malfunctions of printer heads can be effectively prevented. The properties of microemulsions, such as particle sizes, viscosity and surface tension, can be well adjusted by controlling diameters of the water cores, which are closely linked to the water-to-surfactant ratio [72].

To make oil-in-water (o/w) microemulsions, water-insoluble organic components must be used as colorants [73]. Therefore, they are not suitable for inorganic pigments typically used for ceramic decorations. To tackle this problem, reverse microemulsions have been proposed for inorganic materials. In this case, water is spontaneously dispersed into the continuous oil phase at the nanosized scale. The particles of water serve as reaction vehicles, in which nanosized ceramic particles are formed.

Nevertheless, reverse microemulsions have a general problem, i.e., low water solubility, which is insufficient for ceramic inks to ensure desired solid loading levels in practical printing. To address this, it is necessary to increase the water dissolving levels, through selection of suitable reverse microemulsions, utilization of effective surfactants and optimization of processing parameters [74, 75]. Gemini surfactants consist of two similar hydroplethobes, which are connected to two head groups, via a spacer moiety, thus having the characteristics of amphiphiles. For example, didodecyldiphenylether disulfonate (C12-DADS) is a typical Gemini surfactant, which can be used to prepare oil-in-water (o/w) and water-in-oil (w/o) microemulsions [73, 76].

A reverse microemulsion was developed with octyl phenol ethylene oxide condensate (TX-100), n-hexanol and cyclohexane as emulsifier, co-emulsifier and oil phase, respectively, aiming to prepare ZrO_2 ceramic inks for inkjet printing [74]. The maximum content of water was achieved in the optimal system consisting of 19.1% Triton X-100, 12.8% n-hexanol, 23.7% cyclohexane and 44.4% water, as revealed in the quasi-ternary phase diagram. The ZrO_2 ceramic ink was made by mixing the two reverse microemulsions, where water was replaced by zirconium oxychloride solution and ammonia solution. The ink exhibited high stability and uniform dispersion.

A ceramic pigment with the composition of $Ti_{0.97}Cr_{0.015}Sb_{0.015}O_2$ was obtained by using a solvothermal process at 180 °C from microemulsions for inkjet printing [77]. During the preparation, 10 ml Triton X-100 (surfactant), 3 ml n-hexanol (co-surfactant), and 16 ml cyclohexane were mixed with the aid of magnetic stirring to form the oil phase. Then, 2 ml $TiCl_4$ solution in 4 M HCl was mixed with 2 ml DI water, together with $CrCl_3 \cdot 6H_2O$, $SbCl_3$ and urea with designed quantities to form the active phase.

3.3.2.3 Polyols processes

Polyol process is a feasible technology to synthesize various metal and oxide nanosized particles for different applications [78–82]. To obtain nanosized particles, solutions were prepared by mixing corresponding precursors in polyols with high boiling points, which are heated at temperatures below the boiling point at ambient pressure. Pigments prepared via traditional processes usually require high processing temperatures, thus having relatively large particle sizes of >500 nm. Meanwhile, the colors of pigments are highly dependent on the level of phase crystallinity.

In practice, the pigments used to prepare ceramic inks should have relatively small particle sizes (<500 nm), while the crystallinity of the materials should be sufficiently high. In this regard, polyol processes are a promising technique for the synthesis of pigments with nanosized particles and high quality of colors.

In the polyol process, the polyols serve as solvent and reduction agents simultaneously. Due to the high boiling points of polyols, the reaction solutions can withstand high temperatures. Meanwhile, polyols would form complexes with the particles to limit their growth into larger sizes [83, 84]. Ceramic inks for inkjet printing have been developed by using the polyol processes, with solid loading levels of up to 20 wt%.

Feldmann reported a polyol process to develop colloidal nanosized particles dispersed in diethylene glycol [85]. The suspensions, with nanosized particles of $CoAl_2O_4$, Cr_2O_3, $ZnCo_2O_4$, $(Ti_{0.85}Ni_{0.05}Nb_{0.10})O_2$, Fe_2O_3, $Cu(Cr,Fe)O_4$ and so on, have been demonstrated in the study. The corresponding pigment powders could be derived from the suspensions through centrifugation. The powders were most likely crystalline at the reaction temperatures of 180 °C–240 °C. For instance, to obtain α-Fe_2O_3, instead of γ-Fe_2O_3, the processing temperature should be higher than the phase transition temperature. Average sizes of the particles were in the range of 50–100 nm, which were adjustable through controlling the experimental parameters.

To synthesize $ZnCo_2O_4$, diethylene glycol (DEG, 50 ml) was filled in a 250 ml round-bottomed flask, with a reflux condenser, under strong stirring. Zinc(II) acetatedihydrate (4.9 mM) and cobalt(II) acetate tetrahydrate (8.4 mM) were added into the flask. The mixture was then heated at 140 °C for the precursor salts to dissolve. Once a solution was formed, water (2 ml) was added, while the temperature was raised to 180 °C. After reacting for 2 h, the suspension containing $ZnCo_2O_4$ became green in color. The solid pigment powder was collected after centrifugation.

Submicron $CoAl_2O_4$ particles, with sizes of 50–200 nm and spherical morphology, were synthesized by using the polyol process [86]. The average sizes were controllable through adjusting the concentration of the precursors. The suspensions in diethylene glycol contained 10 wt% $CoAl_2O_4$ particles, with sufficiently high stability without agglomeration. A typical recipe for the preparation of sub-micrometer $CoAl_2O_4$ particles with the polyol method is as follows. In typical experiment, 8.05 g $AlOH(CH_3COO)_2$ and 5.63 g $Co(CH_3COO)_2 \cdot 4H_2O$ were added to a round-bottomed flask equipped with a reflux condenser. 50 ml diethylene glycol was filled into the flask, followed by stirring for 15 min. After that, 1.0 ml DI water was added with the aid of stirring. The mixture was heated at 140 °C for 1 h and then at 180 °C for 2 h. The $CoAl_2O_4$ powder was retracted through centrifugation. The as-obtained powder was reddish blue, which became deep blue after calcining at 600 °C for 15 min.

3.3.2.4 Hydrothermal and solvothermal reactions

Hydrothermal processes have been employed to synthesize ceramic powders with various compositions, taking place at high temperatures and high pressures, in autoclaves [87–90]. The precursors could be in different forms, including metallic salts, oxides, hydroxides or even metallic powders. The reactant solutions or suspensions are in liquid media. If the liquid media is water, the reaction is called a hydrothermal process, while it is known as a solvothermal process if the liquid media is an organic solvent. The products can be varied from single crystals to amorphous materials. This method has various advantages, such as controllable particle size, adjustable morphology, and diverse crystallinity. Meanwhile, hydrothermal or solvothermal processes are conducted at relatively low temperatures. In some cases, the calcination step can be skipped, thus avoiding unnecessary growth of particles.

Monodispersed nanosized α-Fe_2O_3 (hematite) particles were synthesized by using a hydrothermal process, with 0.018 M $FeCl_3 \cdot 6H_2O$ and 0.01 M HCl solutions at 100 °C–160 °C [91]. Two methods, i.e., microwave-enhanced hydrothermal and conventional hydrothermal processes, were studied and compared. Acicular yellow β-FeOOH (akaganite) rod-shaped particles, measuring 300 nm in length and 40 nm in thickness, were obtained at 100 °C after reaction for 2–3 h, while spherical α-Fe_2O_3 particles with sizes of 100–180 nm were produced after reaction for 13 h, when using the conventional hydrothermal process. In comparison, the microwave-enhanced hydrothermal process resulted in monodispersed red α-Fe_2O_3 nanosized particles with diameters of 30–66 nm after reaction at 100 °C for 2 h.

A microwave-assisted hydrothermal process was used to prepare TiO_2 suspensions for inkjet printing [92]. Ti^{4+} aqueous precursor solutions were derived from titanium isopropoxide with EDTA and triethanolamine or tetraethylammonium hydroxide, which were treated at temperatures of 100 °C–140 °C. The sample treated at a low temperature of 150 °C exhibited promising photocatalytic effect. In addition, a complexing stabilizer is usually employed to optimize the properties of the precursor solutions. The effect of a complexing stabilizer consisting of ethylenediaminetetraacetic acid and triethanolamine on the formation of a TiO_2 precursor solution for inkjet printing was demonstrated [93].

Highly dispersed nanosized $CoAl_2O_4$ particles were synthesized by using hydrothermal process aided with ultrasonic irradiation for inkjet printing [94]. The incorporation of ultrasonic irradiation promoted the formation and crystallization of the $CoAl_2O_4$ phase. The powder had square morphology, with average particle sizes of <100 nm and a relatively narrow size distribution profile. Aqueous Co-Al nitrate solutions were made of $Co(NO_3)_2 \cdot 6H_2O$ and $Al(NO_3)_3 \cdot 9H_2O$, with a molar ratio of 1:2. With magnetic stirring at 150 rpm, a 3 M NaOH solution was added droplet-wise to titrate the Co-Al precursor solutions with a volume of 200 ml, until the pH value was 8.5. After that, 50 ml feedstock was transferred into a Teflon-lined autoclave for a hydrothermal reaction at 250 °C for 24 h. Ultrasonication was conducted at 20 °C for 30 s at a power of 0.3 kW, with an interval of 30 s. The reaction products were collected and washed with DI water, followed by drying at 80 °C.

3.3.2.5 Other processes

High intensity focused ultrasound (HIFU) treatment was used to reduce the particle size of ferroelectric PZT nanoparticles [95]. The as-synthesized PZT powder had particle sizes in the range of 0.5–1.0 μm, which decreased to 10–20 nm after the HIFU treatment, while crystallinity of the PZT phase was well retained, as evidenced by the XRD characterization results. At the same time, the HIFU treated

PZT nanoparticles exhibited the nearly same particle size and morphology, according to TEM observation. The particle size reduction caused by the shockwave was different from that when using conventional processes.

Capping agents were utilized to control the particle sizes of various pigments. For instance, nanosized $CoAl_2O_4$ pigments with an average particle size of 45 nm were obtained by controlling the concentration of cetyltrimethylammonium bromide (CTAB) and polyvinylpyrrolidone (PVP) as double capping agents, when using chemical co-precipitation process with cobalt nitrate (Co $(NO_3)_3 \cdot 9H_2O$) and aluminum nitrate ($Al(NO_3)_3 \cdot 9H_2O$) as the precursors [96]. The synthesized $CoAl_2O_4$ nanoparticles were homogeneously suspended in the mixture of itaconic acid-co-acrylic acid, DI water, diethylene glycol (DEG) and ethanol, forming inks with a solid loading level of 8 wt%. The particle size of the $CoAl_2O_4$ pigment could be well controlled by adjusting the concentrations of CTAB and PVP. PVP interacted with the Co^{2+} and Al^{3+} ions on N or C = O with the lone electron pairs, through the hydroplethilic polar groups.

$BaTiO_3$ inks were prepared from the precursor solution made of barium isopropoxide, titanium tetraisopropoxide and dehydrated ethanol as starting materials, which were used to fabricate $BaTiO_3$ microdots by inkjet printing [97]. To prevent the precipitation of the precursors due to the hydrolysis reaction, amino acid L-proline, with a strong capability to form complex compounds with metallic ions, was introduced for the stabilization of Ba and Ti compounds. Firstly, l-proline was dissolved in dehydrated ethanol (100 ml) and then 0.01 mol $Ti((CH_3)_2CHO)_4$ and 0.01 mol $Ba((CH_3)_2CHO)_2$ were added. With the aid of stirring, a precursor solution with 0.10 M cathodic ions was obtained. The effect of the concentration of L-proline was evaluated, in the range of 0.025–0.10 M for optimization. Eventually, the solution was diluted with a mixed solution of 1:1 dehydrated ethanol and 2-(2-butoxyethoxy) ethanol by about 10 times.

This strategy was extended to the synthesis of other ceramic pigments, including $Cr:YAlO_3$ (YC), $Pr:ZrSiO_4$ (ZP), and $V:ZrSiO_4$ (ZV) [98]. Most significantly, the processing time was shortened from several hours to minutes. Mixtures containing the precursors of Al_2O_3, SiO_2, Y_2O_3, and ZrO_2 in stoichiometric compositions, dopants of 1.1 wt% Cr, 1.7 wt% Pr, and 2.5 wt% V, together with mineralizers of NaF, NaCl, CaF_2, and $BaCl_2$, with total contents of 4 wt% in YC and ZP and 8 wt% in ZV, were prepared and thoroughly mixed. The mixtures were pressed into pellets for microwave-assisted reactions. The pellets were pre-calcined in a conventional electric furnace at temperatures of 300 °C–1000 °C for 5 min, followed by reaction in a microwave furnace at different powers for different time durations, leading to ceramic pigment powders.

3.4 Inkjet printing of ceramic materials

Oh *et al* studied inkjet-printing efficiencies of Al_2O_3 nanoparticle inks on solid substrates [99]. Various patterns could be deposited with the inks by using a suitable drying agent in the ink solvent, leading to a mixed solvent ink system. Since circulating flow was realized for the ink droplets during drying process, uniform deposition of the nanoparticles was achieved. With optimization, the spreading conditions of the Al_2O_3 ink droplets on solid surfaces were identified for inkjet printing of straight lines. Eventually, uniform Al_2O_3 films could be printed, through the adjustment of the line-to-line pitches.

Al_2O_3 powder with particle size $D_{50} = 200$ nm was used to prepare Al_2O_3 inks, with the solid loading level set to be 4 vol%. The Al_2O_3 inks were prepared with either water (boiling point = 100 °C, surface tension = 72.8 dyn · cm^{-1}) or a mixture of 90 vol% water and 10 vol% drying control agent. Among various solvents, N,N-dimethylformamide (DMF), with boiling point = 153 °C and surface tension = 40.4 dyn · cm^{-1}) was selected as the drying agent. Al_2O_3 powder was dispersed in the solvents by ball milling for 48 h, followed by high-speed mixing at 2000 rpm for 8 min. The Al_2O_3 inks were then filtered through 6 μm nylon mesh to filter out agglomerations.

Pt coated Si wafers, with [100] orientation, were employed as the substrates for inkjet printing. Before printing, the substrates were repeatedly cleaned with acetone and ethanol. The UJ 200 inkjet printing unit (Unijet) was used in the printing experiment. A piezoelectric nozzle with a 50 μm orifice was equipped with a printer. The ejected ink droplets were monitored by using a charge-coupled device (CCD) camera. The distance between the nozzle and the surface of the substrate was set to be 1 mm. The volume of the ink droplet was kept at 150 pL for all experiments. The ink was ejected at speeds of 2.5–3.0 m · s^{-1}.

Figure 3.4 shows SEM images of the dots from the Al_2O_3 inks in water and mixed solvent. As seen in figure 3.4(a), the Al_2O_3 particles were accumulated at the edge of the dots, owing to the outward flowing, because the water was preferentially evaporated from the contact line of the ink droplets. This induced radial outward flow from the center of the ink droplets. As a result, accumulation of Al_2O_3 nanoparticles occurred towards the contact line of the droplets. The Al_2O_3 ring had a width of 20 μm and a height of 4 μm.

In order to remove this coffee ring phenomenon, a co-solvent ink system was designed by mixing a drying agent into the main ink solvent. The drying agent should have a higher boiling point and lower surface tension than the main solvent. In the mixed solvent inks, the main solvent with a lower boiling point evaporated

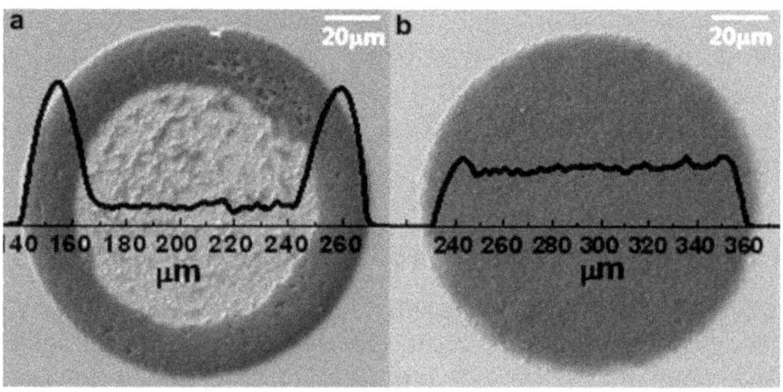

Figure 3.4. SEM images and surface profiles of the Al_2O_3 ink droplets: (a) water single-solvent ink and (b) DMF + water mixed solvent ink. Reprinted from [99], Copyright (2011), with permission from Elsevier.

preferentially at the contact line, thus resulting in outward flow. However, the outward flow was gradually slowed down, since the composition of the solvent on the contact line was varied, leaving more drying agent. As a result, a surface tension gradient across the ink droplet was developed, because high surface tension water was still rich at the center of the ink droplets. In this case, an inward flow toward the center of the ink droplets occurred. The outward and inward flows facilitated the uniform distribution of the nanosized ceramic particles over the ink droplets. DMF was used as drying agent, which has a higher boiling point and lower surface tension than water. Therefore, the Al_2O_3 nanoparticles could be uniformly distributed over the entire contact area of the ink droplets, as demonstrated in figure 3.4(b).

The range of the ink droplet pitches with the printed straight contact lines was in good agreement with the theoretical results. Figure 3.5 shows images of the Al_2O_3 lines printed at different pitches. As Al_2O_3 ink droplets were separated by less than 70 μm, bulges were present along the printed lines, as seen in figure 3.5(a), which is a common phenomenon, especially when the ink droplet pitches have relatively small sizes [100, 101]. As the ink droplet pitches were greater than 100 μm, rounded lines with scallop patterns were formed, as illustrated in figure 3.5(c).

Straight lines can be printed when the pitches are in the range of 70–100 μm, using Al_2O_3 inks with a mixed solvent. In this case, uniform Al_2O_3 films can be developed by adjusting the pitches of the straight lines, with line-to-line pitches narrowed to 25–50 μm. Within this range, the narrower the pitch, the smoother the surface of the films would be. For instance, the surface roughness of the Al_2O_3 film printed with a 25 μm pitch was 0.8 μm, while that of the film with a 50 μm pitch was 1.2 μm. The surface optical and SEM images of the inkjet-printed Al_2O_3 films are depicted in figures 3.6(a) and (b), while the cross-sectional SEM image is presented in figure 3.6 (c), with a thickness of 12 μm.

Freestanding Al_2O_3 ceramic microbeams have been fabricated by using inkjet printing with aqueous colloidal suspensions containing 23 vol% Al_2O_3 [102]. The microbeams exhibited a flexural strength of 920 MPa and a stiffness of 400 GPa, due to the high densification and low microstructural defects. Al_2O_3 powder with a

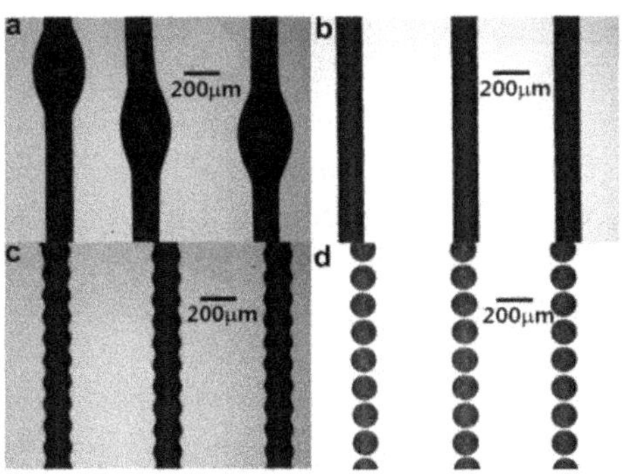

Figure 3.5. Profiles of the ceramic lines inkjet printed with different ink droplet pitches: (a) 50 μm showing line bulges, (b) 90 μm showing straight and uniform line, (c) 120 μm showing non-uniform scallop patterns, and (d) 140 μm showing disconnected dot lines. Reprinted from [99], Copyright (2011), with permission from Elsevier.

Figure 3.6. Morphologies of the inkjet-printed Al_2O_3 ceramic film: (a) optical image, (b) SEM image, and (c) cross-sectional SEM. Reprinted from [99], Copyright (2011), with permission from Elsevier.

particle size of 100–200 nm, polyacrylic acid dispersant and ethylene glycol as rheology modifier were used to prepare the printing inks. The alumina powder was dispersed in a mixture of dispersant (0.8 $mg^{-1} \cdot m^2$ Al_2O_3)) and ethylene glycol (9 wt% after the addition of Al_2O_3), followed by continuous stirring. Three groups of Al_2O_3 ceramic inks, with solid loading levels of 12.6, 17.4 and 23.2 vol%, corresponding to 36, 45 and 54 wt%.

To ensure homogeneity of the ink slurries, the suspensions were ball milled at a relatively low energy with ZrO_2 milling media for 24 h. To eliminate agglomerates, a sedimentation protocol was employed as follows. After that, the inks were kept in glass columns for 24 h to allow larger particles ($\geqslant$5 μm) to settle and low-density organic additives to float. After discarding the precipitants and floating organic materials, the slurries were filtered through 5 μm membranes and then filled into the cartridges of the inkjet printer.

With an image file having multiple copies of a single pixel line pattern, from which layers of ink from a single nozzle could be formed for an arbitrary number of passes, microbeams could be printed. A discrete or non-overlapping printing point for each successful print pass was included to record success or failure of each print pass for each nozzle (figure 3.7). Commercial Al_2O_3 plates were used as the substrates for printing, which were precoated with five layers of graphite aerosol to ensure easy release of the printed items.

The as-printed microbeam samples were cold isostatically pressed (CIP) together with the substrates at 324 MPa and then sintered at 1400 °C for 2 h in vacuum of $<10^{-3}$ Pa. The vacuum sintered microbeams were sandwiched between ZrO_2 plates to avoid warpage and sintered in air for 2 h at 1400 °C. Freestanding microbeams with lengths of >50 mm and sufficiently high mechanical strength were finally obtained.

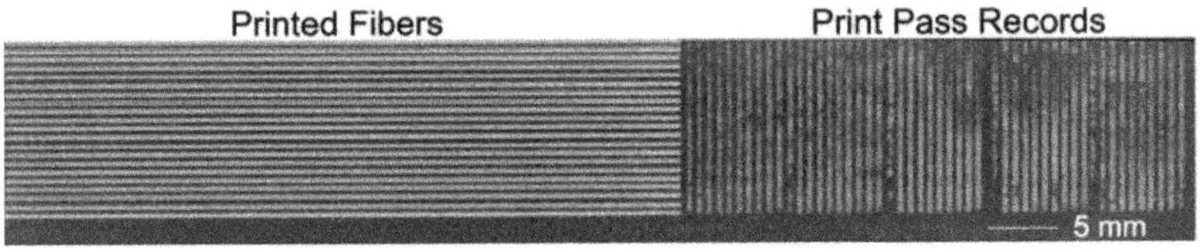

Figure 3.7. Images of the as-printed ceramic microbeams: (left) beams consisting of 50 layers from the 17 vol% Al_2O_3 ink and (right) stripes recording success in printing for all nozzles at each print pass. [102] John Wiley & Sons. © 2012 The American Ceramic Society.

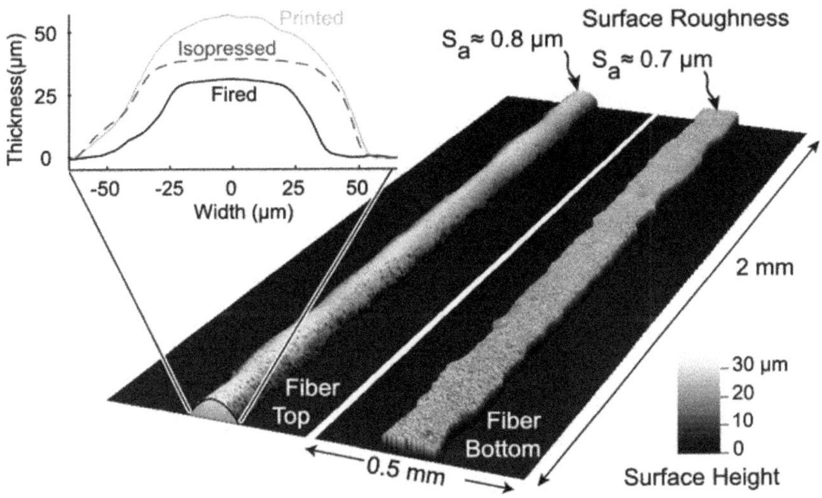

Figure 3.8. Surface map and virtual cross-sections of representative printed beams obtained with interferometry. [102] John Wiley & Sons. © 2012 The American Ceramic Society.

Figure 3.8 shows surface profiles from interferometric measurement, revealing uniform shape and relatively low surface roughness, corresponding to arithmetic an average 3D roughness of <1 μm. Both the top and bottom surfaces were quite smooth, implying that the inkjet-printed items were free of flaws with sizes of >5 μm. Virtual cross-sections extracted from the topographies are illustrated in the inset in figure 3.8, evidencing effective densification of the printed ceramic lines.

A representative polished cross-sectional SEM image is shown in figure 3.9(a), revealing high density and low residual porosity. The volume fraction of pores was $\varphi = 0.03$, consisting of small pores with $D_{90} = 0.25$ μm, which are uniformLy distributed in the ceramics. The grains of the Al_2O_3 ceramics were equiaxed, with an average size of 0.5 μm. ZrO_2 grains with a small grain size of <1 μm were present in the Al_2O_3 matrix, which was attributed to the abrasion from the milling media, appearing as bright dots in the SEM image. Fracture surfaces of the bending tested sample confirmed the uniform grain size and dense microstructure of the printed Al_2O_3 ceramic lines, as evidenced in figure 3.9(b).

A strategy was reported to make Al_2O_3 coating on stainless steel microchannels by using DOD inkjet printing technology [103]. The effects of solid loading level,

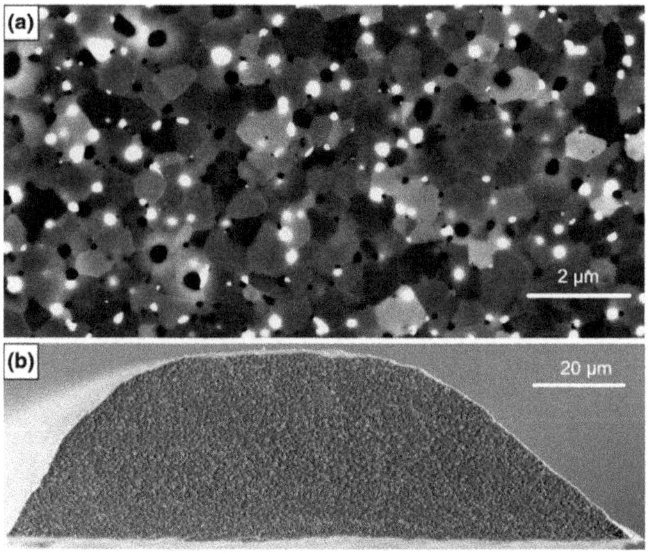

Figure 3.9. Microstructure of the printed Al_2O_3 ceramic beams: (a) high magnification SEM image showing equiaxed grains and about 3% porosity (in black) and ZrO_2 grains (in white) and (b) primary fracture surface low magnification SEM image of the printed microbeam. [102] John Wiley & Sons. © 2012 The American Ceramic Society.

type of co-solvents, hydro-soluble polymers, viscosity and surface tension on printing efficiency of the ceramic inks were systematically studied. Droplet size and velocity of the microdroplets could be adjusted through the piezoelectric activation parameters of the printer. Three shapes of coated films were formed in rectangular microchannels when using inks with different compositions.

Commercially available aqueous Al_2O_3 colloidal suspension, with solid loading of 20 wt% and average ceramic particle size of 100 nm, was used to prepare printing inks. A Microdroplet droplet-on-demand inkjet printer was employed for inkjet printing, with a single head equipped with a nozzle of 100 μm in diameter. The printer was computerized, allowing movement in the *x*-, *y*, and *z*-directions. The microdroplet formation behaviors, directional stability and positional accuracy were monitored by using CCD cameras in the printer system.

Bipolar voltages were applied for the experiment, where magnitudes of the positive voltage and negative voltage were identical, while pulse width of the negative voltage was twice that of the positive voltage. The ink reservoir had a volume of 5 ml, supplying the ink to the head of the printer. To avoid the leakage of ink from the nozzle, pressure was applied, with values between −10 and −14 mbar. Rectangular microchannels on stainless steel foils had channel dimensions of 200 μm × 200 μm × 5 cm. To ensure printing resolution, the velocity of the printer head was set to be 6×10^{-4} m · s^{-1}. After printing, the foils were dried at 70 °C for 20 h and then calcined at 550 °C. Figure 3.10 shows sequential photographs for different inks at given time intervals during the generation of microdroplets. The formation of stable droplets experienced three stages. The initial stage included ejection of fluids

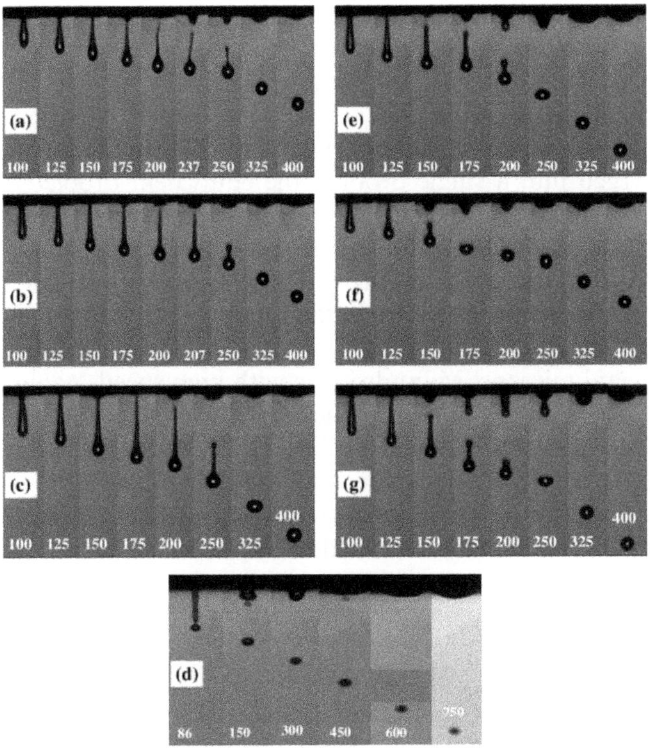

Figure 3.10. Sequent photographs for different ceramic inks at different time intervals (μs) during the formation of inkjet printed droplets: (a) Ink3 at voltage of 100 V with $Z = 3.5$ and pulse width of 40 μs, (b) Ink3 at voltage of 90 V with $Z = 3.5$ and pulse width of 35 μs, (c) Ink3 at voltage of 80 V with $Z = 3.5$ and pulse width of 30 μs, (d) Ink1 at voltage of 42 V with $Z = 30.5$ and pulse width of 20 μs, (e) Ink5 at voltage of 70 V with $Z = 5.9$ and pulse width of 32 μs, (f) Ink5 at voltage of 60 V with $Z = 5.9$ and pulse width of 30 μs and (g) Ink5 at voltage of 80 V with $Z = 5.9$ and pulse width of 36 μs. Reproduced from [103], with permission from Springer Nature.

and neck formation of the fluid ligaments. Secondly, the fluid pinched off from the nozzle. At this stage, the maximum length of the ligament at the time for the fluid to pinch off was determined by the Z value of the inks, as well as magnitude of the applied voltage and width of the pulse. The necking speed of the ligament was dependent on rheology of the ink, magnitude of the applied voltage and width of the pulses. The smaller the Z value, the longer the ligament would be. For example, the times for the ligaments to pinch off were in the ranges of 237–200, 150–155 and 86–100 μs, for the inks with Z values of 3.5, 5.9 and 30.5, respectively. Finally, the ligaments shrunk to form stable droplets. Similarly, the Z value, the applied voltage, and the pulse width had influences on the time and distance for the formation of stable droplets. For instance, stable droplets were formed at 325, 175–250 and 110–125 μs, for the inks with Z values of 3.5, 5.9 and 30.5, respectively.

The length and distance of ligament for the formation of stable droplets had strong influence on the minimum distance of standoff, i.e., the distance between nozzle and the target, which eventually determined the quality of printing [104, 105]. As the standoff distance is too short, interaction of target and nozzle will be triggered through the ligaments of fluid, which will damage the resolution of printing. The minimum standoff distance could be estimated according to the idea of equivalent droplet length in jetting direction [106]. If the ligament shrunk too quickly and merged into a single droplet, the equivalent droplet length would be equal to the distance traveled by the ejected fluid to form a single stable droplet, i.e., the distance would be the shortest length reached by the ligament.

With applied voltages, pulse width, and rheological properties of the inks, the equivalent lengths of the droplet were in the range of 150–650 μm. The speed of the stable microdroplets varied, showing a sharp decrease first and then stabilizing at a value within several microns. Microdroplets with high speeds could be made of the ceramic inks with relatively low Z values, as the voltage and pulse width were proper. The maximum droplet speed of 2.9 $m \cdot s^{-1}$ was achieved by using the ink with $Z = 3.5$.

To examine the effect of the drying process on profiles of the thin films in the microchannels, the ceramic inks were evaluated from the point of view of the coffee ring effect. The microdroplets were deposited on stainless steel plates as single dots. Figure 3.11 shows SEM images of the printed dots after drying at 70 °C. Only two inks exhibited the coffee ring phenomenon, i.e., Ink2 ($Z = 7.7$) and Ink6 ($Z = 11.5$). For the droplets from the inks with single solvent, evaporation occurred at the edge during the drying process, owing to the pinning effect of the contact lines, as stated earlier. The coffee ring effect was effectively suppressed by using mixed solvents to prepare the ceramic inks.

According to the top view and cross-sectional view of the microchannels after coating and calcination, there were three groups of patterns. The type 1 coating was present once the calculation of the offset was error-free. However, for Ink1 ($Z = 30.5$), coating morphologies of type 2 and type 3 were nearly always absent, even though the offset on the surface of the plate was precisely corrected. This is similar to the observations in the case of circular microwells, which were attributed to the hydroplethobic characteristics of the surface and the size of the droplets [107]. Therefore, a small error in positioning could induce the droplets to spread along the side wall of the microwells.

Ink2 ($Z = 7.7$) led to a type 2 pattern, in which the Al_2O_3 nanoparticles were accumulated at the two side edges of the microchannels, while no particles were deposited at the center. This observation was equivalent to the coffee ring effect present on a flat surface, which suggests that the coffee ring effect can be prevented in confined channel structures. The inks without the coffee ring effect resulted in a type 3 pattern. Figure 3.12 depicts cross-sectional SEM images of the microchannels deposited with Ink4 ($Z = 9.78$) and Ink5 ($Z = 5.9$).

Attempts have been made to fabricate defect-free porous Al_2O_3 ceramic multilayers by using inkjet 3D printing techniques [108]. The effect of printing parameters, especially the overlap distance of splats, on porosity and hence mechanical properties, was studied. The Al_2O_3 ceramic multilayers had excellent coherent layer-to-layer and layer-substrate connections. A commercial Al_2O_3 ink and Ceraprinter

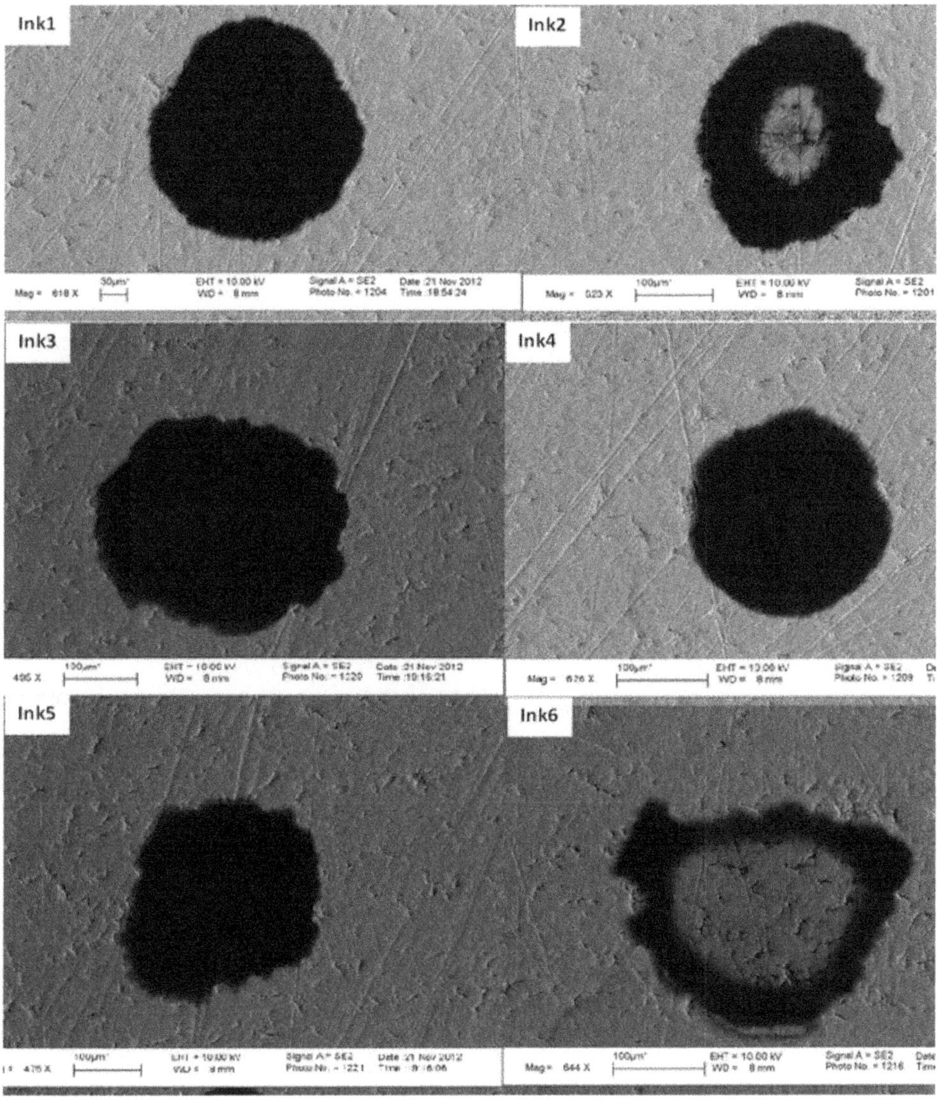

Figure 3.11. SEM images of the dried microdroplets from different inks, with Z values for Ink1, Ink2, Ink3, Ink4, Ink5 and Ink6 to be 30.5, 7.7, 3.5, 9.78, 5.9 and 11.5, respectively. Reproduced from [103], with permission from Springer Nature.

X-series inkjet printer, both from Ceradroplet in France, were used for the printing experiments. The Fuji DIMATIX series Sapphire QS-256/30 AAA multi-nozzle print head (Fujifilm Dimatix, USA) was equipped with a line of 256 piezo-controlled nozzles, arranged at 100 dots-per-inch spacing, with a minimum droplet size of 30 picoliter. The multilayers were printed on commercial alumina ceramic substrates.

The printing pattern was a squared lattice, with four droplets jetted onto the substrate, as schematically shown in figure 3.13. The splat diameter (d) is correlated

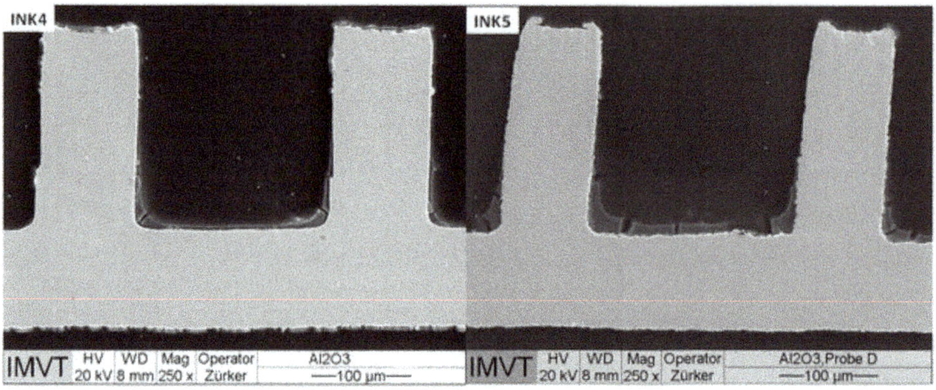

Figure 3.12. Cross-sectional SEM images of the microchannels coated with Ink4 ($Z = 9.78$) and Ink5 ($Z = 5.9$). Reproduced from [103], with permission from Springer Nature.

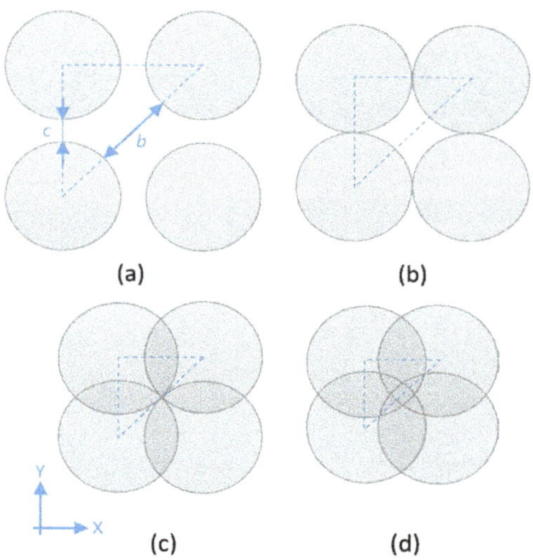

Figure 3.13. Schematic diagram of the printed splat layout as squared lattice pattern with different values of overlapping: (a) $b > 0$, $c > 0$, (b) $b > 0$, $c = 0$, (c) $b = 0$, $c < 0$ and (d) $b < 0$, $c < 0$. Reproduced from [108]. CC BY 4.0.

with the axial overlap distance (c) and the diagonal overlap distance (b). The values of a were in the ranges between 80 and 20 μm, while those of b were between 33.2 and 37.6 μm. To optimize the overlap condition, square samples with dimensions of 1×1 cm^2 were printed with 1–12 layers, alternating in the x and y-directions. Initially, extremely large values of overlap distance parameters, $c = 33.2$ and $b = 80$ μm, were tested. Therefore, no overlap occurred, so that single splat diameter could be readily identified, as depicted in figure 3.14. The splat diameter was on average 80 μm.

Each layer after printing was dried at 100 °C for 2 min to ensure that it was sufficiently strong for the next layer. After the desired number of layers were printed, the samples were finally dried at 100 °C for 24 h. The average thickness of each layer

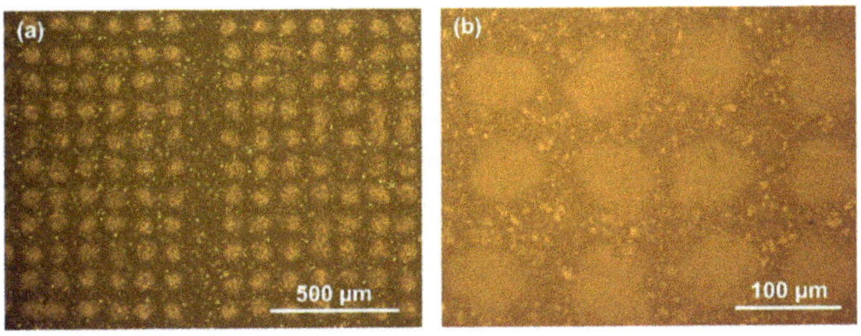

Figure 3.14. Optical microscopic images showing the splat diameter and overlap distance of the printed patterns at different magnifications (b) = 80 μm, (c) = 33.2 μm. Reproduced from [108]. CC BY 4.0.

after it completely dried was about 10 μm. The dried samples were debinded at 400 °C for 2 h and sintered at temperatures of 1200 °C–1500 °C for 4 h, at heating and cooling rates of 5 °C · min^{-1}.

Surface and cross-sectional SEM images of the samples with 10 printing layers after sintering at different temperatures are illustrated in figures 3.15 and 3.16. Obviously, all samples were free of cracks. The samples after sintering at 1200 °C and 1300 °C displayed uniform microstructure. However, high sintering temperatures triggered abnormal grain growth (AGG), with coarse grains with sizes of up to 100 μm, similar to those reported in the literature [109–111].

Nanoindentation tests were conducted at loads in the range of 10–500 mN, corresponding to maximum indentation depths of 0.8–6 μm. Figure 3.17 shows calculated elastic modulus and indentation hardness of the samples sintered at different temperatures, as a function of indentation depth. The variation trends in the mechanical properties were closely associated with the microstructural profiles of the samples. The samples sintered at 1200 °C and 1300 °C readily reached plateau moduli and hardness, because they had homogeneous microstructures. The relatively large deviation in moduli and hardness at low indentation depths was mainly ascribed to the slightly large surface roughness. In comparison, the sintered at 1400 °C and 1500 °C exhibited much larger deviations, irrespective to the indentation depth, since they had high surface roughness caused by the AGG behavior, where the grain sizes were comparable with the size of the indenter. Moreover, the large grains were responsible for the inhomogeneous microstructures of the samples, thus leading to the absence of a well-defined plateau in moduli and hardness.

Al_2O_3 ceramic structures were fabricated by using inkjet printing techniques with different printed paths distributions, including spiral printed path, round trip straight printed path and ladder lap printed path [112]. The effects of the inkjet-printed path on the densification behavior of the green bodies and the thermal shock resistance of the Al_2O_3 ceramics were studied. The green bodies with the ladder lap printed path exhibited the strongest densification behavior. After sintering at 1550 °C, the sample had a density of 3.73 g · cm^{-3} and porosity of 10.8%. The sample with the step-printed path had the highest thermal shock resistance, reaching 11 times.

The Al_2O_3 powder used for the preparation of the ceramic inks had an average particle size of 130 nm. The mixed solution for dispersing the Al_2O_3 powder was

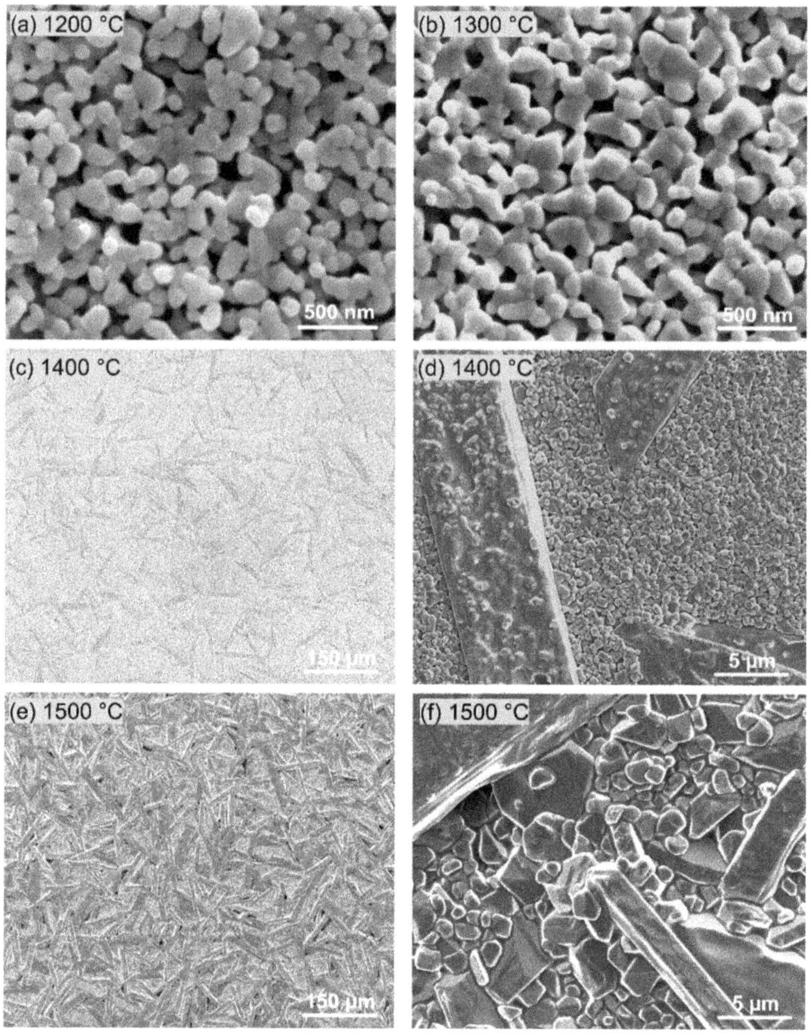

Figure 3.15. Surface SEM images of the samples sintered at different temperatures: (a) 1200 °C, (b) 1300 °C, (c and d) 1400 °C and (e and f) 1500 °C. Reproduced from [108]. CC BY 4.0.

composed of ethylene glycol (9 wt%), acrylic acid (2 wt%) and deionized water (3 wt%). Polyethylene glycol (10 wt%) was used as a dispersant. The ceramic suspensions were ball milled for 3 h, at rotation speed of 500 rpm, leading to ceramic inks with a viscosity of 11 000 MPa · s.

3DS max software was employed to generate 3D models for section slicing. The final data were input to the inkjet printer (Micro make B1, China). A triaxial (x, y, z) micro-positioning device was utilized to deposit the ink to form layered and patterned structures. The samples were built in a layer-by-layer manner along the preset print paths. After every 2D patterned layer was printed, the nozzle was raised in the z-direction, allowing the new layer to be printed until desired thickness was

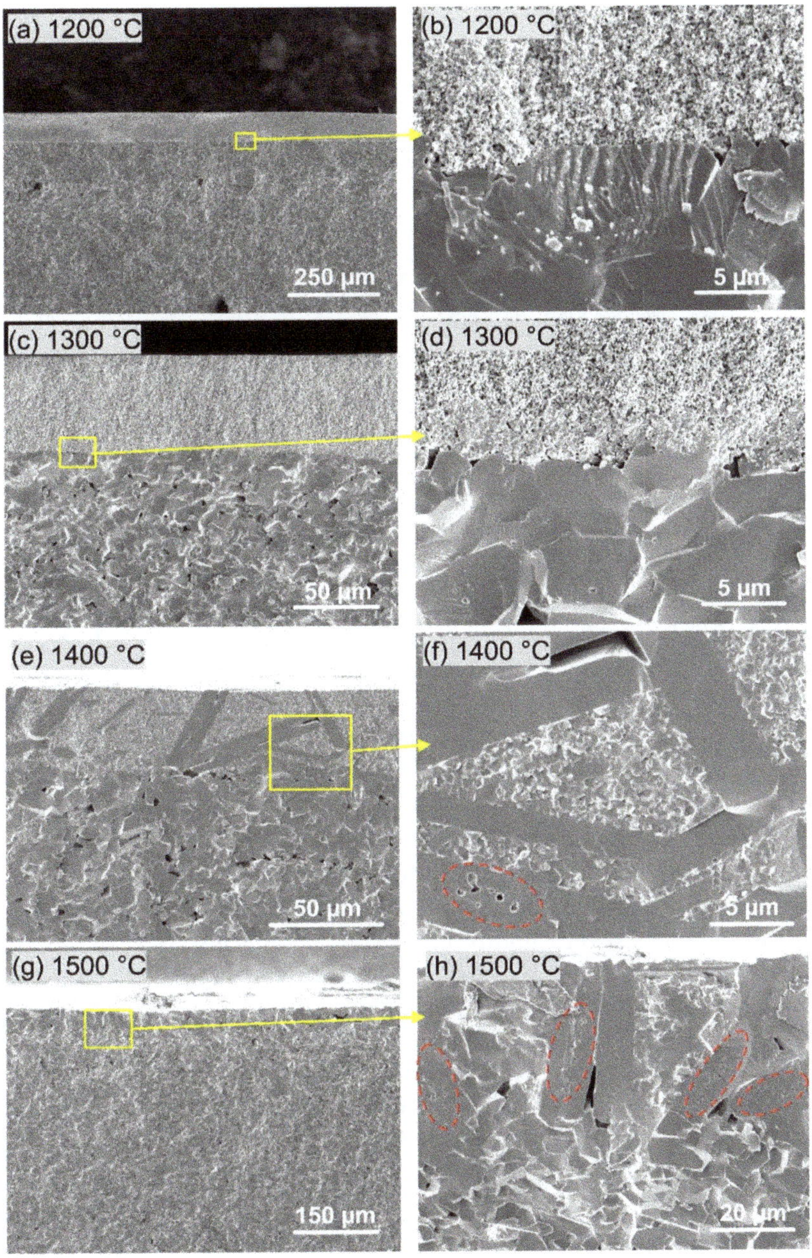

Figure 3.16. Cross-sectional SEM images of the samples sintered at different temperatures: (a and b) 1200 °C, (c and d) 1300 °C, (e and f) 1400 °C and (g and h) 1500 °C. Reproduced from [108]. CC BY 4.0.

achieved. The green bodies were dried and then sintered at 1350 °C, 1450 °C and 1550 °C, for 3 h in air. Figure 3.18 shows photographs of the Al_2O_3 green bodies made with different printed paths.

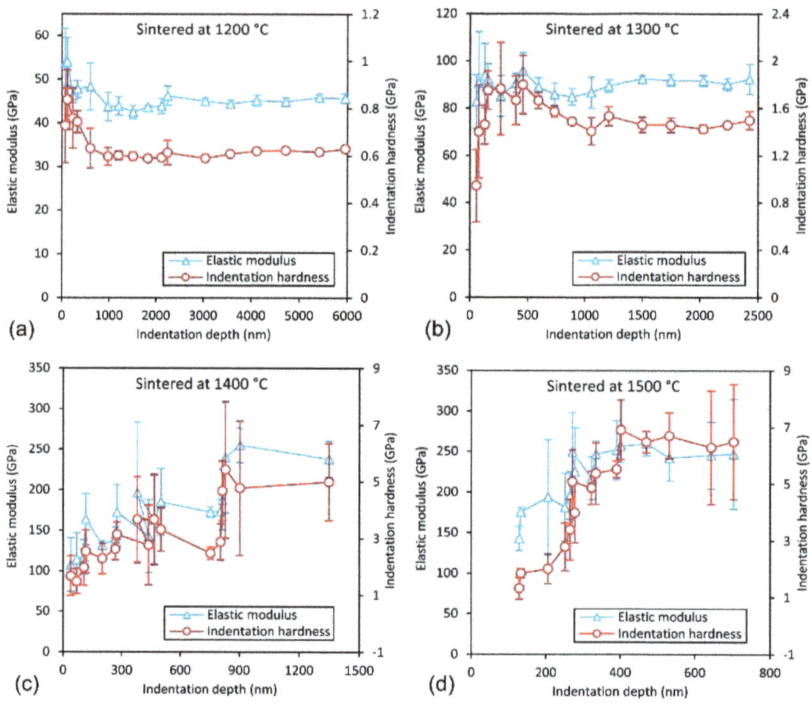

Figure 3.17. Elastic modulus and indentation hardness of the samples sintered at different temperatures: (a) 1200 °C, (b) 1300 °C, (c) 1400 °C and (d) 1500 °C. Reproduced from [108]. CC BY 4.0.

Figure 3.19 shows SEM images of sintered samples with different printed paths. The samples made with the spiral printed path (sample 1) and round trip straight printed path (sample 2) were highly porous after sintering at 1550 °C, as seen in figures 3.19(a) and (b). In contrast, the sample printed using ladder lap path (sample 3) had a much denser microstructure with relatively low porosity, as demonstrated in figure 3.19(c). In all samples, the pores were mainly located at the grain boundaries, as observed in figures 3.19(d)–(f). In addition, the porous microstructures of samples 1 and 2 were further evidenced in figures 3.19(d) and (e), while irregular grains with micron-sized Al_2O_3 particles were present. Their densities were 3.24 and 3.51 $g \cdot cm^{-3}$. Among the three samples, the one with the stepped printed path displayed the highest densification rate and the lowest porosity.

The thermal shock resistance of the sintered samples increased with increasing the complexity of the printing path. After sintering at 1550 C, sample 3 had the highest thermal shock resistance. When brittle ceramics are subject to thermal shock, the generation and propagation of micro-cracks are closely related to the elastic strain energy stored in the materials and the cracked surface energy of crack propagation [113, 114]. As the elastic strain energy stored in a material is not very high, propagation of the original cracks will not be significant. When the crack surfaces have sufficiently high fracture surface energy, crack propagation is suppressed,

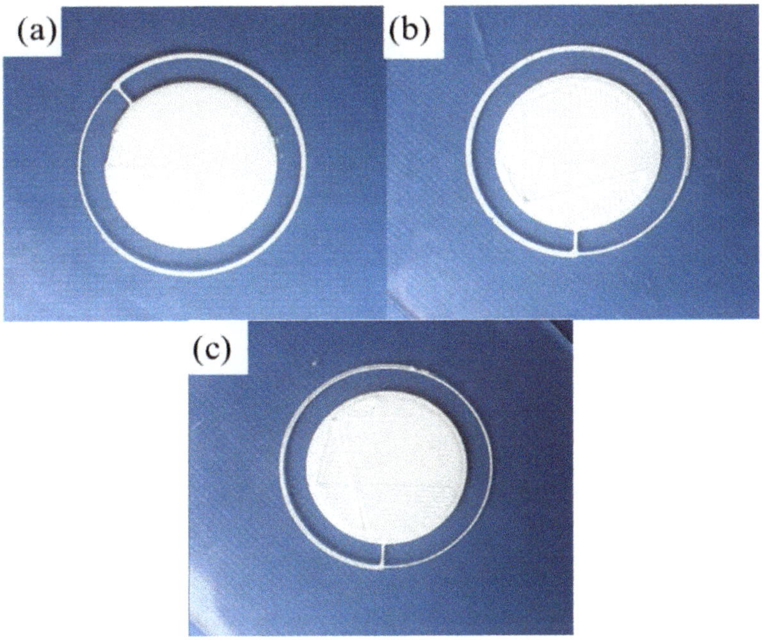

Figure 3.18. Photographs of the Al_2O_3 green bodies made with different printed paths: (a) spiral, (b) round trip straight and (c) ladder lap. Reprinted from [112], Copyright (2018), with permission from Elsevier.

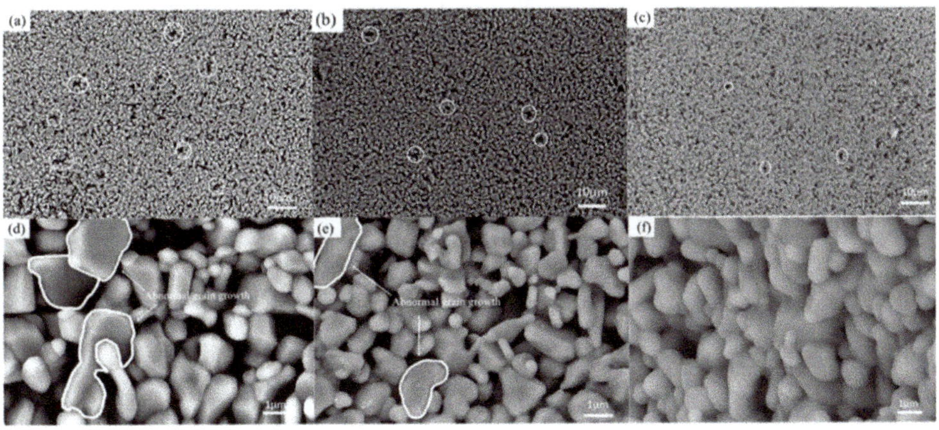

Figure 3.19. SEM images of the samples sintered at 1550 °C with different printed paths: (a and d) spiral, (d and e) round trip straight and (c and f) ladder lap. Reprinted from [112], Copyright (2018), with permission from Elsevier.

hence the materials have high thermal shock resistance. As a result, the thermal shock resistance is proportional to the fracture surface energy and inversely proportional to the strain energy release rate.

Sample 1 and sample 2 experienced AGG, which contributed to crack propagation under thermal stress. Meanwhile, due to their relatively low density, these

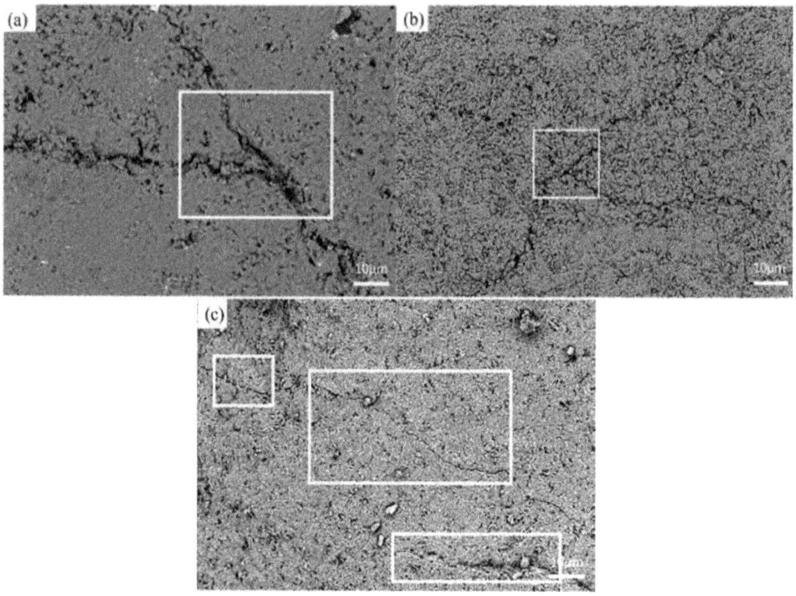

Figure 3.20. Surface SEM images of the polished of samples 1, 2 and 3 sintered at 1550 °C after thermal shock experiment. Reprinted from [112], Copyright (2018), with permission from Elsevier.

samples possessed relatively low mechanical strength. Figure 3.20 shows surface SEM images of the three samples. The cracks in sample 1 and sample 2 were wider and longer than those in sample 3. In sample 1 and sample 2, the thermal stress could not be offset through deformation, as they were subjected to thermal stresses. In this case, the cracks strongly tended to spread. As a consequence, they had lower thermal shock resistance.

Esposito *et al* used a HP Deskjet 1000 inkjet printer to prepare a highly dense gas-tight ZrO_2 electrolyte for solid oxide fuel cells (SOFC), with a thickness of 1.2 μm and a dimension of 16 cm^{-2} [115]. The printing ink was made of yttria stabilized zirconia (YSZ) powder, with an average particle size of 50 nm, dispersed in a water-based medium. The electrolyte processed with optimized sintering procedures exhibited promising adhesion and densification behavior, showing a peak power density of 1.5 $W \cdot cm^{-2}$ at an operating temperature of 800 °C.

Commercial 8YSZ (8 mol% yttria, Tosoh) nanosized powder was dispersed in a mixture of 80 wt% water and 20 wt% ethanol. Two suspensions, with YSZ concentrations of 3.7 vol% (concentrated ink) and 0.9 vol% (dilute ink), were prepared. PVP was used as a dispersant, which was dissolved in ethanol at a weight percentage of 53%. The PVP ethanol solution was added into the YSZ suspensions at about 8 $mg \cdot m^{-2}$ with respect of the solid surface area. All constituents were blended by using ball milling with a rotational mill in a PET flask at a speed of 100 rpm for 10 days, using zirconia milling balls.

A SOFC device was constructed with a thin NiO/YSZ functional anode layer that was deposited on a thicker support of NiO/YSZ anode. The volume ratio of Ni to

YSZ was 40:60, which acted as the support layer and active electrode layer [116]. 8YSZ was used as the anode layer, while 3YSZ was employed at the support layer. These layers were fabricated by using tape-casting and co-laminated, as green materials at about 150 °C. The substrates consisted of a Ni/YSZ cermet anode with thicknesses of 10–15 μm and a Ni/YSZ support layer with a thickness of 300 μm.

The HP printer was modified to print layers on thick and stiff substrates at different print head/substrate distances. An HP 301 black cartridge with a resolution of 600 × 300 dpi was used to fill in the ceramic inks with syringes. Two printing procedures were conducted, i.e., 'Single Droplets' (SD) printing and 'Continuous Printing' (CP). SD was used to print isolated droplets in the form of a square-chess-like pattern, aimed at controlling the quality of the single droplet or avoiding the presence of flaws in the DoD printing. CP was used to deposit sequential droplets to form lines or continuous layers. Each layer was printed in about 15 s after the previous layer was finished, in order to ensure complete evaporation of the solvents and additives in the inks.

The half-cell, consisting of the support, anode and electrolyte layers, was debinded and sintered in air. The debinding procedure was conducted at temperatures of <700 °C for 48 h to eliminate all the organic components. The anode and the printed electrolyte were sintered at temperatures of 1000 °C–1300 °C, for durations of 0.1–6 h. Cathode ink was coated on the electrolyte side of the sintered half-cell, forming a structure of anode support/anode/electrolyte, using screen printing. The cathode ink was a mixture of $La_{0.75}Sr_{0.25}MnO_{3-\delta}$ (LSM) and YSZ, with LSM/YSZ = 50/50 vol%. The printing speed was 60 mm · s^{-1}, with a printing gap of 1 mm and squeegee pressure of 7 bar. After printing, the cathode layer was sintered at 1050 °C for 2 h. An LSM cathode contact layer was then screen printed on the cathode layer similarly, followed by sintering at 1000 °C for 5 h.

The two inks exhibited Newtonian behavior, with the viscosity observed to be constant regardless of the shear rate. The viscosity values were 4.28 and 2.63 mPa · s for the 3.7 vol% ink and the 0.9 vol% ink, respectively. Their surface tension was 36 mN · m^{-1}. Figure 3.21 shows the *We–Re* numbers diagram, which is commonly utilized to describe the printability of inkjet printing inks [117]. According to the *We* and *Re* values, the two YSZ inks were well within the printable regime.

Flaws or defects could be left in printed ceramic structures after sintering through continuous layer printing, especially in the case of droplet-by-droplet deposition, in which discontinuity cannot be fully prevented, due to errors in alignment or missing of droplets. Furthermore, the packing density of ceramic particles cannot be guaranteed when using diluted inks, thus leaving pores or pinholes after sintering. Figure 3.22 shows representative SEM images of the printed YSZ ceramic layers from diluted inks after sintering at different conditions. As observed in figure 3.22(a), misalignment in droplets was present in the layer after sintering at 1300 °C for 6 h, as the ink with 0.9 vol% of YSZ was used. This observation was mainly ascribed to the low degree of precision of the mechanical movement system in the printer.

In the continuously printed (CP) single layer with the concentrated (low dilution) ink, missing droplets were present in some areas with different sizes and shapes, as revealed in figure 3.22(b). There were large regions with sizes of 100 μm. This size is

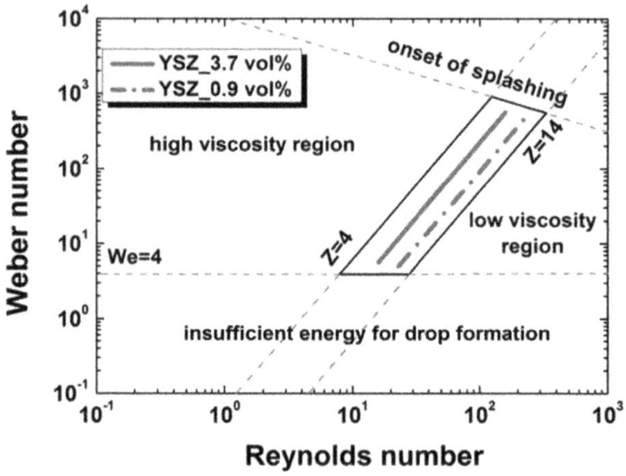

Figure 3.21. Weber–Reynolds numbers diagram defining the regime with promising properties for printability of inks to ensure printing results [118]. Reprinted from [115], Copyright (2015), with permission from Elsevier.

very close to that of the droplets printed with SD. The lack of printing was most likely linked to defective droplets formed from the nozzle or the imprecise impact of the droplets on the surface of the substrate. There were also small-sized defects with irregular shapes in the CP samples, as illustrated in figure 3.22(b), which could be related to the imperfect surface of the substrates, thus having a negative effect on the settling of the ceramic particles after printing.

An SEM image of the circular flaw in the sample sintered at 1300 °C for 6 h is shown in figure 3.22(c). The porous edge was directly caused by the extremely diluted ink. Nevertheless, the areas within the printed layer were relatively homogeneous and dense due to particle packing. As a result, a much denser microstructure was achieved, after the sample was sintered at 1300 °C. In addition, the coffee ring effect was not present in the samples, confirming the suitability of the ceramic inks for the inkjet printing process. Meanwhile, to ensure the fabrication of a gas-tight electrolyte, it is necessary to use a multiple-printing approach.

A representative SEM image of the 2-layered sample with the 3.7 vol% ink after sintering at 1300 °C for 6 h is depicted in figure 3.22(d). It was observed that defects with relatively small sizes in the first layer were covered by the second layer. The defect area was densified to a certain degree, but with visible pores. This observation further indicates that the defects cannot be fully covered if extremely diluted inks are used. To address this issue, the number of print-passes should be increased.

Figure 3.23 shows SEM images of the electrolyte layer printed with the 3.7 vol% ink, after sintering under different conditions. A short duration allowed for a more effective reflection of the effect of sintering temperature on microstructural evolution of the printed layers. As seen in figure 3.23(a), after sintering at 1000 °C for 6 min, the YSZ particles in the printed layer were characterized by spherical morphology, with an average grain size of 100 nm, without the presence of aggregation. At the same time, voids with larger sizes were present occasionally, which were mainly formed during the drying of the ink after printing.

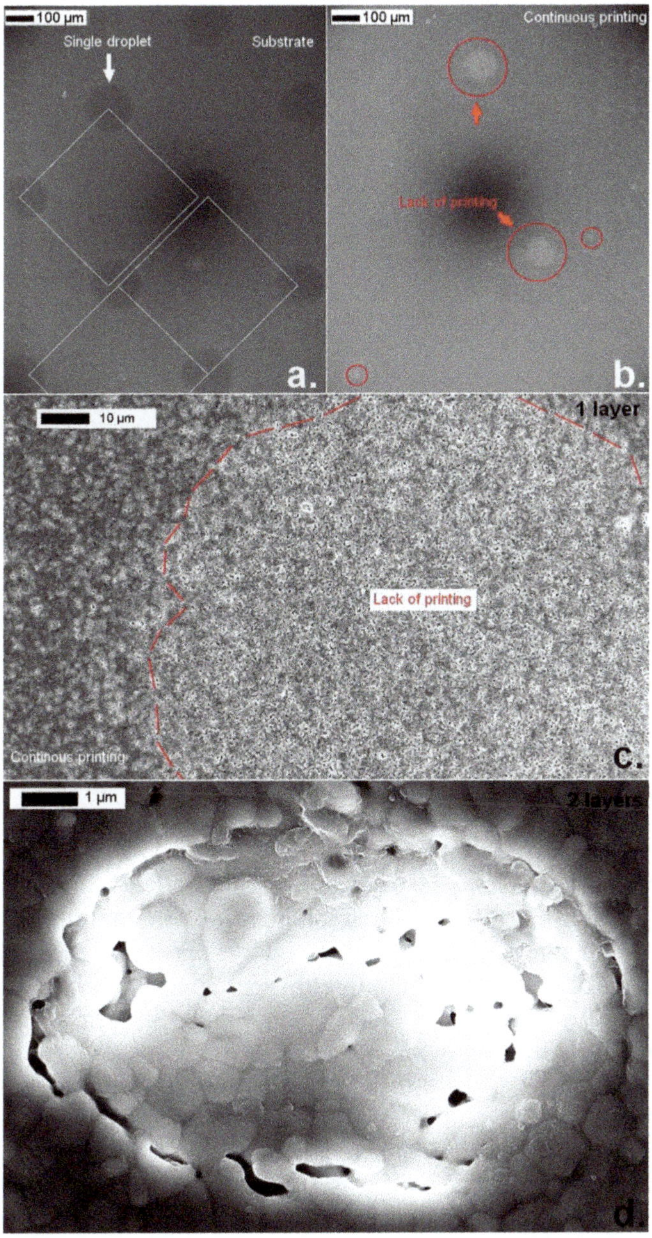

Figure 3.22. SEM images of the printed YSZ layers, after sintering at 1300 °C for 5 min: (a) 0.9 vol% ink printed as single droplets, (b) continuous print using the 3.7 vol% ink, (c) high magnification of printing flaw shown in (b) and (d) typical defects on a two-layer structure. Reprinted from [115], Copyright (2015), with permission from Elsevier.

After sintering at 1150 °C, necking between the YSZ particles was formed, suggesting that the samples experienced initial sintering, with relatively low porosity, as observed in figure 3.23(b). The large voids were similar to those presented in

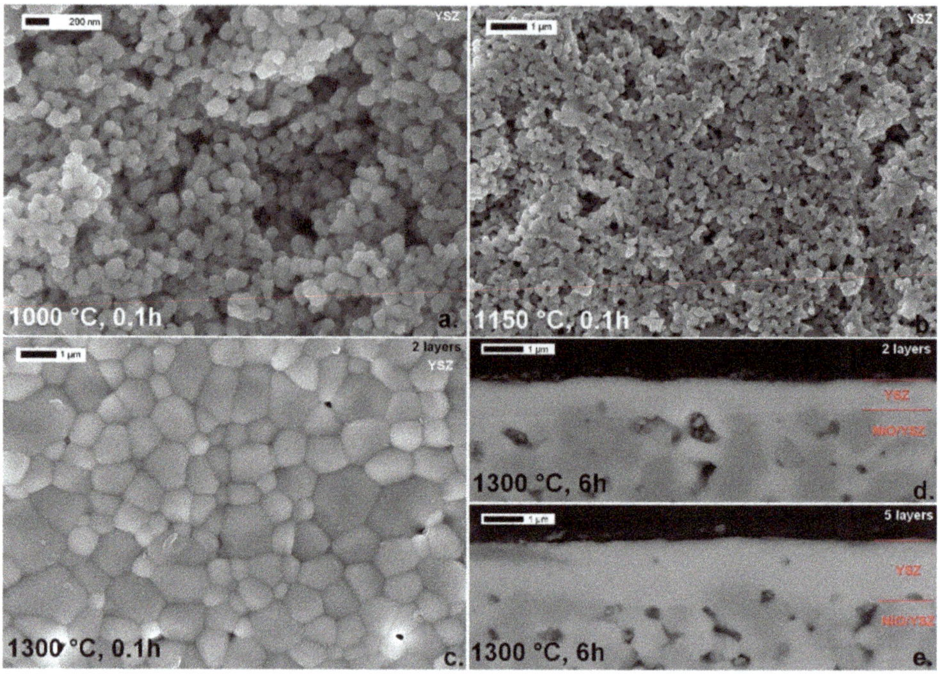

Figure 3.23. SEM images of the printed YSZ layers with the 3.7 vol% ceramic ink: (a) continuous printing after sintering at 1000 °C for 6 min, (b) 1150 °C for 6 min, (c) 1300 °C for 6 min, (d) cross-section view of the half cells made of 2-layer after sintering at 1300 °C for 6 h and (e) 5-layer after sintering at 1300 °C for 6 h. Reprinted from [115], Copyright (2015), with permission from Elsevier.

figure 3.23(b). The sample was nearly fully densified after sintering at 1300 °C, as demonstrated in figure 3.23(c). The grains grew from a nanometer scale to a micrometer scale. In summary, the 3.7 vol% ink was promising for inkjet printing to prepare YSZ electrolyte layers, while the optimal sintering temperature was about 1300 °C. Also, the printing defects could be eliminated by using a multiple-printing process, especially for samples with large areas.

Nanosized YSZ particles were dispersed in water by using a continuous hydrothermal process for the fabrication of thin films via inkjet printing [119]. The YSZ nanosized particles synthesized in supercritical conditions had an average size of 10 nm. The rheology of the suspensions was adjusted to achieve inkjet printability (Z) through the modification of viscosity and surface tension with organic additives. With optimized inks, high-quality YSZ layers with lateral and thickness resolutions of 70 μm and 250 nm, respectively, could be achieved. The YSZ thin films could be sintered at temperatures of $<$1200 °C.

To synthesize YSZ nanosized powders, 0.184 M $Zr(NO_3)_4$ and 0.032 M $Y(NO_3)_3$ mixed solution was used to form precursor streams, at a flow rate of 10 ml · min^{-1}, which was combined with a stream of supercritical water, at a flow rate equivalent to 25 ml · min^{-1} at room temperature, at 397 °C and 270 bar. After reaction, the products were collected, followed by washing and concentrating through centrifugation. For large-scale production of YSZ powders, $Y(NO_3)_3$ and $Zr(Ac)_4$ solutions were utilized as the precursors, which were mixed and introduced into reactors at a flow rate of 20 ml · min^{-1}. The reactions were conducted in supercritical water flow at 375 °C and 241 bar.

PVP K15 with an average molecular weight of 10 000 and PVP with an average molecular weight of 360 000 were used to adjust viscosity of the inks. 2,4,7,9-tetramethyl-5-decyne-4,7-diol ethoxylate (TMDE) was added as a surfactant to control surface tension of the inks, while pH value was regulated with 36 wt% HCl solution. The dispersions were sonicated for 3 min, followed by the addition of PVP, HCl and TMDE.

A Pixdro LP50 inkjet printer equipped with DMC disposable piezoelectric printheads from Dimatix Inks was used for the printing experiments. The printheads had 16 nozzles with a diameter of 21.5 μm, resulting in a droplet volume of 10 pl. The waveform to actuate the piezoelectric elements for ejection of the droplets was optimized for every ink, with a standard trapezoidal pulse. The waveform was characterized by a dwell time of 10 μs and a fall time of 5 μs. For optimization of printing, the filling time of the inks was varied in a range of 4–7 μs, while the maximum voltage was set to be 40–60 V and a jetting frequency of 1 kHz was used for all the tests. The inks were filtered with a syringe filter that had a 700 nm mesh before filling in cartridges of the printer.

The YSZ layers were printed on YSZ/NiO composites made with tape casting after sintering. The droplets were arranged in a square pattern, with different distances, so as to balance the substrate coverage and minimize overlap. In this case, the linear droplet density was optimized to be 500 dpi in both the *x*- and *y*-directions. For multilayer printing, each layer was dried for 2 min. After printing, the YSZ layers were debinded at 600 °C for 4 h with a heating rate of 0.25 °C · min^{-1} and then sintered in air at temperatures of 800 °C, 1000 °C and 1200 °C for 6 h, with a heating rate of 1 °C · min^{-1}. Finally, the samples cooled down to room temperature at a cooling rate of 1.67 °C · min^{-1}.

Because of the relatively low solid loading levels, viscosity (η) and surface tension (σ) of the inks were very close to those of pure water. The as-prepared nanosized YSZ inks had surface tension of 71 mN · m^{-1} and viscosity of 1.2 mPa · s at the shear rate of 1000 s^{-1}, corresponding to a relatively high printability $Z = 33$. The addition of 2 wt% TMDE resulted in the desired value of surface tension, without causing an increase in the viscosity.

Figure 3.24 shows ceramic inks with different Z values in terms of printability. Jetting behaviors of ceramic inks, with droplet volume, speed and shape when printing at optimized waveforms, are described in figure 3.24(A). To examine the effect of nanoparticles on jetting behaviors, experimental results of particle-free inks are demonstrated in figure 3.24(B). Water, PVP and TMDE were mixed to prepare inks to evaluate the printability range, which was similar to the ceramic inks, with components of their liquid phases.

Using ceramic inks with Z of 4.4–20, single round shaped droplets could be printed. As for the inks with $Z = 25$, satellite droplets would be formed, while stable jetting could not be achieved, as the Z values are in the range of 1.6–3.0.

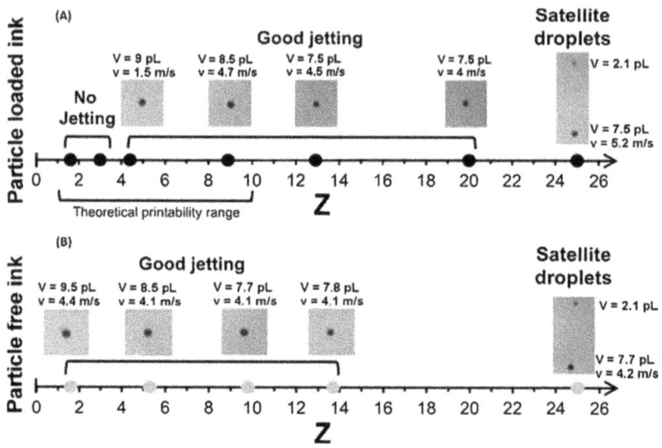

Figure 3.24. (A) Jetting behaviors of ceramic inks with different values of Z. (B) Jetting behaviors of the inks without ceramic particles with similar Z values. In both panels, picture of droplet, volume of droplet (V) and speed (v) for each ink with given Z are presented. Reprinted from [119], Copyright (2019), with permission from Elsevier.

For the inks without ceramic particles with low Z values, which had high viscosity, they allowed for the production of high-quality droplets. The Z values in the range of 1.6–14 offered optimal jetting performance, while the one with $Z = 25$ led to satellite droplets. For these two inks, the volume of droplets increased with increasing viscosity. Therefore, to ensure suitable printability, the Z values of ceramic inks should be in the range of 4.4–20. Inks with too-high viscosities and Z values of <4 cannot be used for printing, while multiple droplets would be formed for the inks with Z values of >20.

The printing experiments were conducted using the ink with $Z = 20$, to optimize the processing parameters to develop high quality patterns and continuous layers. As the droplets hit the surface of substrates, a splat was generated, whose diameter depended on various parameters, such as volume of the droplet and hitting speed, viscosity of the ink, surface tension of the ink and surface energy of the substrate. The diameter of the splat reflected the smallest printable size, corresponding to the highest possible lateral resolution for the given conditions. The splat diameter of dots printed on the YSZ/NiO composite was 70 μm, as seen in figure 3.25(A).

According to the volume of the droplet and diameter of the splat, the deposition of less than 0.5 ng of the ceramic nanoparticles within an area of 0.04 mm^2 could be achieved. Square arrangement of the printed dots was optimal, while linear density was 500 dots per inch in the x- and y-directions. More complicated patterns could also be printed with the set conditions, including honeycomb (figure 3.25(B)) and uniform continuous layers (figure 3.25(C)). In practice, these optimized printing parameters should be modified for different substrates, owing to the relationship between the lateral resolution and the surface energy.

Figure 3.26 shows SEM images of the YSZ samples with ten printing layers after sintering at different temperatures in air. After sintering at 800 °C, particle necking

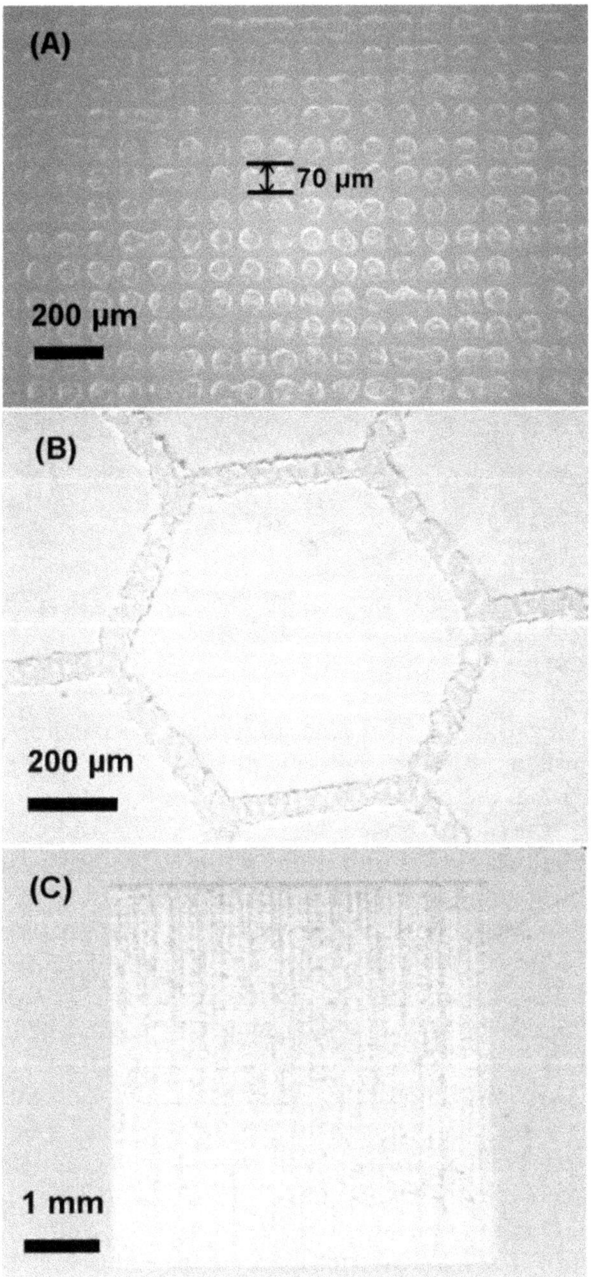

Figure 3.25. Optical images of YSZ samples with different patterns printed with the ink with $Z = 20$ by overprinting 10 times with a linear density of 500 dots per inch: (A) separated splats, (B) honeycomb pattern and (C) continuous squared layer. Reprinted from [119], Copyright (2019), with permission from Elsevier.

and grain growth occurred, as observed in figure 3.26(A). This sample had a porosity of about 9 vol%, while the grain size increased from the initial 10 to 25–30 nm. For the sample sintered at 1000 °C, the porosity was reduced to 2 vol % and the grains

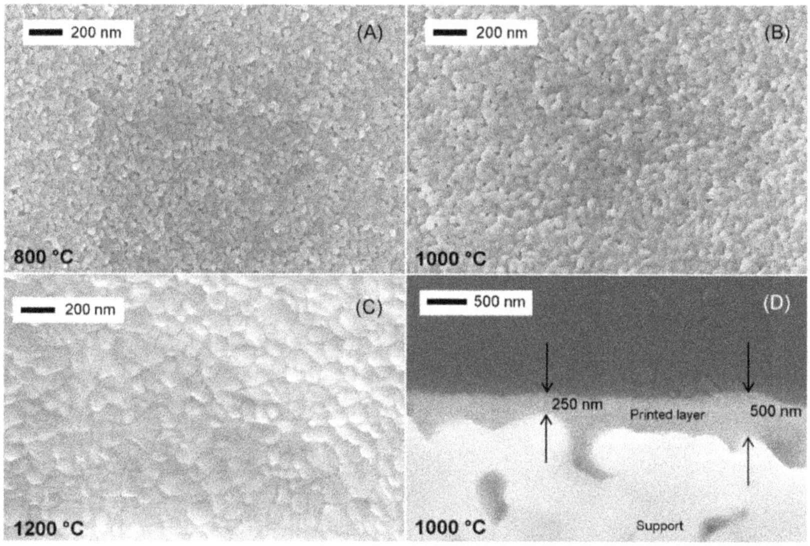

Figure 3.26. SEM images of the YSZ films after sintering at different temperatures: (A) 800 °C, (B) 1000 °C, (C) 1200 °C and (D) cross-section image of the film after sintering at 1000 °C in constrained state. Reprinted from [119], Copyright (2019), with permission from Elsevier.

began to appear, indicating the occurrence of further densification and grain growth, as demonstrated in figure 3.26(B). Sintering at 1200 °C facilitated full densification of the YSZ layer, while the grain size reached 200 nm.

ZrO_2 (3Y-TZP) ceramic teeth were fabricated by using inkjet printing processes [120]. The ceramic ink was made of dental ZrO_2 powder dispersed in a water-based solvent with a solid loading level of 55 vol%. Characteristics of the ZrO_2 ceramic inks and printed filaments were studied to optimize the parameters of printing. After sintering at 1500 °C for 4 h, the ZrO_2 teeth samples had a relative density of 98.5%, hardness of 14.4 GPa, and transverse rupture strength of 520 MPa.

Monomer acrylamide and crosslinking agent N,N′-methylenebisacrylamide, with a weight ratio of 17:1, were dissolved in DI water. Then, 3.0 wt% ammonium citrate ($C_6H_5O_7(NH_4)_3$) was added into the solution as a dispersant. 3Y-TZP dental powder with an average particle size of 0.5 μm was dispersed in the premixed solution under strong stirring. During the stirring process, gradual polymerization of the organic monomers took place, forming a 3D network to coat the particles of the 3Y-TZP powder. 3D inkjet printing of the ZrO_2 ceramic teeth was conducted with computerized control. The ink was extruded from the syringe at a flow rate matching the designed printing speed, by adjusting the applied pressure. The printed green bodies were dried at 80 °C for 12 h, followed by debinding at 400 °C for 1 h and then sintering at 1500 °C for 4 h.

The inks showed a decrease in viscosity with increasing shear rate, demonstrating shear-thinning behavior. The ink was extruded out from the syringe, due to the shear force of air pressure. In this regard, the shear-thinning rheological property of the ceramic inks was advantageous for inkjet printing. The printing speed had a close relation to the pressure of extrusion. As the extrusion pressure was over 0.4 MPa, the quantity of extrusion reached a constant level, at the maximum printing speed of 15 mm · s^{-1}. At this speed, the printed samples had no defect or deformation. After sintering, the shape was well retained, with a shrinkage of 14.7%.

Figure 3.27 shows photographs of the ZrO_2 ceramic teeth printed with the 55 vol% ink after sintering at high temperatures. The shape and size of the samples could be well developed by using the 3D inkjet printing process, which were retained after sintering. Nevertheless, owing to the relatively large diameter of the nozzle, the printed teeth had a roughness of about 10 μm at the side surface. The color of the dental samples after sintering was very close to that of human teeth. Moreover, the printing was finished in a short time of <5 min, thus ensuring rapid prototyping.

Figure 3.28 depicts fractured surface SEM images of the sintered samples. Numerous slender cracks were present in the direction perpendicular to the cross-section surface, as observed in figure 3.28(a). This suggests that the adhesion of the adjacent layers of the green bodies should be further strengthened. As revealed in figure 3.28(b), the gaps and cracks were absent in the sintered samples, while the defects or flaws were almost removed after the sintering at high temperatures.

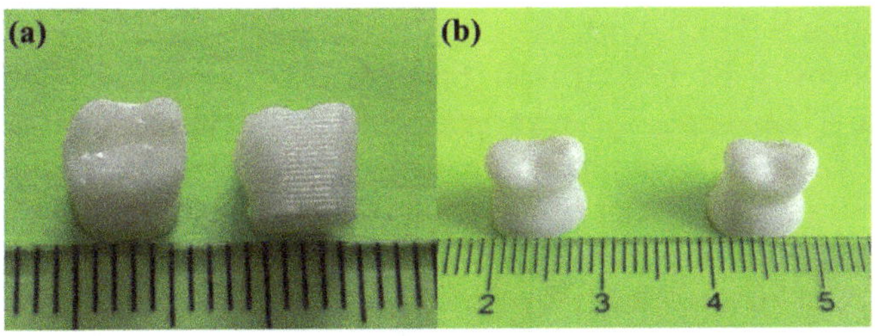

Figure 3.27. Photographs of the 3D inkjet-printed ZrO_2 ceramic teeth after sintering at high temperatures: (a) back lower right tooth and (b) bicuspid tooth. Reprinted from [120], Copyright (2020), with permission from Elsevier.

Figure 3.28. Fractured surface SEM images of the printed samples: (a) green bodies and (b) sintered ceramics. Reprinted from [120], Copyright (2020), with permission from Elsevier.

5YSZ layers have been coated on porous and non-porous substrates by using inkjet printing technology [121]. The dependence of the 5YSZ coatings on characteristics of the substrates was extensively examined. The 5YSZ ceramic inks were obtained by using a high-energy ball milling process. Owing to the requirement of a relatively low sintering temperature, the coatings were mainly of metastable tetragonal phase (*t*-phase) ZrO_2, so they exhibited strong mechanical strengths. When the non-porous substrates had open pores on the surface, the ceramic particles with small sizes would sweep in. To address this problem, coarse and fine particles were mixed to prepare the inks, so that a thin layer was first deposited on the substrates to fill the open pores.

The 5YSZ powder contained t-ZrO_2 and m-ZrO_2, which was dispersed in terpineol solvent at a concentration of 0.4 g ml^{-1}. Meanwhile, a binding agent (ethyl cellulose) was introduced to the suspension at 10 wt%. 5YSZ inks were finally obtained after ball milling for 4.5 h, at 120 and 160 rpm, corresponding to inks of YSZ-1 (coarse) and YSZ-2 (fine), respectively. According to TEM images, the 5YSZ powder consisted of spherical particles, with sizes in the range of 25–35 nm. In contrast, the particle size diameter (PSD) measured by using a dynamic light scattering method was in the range of 0.42–4.5 μm, due to agglomeration.

Stainless steel plates (AISI316L) with a composition of $Fe_{0.64}Ni_{0.36}$ were cut into dimensions of 40 mm × 40 mm × 2.5 mm and polished to achieve a surface roughness of Ra = 2.5 nm. 99.8% α-Al_2O_3 powder was pressed into plates with size of 50 mm × 50 mm × 3 mm, with the samples to be sintered at 1550 °C in air for 2 h, leading to a porosity of 18%. Inkjet printing was carried out with a 3D inkjet printing system (University of Cambridge, UK), equipped with a 16 solenoid Macrojet print head. The ceramic inks were diluted with methanol, under ultrasonication for 10 min before printing. 25 layers of coarse YSZ-1 were first printed and then fine YSZ-2, with an optimal open time interval of 1000 μs, a substrate temperature of 100 °C, and an ink concentration of 30 vol%.

The printed samples on the AISI316L substrates were sintered at 1180 °C for 2 h in vacuum and cooled down in Ar at 2 °C · min^{-1}, so that the 5YSZ coating was maintained to be *t*-phase, while the metal substrates were not oxidized. The samples on the porous Al_2O_3 substrates were also sintered at 1180 °C for 2 h in air. The inkjet printed 5YSZ green bodies on AISI316L and Al_2O_3 were denoted as IJP-SS and IJP-A, respectively, while the sintered samples were named as IJP-SS-S and IJP-A-S.

Figure 3.29 shows SEM images of porous Al_2O_3 substrates and the 5YSZ green bodies on the two substrates. The porous Al_2O_3 substrate had pore sizes in the range of 200–500 nm. The layer printed on the stainless steel substrate displayed no pores, while the one on the porous substrate was slightly porous. Figure 3.30 shows top surface and cross-sectional SEM images and optical micrographs of sintered 5YSZ coatings. The sample on the porous substrate was porous, due to the porous structure of the underlying substrate. The 5YSZ coating for the IJP-SS-S sample exhibited a porous microstructure, with grain sizes in the range of 50–200 nm. The IJP-A-S sample had a grain size of 100–150 nm, along with a pore size of about 100 nm.

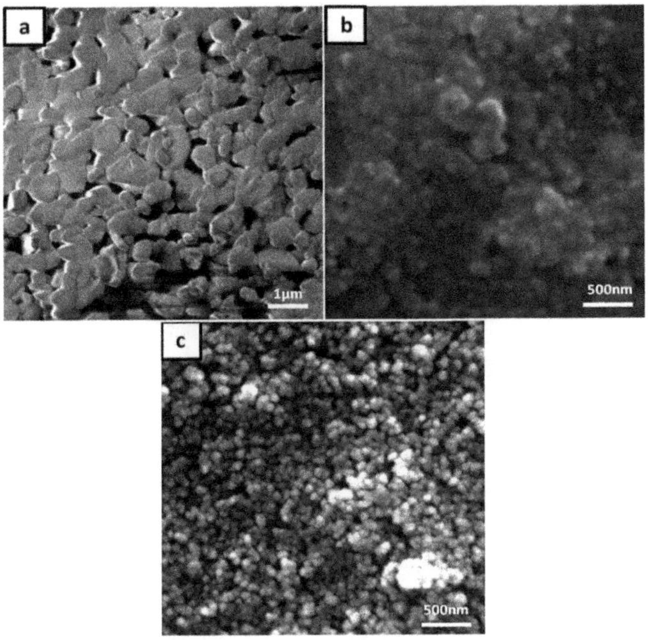

Figure 3.29. SEM images of substrates and green bodies: (a) porous Al_2O_3 substrate, (b) IJP-SS and (c) IJP-A. Reprinted from [121], Copyright (2020), with permission from Elsevier.

Kuscer *et al* reported an approach to develop cost-effective environmentally friendly aqueous inks, made of TiO_2 ceramic nanosized powders, for inkjet printing to fabricate patterns by using piezoelectric inkjet printing technique [122]. Specifically, the TiO_2 powder had particle sizes of <0.6 μm, leading to stable water-based suspensions, which were prepared by using a ball milling process. In order to enable efficient inkjet printing, surface tension and viscous behavior of the inks were finely adjusted, through the incorporation of suitable nonionic amphiphiles and glycerols. The effect of fluidic properties on formation behaviors of the droplets, the interactions between the inks and the substrates and morphologies of the patterns was systematically evaluated. It was shown successfully that uniform, TiO_2 patterns consisting of dots with a regular shape and size of 30 μm on Si/SiO_2 substrates had been printed with a nozzle diameter of 21 μm, at a voltage of 13 V and the frequency of 5 kHz.

The TiO_2 powder had a crystal structure of anatase, with particle sizes of 50–100 nm. The powder was agglomerated, with $D_{50} = 0.6$ μm and $D_{100} = 7$ μm. Nonionic amphiphiles (alcohol ethoxylate genapol UD 50) with a critical micelle concentration (CMC) of 9×10^{-5} M, phenol ethoxylate triton X-100 with a CMC of 2×10^{-4} M, ethoxylated sorbitan ester tween 20 with CMC of 6×10^{-5} M and polyoxoethylene-polyoxopropylene triblock copolymer (ABA-type), pluronic F-127, were employed as additives to prepare the ceramic slurries.

The organic additives and TiO_2 powder with solid loading levels of 1, 5, 10 and 15 vol% were dispersed in distilled water. The distilled water was adjusted to have

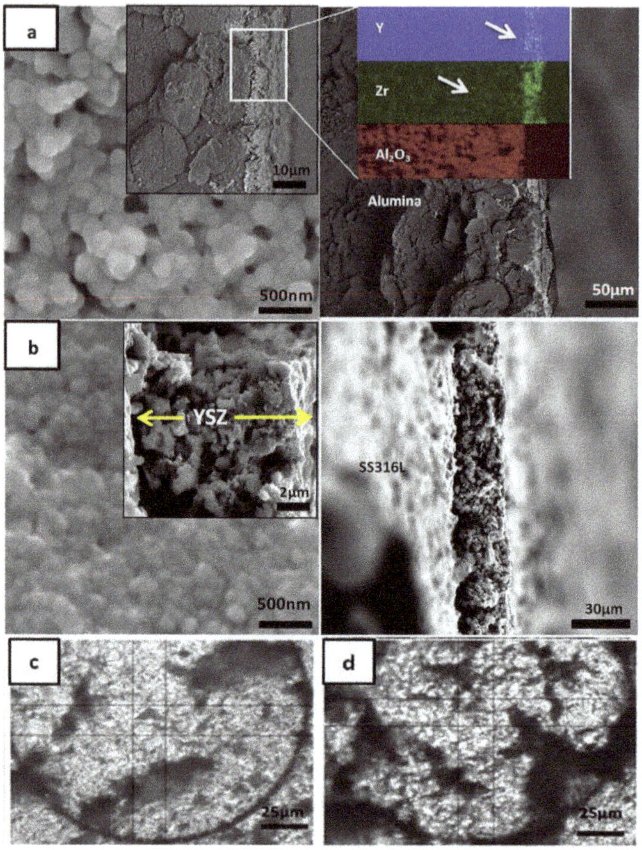

Figure 3.30. Top-surface and cross-sectional SEM images of the sintered samples: (a) IJP-A-S and (b) IJP-SS-S. Optical images showing the hardness test indentation: (c) IJP-SS-S and (d) IJP-A-S. Reprinted from [121], Copyright (2020), with permission from Elsevier.

pH = 7.5 and 11, with ammonium hydroxide solution. As the TiO_2 powder was added, the suspensions were constantly stirred. The mixtures were finally ball milled for 5 h at 1000 rpm with ZrO_2 milling media. The ink with 5 vol% TiO_2 and pH = 11 after ball milling for 5 h was examined for inkjet printing.

Figure 3.31 shows SEM images of the TiO_2 powders taken from the suspensions before and after ball milling for 5 h. Without ball milling, the primary particle was <300 nm, while hard agglomeration was present. The ball milling process effectively broke the agglomerates to be less than 400 nm, making the inks suitable for the inkjet printing through the nozzle of 21 μm.

All inks followed Newtonian behavior with shear rates in the range of 10–100 s^{-1} at room temperature. The viscosity of the inks increased with an increase in the solid loading level of the ceramic powders. The ink with 1 vol% TiO_2 was 0.95 mPa · s at 100 s^{-1}, which was slightly higher than that of water (0.89 mPa · s). The viscosity values of the inks with 5, 10 and 15 vol% TiO_2 at the shear rate of 100 s^{-1} were, respectively, 1.2, 1.5 and 1.8 mPa · s.

Figure 3.31. SEM images of the TiO_2 powders before (a) and after (b) ball milling for 5 h. [122] John Wiley & Sons. © 2011 The American Ceramic Society.

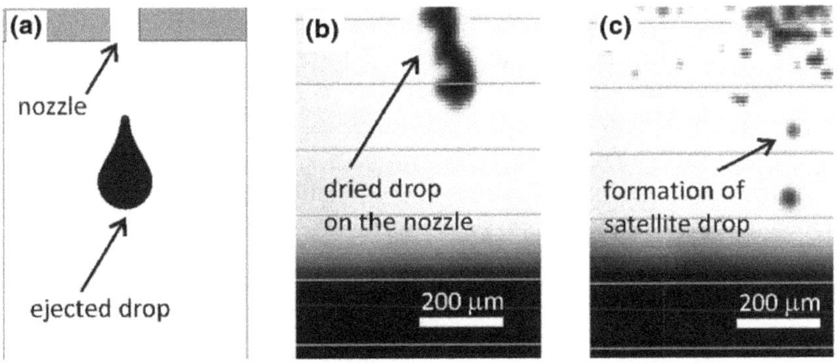

Figure 3.32. Snapshots of the droplets ejected from the nozzle: (a) schematic diagram, (b) ejection of the ink with 15 vol% TiO_2 and 1 wt% Triton X-100 and (c) ejection of the ink with 5 vol% TiO_2 and 1 wt% Triton X-100. [122] John Wiley & Sons. © 2011 The American Ceramic Society.

The inks for inkjet printing should have surface tensions close to 30 $mN \cdot m^{-1}$, while that of the TiO_2 slurries was 68 $mN \cdot m^{-1}$ at room temperature, which was not suitable for the printing process. The above-mentioned surfactants could be used to address this issue, because of their strong self-assembly behaviors to form micelle aggregates. The content of the amphiphiles was in the range of 0.5–1.0 wt%, with respect to the weight of the TiO_2 powder. For example, with the addition of 0.5 wt% genapol UD 50 and triton X-100, the surface tensions were lowered to 22 and 30 $mN \cdot m^{-1}$, respectively, both of which met the requirement of inkjet printing.

The ink made at pH = 11, with 15 vol% TiO_2 and 1 wt% Triton X-100, possessed a surface tension of 32 $mN \cdot m^{-1}$ and a viscosity of 1.8 $mPa \cdot s$ at a shear rate of 100 s^{-1} and room temperature, while the Z value was 17, which was slightly large for printing. In this case, the droplets could not be effectively produced. The inks even could not be squeezed through the nozzle of the printer. Figure 3.32 shows a schematic diagram and a snapshot of the droplets ejected from the nozzle. The jetted ink droplets quickly dried a short while later. As a consequence, droplets could not be well formed on the substrate.

The ink with 5 vol% TiO_2 was used to develop strategies to reduce the drying rate of slurries. The ink had a Z value of 23. Although droplets were readily formed on the Si/SiO_2 substrate, small satellite droplets were present, as seen in figure 3.32(c). The presence of the satellite droplets had a negative effect on the accuracy of the printing. The average diameter of the droplets was 40 μm, with a spacing of 100 μm, as demonstrated in figure 3.33. The drying behavior of liquid droplets could be described by the coffee-ring effect [123]. A droplet dries due to the evaporation of liquid, which is accompanied by shrinking. In the case that the edge line is pinned, shrinkage cannot occur. Therefore, the liquid evaporates at the edge of the droplet, which will be filled by that from the center, resulting in accumulation of solids at the edge, hence forming a ring-like structure. The authors found that glycerol was a useful additive to modify the drying behavior of the TiO_2 inks.

Yang *et al* fabricated TiO_2 thin film photoanodes from colloidal TiO_2 inks through inkjet printing, for the oxidation of organic compounds in aqueous solutions [124]. The device reached a linear response in the range of up to 120 mg $\cdot$ l^{-1} of O_2 and a relatively low detection limit of 1 mg $\cdot$ l^{-1} of O_2. An aqueous TiO_2 colloid was prepared through hydrolysis of titanium butoxide, leading to printing inks with 60 g $\cdot$ l^{-1} TiO_2, into which 1.8 wt% carbowax was dissolved. A piezoelectric inkjet printer (Epson R290) was used to print TiO_2 thin films. The as-printed photoanodes were calcined at 700 °C for 2 h.

Figure 3.34 shows photographs of the inkjet-printed electrode, together with one that was made by using dip-coating. Obviously, the inkjet-printed sample was more homogeneous than the dip-coated one. The dip-coated electrode exhibited light diffraction patterns, because surface tension attracted the liquid layer to the center during the drying process, after the substrates were dipped and withdrawn from the colloidal solutions. Therefore, the dip-coated TiO_2 films were thicker at the center than at the edge. In contrast, the inkjet-printed electrode had no diffraction patterns, since the microdroplets on the surface of the electrode evaporated quickly, without the effect of surface tension on the drying process.

Figure 3.35 shows a representative cross-sectional SEM image of the inkjet-printed TiO_2 thin film. The thin film consisted of TiO_2 particles with sizes in the

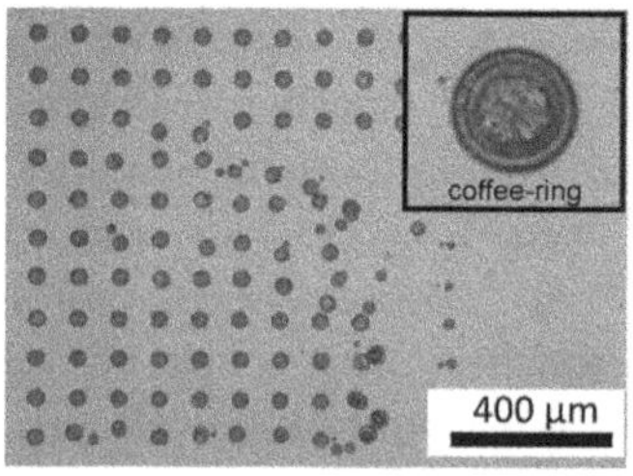

Figure 3.33. Photograph of the as-printed droplets on SiO_2/Si substrate through inkjet printing with the ink containing 5 vol% TiO_2 and 1 wt% Triton X-100. The inset shows an enlarged droplet. [122] John Wiley & Sons. © 2011 The American Ceramic Society.

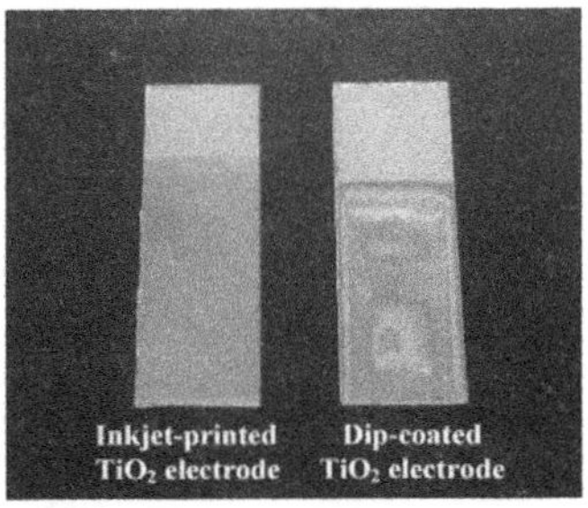

Figure 3.34. Photographs of the inkjet-printed TiO_2 electrode (left) and dip-coated electrode (right). Reprinted from [124], Copyright (2010), with permission from Elsevier.

Figure 3.35. Cross-sectional SEM image of the TiO_2 thin film electrode inkjet printed with 10 layers. Reprinted from [124], Copyright (2010), with permission from Elsevier.

range of 20–40 nm. This observation indicated that calcination at 700 °C caused agglomeration of the TiO_2 nanoparticles (10 nm) in the TiO_2 colloids. The 10-layer TiO_2 film had a thickness of 400 nm. In addition, the inkjet-printed TiO_2 film had a uniform and porous microstructure, without spacing between the adjacent layers, owing to the proper viscosity and surface tension of the colloidal TiO_2 inks.

Arin *et al* developed photocatalytically active TiO_2 coatings on glass substrates with aqueous chemical solutions through inkjet printing, for degradation of methyl orange [67]. Aqueous TiO_2 precursor solutions were derived from tetrabutyl orthotitanate (TNBT) as the Ti source, along with citric acid (CA), acetylacetone and triethanolamine (TEA) as complexing agents. The metal alkoxide was dissolved in ethanol (EtOH), to control the degree of hydrolysis when mixing with water.

For the inkjet printing experiment, a Domino Macrojet electromagnetic nozzle with a 90 μm diameter jewel orifice was equipped on a Roland *x*–*y* plotter. The nozzle could be used to deliver non-magnetic inks with high viscosities of up to 1000 cP, producing droplets of >1 nl at frequencies of up to 1 kHz. The pressure was set to be in the range of 0.3–0.8 bar, while the inkjet printing was conducted at room temperature. The droplets ejected from the nozzle were arranged as a hexagonal droplet array. The inter-droplet distance in the array was optimized to maximize

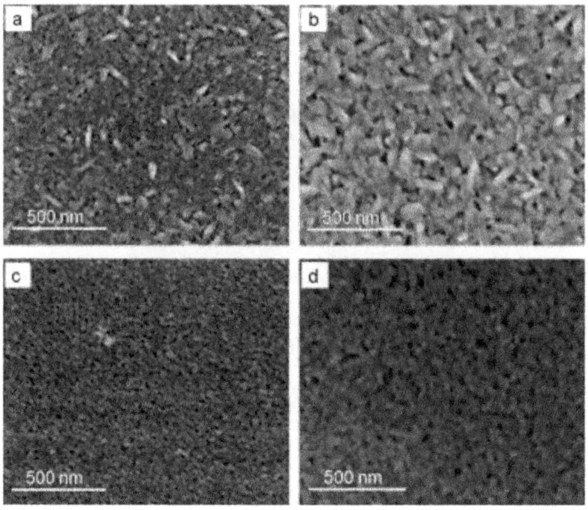

Figure 3.36. Surface SEM images of the TiO_2 films on glass substrates from the Ti–CA (a and b) and Ti–TEA (c and d) solutions after calcination for 1 h at 500 °C (a and b) and 600 °C (b and d). Reprinted from [67], Copyright (2011), with permission from Elsevier.

coverage and overlapping of the droplets, to ensure desired homogeneity of the films. The as-printed samples were dried in air at 60 °C for 3–4 h, followed by calcination at temperatures of 500 °C–650 °C for 1 h in O_2, at a heating rate of 5 °C · min^{-1}.

Figure 3.36 shows representative SEM images of the surface of TiO_2 thin films, from different solutions, after calcination at different temperatures. The films from the solution of Ti–CA exhibited grains with larger size and aspect ratio. In comparison, the TiO_2 film from the solution of Ti–TEA possessed smoother and more porous surfaces. Both the surface roughness and grain size of the two films increased with increasing calcination temperature.

Lead zirconate titanate ($Pb(Zr_xTi_{1-x})O_3$ or PZT) is one of the most important perovskite-type ferroelectric materials, with outstanding dielectric, piezoelectric, and ferroelectric properties [125–127]. PZT particles and a PZT precursor were synthesized by using a chemical process. The PZT particles were dispersed in solvents to form inks with desired rheological properties, while the precursor was directly utilized for inkjet printing [128]. The inks were deposited on Al_2O_3 and steel substrates, which were then sintered with a pulsed laser at a wavelength of 1064 nm.

Organometallic compounds of lead, zirconium and titanium were mixed to synthesize PZT nanoparticles. Water was added to the mixture of the organometallic compounds to form gels. The gels were dried and calcined to obtain PZT nanosized powder with a perovskite crystal structure. Due to agglomeration, the PZT powders were subject to high-energy ball milling. For precursor-based ink, sols were prepared from the mixed organometallic compounds, without the formation of gels. Both the PZT nanosized particles and precursor were blended with vehicles and binders, leading to inks with viscosity of $\leqslant$30 cps and surface tension of $\leqslant$70 N · m^{-1}, as required for inkjet printing. The as-prepared PZT particle inks were filtered with 5 μm Whatman and Cameo syringe filters, to prevent clogging of the nozzle by the agglomerated PZT particles. Although the PZT precursor inks were free of particles, filtering was recommended, in case gel agglomerates were formed from the precursor sols.

The PZT particle inks were printed on both commercial Al_2O_3 and stainless-steel substrates, while the PZT precursor inks were printed only on stainless steel substrates. To ensure the formation of ink droplets, a piezo voltage of 43 V and a maximum jetting frequency of 1 kHz were utilized for printing. Inkjet printing parameters, such as velocity and spacing between droplets, were optimized for different substrates and inks. The printed layers were heated on a hot plate at 40 °C for 30 min to vaporize the vehicle. The samples were then sintered with 1064 nm wavelength pulsed laser at frequency 10 kHz, with maximum scanning velocity of $1\ mm \cdot s^{-1}$ and laser pulse of 200 ns. The sintered samples were immersed in silicon oil at 120 °C for 30 min, which were subject to an electrical current of <1 mA. Specimens with a height of 10 μm were poled using a voltage per unit length between 5 and $7.5\ kV \cdot mm^{-1}$.

Figure 3.37 shows SEM images of the PZT layers printed on the Al_2O_3 substrates at different laser powers, while those on the stainless-steel substrates are depicted in figure 3.38. After laser sintering at low power of 200 mW, organic binders were retained, as the phase with dark grey color on the surface of the particles marked with an arrow, while the necking behavior was less developed, as seen in figures 3.37 (a) and (b). With increasing laser power, neck formation between particles and grain growth occurred, as illustrated in figures 3.37(c)–(f). In addition, too-high laser power was not recommended, because PZT decomposition occurred due to the strong laser irradiation, as observed in figures 3.37(g) and (h). A similar trend was observed for the samples printed on stainless steel substrates.

For the films printed with the precursor inks, according to XRD results, the phase composition was different, as the laser power increased from 400 to 1200 mW. Perovskite phase PZT started to be present after sintering at a laser power of 400 mW, where the sample was nearly totally in an amorphous state, together with minor crystalline phases, at the laser sintering power of 1200 mW. The three crystalline phases included PZT perovskite, PbO and ZrO_2, where PbO and ZrO_2 were the decomposition products of PZT.

Figure 3.39 shows SEM images of the films printed on stainless steel substrates, with the precursor ink. The laser sintering caused thermal stresses and volume variation, along with burnout of the organic additives and phase crystallization. As a result, pores and cracks were formed, thus leading to fragmentation of the PZT films. In most serious cases, the films were broken into small pieces, as indicated in figures 3.39(c), (e), and (g) labeled with *F* and *S*.

Comparatively, the films on stainless steel substrates after sintering at laser powers of 400–800 mW with the particle PZT ink were more integrated, although PZT phase decomposition was still observed. It was found that printing speed had a significant effect. For example, at laser powers of ⩾400 mW and stage speeds of 0.1–$0.5\ mm \cdot s^{-1}$, the degradation of PZT could not be avoided. Therefore, sufficiently low laser powers (200–300 mW) were necessary to ensure the formation of uniform

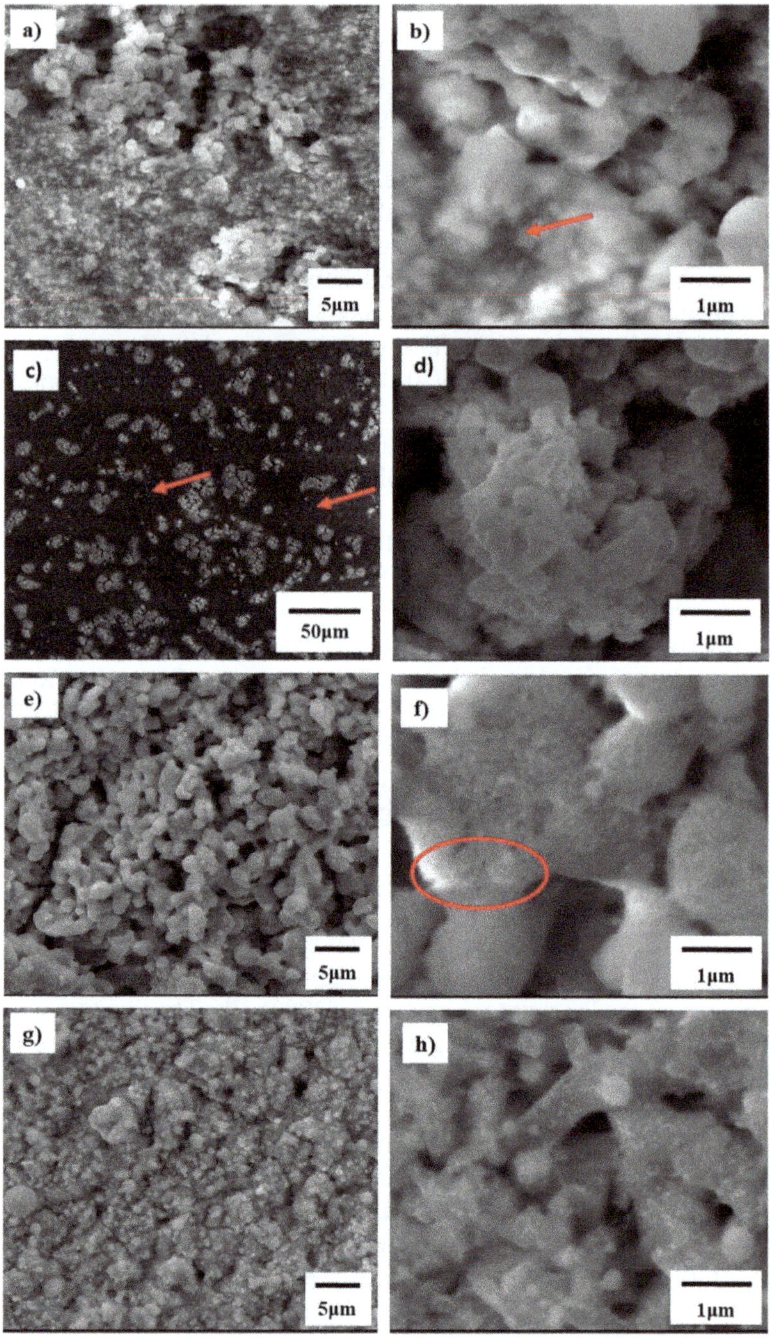

Figure 3.37. SEM images of the nanoparticles on alumina substrate laser sintered at different powers: (a and b) 200 mW, (c and d) 600 mW, (e and f) 2000 mW and (g and h) 3000 mW. Reprinted from [128], Copyright (2018), with permission from Elsevier.

Figure 3.38. SEM images of the nanoparticles on steel substrate laser sintered at different powers: (a and b) 200 mW, (c and d) 600 mW and (e and f) 1000 mW. Reprinted from [128], Copyright (2018), with permission from Elsevier.

and continuous PZT layers. This was because the organic additives have sufficient time to be burned out from the PZT matrix at lower laser powers.

Part of this text has been reproduced from [129]. CC BY 3.0.

$CuFe_2O_4$ inks were developed for inkjet printing to fabricate cathodes of low temperature fuel cells [129]. High-purity commercially available $CuFe_2O_4$ powder, with an average particle size of 3 μm, was used to prepare the ceramic inks. The dispersing medium was terpineol, while ethyl cellulose (EC) and 1,5-pentandiol were utilized as the dispersant and humectant/surfactant. Three dispersions, with 3.5 wt% $CuFe_2O_4$ and 0.03 wt% EC, as well three levels of 1,5-pentandiol, including 0, 15

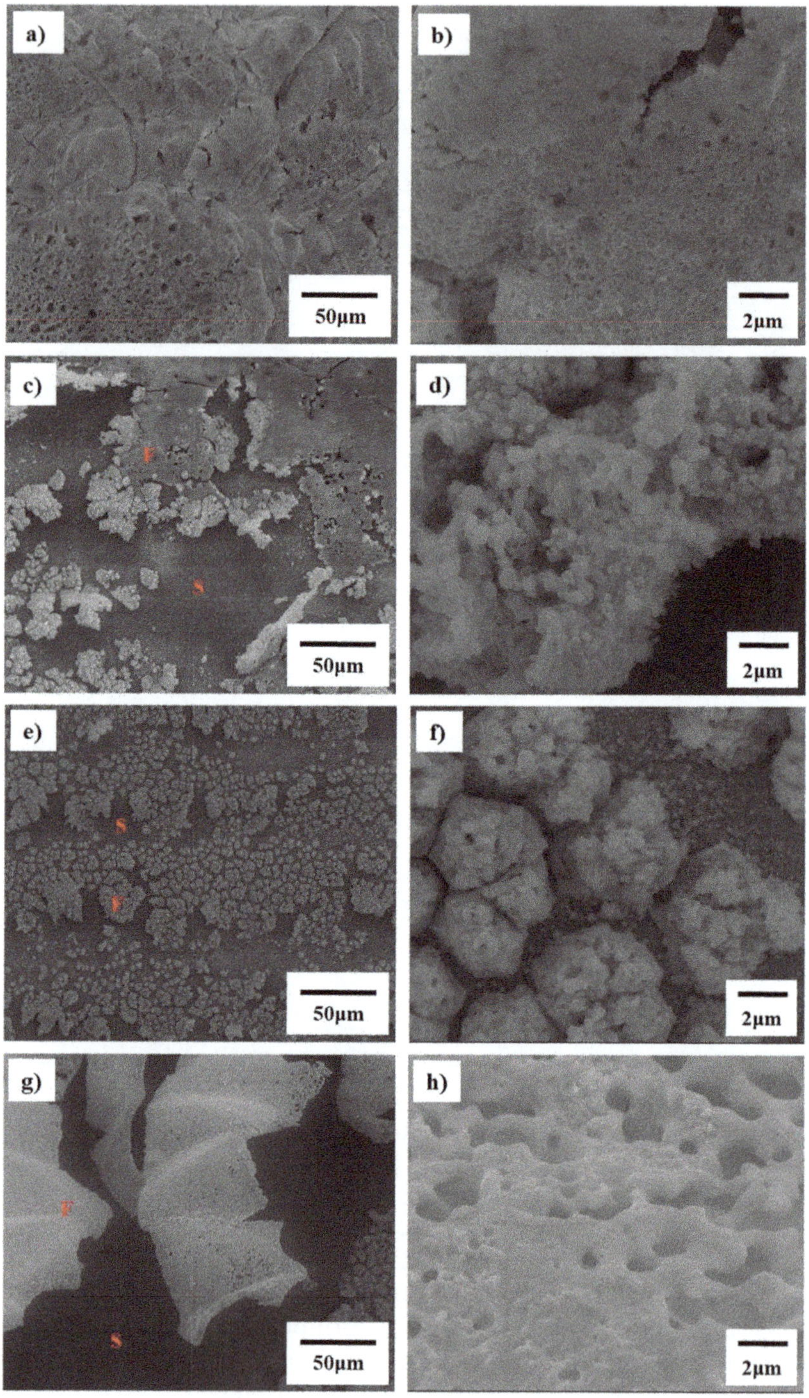

Figure 3.39. SEM images of the PZT precursor on steel substrates laser processed at different powers: (a and b) 400 mW, (c and d) 600 mW, (e and f) 800 mW and (g and h) 1200 mW. Reprinted from [128], Copyright (2018), with permission from Elsevier.

and 20 wt%, were formulated to study the effect of 1,5-pentandiol on printability of the $CuFe_2O_4$ inks and electrochemical performance of the final cathodes. These three inks were denoted as Ink (0), Ink (1) and Ink (2).

The components were blended and mixed by using a ball milling process, with small, medium and large sizes of zirconia balls, in a zirconia vial, at rotation speed of 300 rpm for 2 weeks. Finally, the inks were vacuum filtered through glass microfiber filters. The flowability of Ink (0) was too low for vacuum filtration. Ink (2) was modified by adding cyclopentanone (Sigma–Aldrich, Germany), at a volume ratio of 3:1 (Ink (2): Cyclopentanone), denoted as Ink (2)—Samba, to reduce the viscosity.

The nanocomposite electrolyte substrate had a porous-dense-porous three-layer structure, with the dense and porous layers both consisting of NLK-GDC. The porous layers in the electrolyte served to provide capillary forces, regulating the stage of droplet dispersal and preventing the coffee-string effect. The NLK-GDC powder for the dense layer was prepared by using a solid state reaction process. Lithium, sodium and potassium carbonates, with eutectic composition of Li:Na:K = 32.1:33.4:34.5, were mixed with GDC powder at a weight ratio of 30:70, through ball milling for 24 h at 300 rpm using zirconia milling media. For the porous layer, the NLK-GDC nanocomposite powder was mixed with 25 wt% EC, followed by intense grinding.

The three-layer porous electrolyte substrates were fabricated by cold pressing with a 13 mm die, with 0.1 g porous, 0.3 g dense and 0.1 g porous NLK-GDC powders, for 2 min under a pressure of 250 MPa. The pellets were then heated at $1\ °C \cdot min^{-1}$ to 350 °C and $2\ °C \cdot min^{-1}$ to 700 °C for 1 h. The sintered substrate had a diameter of 12.2 mm and a thickness of 1.22 mm, with the dense layer and porous layers measuring 0.73 and 0.25 mm in thickness, respectively.

To prepare the anode, Ni_2O_3 and GDC powders were mixed at a weight ratio of NiO:GDC = 60:40, leading to NiO-GDC. After that, EC was added to the NiO-GDC mixture at a weight ratio of NiO-GDC: EC = 90: 10. The mixture was thoroughly ball milled for 1 h at 300 rpm. 0.4 g NiO-GDC and 0.3 g GDC were used to fabricate two-layered anode-supported substrates, by using cold co-pressing. The anode-supported substrate pellets were sintered at 1500 °C with a heating rate of $1\ °C \cdot min^{-1}$ for 4 h.

Porous NLK-GDC nanocomposite electrolytes were inkjet-printed by using a Dimatix DMP-2800 (Fujifilm, Santa Clara, USA) inkjet printer, with a piezoelectric transducer and 10 pl Legacy DMP and Samba cartridges. The Legacy DMP and Samba cartridges had 16 and 12 nozzles, respectively, both with a diameter of 21.5 μm, while the reservoir had a capacity of 1.5 ml. The piezoelectric actuators in the printhead nozzles were driven by an electrical signal, which was regulated through a single waveform. The printing was conducted at a cartridge temperature of 45 °C, a jetting voltage of 35 V, and a jetting frequency of 30 kHz, with the dwell and falling times of 4 and 20 μs, respectively. A delay of 10 s was set between printing individual layers, while the printer's vacuum plate was kept in an active state.

Firstly, 50 layers of porous NLK-GDC nanocomposite electrolytes were inkjet-printed with Ink (1) and Ink (2) to determine the optimal concentration of 1,5-pentandiol for the $CuFe_2O_4$ inks. After both sides were printed, the sample was dried at 130 °C for 1 h and then sintered at 600 °C for 1 h with a heating rate of 1 °C · min^{-1}. Using Ink (2), 30–200 layers were printed to find the optimal number of layers. Finally, Ink (2)—Samba and 100 layers were identified as the optimal inkjet printing parameters.

Symmetric cells were also made with a droplet-casting deposition process, with Ink (2) and Ink (2)—Samba, as compared with the inkjet-printed samples. 0.05 ml ink was dropleted onto the porous NLK-GDC electrolyte, which was heated on hot plate at 60 °C. The drying and sintering processes were conducted in a similar manner. The two sides of the electrolyte were coated alternatively. To achieve more effective comparison, the same loading of the cathode ink was used for the inkjet-printed and droplet-cast samples. The 100-layer inkjet-printed and droplet-cast samples were estimated to have thicknesses of 14 and 45 μm, respectively.

Nyquist plots of inkjet-printed symmetric cells with 40–200 layers with Ink (2), which were measured at 550 °C, are shown in figure 3.40(a). The insets depict the enlarged Nyquist plots for the cells in the high frequency region of 1.07–100 kHz and the corresponding equivalent circuit. The area specific resistance (ASR) decreased with the increasing number of printed layers, with the 100-layer inkjet-printed symmetric cell reaching the lowest value of 9.91 Ω · cm^2.

In the device with a 30-layer electrode, the active materials were insufficient to provide enough active sites for the oxygen reduction reaction, so the sample could not be measured. As the number of inkjet-printed layers increased to 40 and 50, the high content of $CuFe_2O_4$ and the increased thickness provided more active sites for the electrode reactions. In the 100-layer device, the site effect was maximized. Further increasing the number of printed layers to 200 caused the electrode to become too thick and less porous, thus blocking oxygen diffusion and increasing resistance.

Meanwhile, R_{Ohm} was inversely proportional to the number of inkjet-printed electrodes in the whole range of 40–200 layers. The monotonic decrease in R_{Ohm} could be attributed to the enhanced contact at the electrolyte/electrode interface, owing to the presence of more active materials and denser packing of the ceramic particles. ASR and R_{Ohm} of inkjet-printed symmetric cells, with 40, 50, 100 and 200 layers, measured at 400 °C–550 °C, are illustrated in figure 3.40(b). In summary, the 100-layer inkjet-printed symmetric cell was optimal, owing to the lowest level of ASR and the acceptable value of R_{Ohm}.

The inkjet printing led to electrode materials with more hierarchical porous microstructures, thus offering more reactive sites. In this case, the ASR value of the inkjet-printed sample was 2.3 times lower than that of the droplet-cast sample at the same measurement temperature of 550 °C. This phenomenon was observed for both the Ink (2) and Ink (2)–Samba.

Figure 3.41 shows surface SEM images of the 100-layer inkjet-printed and droplet-cast samples with Ink (2). $CuFe_2O_4$ particles were uniformLy distributed

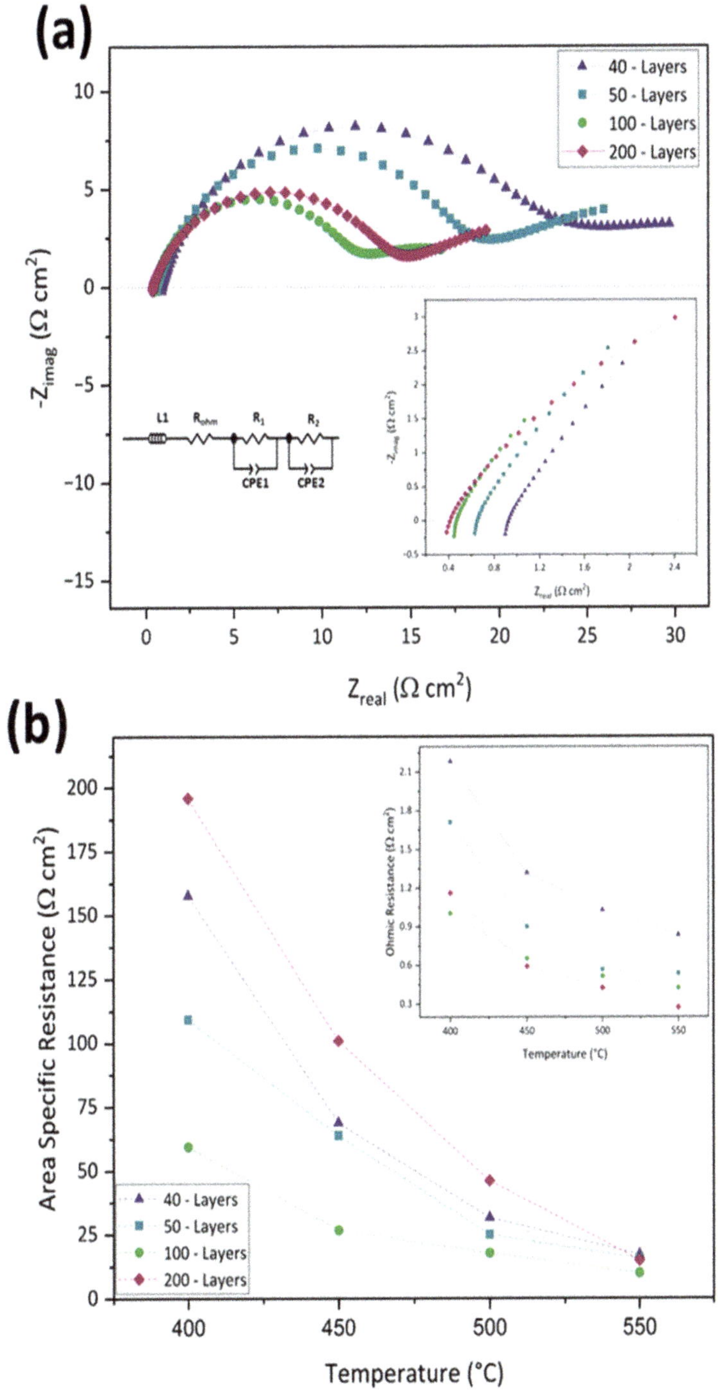

Figure 3.40. (a) Electrochemical performances of the symmetric inkjet printed cells with Ink (1) and Ink (2) at 550 °C. The insets are enlarged Nyquist plots in the high frequency region of 1.07–100 kHz and the equivalent circuit. (b) Area specific resistance (ASR) of inkjet-printed symmetric cells with different numbers of layers at 400 °C–550 °C. The inset depicts R_{Ohm} curves of the cells. Reproduced from [129]. CC BY 3.0.

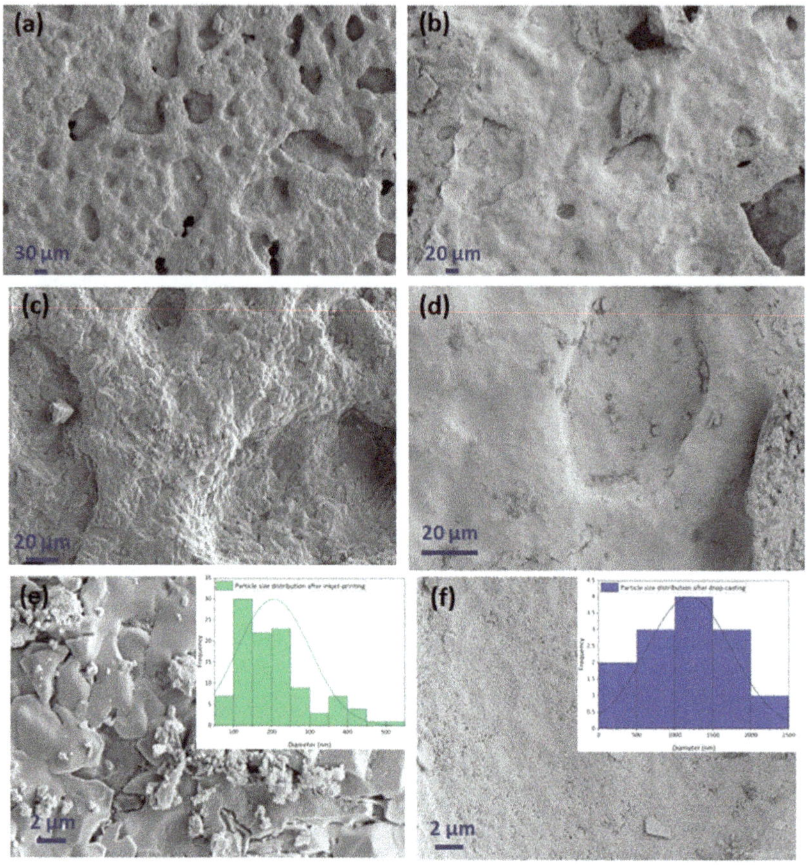

Figure 3.41. Surface SEM images of the NLK-GDC symmetric cells with 100-layer porous cathode: (a, c, and e) inkjet printing and (b, d, and f) droplet-casting. The insets of (e) and (f) panels show the particle size distribution profiles of $CuFe_2O_4$ particles on surface of the two samples analyzed with ImgaeJ software. Reproduced from [129].

on the surface of the inkjet-printed layers, as revealed in figures 3.41(a), (c), and (e). In comparison, the droplet-cast samples exhibited fewer active materials on the surface. As seen in figures 3.41(c) and (e), the $CuFe_2O_4$ particles were present with a hierarchical microstructure, which was beneficial to charge transportation during the electrochemical reaction. Moreover, the hierarchical microstructure of the inkjet-printed samples ensured an environment for the oxygen reduction reaction (ORR) in the cathode. The droplet-cast samples displayed no such structure, as observed in figures 3.41(d) and (f).

According to the particle size distribution profiles of the inkjet-printed and droplet-cast layers derived using the ImageJ software, as demonstrated in figures 3.41 (e) and (f), the inkjet-printed sample had an average particle size of 142 nm, while the size of the droplet-cast sample was 1.16 μm. This observation confirmed that the inkjet-printed cells possessed much higher specific surface area, thus having more reactive sites and hence stronger surface reactions [130–132]. Moreover, the high-speed ink ejection would help reduce the particle size during the inkjet printing process.

Figure 3.42 shows cross-sectional SEM images of the interface area of the inkjet-printed and droplet-cast samples at different magnifications. The electrode/electrolyte interface of the inkjet-printed sample was entirely coated with $CuFe_2O_4$ nanosized particles, as shown in figures 3.42(c) and (e), suggesting that the ceramic ink infiltrated into the porous layer of the substrates. In contrast, in the interface area of the droplet-cast sample, $CuFe_2O_4$ particles were agglomerated, forming a structure with scattered islands. Therefore, the enhanced electrochemical performance of the inkjet-printed cells was closely linked to the unique interface microstructure.

Representative SEM images of the interface area of the symmetric cells coated with Ink (2)—Samba are shown in figure 3.43. Obviously, the inkjet-printing process effectively enabled the homogeneous distribution of the $CuFe_2O_4$ particles in the interfacial area, as compared with the droplet-cast technique. Similar results were observed in the BET measurement of the two samples.

YAG/10at% Yb:YAG/YAG transparent ceramic planar waveguide (PWG) gain medium was fabricated by using inkjet printing, which was compared with the dry pressing process [133]. The composition and rheological property of the ceramic inks were studied to optimize the conditions to increase printing accuracy and the quality

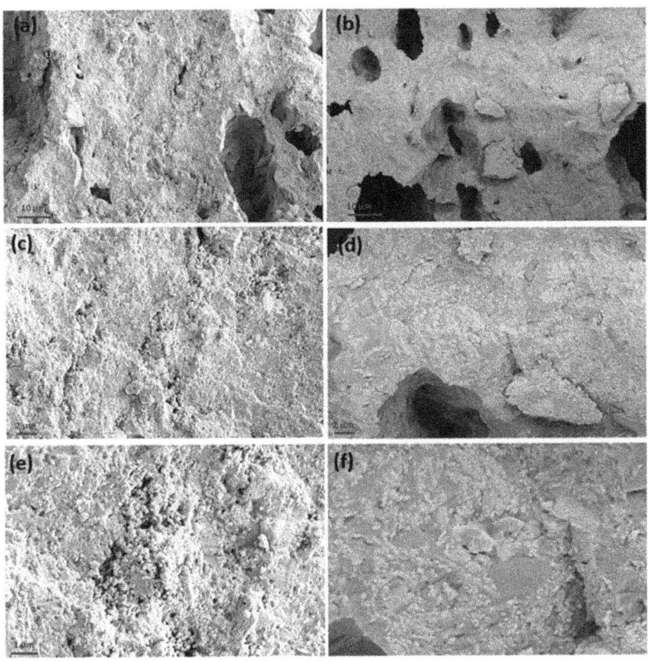

Figure 3.42. Cross-sectional SEM images of the 100-layer porous NLK-GDC symmetric cells at different magnifications: (a, c, and e) inkjet printing and (b, d, and f) droplet-casting. Reproduced from [129]. CC BY 3.0.

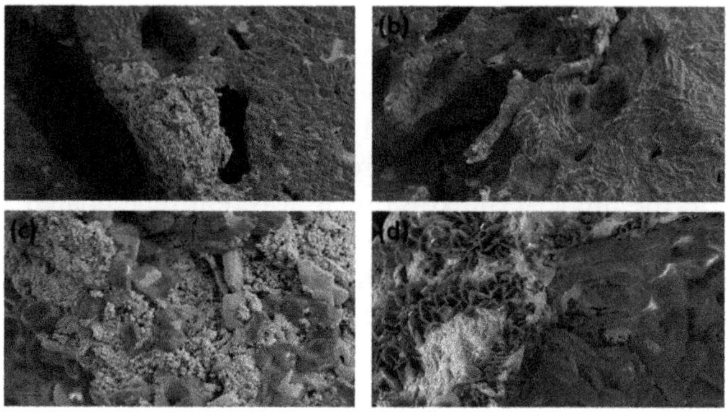

Figure 3.43. Cross-sectional SEM images of the 100-layer porous NLK-GDC symmetric cells made with Ink (2)-Samba: (a and c) inkjet printing and (b and d) droplet-casting. Reproduced from [129]. CC BY 3.0.

of the final products. The PWG had an in-line transmittance of 81.7% at a wavelength of 1030 nm, while the ceramics had an average grain size of 2.3 μm. The diffusion coefficient and mean diffusion distance of ^{172}Yb ions across the interface between the cladding YAG layer and the core Yb:YAG layer were examined. When pumped with a single end by using a 940 nm laser diode, the Yb:YAG PWG oscillator was able to generate a continuous wave laser at the wavelength of 1030 nm, corresponding to the highest power of 3.8 W and highest absorbed-output slope efficiency of 64.6%.

High purity Al_2O_3 and Y_2O_3 were mixed with a stoichiometric composition to synthesize $Y_3Al_5O_{12}$ (YAG) through the conventional solid state reaction process. The mixture was used to fabricate the YAG cladding layer with dry pressing. 10at% ytterbium oxide (Yb_2O_3) was mixed with the Al_2O_3 and Y_2O_3 mixture of YAG to prepare the Yb:YAG core by using inkjet printing (IJP) technique.

The powder mixture used for preparing the Yb:YAG core was suspended in a water solvent with 15–30 vol% solid loading. Then, dispersant Dolapix CE-64 (0.4 wt%, 65 wt% solution, Zschimmer & Schwarz Chemical (Foshan) Co., Ltd, China) was added to obtain aqueous Yb:YAG ink. Ammonia solution (AR, Sinopharm Chemical Reagent Co., Ltd, China) was used to control the pH (adjusted to 11) to increase the dispersing efficiency. At the same time, diethylene glycol (10 vol %, AR, Sinopharm Chemical Reagent Co., Ltd, China) was introduced to maintain moisture, while *N*-(2-hydroxyethyl)ethylenediamine (5 vol%, CP, Sinopharm Chemical Reagent Co., Ltd, China) was utilized to stabilize the pH value.

After that, the suspensions were subjected to ball milling for 24 h, with Al_2O_3 balls at a mass ratio of powder to balls of 5:11, at a rotation speed of 100–150 rpm. Then, polyethylene glycol 400 (0.1 wt%, CP, Sinopharm Chemical Reagent Co., Ltd, China) and dimethyl silicone oil (Shanghai Macklin Biochemical Co., Ltd, China) were incorporated as the binder and defoamer. Eventually, the suspensions were treated with ultrasonication and defoaming using a planetary defoaming mixer (Thinky), leading to ceramic inks for inkjet printing.

In the PWG structure, the lower YAG cladding was molded through dry pressing with a dimension of 66.5 × 44.5 mm^2. A layer of Yb:YAG with a size of 50 × 30 mm^2 was printed on the YAG cladding. Then, the printed sample was put back into the dry pressing mold, after which the YAG powder was filled, followed by a second round of dry pressing to form the upper YAG cladding. The pressed samples were cold isostatic pressed (CIP) at 200 MPa to further increase the density of the green bodies, thus ensuring the optical properties of the final sintered products.

The CIPed samples were calcined at 800 °C for 2 h to burn out all organic additives. The debound samples were sintered at 1680 °C for 14 h in vacuum at a vacuum level of 5×10^{-3} Pa, followed by hot isostatic pressing (HIP) sintering at 1650 °C for 6 h at 190 MPa in Ar. After that, the samples were annealed at 1270 °C for 80 h for oxidation of Yb^{2+} to Yb^{3+}. After sintering, the samples were made into dimension of 13.5 × 8.0 × 1.8 mm^3 for laser characterization. Figure 3.44 shows the fabrication process of the PWG structure.

Densities of the samples, derived from the inks with solid loading levels of 10, 20 and 30 vol%, were 1.29, 1.56 and 1.63 $g \cdot cm^{-3}$, respectively. Figure 3.45 shows vertical section views of the droplets. The average contact angle of the droplets decreased with increasing solid loading level. Theoretically, Z values of the inks with 10, 20and 30 vol% solid were 7.60, 2.32 and 8.48, respectively, suggesting that the inks were all suitable for the IJP experiment.

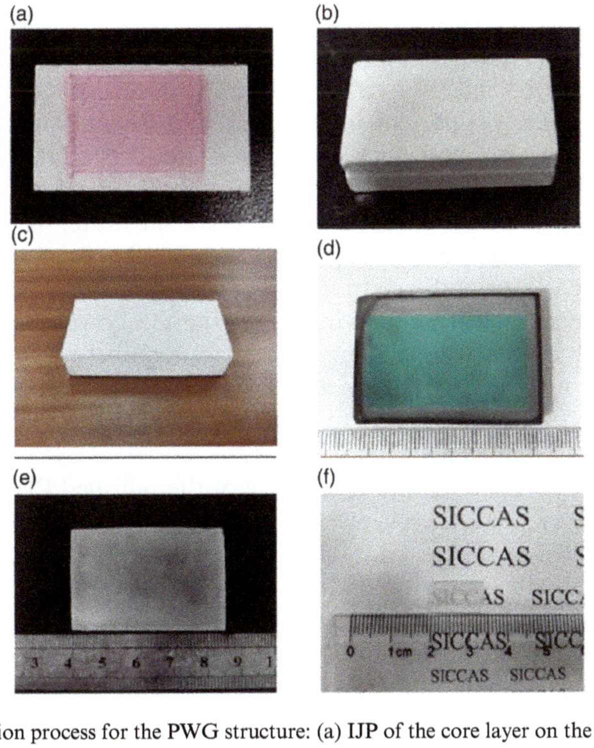

Figure 3.44. Preparation process for the PWG structure: (a) IJP of the core layer on the dry-pressed substrate, (b) PWG green body after CIP, (c) vacuum sintered sample, (d) sample after HIP, (e) annealed sample, and (f) machined and polished sample. Reproduced from [133]. CC BY 4.0.

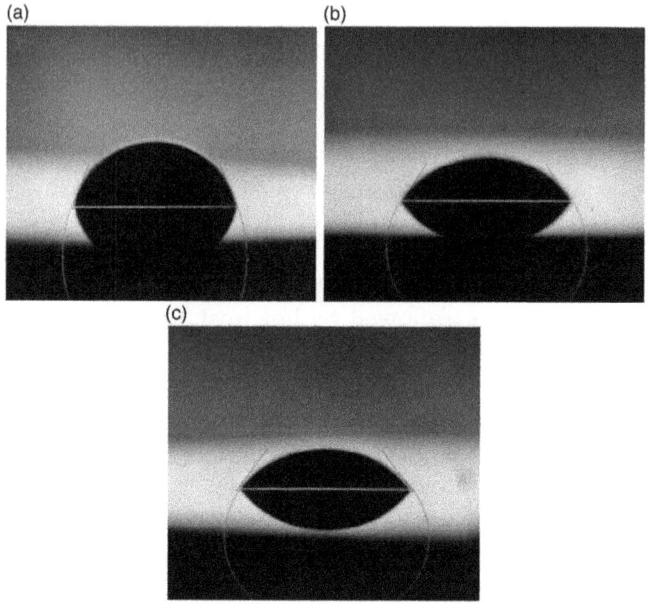

Figure 3.45. Images depicting the vertical sections of the droplets for the three inks with different solid loading levels: (a) 10 vol%, (b) 20 vol% and (c) 30 vol%. Reproduced from [133]. CC BY 4.0.

Figure 3.46 shows polished cross-sectional SEM images and the grain size distribution profile of the PWG structure with a core layer of 10at% Yb:YAG. The core layer was clearly present in the middle of the structure. One side of the interface between the core and the cladding was smooth, while the other side was relatively rough. The core layer had thicknesses in the range of 190–215 μm. Because the core layer was printed on the dry-pressed YAG substrate, from ceramic ink in a point-to-line-to-plane manner, through the nozzle, the flatness of the interface between the printed layer and substrate was ensured. In this case, the printed lines would not be sufficiently diffused, owing to the limited flowability of the inks. As a result, the surface of the printed core layer was relatively rough. Therefore, the bottom interface was not as flat as the upper one. The ceramics had a dense microstructure and uniform grain size distribution, with an average grain size of 2.3 μm, as shown in figures 3.46(b) and (c).

Figure 3.47 depicts the transmittance curve of the polished PWG with the 10at% Yb:YAG core layer, demonstrating an in-line transmittance of 81.7% at the wavelength of 1030 nm. The excitation light source of the emission spectra was a laser at the wavelength of 915 nm. The strongest absorption peak was present at about 940 nm, corresponding to the extranuclear transition of electrons in Yb^{3+} from the lowest stark energy level of the ground state ($^2F_{7/2}$) to the middle stark energy level of the excited state ($^2F_{5/2}$). The strongest emission peak was at the wavelength of 1030 nm, representing the extranuclear transition of electrons in Yb^{3+} from the lowest stark energy level of the excited state ($^2F_{5/2}$) to the middle stark

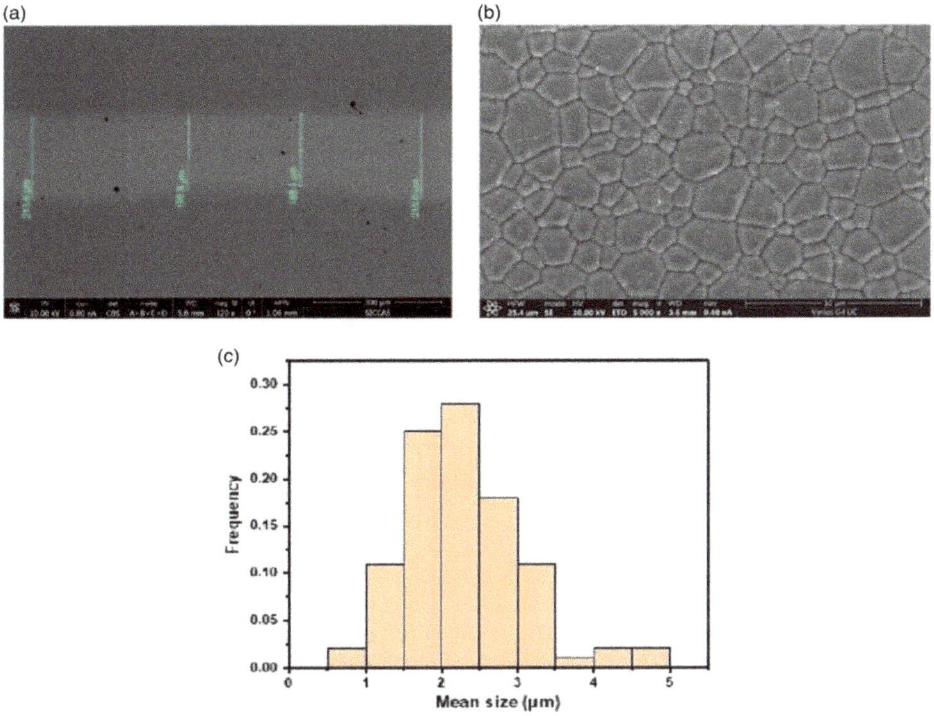

Figure 3.46. SEM images and grain size of the PWG: (a) polished cross-sectional SEM image, (b) polished and etched surface SEM image and (c) distribution profile of grain size calculated according to panel (b). Reproduced from [133]. CC BY 4.0.

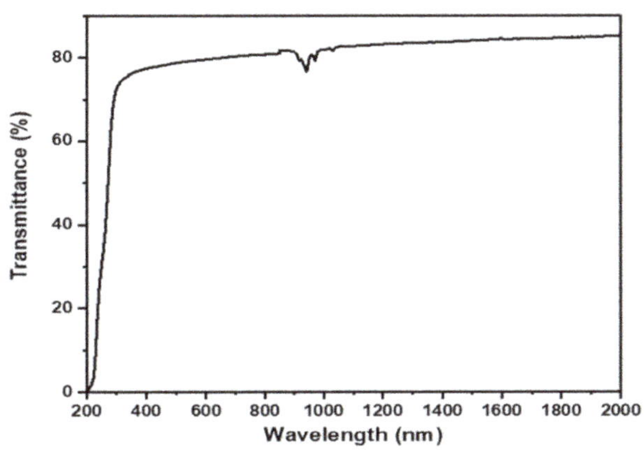

Figure 3.47. Transmittance curve of the PWG structure. Reproduced from [133]. CC BY 4.0.

energy level of the ground state ($^2F_{7/2}$). In addition, by comparing the absorption spectrum and emission spectrum, it was found that the absorption and emission peaks were at the wavelengths of 941, 969 and 1030 nm for the PWG, suggesting

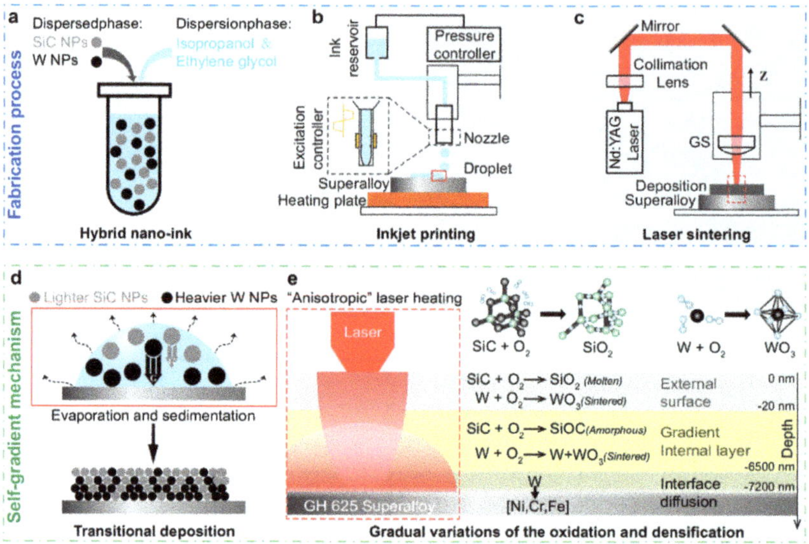

Figure 3.48. (a–c) Schematic diagram describing preparation process of the SiOC–WO_x gradient ceramic matrix composite (CMC) thin films using inkjet printing and laser sintering. (d and e) Formation mechanisms of the self-gradient microstructure. During the inkjet printing, the SiC and W nanoparticles were deposited in a transitional style in the depth direction, due to the difference in density between the two components. In the laser sintering process, gradient oxidation and densification of the CMC occurred along the depth direction. Reproduced from [141]. CC BY 4.0.

that the Yb:YAG PWG possessed self-absorption characteristics at these three wavelengths.

Ceramic matrix composites (CMCs) have a wide range of applications in different engineering fields [134–140]. However, CMCs that have simultaneously high toughness and stiffness are scarcely available in nature. As an example, SiOC composite films that were reinforced with W-based nanosized particles, with gradient structures and properties, were developed by using integrated hybrid nanoparticle inkjet printing, combined with selective laser sintering [141].

The resultant SiOC–WO_x films displayed strong enhancements in both stiffness and toughness, where the hardness and modulus were improved by two times, while the fracture toughness was enhanced by nearly four times, as compared with those of the matrix materials. Furthermore, the SiOC–WO_x films exhibited high interfacial coherence, with a maximum bonding strength of 86.6 MPa and a working temperature of 1050 °C. The enhancement in mechanical properties was ascribed to the presence of the metal–ceramic compositional gradient and the homogeneously distributed self-assembled reinforcement nanosized particles.

Figure 3.48 shows the experimental details. The SiC and W inks were prepared by dispersing their respective precursor in a mixed solvent of ethylene glycol (99.9% purity, Macklin, China) and isopropyl alcohol (99.9% purity, Macklin, China), with a volume ratio of 3:5. The SiC and W nanoparticles (Deke Daojin, China) had an

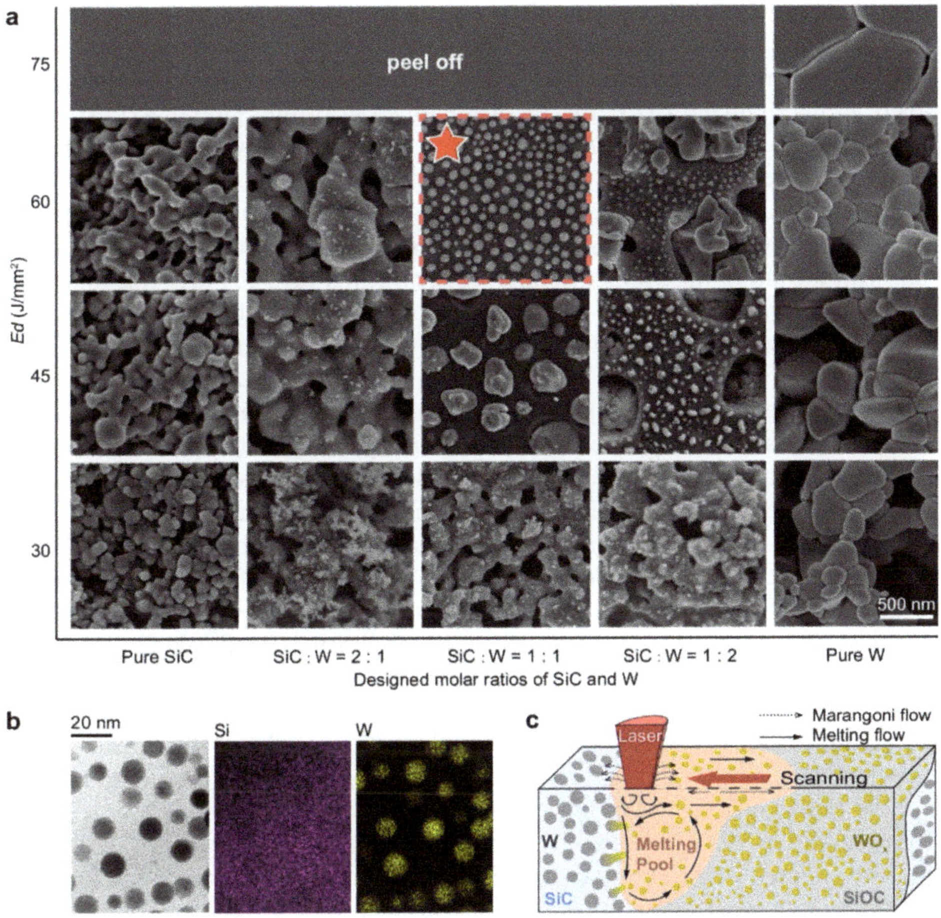

Figure 3.49. (a) Micro-morphological morphologies of the samples with different ratios of SiC:W after sintering at different laser energy densities (*Ed*). With increasing content of W nanoparticles, the porosity of the sintered body decreased, confirming the laser absorption enhanced sintering effect of W. (b) Microstructure characterized by the 'rock-embedded' dense pattern of the sample with SiC:W = 1:1, after sintering at laser energy densities of $Ed = 45–60\ \mathrm{J \cdot mm^{-2}}$. The enrichment in W in the spherical 'rock' was confirmed by the EDS result. (c) Schematic diagram showing the formation mechanism of the 'rock-embedded' microstructure through the synergistic effect of compositional gradient and laser sintering. Reproduced from [141]. CC BY 4.0.

average size of 50 nm. Meanwhile, SiC–W hybrid inks with different compositions were prepared in a similar way, as illustrated in figure 3.49(a).

Solid loading level of all the inks was controlled to be $250\ \mathrm{mg \cdot ml^{-1}}$. Polyvinylpyrrolidone, with a molecular weight of 58 000 (99.99% purity, Macklin, China), at a concentration of $2.5\ \mathrm{mg \cdot ml^{-1}}$, was used to stabilize the ink suspensions. The suspensions were treated with an ultrasonic crusher (Scientz-IID, Scientz, China) at 150 W for 300 min, to obtain homogeneous inks. They were kept at 10 °C in a water bath. Before being used for printing, the inks were filtered through a polytetrafluoroethylene filter membrane with a pore size of 0.45 μm, so that all large aggregated particles were removed.

The SiC–W inks were coated on a flat substrate of Inconel 625 superalloy as droplets with a diameter of 20 μm, by using an inkjet printer (Jetlab 4xl, MicroFab, America). Inconel 625 superalloy was used as the substrate, owing to its outstanding high-temperature mechanical strength, since the SiOC–WO_x thin films were treated at high temperatures. To ensure high quality inkjet printing, the superalloy substrates were washed with isopropyl alcohol, ethanol, and ultrapure water for 3 min, with the aid of ultrasonication. During printing, the substrate was maintained at a temperature of 160 °C, to ensure rapid evaporation of the solvent and development of uniform thin films. The films had a rectangular dimension of 3 × 30 mm^2.

A single-mode pulsed laser (P50QB, Raycus, China), with a wavelength of 1064 nm, was used to sinter the printed thin films, through the generation of localized high temperatures at the surface of the layers. The laser had a pulse width of 130 ns, at a repetition frequency of 80 kHz. A programmable Galvanometer Scanner (GS) system (RC1001-R, Jinhaichuang, China) was utilized to control the laser beam.

The mechanical and electrical properties of the SiOC–WO_x thin films were dependent on the composition of the inks and the degree of densification. The presence of SiC guaranteed the films to have high hardness and electrical resistance, while making it difficult for the samples to be sintered, by using the Nd:YAG laser, because of its relatively low fracture toughness and low absorption of the 1064 nm laser. To address this problem, W nanoparticles were introduced to increase the laser absorption. Optimal parameters included a laser energy density of 60 $J \cdot mm^{-2}$ and a SiC-to-W molar ratio of 1:1.

Representative SEM images of the thin films from the inks with different SiC/W ratios and sintered at different laser energy densities are demonstrated in figure 3.49 (a). At low laser energy density, e.g., $Ed = 30 \ J \cdot mm^{-2}$, the ceramic particles were diffused to form necks. In other words, the sintering process was at the initial sintering stage, so the sample had a relatively high porosity. Importantly, the addition of *W* nanoparticles promoted the densification process of the composite thin films, owing to the enhanced laser absorption [142].

When the energy density was increased to $Ed = 45 \ J \cdot mm^{-2}$, the necking effect of the SiC sample was more pronounced. Meanwhile, the presence of W nanoparticles promoted both densification and grain growth. For the sample with the composition of SiC:W = 1:1, nearly full densification was achieved, accompanied by the formation of a 'rock-embedded' microstructure. The 'rocks' acted as the reinforcement agent, thus leading to an enhancement in mechanical strength. Such microstructure was absent in the sample with a higher content of W nanoparticles. At the energy density of $Ed = 60 \ J \cdot mm^{-2}$, the size of the 'rocks' was reduced, which were present in smaller spherical particles and uniformly distributed in the matrix. In comparison, at too-high energy density increases, i.e., $Ed = 75 \ J \cdot mm^{-2}$, over-sintering occurred and the layers were peeled off from the substrates.

The sample with the 'rock-embedded' microstructure was sliced and characterized with TEM and EDS. The distribution of Si was uniform in the composite, while W was only present in the 'rock', as observed in figure 3.49(b). Therefore, it was concluded that the matrix was rich in Si, while W was inside the reinforcement component. The formation of the 'rock-embedded' microstructure could be explained schematically in figure 3.49(c). During sintering with laser irradiation, SiC was molten due to its low melting point, while W was in the solid state owing to the high melting temperature.

In addition, since the beam of the laser had a Gaussian profile, both a temperature gradient and a surface tension gradient could be generated from center to edge in the melting area [143–145]. In this case, an internal melt flow was formed, which facilitated rearrangement of the nanosized particles, leading to their uniform distribution in the matrix. Also, the increase in the laser intensity resulted in a further increase in the temperature gradient. As a consequence, the reinforcement agent was present in more uniform and homogeneous distribution.

Phase compositions corresponding to the elements were characterized by using XPS, SIMS and so on. The atom ratio of W:Si revealed by XPS results was 3:7, which was different from that of 1:1 in the ink, confirming the gradient deposition of SiC and W nanoparticles during the inkjet printing process. Element distribution profiles of Si and W in the depth direction were examined from the surface down to a depth of 60 nm. At surface, the Si–C bond was not observed, while Si–O was present, implying that SiC had been oxidized. The content of Si–O was decreased to 83.6% at the depth of $h = -20$ nm, after which the ratio of Si–C:Si–O remained nearly unchanged. Similarly, W was fully oxidized to WO_3 at the surface. However, the ratio of W^{6+}:W^{4+}:W continuously varied with increasing depth in the range of 0–60 nm. As the depth was from $h = -20$ to $h = -40$ nm, the content of W was increased from 9.8% to 18.6%, while that of W^{6+} was decreased from 67.2% to 53.9%.

According to the SIMS results, a turning point at $h = -20$ nm was present for all the five ions, suggesting that the oxidation occurred within the thin surface layer. The percentage of Si–O and the valence state of W were abruptly changed at the depth of $h = -20$ nm. The contents of Si^+ and SiO^+ leveled off, so that the percentages of Si–C and Si–O were not varied, as the depth was $\geqslant$20 nm. The matrix was amorphous SiOC with the composition of $SiO_{4-x}C_x$, due to the diffusion of external O atoms into the SiC lattice. According to the XPS results, the SiOC phase had a composition of $SiO_{(1.32-1.48)}C_{(0.26-0.34)}$ [146, 147]. Since the intensity of WO^{2+} or WO^+ decreased, the degree of W oxidation slowed down at certain depth.

Gradient characteristics of the SiOC–WO_x composite thin films, as described in aspects, Si/W atomic ratio, degree of oxidation, and grain size, are depicted in figure 3.40. As demonstrated in figure 3.40(a), the content of W increased, while that of Si decreased along the depth direction. This observation was ascribed to the difference in settling behaviors between the two components, owing to the difference in their densities. Variations in oxidation degrees of Si and W are shown in figure 3.40(b). The oxidation of Si abruptly stopped at the depth of about 20 nm, while that of W gradually decreased. The decreased degree of oxidation was related to the reduction in oxygen partial pressure inside the thin films.

Grain size of the WO_x reinforcement agent gradually decreased with increasing depth, as presented in figure 3.40(c), which was closely linked to the variations in sintering temperature and degree of oxidation. The thin films were sintered at higher temperatures at the bottom part, because the superalloy substrate had strong laser absorption [148]. Therefore, the grains of the corresponding phases were effectively refined [149]. The oxidation of the metals was realized by capturing oxygen atoms into the lattice, thus leading to an increase in grain size [150]. In other words, the lower degree of oxidation in the inner layers might be responsible for the decrease in the grain size of the components.

As revealed by the high-resolution TEM (HRTEM) images and the selected area electron diffraction (SAED) patterns, W particles were nanosized monocrystals. Meanwhile, WO_3 and W monomers coexisted in some circular areas. The gradient properties of the SiOC–WOx thin films could be described in three aspects, including Si/W atomic ratio, degree of oxidation and grain size. In the depth region of $\leqslant$20 nm, the oxidation degree of W decreased gradually, while that of Si decreased sharply. The gradual reduction in the grain size of WO_x reinforcement agent versus depth was due to the variation in the localized sintering temperature and hence the degree of oxidation.

XRD patterns of the substrates with and without the composite thin film are shown in figure 3.41(a). Different peaks of WO_3 were present, suggesting that W diffused into the superalloy during the cleaning and etching process and then was oxidized. Meanwhile, the main peaks of the superalloy substrate shifted to a lower angle with the deposition of the gradient thin film, as seen in the inset in figure 3.41 (a), indicating that W was doped into the Ni–Cr–Fe superalloy. In addition, the relative intensities of the main peaks of [Ni,Fe] and [Cr,Fe] were altered, implying that solid solution had probably formed between the two components during the laser sintering process.

Cross-sectional SEM images of the SiOC–WO_x composite thin films are depicted in figure 3.41(b). The coherent film-substrate interface was evidenced by cross-sectional SEM images. The thin films exhibited surface hardness of up to 20 GPa, which was close to that of amorphous silica and SiC ceramics, but much higher than that of Inconel 625 superalloy [151]. The high surface hardness resulted from the synergistic effect of WO_3 reinforcement and laser densification.

Electrical resistance of the SiOC–WO_x composite thin film was measured at temperatures in the range of 20 °C–1050 °C. Generally, with increasing temperature, the resistance of insulating materials decreases, because of the increased energy of conduction carriers and the activation of defects. In contrast, the resistance of the SiOC–WO_x composite thin films slightly increased with temperature till 400 °C. It could be understood that the laser sintered films had relatively low electrical activation energy, owing to the polariton conduction mechanism [152, 153]. At temperatures of $\geqslant$400 °C, the SiOC–WO_x films exhibited a reduction trend in resistance, with promising insulation efficiency and temperature stability (figures 3.50 and 3.51).

Silicon nitride (Si_3N_4) ceramics were prepared by using direct inkjet printing process, with aqueous suspensions containing organic additives [154]. Thin layers at micron-sized scale were printed to form 3D structures, with shapes of gearwheels and engineering parts, followed by pressureless sintering to obtain ceramic products.

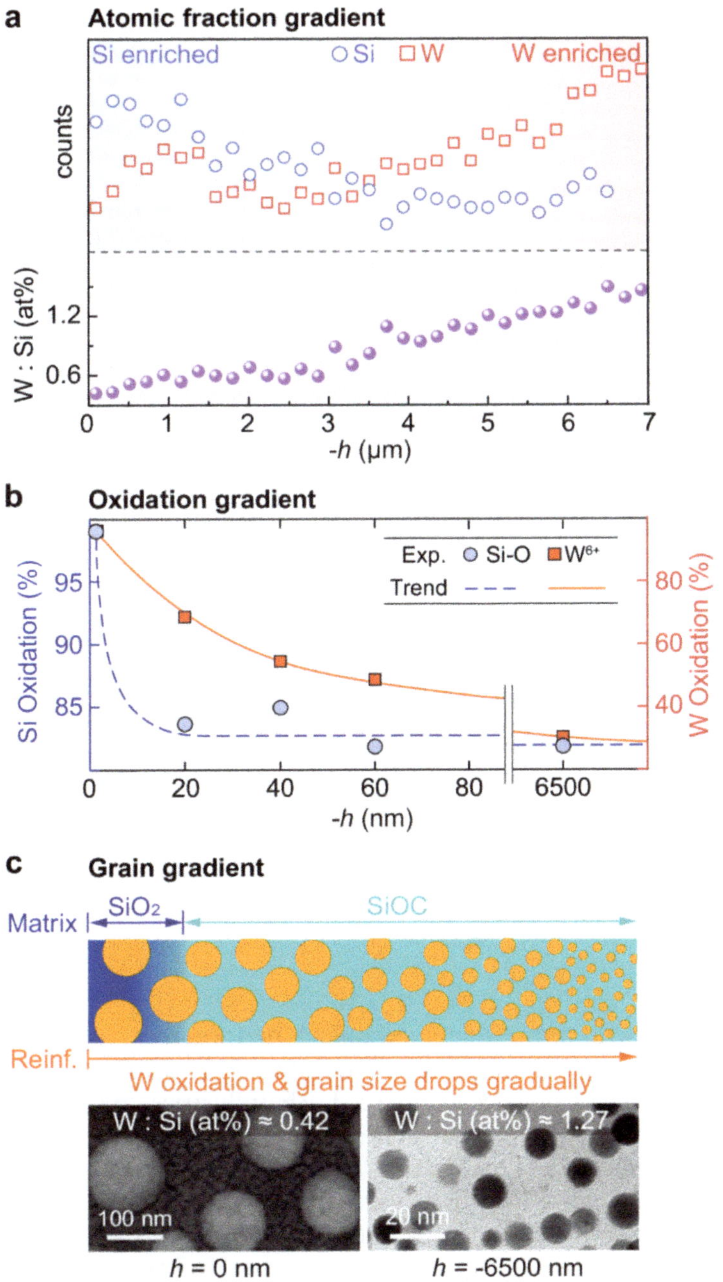

Figure 3.50. (a) Concentrations of Si and W versus depths, measured with FIB-EDS. A higher ratio of W:Si was present in the deeper region. (b) Degrees of oxidation of Si and W versus depth. (c) Morphology and grain size of WO_x as the reinforcement agent at different depths. Reproduced from [141]. CC BY 4.0.

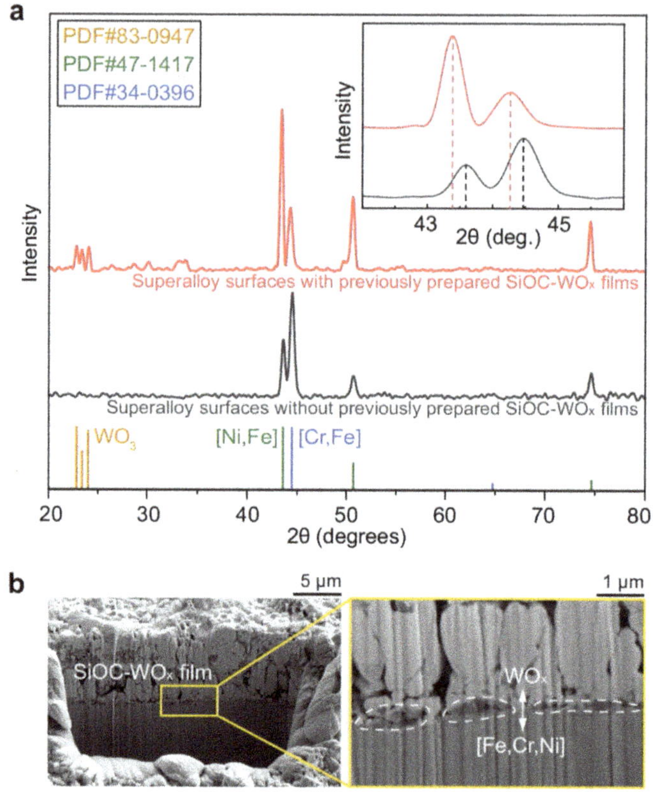

Figure 3.51. (a) XRD patterns of the samples with and without the SiOC–WO_x composite thin film, demonstrating that W in the film diffused into the Inconel 625 superalloy substrate, evidenced by the presence of the WO_3 peak and the shift to lower diffraction angle of the peaks of [Ni,Fe] and [Cr,Fe]. (b) Cross-sectional SEM images of the SiOC–WO_x composite thin films together with substrate, showing strongly coherent interfaces with permeable grayscale to confirm the diffusion of elements. Reproduced from [141]. CC BY 4.0.

Dense structures with promising mechanical performances were achieved. The ceramics showed no obvious delamination or any other defects.

Commercial α-Si_3N_4 powder (SN-E10, UBE Industries, Japan), with a specific surface area of 10 $m^2 \cdot g^{-1}$ and an average particle size of 0.5 μm, was utilized to prepare the ceramic inks. Powder of yttrium aluminium garnet powder (YAG, Sintertechnik, Pretzfeld, FRG) was used to promote densification of the Si_3N_4 ceramics. 30.2 vol% Si_3N_4 and 2.3 vol% YAG were dispersed in water, with dispersants consisting of polyacrylic and carboxylic acids. The suspensions were adjusted by adding ethanol, ethylene glycol (Merck KGaA, Darmstadt, Germany), binder (PEG 400, Merck KGaA, Darmstadt, Germany) and organic additive to prevent the occurrence of flocculation. A modified droplet-on-demand HP DeskJet 930c system was adopted as the printer. After printing, the green bodies were calcined at 500 °C for 2 h and the sintered at 1780 °C for 2 h in vacuum of 0.1 MPa with flowing N_2.

D_{50} value of the powder after attrition milling was reduced to 0.4 μm, which met the requirement that the agglomerated size should be <1 μm to prevent nozzle clogging. Figure 3.52 shows the surface SEM image of the green body after printing for 10 layers, which were printed in the direction from left to right. The arrayed droplet lines in the horizontal direction were clearly present in the SEM image. The printed layers had a uniform surface, without the observation of any process-related defects or flaws. The Si_3N_4 particles were packed with a high density.

Figure 3.53 depicts the SEM image of the sintered sample with a gearwheel geometry. The surface was flaw-free with relatively high accuracy. The outer contour of the cogs was somehow distorted, owing to data quality degradation during the transition from the 3D file to the pictures for printing. After sintering, the density of the ceramics reached 3.18 g · cm^{-3}. Meanwhile, α-Si_3N_4 was entirely transferred into β-Si_3N_4 phase. In addition, secondary phases, including silicon

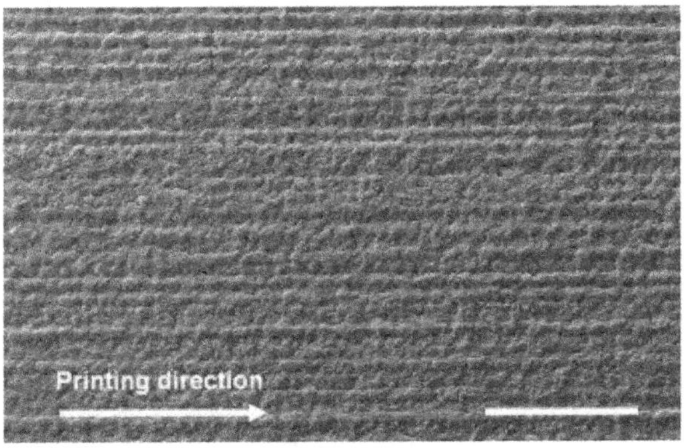

Figure 3.52. Surface SEM image of the as-printed green body with 10 printing layers, with scale bar to be 1 mm. Reprinted from [154], Copyright (2008), with permission from Elsevier.

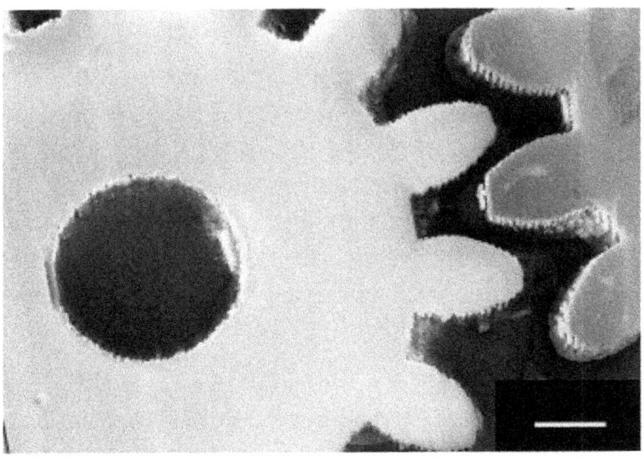

Figure 3.53. SEM image of the sintered Si_3N_4 ceramic structure with gearwheels geometry, with the scale bar to be 1 mm. Reprinted from [154], Copyright (2008), with permission from Elsevier.

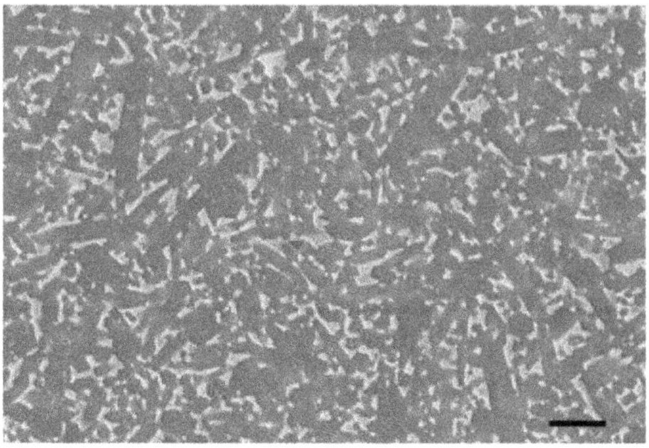

Figure 3.54. SEM image of the final ceramics after plasma etching, with scale bar to be 2 μm. Reprinted from [154], Copyright (2008), with permission from Elsevier.

oxynitride ($Si_3Al_7O_3N_9$), yttrium silicate ($Y_4Si_3O_{12}$), and a glassy phase, were detected in the XRD patterns. Also, a small quantity of silicon was observed, which was attributed to the decomposition of Si_3N_4 during the sintering process. The Si_3N_4 displayed a homogeneous microstructure, as illustrated in figure 3.54.

3.5 Concluding remarks

Inkjet printing has been demonstrated as a powerful manufacturing technology for the fabrication of structural and functional ceramics. In this process, ceramic inks can be deposited precisely on different substrates, with various special configurations designed by customers. Through controlling sizes, shapes and placement of the droplets, it is possible to print high-resolution patterns, structures and geometries that cannot be achieved with the conventional ceramic processing technique, thus opening up new avenues for the production of ceramic materials. For instance, decorative ceramic tiles with enhanced imagery, customized ceramic art products, and functionalized ceramic components with unique surface patterns, textures, and compositions, have been created.

On one hand, from a material perspective, the development of ceramic inks with desired rheological performance, long-time stability, and high solid loading levels cannot be overemphasized to guarantee the quality of the printed items. On the other hand, the invention and construction of sophisticated inkjet printers are equally important to ensure the efficiency and effectiveness of printing. In addition, the capabilities to print multi-materials should be explored, thus allowing for the creation of ceramic composites with currently unavailable functionalities.

References

[1] Lv X Y, Ye F, Cheng L F, Fan S W and Liu Y S 2019 Binder jetting of ceramics: powders, binders, printing parameters, equipment, and post-treatment *Ceram. Int.* **45** 12609–24

[2] Pan Z D, Wang Y M, Huang H N, Ling Z Y, Dai Y G and Ke S J 2015 Recent development on preparation of ceramic inks in ink-jet printing *Ceram. Int.* **41** 12515–28

[3] Arin M, Lommens P, Vandeput D, Van Acker J and Van Driessche I 2014 Durability and efficiency of ink-jet printed TiO_2 coatings: Influence of processing temperature *Thin Solid Films* **556** 160–7

[4] Zhou X Q, Li W, Wu M L, Tang S and Liu D Z 2014 Enhanced dispersibility and dispersion stability of dodecylamine-protected silver nanoparticles by dodecanethiol for ink-jet conductive inks *Appl. Surf. Sci.* **292** 537–43

[5] Von Hagen R, Sneha M and Mathur S 2014 Ink-jet printing of hollow SnO_2 nanospheres for gas sensing applications *J. Am. Ceram. Soc.* **97** 1035–40

[6] Guo Y, Patanwala H S, Bognet B and Ma A W K 2017 Inkjet and inkjet-based 3D printing: connecting fluid properties and printing performance *Rapid Prototyp. J.* **23** 562–76

[7] Calvert P 2001 Inkjet printing for materials and devices *Chem. Mater.* **13** 3299–305

[8] Chung S, Cho K and Lee T 2019 Recent progress in inkjet-printed thin-film transistors *Adv. Sci.* **6** 201801445

[9] Derby B 2015 Additive manufacture of ceramics components by inkjet printing *Engineering* **1** 113–23

[10] Joglekar-Athavale A and Shankarling G S 2022 Review: development of inkjet printing colorants in ceramics *Pigm. Resin Technol* **51** 273–89

[11] Lohse D 2022 Fundamental fluid dynamics challenges in inkjet printing *Annu. Rev. Fluid Mech.* **54** 349–82

[12] Wijshoff H 2018 Drop dynamics in the inkjet printing process *Curr. Opin. ColloidInterface Sci.* **36** 20–7

[13] Martin G D, Hoath S D and Hutchings I M 2006 Inkjet printing–the physics of manipulating liquid jets and drops *Conf. on Engineering in Physics—Synergy for Success* (London, England: IOP Publishing) p 012001

[14] Jang H W, Kim J, Kim H T, Yoon Y, Lee S N, Hwang H *et al* 2010 Fabrication of nonsintered alumina-resin hybrid films by inkjet-printing technology *Jpn. J. Appl. Phys.* **49** 071501

[15] Polsakiewicz D A and Kollenberg W 2011 Highly loaded alumina inks for use in a piezoelectric print head *Materialwiss. Werkstofftech.* **42** 812–9

[16] Tay B Y, Rashid H and Edirisinghe M J 2000 On the preparation of ceramic ink for continuous jet printing *J. Mater. Sci. Lett.* **19** 1151–4

[17] Tomov R I, Krauz M, Jewulski J, Hopkins S C, Kluczowski J R, Glowacka D M *et al* 2010 Direct ceramic inkjet printing of yttria-stabilized zirconia electrolyte layers for anode-supported solid oxide fuel cells *J. Power Sources* **195** 7160–7

[18] Song J H, Edirisinghe M J and Evans J R G 1999 Formulation and multilayer jet printing of ceramic inks *J. Am. Ceram. Soc.* **82** 3374–80

[19] Ebert J, Özkol E, Zeichner A, Uibel K, Weiss Ö, Koops U *et al* 2009 Direct inkjet printing of dental prostheses made of zirconia *J. Dent. Res.* **88** 673–6

[20] Kim S and McKean D E 1998 Aqueous TiO_2 suspension preparation and novel application of ink-jet printing technique for ceramics patterning *J. Mater. Sci. Lett.* **17** 141–4

[21] Blazdell P 2003 Solid free-forming of ceramics using a continuous jet printer *J. Mater. Process. Technol.* **137** 49–54

[22] Cappi B, Ebert J and Telle R 2011 Rheological properties of aqueous Si_3N_4 and $MoSi_2$ suspensions tailor-dade for direct inkjet printing *J. Am. Ceram. Soc.* **94** 218–23

[23] Windle J and Derby B 1999 Ink jet printing of PZT aqueous ceramic suspensions *J. Mater. Sci. Lett.* **18** 87–90

[24] Zhou Z J, Yang Z F and Yuan Q M 2008 Barium titanate ceramic inks for continuous ink-jet printing synthesized by mechanical mixing and sol–gel methods *Trans. Nonferrous Met. Soc. China* **18** 150–4
[25] Tseng W J, Lin S Y and Wang S 2006 Particulate dispersion and freeform fabrication of $BaTiO_3$ thick films via direct inkjet printing *J. Electroceram.* **16** 537–40
[26] Arcos D and Vallet-Regí M 2010 Sol–gel silica-based biomaterials and bone tissue regeneration *Acta Biomater.* **6** 2874–88
[27] Choi G, Choi A H, Evans L A, Akyol S and Ben-Nissan B 2020 A review: recent advances in sol–gel-derived hydroxyapatite nanocoatings for clinical applications *J. Am. Ceram. Soc.* **103** 5442–53
[28] Kajihara K, Kanamori K and Shimojima A 2022 Current status of sol–gel processing of glasses, ceramics, and organic–inorganic hybrids: a brief review *J. Ceram. Soc. Jpn.* **130** 575–83
[29] Li L, Liu X L, Wang G, Liu Y L, Kang W M, Deng N P *et al* 2021 Research progress of ultrafine alumina fiber prepared by sol–gel method: a review *Chem. Eng. J.* **421** 127744
[30] Schileo G 2013 Recent developments in ceramic multiferroic composites based on core/shell and other heterostructures obtained by sol–gel routes *Prog. Solid State Chem.* **41** 87–98
[31] Secu M, Secu C and Bartha C 2021 Optical properties of transparent rare-earth doped sol–gel derived nano-glass ceramics *Materials* **14** 6871
[32] Song X Z, Segura-Egea J J and Díaz-Cuenca A 2023 Sol–gel technologies to obtain advanced bioceramics for dental therapeutics *Molecules* **28** 6967
[33] Vinogradov A V and Vinogradov V V 2014 Low-temperature sol–gel synthesis of crystalline materials *RSC Adv.* **4** 45903–19
[34] Walcarius A, Mandler D, Cox J A, Collinson M and Lev O 2005 Exciting new directions in the intersection of functionalized sol–gel materials with electrochemistry *J. Mater. Chem.* **15** 3663–89
[35] Zhang J, Zhang J X, Sun Q L, Ye X L, Ma X M and Wang J 2022 Sol–gel routes toward ceramic nanofibers for high-performance thermal management *Chemistry* **4** 1475–97
[36] Xiao Z H and Luo W Y 2011 Nano-size beta-spodumene glass-ceramic powders synthesized by sol–gel route *Appl. Mech. Mater* **110–6** 3844–8
[37] Krasovec U O, Vidmar T, Gunde M K, Korosec R C and Perse L S 2022 In-depth rheological characterization of tungsten sol–gel inks for inkjet printing *Coatings* **12** 112
[38] Yang J M, Wang H Q, Zhou B, Shen J, Zhang Z H and Du A 2021 Versatile direct writing of aerogel-based sol–gel inks *Langmuir* **37** 2129–39
[39] Gurauskis J, Gil V, Lin B and Einarsrud M A 2022 Pilot scale fabrication of lanthanum tungstate supports for H_2 separation membranes *Open Ceram.* **9** 100226
[40] Nabavi M S, Mohammadi T and Kazemimoghadam M 2014 Hydrothermal synthesis of hydroxy sodalite zeolite membrane: separation of H_2/CH_4 *Ceram. Int.* **40** 5889–96
[41] Van Gestel T, Hauler F, Bram M, Meulenberg W A and Buchkremer H P 2014 Synthesis and characterization of hydrogen-selective sol–gel SiO_2 membranes supported on ceramic and stainless steel supports *Sep. Purif. Technol.* **121** 20–9
[42] Ghaffari M, Barzegar A, Janghorban K and Javidi M 2011 Investigation of effective parameters in improving corrosion resistance of silica coated stainless steel via sol–gel method *Corros. Eng. Sci. Technol* **46** 605–10
[43] Karata S, Kizilkaya C, Kayaman-Apohan N and Güngör A 2007 Preparation and characterization of sol–gel derived UV-curable organo-silica-titania hybrid coatings *Prog. Org. Coat.* **60** 140–7

[44] Khelifa F, Druart M E, Habibi Y, Bénard F, Leclère P, Olivier M *et al* 2013 Sol–gel incorporation of silica nanofillers for tuning the anti-corrosion protection of acrylate-based coatings *Prog. Org. Coat.* **76** 900–11

[45] Shange M G, Khumalo N L, Mohomane S M and Motaung T E 2024 Factors affecting silica/cellulose nanocomposite prepared via the sol–gel technique: a review *Materials* **17** 1937

[46] Xiao Z H, Luo W Y and Wang S L 2010 Thermal expansion property of P_2O_5 doped lithium aluminosilicate glass-ceramic synthesized by sol–gel process *Mater. Sci. Forum* **663–5** 1281–4

[47] Chouiki M and Schoeftner R 2011 Inkjet printing of inorganic sol–gel ink and control of the geometrical characteristics *J. Sol–Gel Sci. Technol.* **58** 91–5

[48] Fasaki I, Siamos K, Arin M, Lommens P, Van Driessche I, Hopkins S C *et al* 2012 Ultrasound assisted preparation of stable water-based nanocrystalline TiO_2 suspensions for photocatalytic applications of inkjet-printed films *Appl. Catal.* A **411** 60–9

[49] Kignelman G and Thielemans W 2021 Synergistic effects of acetic acid and nitric acid in water-based sol–gel synthesis of crystalline TiO_2 nanoparticles at 25 °C *J. Mater. Sci.* **56** 16877–86

[50] Wan X J, Tan G F, Cai L and Zhang Y C 2023 Preparation and characterization of Li_2TiO_3 ceramic pebbles by modification of the water-based sol–gel method *Int. J. Appl. Ceram. Technol.* **20** 2760–71

[51] Kumar R S, Erkulla S, Khanra A K and Johnson R 2020 Aqueous sol–gel processing of precursors and synthesis of aluminum oxynitride powder therefrom *J. Sol–Gel Sci. Technol.* **93** 100–10

[52] Mahy J G, Lejeune L, Haynes T, Lambert S D, Marcilli R H M, Fustin C A *et al* 2021 Eco-friendly colloidal aqueous sol–gel process for TiO_2 synthesis: the peptization method to obtain crystalline and photoactive materials at low temperature *Catalysts* **11** 768

[53] Malengreaux C M, Pirard S L, Léonard G, Mahy J G, Herlitschke M, Klobes B *et al* 2017 Study of the photocatalytic activity of Fe^{3+}, Cr^{3+}, La^{3+} and Eu^{3+} single-doped and co-doped TiO_2 catalysts produced by aqueous sol–gel processing *J. Alloys Compd.* **691** 726–38

[54] Atkinson A, Doorbar J, Hudd A, Segal D L and White P J 1997 Continuous ink-jet printing using sol–gel 'ceramic' inks *J. Sol–Gel Sci. Technol.* **8** 1093–7

[55] Mougenot M, Lejeune M, Baumard J F, Boissiere C, Ribot F, Grosso D *et al* 2006 Ink jet printing of microdot arrays of mesostructured silica *J. Am. Ceram. Soc.* **89** 1876–82

[56] Lim J, Jung H, Baek C, Hwang G T, Ryu J, Yoon D *et al* 2017 All-inkjet-printed flexible piezoelectric generator made of solvent evaporation assisted $BaTiO_3$ hybrid material *Nano Energy* **41** 337–43

[57] Thuy T T, Hoste S, Herman G G, Van de Velde N, De Buysser K and Van Driessche I 2009 Novel water based cerium acetate precursor solution for the deposition of epitaxial cerium oxide films as HTSC buffers *J. Sol–Gel Sci. Technol.* **51** 112–8

[58] Gallage R, Matsuo A, Fujiwara T, Watanabe T, Matsushita N and Yoshimura M 2008 On-site fabrication of crystalline cerium oxide films and patterns by ink-jet deposition method at moderate temperatures *J. Am. Ceram. Soc.* **91** 2083

[59] Acosta M, Novak N, Rojas V, Patel S, Vaish R, Koruza J *et al* 2017 $BaTiO_3$-based piezoelectrics: fundamentals, current status, and perspectives *Appl. Phys. Rev.* **4** 041305

[60] Gao J H, Xue D Z, Liu W F, Zhou C and Ren X B 2017 Recent progress on $BaTiO_3$-based piezoelectric ceramics for actuator applications *Actuators* **6** 24

[61] Jain A, Wang Y G and Shi L N 2022 Recent developments in $BaTiO_3$ based lead-free materials for energy storage applications *J. Alloys Compd.* **928** 167066
[62] Zhang Q C, Jia Y M, Wu W W, Pei C J, Zhu G Q, Wu Z S *et al* 2023 Review on strategies toward efficient piezocatalysis of $BaTiO_3$ nanomaterials for wastewater treatment through harvesting vibration energy *Nano Energy* **113** 108507
[63] Ardo S and Meyer G J 2009 Photodriven heterogeneous charge transfer with transition-metal compounds anchored to TiO_2 semiconductor surfaces *Chem. Soc. Rev.* **38** 115–64
[64] Liu H F, Dao A Q and Fu C Y 2016 Activities of combined TiO_2 semiconductor nanocatalysts under solar light on the reduction of CO_2 *J. Nanosci. Nanotechnol.* **16** 3437–46
[65] Navidpour A H, Abbasi S, Li D H, Mojiri A and Zhou J L 2023 Investigation of advanced oxidation process in the presence of TiO_2 semiconductor as photocatalyst: property, principle, kinetic analysis, and photocatalytic activity *Catalysts* **13** 232
[66] Ohya T, Nakayama A, Shibata Y, Ban T, Ohya Y and Takahashi Y 2003 Preparation and characterization of titania thin films from aqueous solutions *J. Sol–Gel Sci. Technol.* **26** 799–802
[67] Arin M, Lommens P, Avci N, Hopkins S C, De Buysser K, Arabatzis I M *et al* 2011 Inkjet printing of photocatalytically active TiO_2 thin films from water based precursor solutions *J. Eur. Ceram. Soc.* **31** 1067–74
[68] Klier J, Tucker C J, Kalantar T H and Green D P 2000 Properties and applications of microemulsions *Adv. Mater.* **12** 1751–7
[69] Komura S 2007 Mesoscale structures in microemulsions *J. Phys.: Condens. Matter* **19** 463101
[70] López-Quintela M A, Tojo C, Blanco M C, Rio L G and Leis J 2004 Microemulsion dynamics and reactions in microemulsions *Curr. Opin. Colloid Interface Sci.* **9** 264–78
[71] Wolf S and Feldmann C 2016 Microemulsions: options to expand the synthesis of inorganic nanoparticles *Angew. Chem. Int. Ed* **55** 15728–52
[72] Debuigne F, Jeunieau L, Wiame M and Nagy J B 2000 Synthesis of organic nanoparticles in different W/O microemulsions *Langmuir* **16** 7605–11
[73] Magdassi S and Ben Moshe M 2003 Patterning of organic nanoparticles by ink-jet printing of microemulsions *Langmuir* **19** 939–42
[74] Guo R, Qi H, Guo D, Chen X, Yang Z and Chen Y 2003 Preparation of high concentration ceramic inks for forming by jet-printing *J. Eur. Ceram. Soc.* **23** 115–22
[75] Guo R, Qi H, Chen Y and Yang Z 2003 Reverse microemulsion region and composition optimization of the AEO_9/alcohol/alkane/water system *Mater. Res. Bull.* **38** 1501–7
[76] Magdassi S, Ben Moshe M, Talmon Y and Danino D 2003 Microemulsions based on anionic gemini surfactant *Colloids Surf. A: Physicochem. Eng. Asp* **212** 1–7
[77] Jovaní M, Domingo M, Machado T R, Longo E, Beltrán-Mir H and Cordoncillo E 2015 Pigments based on Cr and Sb doped TiO_2 prepared by microemulsion-mediated solvothermal synthesis for inkjet printing on ceramics *Dyes Pigm.* **116** 106–13
[78] Fiévet F, Ammar-Merah S, Brayner R, Chau F, Giraud M, Mammeri F *et al* 2018 The polyol process: a unique method for easy access to metal nanoparticles with tailored sizes, shapes and compositions *Chem. Soc. Rev.* **47** 5187–233
[79] Guan P Y, Xu Z M, Lin X, Chen N, Tong H, Ha T J *et al* 2018 Recent progress in silver nanowires: synthesis and applications *Nanosci. Nanotechnol. Lett.* **10** 155–66

[80] Kimberly T Q, Frasch M H and Kauzlarich S M 2024 Colloidal synthesis of two-dimensional nanocrystals by the polyol route *Dalton Trans.* **53** 13280–97
[81] Varanda L C, Souza C G S, Moraes D, Neves H R, Souza J B Jr, Silva M F *et al* 2019 Size and shape-controlled nanomaterials based on modified polyol and thermal decomposition approaches. A brief review *An. Acad. Bras. Cienc.* **91** 04
[82] Willard M A, Kurihara L K, Carpenter E E, Calvin S and Harris V G 2004 Chemically prepared magnetic nanoparticles *Int. Mater. Rev.* **49** 125–70
[83] Feldmann C 2003 Polyol-mediated synthesis of nanoscale functional materials *Adv. Funct. Mater.* **13** 101–7
[84] Feldmann C and Jungk H O 2001 Polyol-mediated preparation of nanoscale oxide particles *Angew. Chem. Int. Ed* **40** 359–62
[85] Feldmann C 2001 Preparation of nanoscale pigment particles *Adv. Mater.* **13** 1301–3
[86] Merikhi J, Jungk H O and Feldmann C 2000 Sub-micrometer $CoAl_2O_4$ pigment particles: synthesis and preparation of coatings *J. Mater. Chem.* **10** 1311–4
[87] Darr J A, Zhang J Y, Makwana N M and Weng X L 2017 Continuous hydrothermal synthesis of inorganic nanoparticles: applications and future directions *Chem. Rev.* **117** 11125–238
[88] Hayashi H and Hakuta Y 2010 Hydrothermal synthesis of metal oxide nanoparticles in supercritical water *Materials* **3** 3794–817
[89] Ruilin L , Yee K Y, Salleh N A, Deghfel B, Zakaria Z, Yaakob M K *et al* 2024 Hydrothermal synthesis of manganese doped zinc oxide wurtzite nanoparticles for supercapacitors: a brief review *Inorg. Chem. Commun.* **170** 113311
[90] Rajamathi M and Seshadri R 2002 Oxide and chalcogenide nanoparticles from hydrothermal/solvothermal reactions *Curr. Opin. Solid State Mater. Sci.* **6** 337–45
[91] Katsuki H and Komarneni S 2001 Microwave-hydrothermal synthesis of monodispersed nanophase α-Fe_3O_3 *J. Am. Ceram. Soc.* **84** 2313–7
[92] Arin M, Lommens P, Hopkins S C, Pollefeyt G, Van der Eycken J, Ricart S *et al* 2012 Deposition of photocatalytically active TiO_2 films by inkjet printing of TiO_2 nanoparticle suspensions obtained from microwave-assisted hydrothermal synthesis *Nanotechnology* **23** 165603
[93] Arin M, Watté J, Pollefeyt G, De Buysser K, Van Driessche I and Lommens P 2013 Low temperature deposition of TiO_2 layers from nanoparticle containing suspensions synthesized by microwave hydrothermal treatment *J. Sol–Gel Sci. Technol.* **66** 100–11
[94] Kim J H, Son B R, Yoon D H, Hwang K T, Noh H G, Cho W S *et al* 2012 Characterization of blue $CoAl_2O_4$ nano-pigment synthesized by ultrasonic hydrothermal method *Ceram. Int.* **38** 5707–12
[95] Wang W, Zhang Q, Liu Z X and Libor Z 2010 Highly efficient size reduction of nanoparticles by the shock wave method *Funct. Mater. Lett.* **3** 299–302
[96] Peymannia M, Soleimani-Gorgani A, Ghahari M and Najafi F 2014 Production of a stable and homogeneous colloid dispersion of nano $CoAl_2O_4$ pigment for ceramic ink-jet ink *J. Eur. Ceram. Soc.* **34** 3119–26
[97] Wagata H, Taniguchi T, Gallage R, Subramani A K, Sakamoto N, Watanabe T *et al* 2010 Fabrication of $BaTiO_3$ thin films and microdot patterns by halide-free nonaqueous solution route *J. Am. Ceram. Soc.* **93** 381–6
[98] Blosi M, Dondi M, Albonetti S, Baldi G, Barzanti A and Zanelli C 2009 Microwave-assisted synthesis of Pr-$ZrSiO_4$, V-$ZrSiO_4$ and Cr-$YAlO_3$ ceramic pigments *J. Eur. Ceram. Soc.* **29** 2951–7

[99] Oh Y, Kim J, Yoon Y J, Kim H, Yoon H G, Lee S N *et al* 2011 Inkjet printing of Al_2O_3 dots, lines, and films: from uniform dots to uniform films *Curr. Appl Phys.* **11** S359–S63

[100] Soltman D and Subramanian V 2008 Inkjet-printed line morphologies and temperature control of the coffee ring effect *Langmuir* **24** 2224–31

[101] Tekin E, Smith P J, Hoeppener S, van den Berg A M J, Susha A S, Rogach A L *et al* 2007 Inkjet printing of luminescent CdTe nanocrystal-polymer composites *Adv. Funct. Mater.* **17** 23–8

[102] Philips N R, Compton B G and Begley M 2012 High strength alumina micro-beams fabricated by inkjet printing *J. Am. Ceram. Soc.* **95** 3016–8

[103] Mogalicherla A K, Lee S, Pfeifer P and Dittmeyer R 2014 Drop-on-demand inkjet printing of alumina nanoparticles in rectangular microchannels *Microfluid. Nanofluid.* **16** 655–66

[104] Hoath S D, Hutchings I M, Martin G D, Tuladhar T R, Mackley M R and Vadillo D 2009 Links between ink rheology, drop-on-demand jet formation, and printability *J. Imaging Sci. Technol.* **53** 041208

[105] Tekin E, Smith P J and Schubert U S 2008 Inkjet printing as a deposition and patterning tool for polymers and inorganic particles *Soft Matter* **4** 703–13

[106] Kwon K S 2010 Experimental analysis of waveform effects on satellite and ligament behavior via *in situ* measurement of the drop-on-demand drop formation curve and the instantaneous jetting speed curve *J. Micromech. Microeng.* **20** 115005

[107] Lu G, Tan H Y, Neild A, Liew O W, Yu Y and Ng T W 2010 Liquid filling in standard circular well microplates *J. Appl. Phys.* **108** 124701

[108] Chen Z W and Brandon N 2016 Inkjet printing and nanoindentation of porous alumina multilayers *Ceram. Int.* **42** 8316–24

[109] Fisher J G, Choi S Y and Kang S J L 2006 Abnormal grain growth in barium titanate doped with alumina *J. Am. Ceram. Soc.* **89** 2206–12

[110] Hong S H and Kim D Y 2001 Effect of liquid content on the abnormal grain growth of alumina *J. Am. Ceram. Soc.* **84** 1597–600

[111] Kim Y M, Hong S H and Kim D Y 2000 Anisotropic abnormal grain growth in TiO_2/SiO_2-doped alumina *J. Am. Ceram. Soc.* **83** 2809–12

[112] Peng Z J, Luo X D, Xie Z P, An D and Yang M M 2018 Effect of print path process on sintering behavior and thermal shock resistance of Al_2O_3 ceramics fabricated by 3D inkjet-printing *Ceram. Int.* **44** 16766–72

[113] Chang D M and Wang B L 2011 Thermal shock resistance of brittle ceramic materials with embedded elliptical cracks *Philos. Mag. Lett.* **91** 648–55

[114] Collin M and Rowcliffe D 2000 Analysis and prediction of thermal shock in brittle materials *Acta Mater.* **48** 1655–65

[115] Esposito V, Gadea C, Hjelm J, Marani D, Hu Q, Agersted K *et al* 2015 Fabrication of thin yttria-stabilized-zirconia dense electrolyte layers by inkjet printing for high performing solid oxide fuel cells *J. Power Sources* **273** 89–95

[116] Hauch A and Mogensen M 2010 Ni/YSZ electrode degradation studied by impedance spectroscopy: effects of gas cleaning and current density *Solid State Ion* **181** 745–53

[117] Derby B and Reis N 2003 Inkjet printing of highly loaded particulate suspensions *MRS Bull.* **28** 815–8

[118] Derby B 2010 Inkjet printing of functional and structural materials: fluid property requirements, feature stability, and resolution *Annu. Rev. Mater. Res.* **40** 395–414

[119] Rosa M, Gooden P N, Butterworth S, Zielke P, Kiebach R, Xu Y *et al* 2019 Zirconia nano-colloids transfer from continuous hydrothermal synthesis to inkjet printing *J. Eur. Ceram. Soc.* **39** 2–8

[120] Shi Y L and Wang W Q 2020 3D inkjet printing of the zirconia ceramic implanted teeth *Mater. Lett.* **261** 127131

[121] Venkatesh S, Rahul S H and Balasubramanian K 2020 Inkjet printing yttria stabilized zirconia coatings on porous and nonporous substrates *Ceram. Int.* **46** 3994–9

[122] Kuscer D, Stavber G, Trefalt G and Kosec M 2012 Formulation of an aqueous titania suspension and its patterning with ink-jet printing technology *J. Am. Ceram. Soc.* **95** 487–93

[123] Deegan R D, Bakajin O, Dupont T F, Huber G, Nagel S R and Witten T A 1997 Capillary flowasthe cause of ring stains from dried liquid drops *Nature* **389** 827–929

[124] Yang M, Li L H, Zhang S Q, Li G Y and Zhao H J 2010 Preparation, characterisation and sensing application of inkjet-printed nanostructured TiO_2 photoanode *Sensors Actuators* B **147** 622–8

[125] Ichangi A, Bergamini A and Clemens F 2024 Chemical modification of lead zirconate titanate piezoceramics through cold-sintering process *J. Alloys Compd.* **988** 174282

[126] Liu W B, Zhang F P, Zheng T, Li H J, Ding Y, Lv X *et al* 2024 Ultra-broad temperature insensitive Pb(Zr, Ti)O_3-based ceramics with large piezoelectricity *J. Mater. Sci. Technol* **192** 19–27

[127] Niu L, Han X T, Wang H T, Zhou Y, Zhang X R and Li J H 2024 Flash sintering of lead zirconate titanate ceramics under high voltage at room temperature *J. Eur. Ceram. Soc.* **44** 6169–77

[128] Fraile I, Gabilondo M, Burgos N, Azcona M and Castro F 2018 Laser sintered ceramic coatings of PZT nanoparticles deposited by inkjet printing on metallic and ceramic substrates *Ceram. Int.* **44** 15603–10

[129] Golkhatmi S Z, Lund P D and Asghar M I 2024 A novel $CuFe_2O_4$ ink for the fabrication of low-temperature ceramic fuel cell cathodes through inkjet printing *Mater. Adv* **5** 143–58

[130] Ahn M, Lee J and Lee W 2017 Nanofiber-based composite cathodes for intermediate temperature solid oxide fuel cells *J. Power Sources* **353** 176–82

[131] Fan L Q, Xiong Y P, Liu L B, Wang Y W and Brito M E 2013 Preparation and performance study of one-dimensional nanofiber-based $Sm_{0.5}Sr_{0.5}CoO_{3-\delta}$–$Gd_{0.2}Ce_{0.8}O_{1.9}$ composite cathodes for intermediate temperature solid oxide fuel cells *Int. J. Electrochem. Sci.* **8** 8603–13

[132] Kim S J, Woo D, Kim D, Lee T K, Lee J and Lee W 2023 Interface engineering of an electrospun nanofiber-based composite cathode for intermediate-temperature solid oxide fuel cells *Int. J. Extreme Manuf* **5** 015506

[133] Wang H R, Gao W L, Zhang J, Ma J, Ji H H, Xie M M *et al* 2024 Inkjet printing of Yb: YAG transparent ceramic planar waveguide laser gain medium *Adv. Photonics Res* **5** 202300320

[134] Chen Q, Bai S X and Ye Y C 2023 Highly thermal conductive silicon carbide ceramics matrix composites for thermal management: a review *J. Inorg. Mater* **38** 634–46

[135] Diaz O G, Luna G G, Liao Z R and Axinte D 2019 The new challenges of machining ceramic matrix composites (CMCs): review of surface integrity *Int. J. Mach. Tools Manuf* **139** 24–36

[136] Karadimas G and Salonitis K 2023 Ceramic matrix composites for aero engine applications: a review *Appl. Sci.* **13** 3017

[137] Sommers A, Wang Q, Han X, T'Joen C, Park Y and Jacobi A 2010 Ceramics and ceramic matrix composites for heat exchangers in advanced thermal systems: a review *Appl. Therm. Eng.* **30** 1277–91

[138] Song C K, Ye F, Cheng L F, Liu Y S and Zhang Q 2022 Long-term ceramic matrix composite for aeroengine *J. Adv. Ceram* **11** 1343–74
[139] Sun J X, Ye D R, Zou J, Chen X T, Wang Y, Yuan J S *et al* 2023 A review on additive manufacturing of ceramic matrix composites *J. Mater. Sci. Technol* **138** 1–16
[140] Yang J S, Chen J L, Ye F, Cheng L F and Zhang Y 2022 High-temperature atomically laminated materials: the toughening components of ceramic matrix composites *Ceram. Int.* **48** 32628–48
[141] Chen X Y, Qiu L, Zhang M S, Huang J and Tao Z 2024 Nanoparticle-reinforced SiOC ceramic matrix composite films with structure gradient fabricated by inkjet printing and laser sintering *Commun. Mater* **5** 96
[142] Zhao L, Macías J G S, Douillard T, Li Z H and Simar A 2021 Unveiling damage sites and fracture path in laser powder bed fusion $AlSi_{10}Mg$: comparison between horizontal and vertical loading directions *Mater. Sci. Eng.* A **807** 140845
[143] Zhang J L, Gao J B, Song B, Zhang L J, Han C J, Cai C *et al* 2021 A novel crack-free Ti-modified Al–Cu–Mg alloy designed for selective laser melting *Addit. Manuf* **38** 101829
[144] Wang H Z, Chen P, Shu Z X, Chen A N, Su J, Wu H Z *et al* 2023 Laser powder bed fusion of poly-ether-ether-ketone/bioactive glass composites: processability, mechanical properties, and bioactivity *Compos. Sci. Technol.* **231** 109805
[145] Wang H Z, Chen P, Su J, Chen Z Y, Yang L, Yan C Z *et al* 2023 Isothermal crystallization behavior of poly-ether-ether-ketone/bioactive glass composites and its correlation with scaffold warpage in laser powder bed fusion process *Addit. Manuf* **78** 103852
[146] Peña-Alonso R, Mariotto G, Gervais C, Babonneau F and Soraru G D 2007 New insights on the high-temperature nanostructure evolution of SiOC and B-doped SiBOC polymer-derived glasses *Chem. Mater.* **19** 5694–702
[147] Luo C J, Tang Y S, Jiao T and Kong J 2018 High-temperature stable and metal-free electromagnetic wave-absorbing SiBCN ceramics derived from carbon-rich hyperbranched polyborosilazanes *ACS Appl. Mater. Interfaces* **10** 28051–61
[148] Chen X Y, Zhang M S, Zhu J Q, Tao Z and Qiu L 2023 Laser sintering of Cu nanoparticles deposited on ceramic substrates: experiments and modeling *Addit. Manuf* **69** 103527
[149] Gu D D and Shen Y F 2008 Direct laser sintered WC–10Co/Cu nanocomposites *Appl. Surf. Sci.* **254** 3971–8
[150] Wozniak J, Jastrzebska A, Cygan T and Olszyna A 2017 Surface modification of graphene oxide nanoplatelets and its influence on mechanical properties of alumina matrix composites *J. Eur. Ceram. Soc.* **37** 1587–92
[151] Gao Y and Zhou M Z 2018 Superior mechanical behavior and fretting wear resistance of 3D-printed inconel 625 superalloy *Appl. Sci.* **8** 8122439
[152] Zhang Y, Dalpian G M, Fluegel B, Wei S H, Mascarenhas A, Huang X Y *et al* 2006 Novel approach to tuning the physical properties of organic-inorganic hybrid semiconductors *Phys. Rev. Lett.* **96** 026405
[153] Liu Z C, Liang J S, Su S J, Zhang C Y, Li J, Yang M J *et al* 2021 Preparation of defect-free alumina insulation film using layer-by-layer electrohydrodynamic jet deposition for high temperature applications *Ceram. Int.* **47** 14498–505
[154] Cappi B, Özkol E, Ebert J and Telle R 2008 Direct inkjet printing of Si_3N_4: characterization of ink, green bodies and microstructure *J. Eur. Ceram. Soc.* **28** 2625–8

IOP Publishing

Additive Manufacturing of Ceramics

Ling Bing Kong, Zhuohao Xiao, Bin He and Yin Liu

Chapter 4

Selective laser sintering/melting (SLS/SLM)

Selective laser sintering (SLS) is also known as selective laser melting (SLM), which is within the scope of the powder bed fusion manufacturing technologies. In this case, the materials are directly scanned using laser beams, where the laser irradiated areas are fused, solidified or sintered simultaneously. Ceramics with different types of materials have been fabricated by using this technology, which will be presented and described in this chapter.

4.1 Brief overview

Selective laser sintering (SLS), also known as selective laser melting (SLM), belongs to be one of the powder bed fusion manufacturing technologies, in which a beam of lasers is used as the energy suppliers to sinter or fuse the particles of given powders for different types of materials, including metals, metallic alloys, polymers, ceramics and composites [1–9]. In terms of laser powder bed fusion processes, there are mainly four binding mechanisms, including (i) solid-state sintering, (ii) liquid-phase sintering (structural materials with distinct binders), (iii) partial melting (structural materials without distinct binders) and (iv) full melting [10].

During solid-state sintering process, there is no melting or phase change, while necks between adjacent particles are formed, at sufficiently high temperatures for sufficiently long time durations [11–13]. Rapid melting/cooling for ceramics made with SLM encountered issues, especially regarding the elimination of pores. In comparison, ceramics sintered with solid state sintering process usually exhibit homogeneous microstructure and reasonable grain growth, owing to the slow diffusion of atoms.

Liquid-phase sintering means that liquid phases are formed during the sintering process at certain high temperatures, due to the relatively low melting points of some components [14–16]. The presence of the liquid phases offers smooth paths for mass transportation and diffusion through capillary effect or dissolution-precipitation

doi:10.1088/978-0-7503-4831-7ch4

behavior. Therefore, the liquid phases act as binders for the grains to be combined, thus leading to enhanced densification of the materials.

Partial melting is present as just the shell of ceramic particles is melted, because the input energy is below a certain level [17]. Nevertheless, the molten phase serves as a binder to connect those particles that are not molted. However, ceramics formed through partial melting usually have a relatively low degree of densification and weak mechanical strength. In this regard, particle melting is not expected to have a wide range of applications.

Full melting is applicable to materials with relatively low melting points, which has been extensively employed to process metallic materials with net shape formation, nearly full densification and outstanding mechanical strengths [18, 19]. However, since ceramic materials generally have relatively high melting temperatures, there are limited successful examples reported in the open literature for the processing of ceramics with the full melting approach.

In recent years, various ceramics have been processed by using SLS/SLM. Although reviews in such topic have been available in the literature, they are focused on in one aspect or another [20–23]. This chapter aims to summarize the progress in ceramic materials or structures that have been processed by using SLS and SLM technologies in a more extensive way, covering topics from fundamental considerations to examples of various materials.

4.2 Fundamental considerations

4.2.1 General descriptions

In the additive manufacturing process, three-dimensional (3D) solid pieces are generated from computer aided design/manufacturing (CAD/CAM) data, which are realized by sintering/melting materials in the form of powders, in the layer-by-layer manner, with the aid of lasers as the energy source [24]. A 3D printing facility usually consists of four main parts, including a laser source, powder storage chamber, printing platform and powder spreading kit. Laser source is definitely the most essential component in any 3D printing machines. There are two commercial lasers that have been used for 3D printing, i.e., CO_2 and Nd:YAG lasers?, with wavelengths of $\lambda = 10.6$ μm and $\lambda = 1.064$ μm, respectively.

The printing platform is lowered by one-layer thickness, and then a scraper or a roller drum is driven to spread a certain quantity of the powder over the building platform. When the powder is deposited on the building platform, selected areas of the printed layer are sintered or melted, according to the CAD/CAM data. This layer-by-layer printing process is repeated until the whole structure is fabricated. After printing and cooling, the printed structures are taken out from the powders, while the rest of the powders can be recycled.

In practice, there are various factors that are necessary to consider, such as laser–matter compatibility, printing capacity, thermal gradients and so on. Among them, the key to ensure the success of printing is the matching degree between the laser to be used and the powder to be printed. In other words, the powder should be able to effectively absorb the energy of the laser radiation. Different materials exhibit different responsive behaviors to lasers, depending on their wavelengths [25]. When the ceramic materials to be printed have too low an absorption of laser energy, they can be modified by introducing absorbers to enhance the energy absorbing capacity [26].

Another factor is the relationship between the type of lasers, spot size of the laser beam and energy density for given materials. For example, oxide ceramic powders have different energy absorptions to CO_2 and Nd:YAG lasers, owing to their different wavelengths. For typical ceramics, the absorption of CO_2 laser is much stronger than that of Nd:YAG laser. Meanwhile, due to the large spot size the CO_2 laser, the energy density is not sufficiently high, which is also an issue to be considered.

Unlike metallic and polymeric materials, ceramics are not tolerant thermal gradients to be present between the printed layers. To tackle this problem, SLS/SLM machines should be modified. For instance, by preheating the printing chamber and the powder bed to a temperature below the melting points of the powder materials to be printed, it is possible to reduce the thermal gap [27]. Another way is to apply a post-sintering or debinding step to narrow the thermal gradient ranges. In this case, the green bodies are obtained by just melting the polymeric component in the mixtures at relatively low temperatures. As a result, the green bodies have sufficiently high mechanical strengths. After that, final ceramic structures are developed by eliminating the organic binders. These two processes are known as direct and indirect SLS/SLM.

4.2.2 Properties of the ceramic powders

Properties of powders are an important factor in ensuring the success of SLS/SLM processes. Flowability of powders directly affects the quality of the printed structures. To comprehensively characterize the flow properties of powders and predict their behaviors at different processing conditions, it is necessary to used different analysis techniques [28]. Of course, there is deference in the requirement of flowability for different printing facilities. Other properties of the powders, including particle morphology, particle size and size distribution, are also of critical influences on the quality of both the printed green bodies and final ceramic products. Ideally, all the particles are of spherical morphology. The particle size distribution should be as narrow as possible [29].

4.2.3 Laser–powder interaction

Characteristics of the lasers also play an important role in determining the printing efficiency of SLS/SLM. The influence of the lasers is reflected by the interactions between them and the ceramic materials to be printed. Specifically, the given ceramic powders should be able to absorb the energies of the lasers, so that the printed green bodies will be sintered or melted. Either Nd:YAG laser or fiber laser has been employed as the heating sources in the SLM processes. Oxide powders usually have relatively low absorption efficiency to the laser with wavelength of 1.06 μm, thus causing huge waste of the laser energy and hence leading to low productivities.

In comparison, the absorption to CO_2 laser with wavelength of 10.6 μm is quite high. As a consequence, CO_2 laser has been adopted in most studies.

Interactions between the laser and ceramic powder have a strong influence on propagation behaviors of the laser beams in the powder beds. The ceramic powder beds for the SLM printing are generally loosely packed, without pressing or heating, thus being highly porous, with relatively low density. In this case, the laser beam is multiply scattered or reflected in the ceramic layer, before it is absorbed to trigger densification of the powder layers. In addition, the laser energy absorption by the ceramic powders is also dependent on various other parameters, including particle sizes, size distribution, morphology of the particles, density of the particles, impurity levels and so on, for given materials.

4.2.4 Processing parameters

Laser power and scanning speed are two important processing parameters, which influence the quantity of the energies that are transferred into the ceramic materials to be printed. The laser power is defined as the energy transferred out per unit time (second), while the scanning speed controls the time for the laser beam to focus on a specific area. The two parameters should be balanced, so as to prevent overheating or insufficient heating. For a predesigned scanning profile, the printing rate is dependent on the scanning speed, the hatching distance and the layer thickness. The higher the scanning speed, the higher the productivity obtained. Therefore, the scanning speed should be as high as possible.

4.2.5 Melting and consolidation mechanisms

If partial melting occurs, only the surfaces of the particles that have low melting point are melted by the laser beam. The molten items serve as glue to generate neck-like structures between adjacent particles, thus eventually forming porous structures. As the materials are fully melted, movable molten pools are produced, owing to the moving energy input of the laser. Because of the interactions between the laser beam and the ceramic powder, large temperature gradients and huge temperature differences inside the molten pools are induced.

The density of the final products obtained by using the SLM printing is dependent on viscosity of the ceramic melts. On one hand, the flow and deformation of the melted particles make them to merge as larger melts. On the other hand, once the entrapped gases are unable to diffuse out, air bubbles that are formed tend to escape from the melts. Of course, if the viscosity is too high, the melts cannot flow.

It is well known that surface tension is the driving force of densification of powdery materials. However, high surface tension favors the densification of ceramic materials, while low surface tension is favorite for the bubble escape. Therefore, it is still a challenge to balance the surface tension for the optimization of densification of ceramics processed when using SLM printing. There are three categories of behaviors in SLS/SLM processes, i.e., irregular unstable melt tracks, continuous stable melt tracks and balling effects, which are closely related to the properties of the ceramic powders and the printing parameters.

4.3 Ceramic materials processed with SLS/SLM

4.3.1 Silica

Silica sands were made into structures by using SLS process [30]. Effects of processing parameters, such as power of laser, beam diameter of laser, printing speed, scanning rate and mixture ratio of powder, on dimension accuracy and densification of the green bodies, were studied. It was revealed that lamination and deformation of the sintered products could be minimized by reducing the thickness of slicing and optimizing the processing parameters.

The printing facility was equipped with a cross-flow CO_2 laser with power of 3 kW and a lab made powder layering device, which was aided with computational modeling software and SIEMENS numerical control system. The numerical control system had an accuracy 0.1 mm. Figure 4.1 shows schematic diagram of the SLS printing system and working principle. During printing process, the powder was scanned along the tracks by laser beam, based on the structures of the CAD models. After finishing a layer, the piston was adjusted down by a space equivalent to the thickness of one layer. After that, the powder was scanned over the as-printed layer. Silica sand structures were finally obtained after repeating for designed number of layers.

The powders were mixtures of silica sand and phenol formaldehyde (PF) resin. The silica sands contained 99% SiO_2, 0.22% Al_2O_3 and trace of TiO_2, with melting temperature of 1750 C. PF resin with grain size of 200 meshes and softening points of 105 °C–115 °C was used binder. Length, width and height of the sintered structures are denoted as *L*, *W* and *H*. Laser power, scanning speed, overlapping, laser beam diameter and powder mixture ratio are named as *P*, *F*, η, *D* and Φ. All the

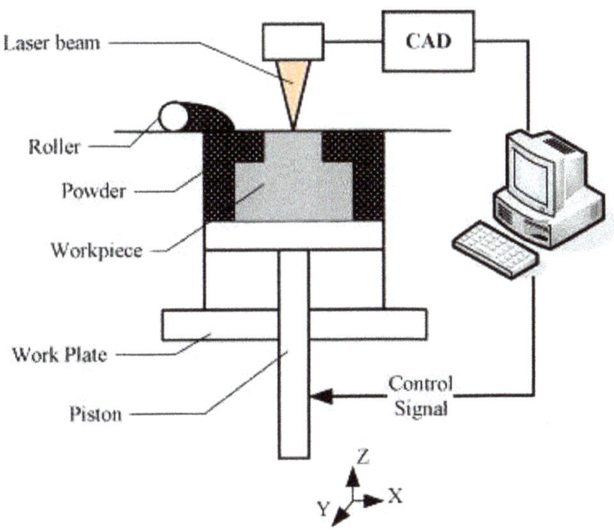

Figure 4.1. Schematic diagram of the SLS system along with the printing principle. Reprinted from [30], Copyright (2007), with permission from Elsevier.

sintered samples were characterized by using KEYENCE three-dimensional microscopy.

Softening temperature of the silica sand was about 110 °C, while the optimal solidification temperature was about 250 °C. In experiment, electric current was used as the reference to represent the laser power. The optimal electric current was about 1.0 A for the PF resin to be completely melted.

Figure 4.2 shows micro morphologies of the samples sintered at different laser powers. Morphology of the pristine silica sands is depicted in figure 4.2(a). The particles of the silica sands were semitransparent crystals, with a milk white and pale yellow appearance. The silica sand particles were wrapped by the polymer binder, forming composite like structures. As shown in figure 4.2(b), the surface of the sample displayed a pale yellow color after irradiation at low laser power, where the binder was not completely melted, so that the strength of the samples was not sufficiently high. After irradiation at medium power, the surface of the sample

Figure 4.2. Micro morphologies of the sintered samples irradiated at different laser powers: (a) pristine silica sand particles, (b) low power, (c) medium power and (d) high power. Reprinted from [30], Copyright (2007), with permission from Elsevier.

became deep brown in color, while silica sand particles were combined by the solid binder. The sample was semitransparent, demonstrating a strong densification effect, as seen in figure 4.2(c). At too high laser power, some areas on surface of the sample presented black brown color, whereas carbonization of the binder occurred, as observed in figure 4.2(d).

The effect of scanning rate on dimension of the sintered structures was demonstrated at given conditions, including laser beam diameter of 3 mm, overlapping width of 0.5 mm, laser power of 12 W and weight ratio of silica sand powder to PF resin of 14:1. With increasing scanning speed, the length, width and height of the samples were all monotonically decreased. The higher the scanning speed, the shorter the dwelling time of the laser beam on the scanning spot would be. Since the laser power was fixed, the energy input was decreased, thus leading to corresponding reduction in the heated area and hence dimension.

With increasing overlapping, the sintering depth was increased. Because the increase in overlapping resulted in increments in the absorption of laser energy, more fraction of the polymer binder was melted, so that the depth of the softened binder increased. Given that other processing parameters were constant, energy density of the laser beam was decreased with increasing beam diameter. As a result, the sintered depth was largely decreased.

The variation in powder mixture ratio Φ had a relatively weak effect on dimensions of the printed structures. With increasing content of binder, the dimensions of the sintered items was just increased to a very low degree. The content of the binder has a direct influence the connecting bridges of the silica sand particles and hence the softened and melted areas, as well as the depth of the powder mixtures. In this regard, the strength of the sintered structures would be proportionally enhanced with the content of the binder. However, if the content of the binder is too high, the deformation and contraction of the samples would be more pronounced, which was a negative effect on dimensional stability of the structures. The optimal powder mixture ratio was 11:1.

Since the as-printed samples had residual powders that were not fully melted in the SLS process, the microstructures of the structures are inhomogeneous, leading to low mechanical strength. Therefore, post-heat treatment was necessary. The post thermal treatment process served as hardening process, through which the congregating behavior of the binder was effectively prevented, while the moisture and organic components were volatilized or burned out. As a consequence, the binder was uniformly distributed in the final structures. The optimal post thermal treatment condition was 250 °C for 0.5 h. In addition, to avoid carbonization, the post-treatment temperature should be below 300 °C. The silica sand structures had sufficiently high mechanical strength to be used for casting.

Noting that the products made with SLS are of laminating structures, the layer thickness and process parameters should be properly controlled and optimized, otherwise the stepping effect would be present. This would have a negative effect on dimension accuracy and surface finish of the final products. Meanwhile, the SLS process usually experiences non-uniform heating and contraction, the printed items tend to have structural deformation. In this aspect, the slicing thickness should be as thin as possible.

However, if the single layer too thin, the production time would be increased. At the same time, too-thin a slicing layer would have more critical requirements for the powders. Also, laser power density is dependent on the laser power and beam dimension, while heating temperature and time duration required for given powders are closely related to the laser power density and scanning speed. Once the power density is too low and the scanning speed is too high, the powders could not be completely melted. In this case, the strength of the ceramics would be sacrificed. If the powder temperature is too high, the effect of the binder would be destroyed. Therefore, the processing parameters should be well controlled and optimized. Figure 4.3 shows photographs of representative silica sand structures SLS processed with optimal processing parameters.

SLS process has been used to fabricate silica ceramic cores [31]. Specifically, the authors combined fine and coarse powders to achieve maximum mechanical performances of the silica ceramics. Flexural strength of the silica green bodies reached a maximum of 3.20 MPa, as the mass ratio of coarse powder (D_{50} = 45.6 μm) to the fine powder (D_{50} = 23.3 μm) was 4:5. The corresponding silica ceramics exhibited the maximum flexural strength of 7.15 MPa, along with a linear shrinkage of 4.76% and an apparent porosity of 47.3%. Compared with that without the introduction of the fine powder, the flexural strength of the final silica ceramics was increased by more than five times.

Two silica powders, 1# and 2#, with particle sizes of D_{50} = 45.6 and 23.3 μm, were used as the raw materials. Epoxy resin (E12, purity~98%) was used as binder, with an average particle size of 8.9 μm. 1# and 2# silica powders were mixed at mass ratios of 1:0, 7:3, 4:5, 3:7 and 0:1, and while 15 wt% epoxy resin E12 was included as binder, with the samples to be denoted as S1, S2, S3, S4 and S5, respectively.

The precursor powders were thoroughly mixed by using a three-dimensional powder rolling machine, at a speed of 150 rpm for 12 h. With increasing content of 2# silica powder, the apparent density of the composite powder was increased first and then decreased, with a peak density 0.89 $g \cdot cm^{-3}$ for S3. The spaces between the coarse particles were filled with the fine particles, leading to increment in bulk density of the mixture. The flowability of the mixed powders was improved after the addition of the 2# silica powder. Nevertheless, too high content of the fine silica powder is not recommended, otherwise, the effect of particle grading would be more or less weakened.

A commercial HK C250 SLS machine was used for the printing experiment. The printing chamber of the machine was a cubic shape 250 mm in length, while the laser had a wavelength of 10.6 μm. Processing parameters included preheating temperature of 45 °C, layer thickness of 100 μm, laser power of 9 W, scanning space of 100 μm and scanning speed of 180 $cm \cdot s^{-1}$. The epoxy resin E12 was completely burned out after calcining at 600 °C for 2 h. Silica ceramics with desired densification were obtained after sintering at 1200 °C for 6 h.

Figure 4.4 shows SEM images of the silica green bodies with different compositions of the two silica powders. The silica particles with spherical morphology were

Figure 4.3. Photographs of representative silica sand structures sintered by selective laser sintering. Reprinted from [30], Copyright (2007), with permission from Elsevier.

Figure 4.4. SEM images of the silica green bodies with different particle compositions: (a) S1, (b) S2, (c, f) S3, (d) S4, (e) S5, (a–e) fracture surface and (f) natural surface. Reprinted from [31], Copyright (2022), with permission from Elsevier.

strongly fused with clear necks, which were formed due the solidification of the melted epoxy resin during the SLS printing process. As observed in figure 4.4(a), there were gaps in between adjacent layers. The laser irradiation induced decomposition of the epoxy resin E12, producing gases that escaped along the printing direction. As a result, the combination of the printed layers was weakened. Comparatively, the green body of sample S3 exhibited a dense microstructure, owing to the proper fraction of the fine particles, as evidenced in figure 4.4(c).

However, as revealed in figure 4.4(d), the green body of sample S4 had numerous pores with small size, which was attributed to the agglomeration of the fine particles. The presence of the small pores led to increase in the apparent porosity, resulting in decrease in the mechanical strength. In other words, the sample S3 possessed optimal content of 2# silica powder. The particle composition had a strong influence on flexural strength of the silica green bodies. With increasing content of 2# silica powder, the flexural strength was increased first and then decreased, reaching a maximum value of 3.20 MPa for sample S3, in agreement with the variation trend in the apparent density.

The particle composition had almost no effect on crystallization of the fused silica. According to XRD results, cristobalite phase was present in the ceramic samples. The fused silica was crystallized to β-cristobalite, known as high-temperature cristobalite, at about 1200 °C [32]. The β-cristobalite was transformed to low-temperature α-cristobalite during the cooling process. Accompanying the crystalline phase transition, there is a variation in volume, thus generating internal stress in the samples. As a consequence, microcracks could be formed in the silica ceramics [32–34]. Since cristobalite is a high-temperature phase, which would block the liquid phase to flow in the silica ceramics, so as to increase the creep resistance and reduce the deflection at high temperatures. It is therefore important to adjust the content of the cristobalite phase to achieve high mechanical strength of the silica ceramics.

Figure 4.5 depicts cross-sectional SEM images of the silica ceramics from the mixtures with different particle compositions. A silica ceramic matrix was formed through the connection of the silica particles, owing to the binder. The particles with small sizes tended to form a liquid phase during the sintering process, due to their high surface energy. The liquid phase flowed into the pores in the silica ceramics, forming sintering necks of the grains.

Meanwhile, the flowing characteristics of the liquid phase promoted rearrangement of particle and mass transfer during the early stage of sintering, speeding up densification of the samples. This is the reason why the densification of the silica ceramics was gradually enhanced in the first three samples. However, as stated earlier, too high a content of fine powder led to the formation of small pores in the ceramics, which broke the connections of the large ceramic grains.

Figure 4.6 shows a schematic diagram illustrating the effect of particle composition on microstructure of the silica ceramics. With an increasing ratio of 2# silica powder, the number of pores in the ceramics was first decreased and then increased. Without the addition of the fine powder, large pores were present among the large particles, so that sintering necks were hardly formed, as presented in figure 4.6(a). As seen in figure 4.6(b), sintering necks were produced for both the small and large

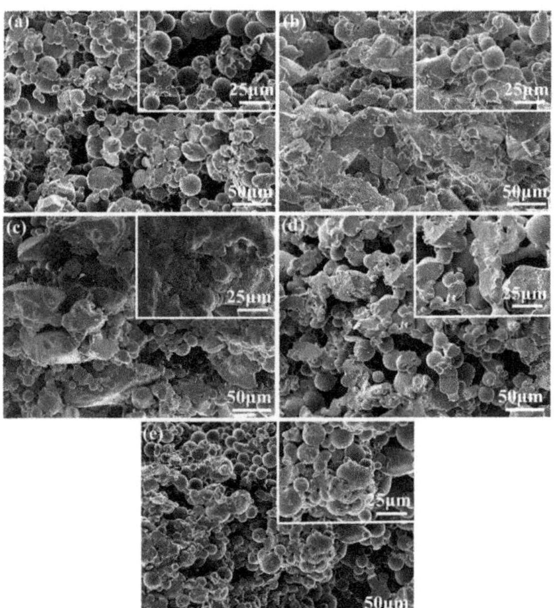

Figure 4.5. Fractured surface SEM images of the silica ceramics with different particle compositions: (a) S1, (b) S2, (c) S3, (d) S4 and (e) S5. Reprinted from [31], Copyright (2022), with permission from Elsevier.

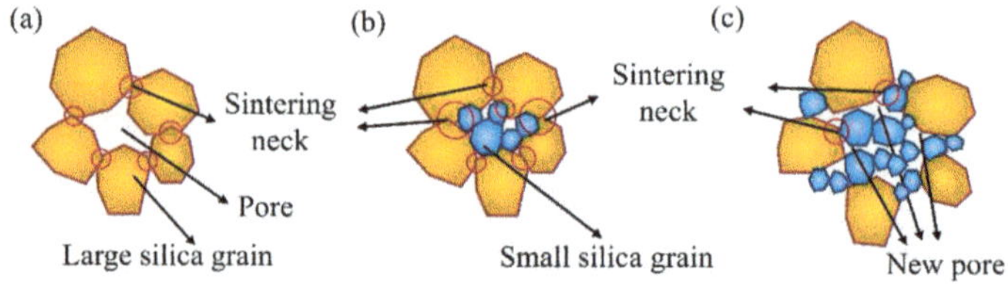

Figure 4.6. Schematic diagram showing effects of particle composition on microstructure of the final silica ceramics. Reprinted from [31], Copyright (2022), with permission from Elsevier.

grains, thus leading to a high degree of densification. With optimal content of 2# silica powder, the pores between large grains could be nearly entirely filled by the small grains, resulting in silica ceramics with high mechanical strength. As the content of 2# silica powder was excessive, the effect of particle grading was suppressed, so that additional pores were formed, which in turn weakened the densification and hence mechanical strength, as demonstrated in figure 4.6(c).

Silica ceramic cores were fabricated by using SLS printing processes, combined with vacuum infiltration (VI) [35]. The optimal SLS processing parameters included, and for the laser power of 11 W, scanning speed of 2000 mm $\cdot$ s^{-1} and hatch space of 100 μm. As the infiltration time was increased to 2 h, room temperature flexural strength of the ceramics was gradually increased to 7.45 MPa, while the linear shrinkage was decreased to 0.6%, owing to the formation of α-cristobalite and the presence of mullite. Moreover, the silica-based ceramics obtained with optimal SLS processing parameters and infiltration time exhibited high temperature creep deformation of 0.31 mm and flexural strength at 1550 °C of 15.04 MPa, making them suitable for the applications as ceramic cores.

The silica-based green bodies were prepared by using composite powders, consisting of 79 wt% spherical fused silica powder with purity >99.8% and average particle size of 23.3 μm, 1 wt% mullite fiber with diameters of 5–15 μm and lengths of 50–100 μm, 15 wt% epoxy resin E12 with average particle size of 52.9 μm and 5 wt% of zirconium silicate powder (99.0% purity). The mixtures of the powders were ball milled for 6 h at 96 rpm at room temperature, with 6-mm diameter zirconia balls. The mixed composite powders were made into silica-based green bodies with SLS. To obtain high-performance silica-based ceramics, SLS parameters, including hatch space, laser power and scanning speed, were optimized through orthogonal experiments.

After printing, the green bodies were infiltrated with nano silica slurry with a concentration of 30 wt% and average particle size of 12 nm. The filtration process was conducted at a given vacuum level, for infiltration times of 30–150 min. The infiltrated samples were dried at 60 °C for 4 h to evaporate the solvent. According to fractured surface SEM images of the composite powders and the silica-based ceramics, the silica powder consisted of spherical particles, while the mullite fibers exhibited rod-like morphology.

Experimental results indicated that flexural strength and bulk density of the ceramics could be optimized by adjusting the laser energy density. When the laser energy density was lower than 0.55 J · mm^{-3}, the printed green bodies had low bulk density and mechanical strength, due to the weak bonding among the particles. If the laser energy density was higher than 0.55 J · mm^{-3}, warpage and excessive sintering occurred, thus lowering the bulk density and mechanical strength. Therefore, it is important to optimize the laser energy density during the SLS process.

The silica-based ceramics after infiltration and sintering at 1200 °C contained phases of cristobalite, mullite and zircon. As mentioned earlier, α-cristobalite was present, owing to the phase transformation of β-cristobalite. During sintering at high temperature, β-cristobalite was produced due to the crystallization of the fused silica. During cooling process, β-cristobalite (stable at high temperature) transformed to α-cristobalite (stable at room temperature). Meanwhile, zircon was stable up to 1600 °C [36]. At the same time, the infiltrated nano silica was transformed to mullite [37]. The two crystalline phases were beneficial to improvement in mechanical strength of the ceramics.

Figure 4.7 shows fractured surface SEM images of the sintered silica-based ceramics with infiltration for different time durations. After infiltration, porosity of the ceramics was significantly decreased, because the micropores of the silica ceramic matrix was filled by the nanosized particles. Meanwhile, the cristobalite crystallized from the infiltrated silica would be transferred to more mullite, which blocked the propagation of microcracks formed due to the phase transition from β-cristobalite to α-cristobalite in the ceramics. Also, the formation of mullite was accompanied by volume expansion, helping to reduce the porosity. Therefore, with prolonging infiltration time, the content of the infiltrated silica also was increased, so that the porosity was decreased accordingly.

Figure 4.8 shows flexural strength and linear shrinkage of the silica-based ceramics as a function of the nano-silica infiltration time. The increase in flexural strength and decrease in linear shrinkage of the silica-based ceramics were credited to the crystallization of cristobalite phase and the formation of additional mullite. After infiltration for 120 min, the flexural strength reached 7.45 MPa, which kept unchanged with prolonged infiltration time. The phase transition from β-cristobalite to α-cristobalite brought out volume reduction of about 5%, which could induce the formation of microcracks, thus leading to reduction in flexural strength and increment in linear shrinkage.

The silica-based ceramics without infiltration had creep deformation of 10.29 mm, which was quickly dropped to 0.31 mm after infiltration for 2 h. Such performance meets the requirement as the ceramic cores for the fabrication of hollow blades. At high temperature of 1550 °C, the β-cristobalite acted as seeds for the crystallization of fused silica into cristobalite. The cristobalite phase distributed on the surface of fused silica grains disturbed the viscous flow of the liquid phase, thus reducing the creep deformation of the silica-based ceramics [38]. However, too long a time infiltration is not necessary, because no more improvement is available, owing to the coagulation of the fused silica with infiltrated silica. Therefore, the optimal infiltration time was 2 h for the silica ceramic cores. Figure 4.9 shows photographs of representative silica-based ceramic cores fabricated with optimized SLS processing parameters and infiltration time.

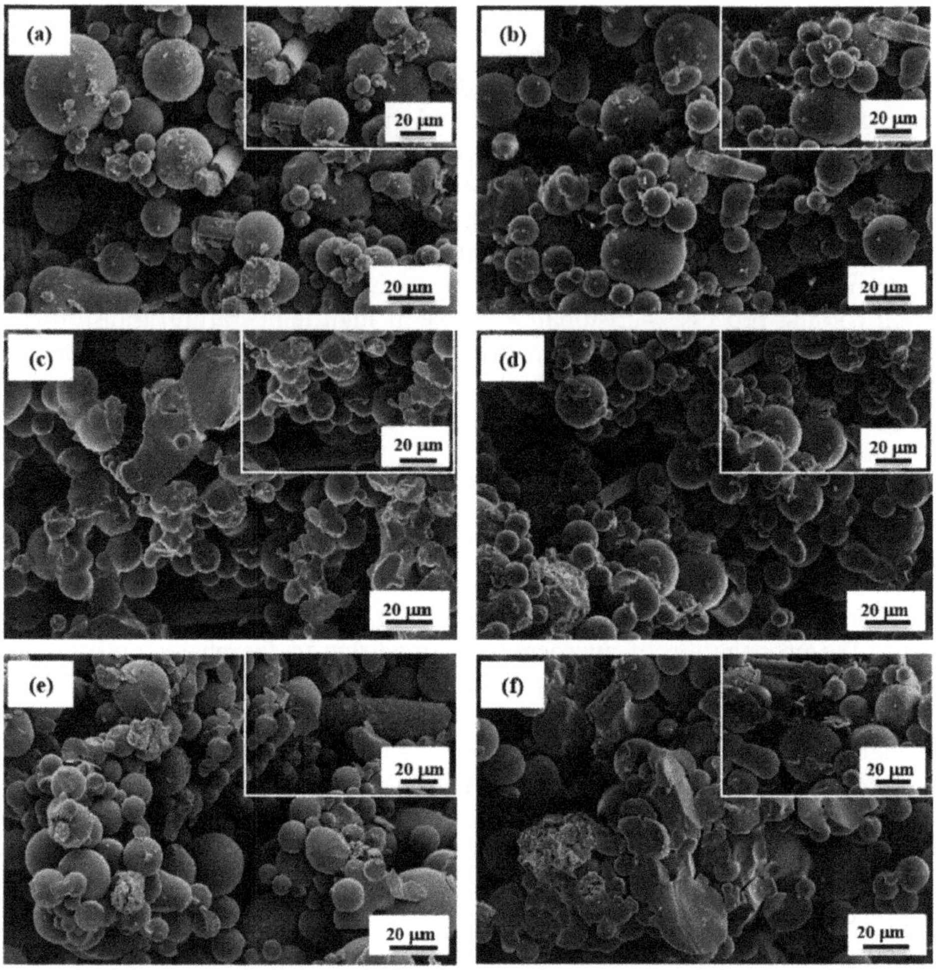

Figure 4.7. Fractured surface SEM images of the sintered silica-based ceramics after infiltration for different times: (a) 0 min (non-infiltrated), (b) 30 min, (c) 60 min, (d) 90 min, (e) 120 min and (f) 150 min. Reprinted from [35], Copyright (2021), with permission from Elsevier.

SLS has been combined with various post-treatment methods, including silica-sol infiltration (SI), vacuum silica-sol infiltration (vSI), debinding (DB) and pressureless sintering (PS), to prepare Al_2O_3–SiO_2 ceramics [39]. Geometric dimension accuracy and surface quality of the final products could be maximized by optimizing the SLS parameters and post-treatment conditions. The optimal SLS processing parameters included hatch spacing of 0.15 mm, laser power of 10 W, layer thickness of 0.1 mm and scanning speed of 1500 mm · s^{-1}. Correspondingly, the SLS/DB/vIN/FS samples exhibited the smallest linear shrinkage ratio of <1), the least warpage degree of <3), and the highest surface quality with surface altitude difference of <170 μm. The materials consisted of corundum, mullite, quartz and cristobalite, with the cristobalite derived from the infiltrated silica-sol. Because the silica-sol had

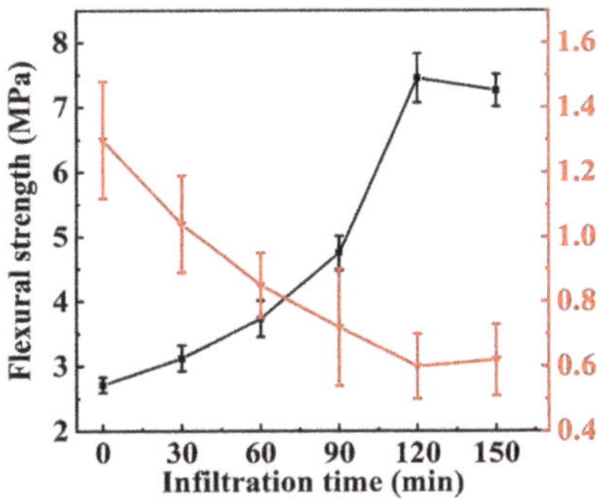

Figure 4.8. Room temperature flexural strength and linear shrinkage of the silica-based ceramics as a function of infiltration time. Reprinted from [35], Copyright (2021), with permission from Elsevier.

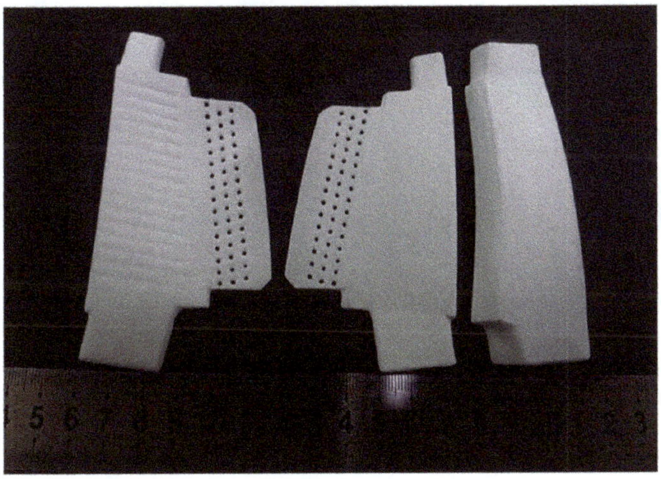

Figure 4.9. Photographs of representative silica-based ceramic cores fabricated by using SLS-VI with optimized SLS processing parameters and infiltration time. Reprinted from [35], Copyright (2021), with permission from Elsevier.

infiltrated into pores of the samples during sintering, the magnitude of shrinkage was reduced and hence the geometric accuracy and surface quality of the finals ceramics were improved.

Sintered corundum (Al_2O_3 >99.7 wt%, D_{50} = 15.6 μm), reactive α-Al_2O_3 micro-powder (Al_2O_3 >99.9 wt%, D_{50} = 2.8 μm, and quartz sand (SiO_2 >93%, D_{50} = 146.1 μm) were used as the raw materials. Epoxy resin E12, with a content of 8 wt%, was used as a binder for the SLS printed structures. The silica-sol of high purity ammonia type contained 32 wt% SiO_2, with room temperature viscosity of 2.0 $mm^2 \cdot s^{-1}$ and particle sizes of 10–21 nm, for infiltration experiment. Figure 4.10 shows SEM images of the raw materials.

The raw material powders were mixed with a drum mill and zirconia balls. The mixing was conducted for 2.5 h, while the rotation speed was 96 rpm. The printing

Figure 4.10. SEM images of the raw materials: (A) calcined Al_2O_3 powder, (B) micro-sized α-Al_2O_3 powder, (C) quartz sand and (D) epoxy resin E12. [39] John Wiley & Sons. © 2019 The American Ceramic Society.

process was carried out with a HK-C250 machine, which was equipped with a CO_2 laser. The laser wavelength was 10.6 μm, while the laser spot size was 0.2 mm. During the SLS printing process, the temperature in the working chamber was set to be about 45 °C. According to the processing parameters, including incident laser power, scanning speed, hatch space and layer thickness, nine groups of samples were prepared.

The as-printed green bodies were infiltrated with silica-sol, followed by thermal debinding and pressureless sintering (PS). Four post-treatment routes were used for the samples. In the first group, no post-treatment was involved, while the samples were only sintered at 1450 °C for 3 h (denoted as PS). The PS process consisted of a slow heating process at 1.0 °C $\cdot$ min^{-1} till 650 °C with a dwell time of 30 min and fast heating process at 5 °C $\cdot$ min^{-1} to 1450 °C with a holding time of 3 h. The resultant sample was named SLS/PS.

In the second group, the samples were first infiltrated with silica-sol in ambient conditions for 2 h (denoted as IN), followed the PS process. The samples were labeled as SLS/IN/PS. In the third group, all samples were infiltrated with silica-sol at vacuum of about 0.005 MPa (denoted as vIN) for 2 h, resulting in samples of SLS/vIN/FS. The fourth group of samples were treated at the same conditions as the third group, while a debinding process was additionally involved (denoted as DB). The DB samples were infiltrated with silica-sol at vacuum and then sintered at 1450 °C for 3 h, and were then called SLS/DB/vIN/PS.

The samples printed with laser energy densities of 0.44, 0.53 and 0.80 J · mm^{-2} exhibited promising macro-appearance and high geometric accuracy. As the incident laser energy density was lower than 0.36 J · mm^{-2}, SLS printed green bodies were of very loose surface and fragile characteristics, due to the weak bonding among the particles. Too high an laser energy density, e.g., >1.28 J · mm^{-2}, led to serious warpage and oversintering of the ceramics. For some samples, even although medium incident laser energy density was adopted, e.g., 0.60–0.67 J · mm^{-2}, oversintering was observed, implying that the laser energy density and incident laser power should be carefully selected.

The SLS/PSed samples displayed the highest bulk density and the lowest apparent porosity, while the SLS/IN/PS, SLS/vIN/PS and SLS/DB/vIN/PSed samples had a lower density and higher apparent porosity. For the samples from SLS/IN/PS to SLS/vIN/PS and eventually to SLS/DB/vIN/PS, their bulk density showed a decreasing trend and the apparent porosity was slightly increased.

Figure 4.11 shows fractured surface SEM images of the samples in the direction perpendicular to the laser processing plane. The sample SLS/PS had the most dense microstructure, while internal voids were observed in the other three samples. The porous microstructure of the SLS/IN/PSed, SLS/vIN/PSed and SLS/DB/vIN/PSed samples was in agreement with the density results. It was indicated that densification of the samples after the infiltration of silica-sol was slowed down.

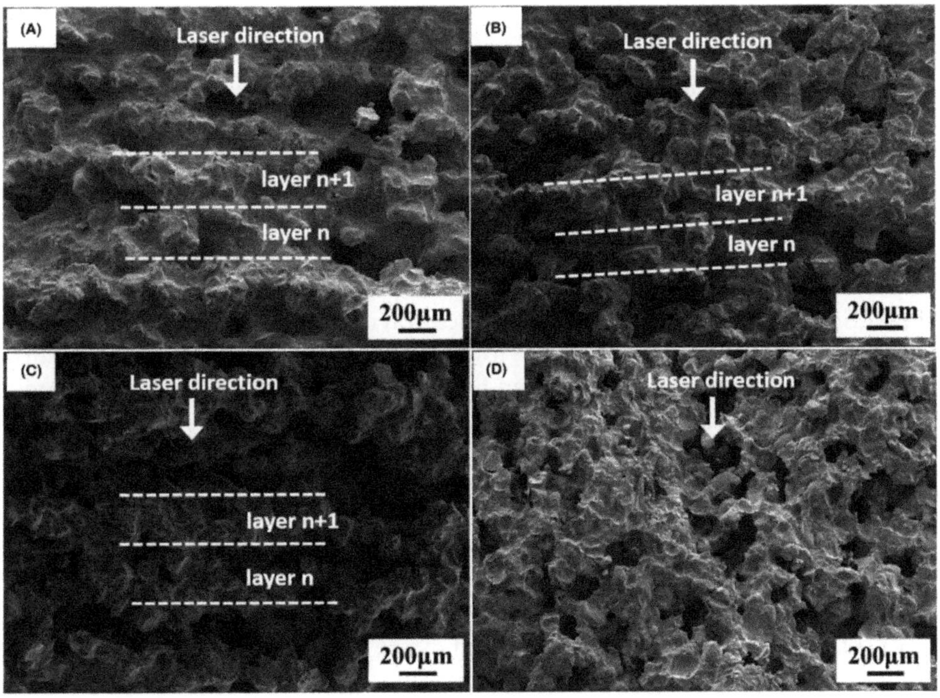

Figure 4.11. Fractured surface SEM images of the samples posttreated with different conditions: (A) SLS/PS, (B) SLS/IN/PS, (C) SLS/vIN/PS and (D) SLS/DB/vIN/PS. [39] John Wiley & Sons. © 2019 The American Ceramic Society.

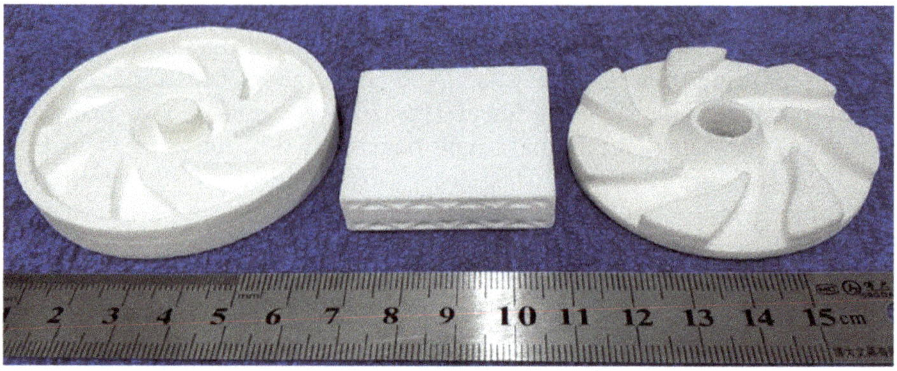

Figure 4.12. Photographs of the Al_2O_3–SiO_2 ceramic items fabricated with optimized SLS parameters and post-treatment process. [39] John Wiley & Sons. © 2019 The American Ceramic Society.

The SLS/IN/PS and SLS/vIN/PS samples exhibited laminated structures, with thickness to be 200–250 μm, corresponding to the thickness of powder layer for the SLS printing process. In comparison, the SLS/DB/vIN/FPed sample had almost no laminated characteristics, when observing from the fractures surface perpendicular to the laser processing plane. In other words, the SLS/DB/vIN/PS post-treatment suppressed the delamination effect of the SLS printed Al_2O_3–SiO_2 ceramics.

In terms of stoichiometric composition of mullite, the quantity of Al_2O_3 was insufficient, since the mass ratio of Al_2O_3 to SiO_2 in the starting materials was about 39:61. Therefore, both the sintered corundum and micro-sized α-Al_2O_3 powder should be used up for the formation of mullite. In fact, they are present in the XRD pattern, because of their low reactivity. In other words, the reaction to form mullite only partially took place during the sintering process.

Figure 4.12 shows photographs of representative Al_2O_3–SiO_2 ceramic structures, obtained with optimized SLS processing parameters and post-treatment route, i.e., SLS/DB/vIN/PS. The left and right ones are the reverse molds of impeller component and the corresponding impeller component, while the middle one is Al_2O_3–SiO_2 ceramics with staggered triangular pores. This confirmed the capability of the SLS/DB/vIN/PS process for the fabrication of complicated ceramic structures. Nevertheless, the issues of relatively low geometric accuracies and poor surface qualities of the ceramic items fabricated by using SLS printing technology should be further improved, in order for large-scale industrial production.

4.3.2 Alumina

Parts of this section have been reprinted from [46], Copyright (2022), with permission from Elsevier.

A similar work was reported by Shahzad *et al.* to fabricate alumina ceramic structures with SLS [40]. High purity α-alumina (grade SM8, Baikowski, France) powder with $D_{50} = 0.3$ μm was used as the raw material, while polyamide 12 (PA 12, grade DuraForm PA, 3D Systems, USA) with $D_{50} = 100$ μm was used as the binder. Dimethyl sulfoxide (DMSO, Merck Co.) was used as solvent to prepare composite microspheres.

Alumina and PA powders (10 vol%) were blended with DMSO (90 vol%), with the aid of strong magnetic stirring. The suspensions were then heated at 140 °C, which was above the dissolution temperature of 135 °C, for 15 min in N_2 to ensure the dissolution of PA in DMSO. After cooling to room temperature, the suspensions were vacuum filtrated to collect the Al_2O_3-PA composite particles from the DMSO. The precipitates were repeatedly washed with ethanol and then dried at 80 °C for 24 h. Two compositions were examined, with 50 vol% (CP50:50) and 40 vol% (CP40:60) alumina.

Green bodies were printed by using a Sinterstation 2000 (DTM Corporation/3D Systems, USA) 3D printer equipped with a 100 W CO_2 laser (f100, Synrad, USA) having laser beam diameter of 400 μm. Powder layers were coated by using a counter current roller, which were irradiated with the laser beam during scanning. To prevent thermal oxidation of PA, the SLS printing experiment was conducted in N_2 protection. The powder bed was heated to 160 °C, i.e., which is between the PA melting of 190 °C and crystallization temperature of 157 °C.

The multilayer green bodies had a square dimension of 15×15 mm^2. Printing parameters included laser power of 3–7 W, scan speed of 100–1257 mm $\cdot$ s^{-1} and scan spacing of 150–300 μm, for optimization. The layer thickness was fixed to be 150 μm. The binder was burned out from the green bodies by calcination to 600 °C in air for 2 h, at a heating rate of 0.1 °C $\cdot$ min^{-1}. The calcined samples were sintered at 1600 °C for 1 h in air, at a heating rate of 5 °C $\cdot$ min^{-1}.

The composite microspheres exhibited outstanding flowing properties, while green bodies with desired quality could be printed over a relatively narrow window of laser energy density. Too low laser energy density led to printed green bodies with loose compaction owing to the incomplete melting of the binder, while too high laser energy density induced decomposition of PA and distortion of the printed bodies. The green bodies printed at the laser energy densities of 0.176–0.37 J $\cdot$ mm^{-3} had sufficiently high mechanical strength.

Figure 4.13 shows fractured surface SEM images of two green bodies. As seen in figure 4.13(a), for the sample CP50:50, spheres of the starting powder were clearly present, without collapse during the SLS printing process, suggesting that no plastic flow occurred for the composite powder. As a result, the green body had a loose compact, due to the presence of residual intersphere spaces. In comparison, the sample CP40:60 experienced polymer and material flow, as illustrated in figure 4.13 (b), so that it had higher density.

Both the green and sintered densities of the sample CP40:60 were higher than those of the CP50:50, despite the 10 vol% higher polymer content of CP40:60. This could be explained in terms of the plastic flow of the polymer during SLS printing process. For the sample CP40:60, there was large intersphere contact area in between the Al_2O_3 particles, ensuring the formation of necks during the sintering process. However, the two samples possessed relatively high residual porosity, as demonstrated in figure 4.14 for the CP40:60 sintered ceramics. Nevertheless, complex shaped structures could be fabricated by using the SLS/debinding/sintering process, as presented in figure 4.15.

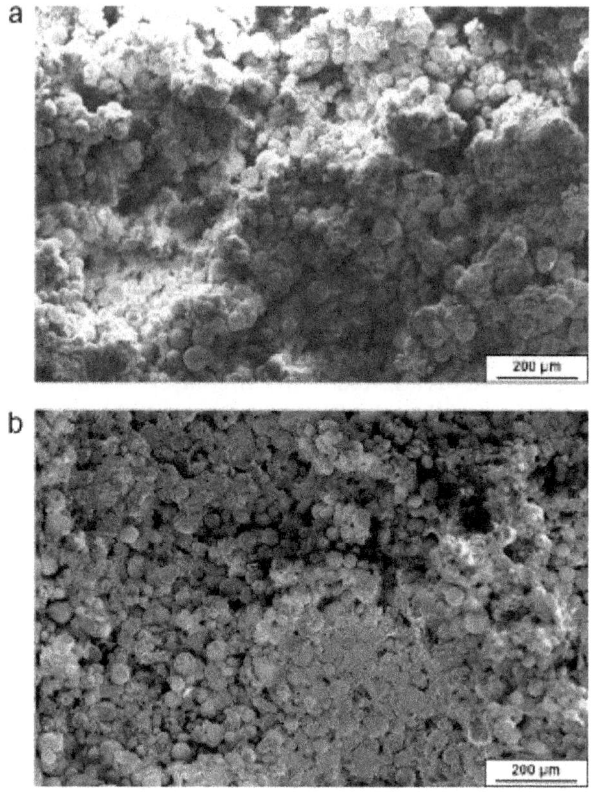

Figure 4.13. Cross-sectional SEM images of the SLS printed green bodies: (a) CP50:50 and (b) CP40:60. Reprinted from [40], Copyright (2012), with permission from Elsevier.

Figure 4.14. Representative SEM image of the sintered alumina ceramics. Reprinted from [40], Copyright (2012), with permission from Elsevier.

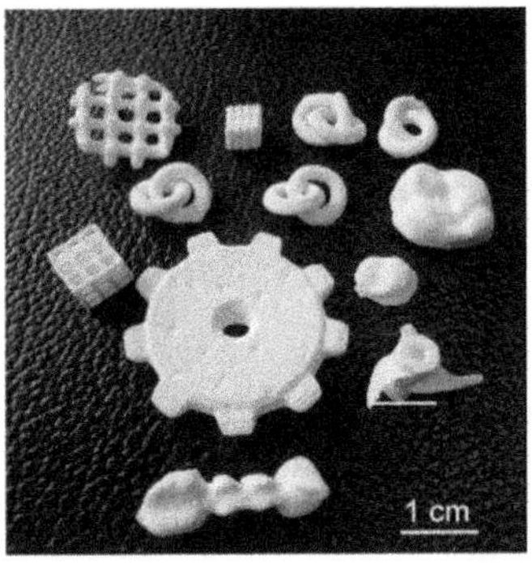

Figure 4.15. Photographs of the alumina ceramics fabricated by using the SLS/debinding/sintering process with different structures. Reprinted from [40], Copyright (2012), with permission from Elsevier.

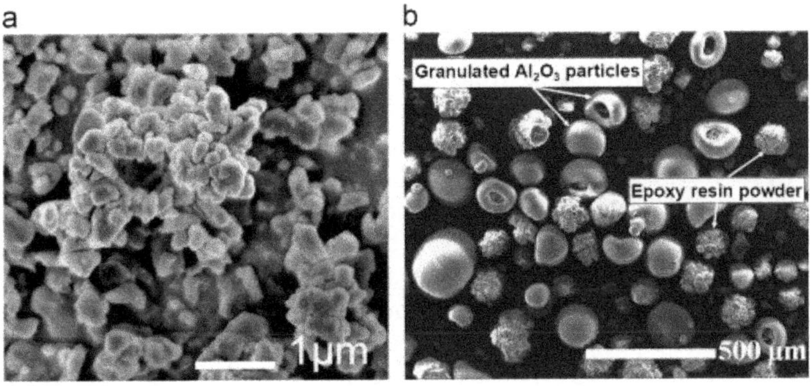

Figure 4.16. SEM images the raw materials: (a) fine α-Al_2O_3 powder and (b) Al_2O_3-epoxy resin E06 composite microspheres. Reprinted from [41], Copyright (2014), with permission from Elsevier.

SLS was combined with cold isostatic pressing (CIP) to fabricate Al_2O_3 ceramics with complex structures [41]. The starting materials included Al_2O_3 powder and polymer additives, i.e., epoxy resin E06 (ER06) and polyvinyl alcohol (PVA). Commercial α-Al_2O_3 powder with purity of 99.7% and $D_{50} = 0.4$ μm was coated with 1.5 wt% PVA through spray drying process. The coated spherical particles with an average size of 80 μm were mixed with ER06 powder with particle size of 50 μm at the concentration of 8 wt% after blending for 24 h. SEM images of the fine α-Al_2O_3 powder and the Al_2O_3-epoxy resin E06 composite microspheres are shown in figures 4.16(a) and (b), respectively.

The SLS machine was equipped with a 55 W CO_2 laser with a laser beam diameter of 200 μm. Processing parameters included laser power of 15–21 W, scanning speed of 1600–2000 mm $\cdot$ s^{-1} and scanning space of 100–140 μm, for experimental optimization. The printed layer thickness was set to be 150 μm, while the sample dimension was $50 \times 10 \times 10$ mm^3 ($L \times W \times H$). Plastic airproof canning was adopted for cold isostatic pressing of the SLS printed Al_2O_3 bars. Liquid RTV-2, $CH_3COOCH_2CH_3$, $Si(OEt)_4$, and $Bu_2Sn(OCOC_{11}H_{23})_2$ were mixed according to the mass ratio of 100:(3–4):(0.5–1):1. After coating with the mixed liquid, the samples were heated at 85 °C for a while.

For debinding, the green bodies were heated at 330 °C, 450 °C and 800 °C or 1000 °C, for 2, 1 and 2 h, at heating rates of 4, 1 and 1 °C · min^{-1}. The debound samples were sintered at 1600 °C for 2 h. The heating rates to 1000 °C and 1600 °C were 10 and 5 °C · min^{-1}.

Figure 4.17 shows SEM images of the Al_2O_3 green bodies. The agglomeration of the composite powder was not varied significantly during the laser sintering process. The spheric morphology of the spray dried Al_2O_3 agglomerates was well retained, while the particles were coated by the melted ER06. The sample had a relatively high porosity, as observed in figure 4.17(a), while sintering necks were formed between the Al_2O_3 agglomerates, as demonstrated in figure 4.17(b). During the SLS printing process, the epoxy absorbed the laser energy was melted, which then covered the Al_2O_3 agglomerates, so that they were bonded together to form the green bodies.

Density of the green bodies was dependent on the SLS processing parameters. Among the various parameters, the effect of laser power was the most pronounced. As the laser power was raised from level 1 to level 3, the relative density was increased from 31% to 34%. A narrow scanning space was beneficial to resin adhesion, because of the widened temperature overlap field. The optimal parameters were laser power of 21 W, scanning speed of 1600 mm · s^{-1} and scanning space of 100 μm, in terms of achieving the highest relative density.

Densification of the SLS Al_2O_3 green bodies displayed a close relation to CIP pressure. As the CIP maximum pressure was increased from 50 to 335 MPa, density of the Al_2O_3 green bodies was increased from 1.25 to 1.64 g · cm^{-3}, corresponding to an increase in relative density from 40% to 54%. At the initial stage of CIP, pores were eliminated and Al_2O_3 particles were rearranged, when the pressure was $<$50 MPa. Epoxy resin necks were damaged at this stage. In this case, the green bodies possessed the highest densification rate.

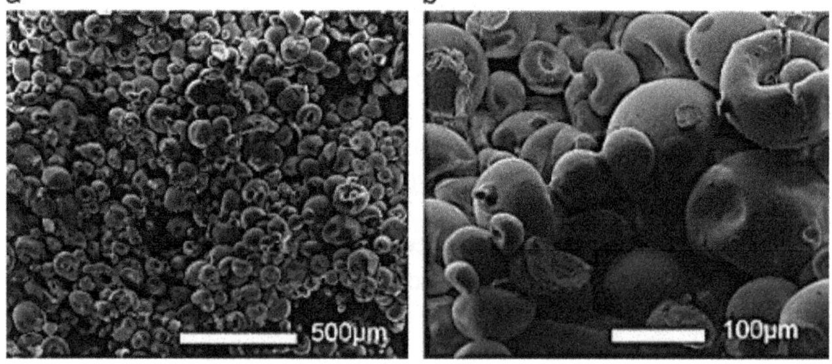

Figure 4.17. Fractured surface SEM images of the SLS samples fabricated from the epoxy resin E06-Al_2O_3 composite powder: (a) low magnification and (b) high magnification showing the bound necks. Reprinted from [41], Copyright (2014), with permission from Elsevier.

At medium stage of CIP, the damage of the bonding necks became more serious, whereas they were crushed into pores in green bodies. The PVA around the Al_2O_3 particles were also isostatic pressed and the contact area increased. The mechanism of densification at this stage was plastic flow of the polymers e. Meanwhile, the presence of PVA led to decrease in friction among in between the spherical particles, thus promoting the plastic flow. At high CIP pressures, both polymer and the PVA was ruptured and filled into the pores in the green bodies. Surface contact area was increased because of the increase in deformation resistance. Therefore, the densification rate was decreased as the pressure was over 200 MPa.

Fractured surface SEM image of the green body after SLS/CIP treatment at 200 MPa is shown in figure 4.18(a). In the debinding process, the binder had completely decomposed, so that the weight of the samples was reduced. Therefore, relative density of the samples could not be further increased. Due to the formation of the sintering necks between the fine Al_2O_3 particles, shape of the green bodies was well retained, with shrinkage in length, width and thickness directions.

As seen in figure 4.18(b), intergranular fracture was the main fracture mode in the ceramics after sintering. After completely debinding, the samples consisted of only fine Al_2O_3 particles. The densification took place through the solid phase sintering, owing to the high surface energy of the fine Al_2O_3 particles. With optimized sintering parameters, the relative density of the sintered Al_2O_3 ceramics was 92%, with dense microstructure as illustrated in figure 4.18(c). The shrinkages of length (L), width

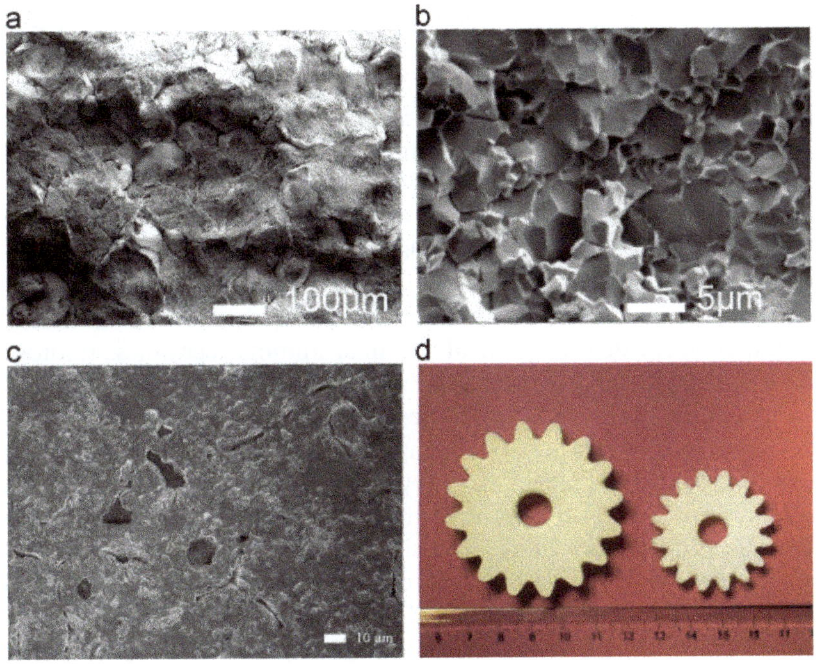

Figure 4.18. Fractured surface SEM images of different samples: (a) green bodies with SLS/CIP treatment at 200 MPa, (b) sintered Al_2O_3 ceramics and (c) Al_2O_3 ceramics. (d) Photographs of the Al_2O_3 green body (left) and sintered Al_2O_3 ceramics (right). Reprinted from [41], Copyright (2014), with permission from Elsevier.

(W), height (H) and volume (V) were 28.3%, 28.7%, 32.5% and 65.5%, respectively. Since the layer thickness was relatively large and the laser re-sintering in the SLS printing direction was limited, shrinkage of H was larger than those of L and W. Photographs of representative samples are shown in figure 4.18(d).

A phase inversion approach was developed to prepare spherical alumina particles, by coating with a thin layer of polystyrene (PS), for indirect SLS printing [42]. High purity α-alumina (SAW-70) powder, with $D_{10} = 40 \mu m$, $D_{50} = 75\ \mu m$ and $D_{90} = 120\ \mu m$, was used as the ceramic material. Expanded polystyrene (EPS) was used as the binder, while dichloromethane (Extra pure) was used as the solvent to prepare the core–shell composite powder.

Phase inversion process (PIP) was used to prepare 6 wt% PS-alumina core–shell composite powders, involving steps of heating, mechanical stirring, cooling, washing, filtration and drying. Alumina and PS powders (30 vol%) were suspended in dichloromethane (70 vol%), with the aid of mechanical stirring. The dissolution temperature of expanded polystyrene in the dichloromethane is ambient temperature, while the boiling point of dichloromethane is 40 °C. The suspension was heated to 38 °C for 10 min for complete dissolution of the EPS dichloromethane. The composite powder precipitated from the suspension after cooling to room temperature. The precipitated powder was collected through vacuum filtration, followed by repeated washing and drying at 90 °C for 4 h. The dried powder was manually crushed and then ball milled at 30 rpm for 40 min, with hardened steel balls as milling media, at balls to powder weight ratio of 2:1.

Green bodies were printed by using a selective laser sintering system equipped with a pulsed Nd:YAG laser, while laser spot diameter was 400 μm. Processing parameters tested included laser power of 2–7 W, pulse frequency of 1–50 kHz, pulse width of 0.5–50 μs and scanning speed of 5–50 $mm \cdot s^{-1}$, while the scan spacing was fixed to be 50 μm. 3D structures of alumina ceramics were printed with optimized processing parameters.

Since the densities of alumina and PS were 3.8 and 1.05 $g \cdot cm^{-3}$, respectively, the theoretical density of the 6 wt% PS coated alumina powder was calculated to be 3.28 $g \cdot cm^{-3}$. The relative densities of the green bodies were in the range of 66.6%–81.3%, with maximum standard deviation of 0.61%. Such green bodies with high density would ensure effective densification of the final alumina ceramics. Photographs of representative samples with complex structures are shown in figure 4.19.

Surface and cross-sectional SEM images of selected sintered alumina ceramics are shown in figures 4.20 and 4.21. The samples were processed at scanning speed of 20 $mm \cdot s^{-1}$, laser power of 6 W, frequency of 10 kHz, pulse width of 10 μs and layer thickness of 400 μm. Strong connections among the particles were observed, with the adjacent particles to be linked via the sintering necks. The uniform PS shell on the alumina particles guaranteed dense compaction of the ceramic powder. The connection necks exhibited thicknessed of 20–70 μm. Noting the layer thickness of 400 μm and the particles size of 125 μm, the single layer contained at least three rows of particles. This implied that the laser radiation penetrated the printed layer in thickness direction, because the PS shell and spherical alumina particles could absorb nearly 80% of the energy at the wavelength of the laser.

Figure 4.19. Photographs of selected green bodies with different complex structures printed by using the indirect SLS, with optimal laser parameters, including frequency of 10 kHz, pulse width of 10 μs, laser power of 6 W and scan speed of 20 mm · s^{-1}. Reprinted from [42], Copyright (2018), with permission from Elsevier.

Figure 4.20. Top view SEM images of the sintered ceramics: (a) low magnification and (b) high magnification. Reprinted from [42], Copyright (2018), with permission from Elsevier.

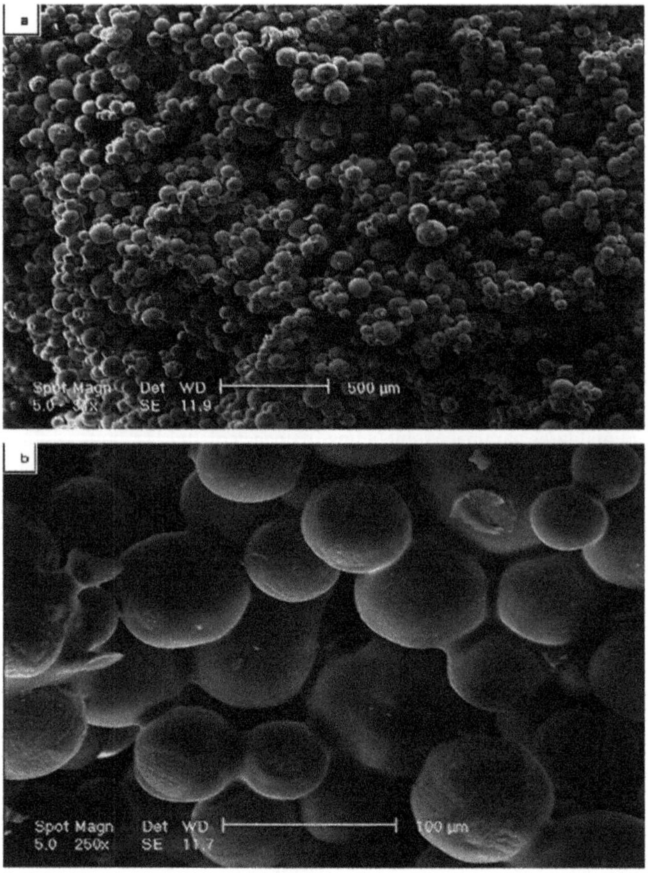

Figure 4.21. Side view SEM images of the sintered ceramics: (a) low magnification and (b) high magnification. Reprinted from [42], Copyright (2018), with permission from Elsevier.

A powder-based selective laser sintering (SLS) approach was reported to fabricate Al_2O_3 ceramic foams with near-zero shrinkage, high porosity and superior mechanical strength [43]. The ceramic foams were derived from the Al_2O_3/Al composite powders through reaction bonding (RB), resulting in coral-like hollow-sphere structures. A near-zero shrinkage of about 1% and a high porosity of about 74% were realized through the Kirkendall effect, owing to the oxidation of Al. At the same time, owing to the reinforced sintering necks and robust bond-bridge connections between hollow spheres and coral-like structures, a high bending strength of about 7.5 MPa was achieved.

Spherical Al_2O_3 powder had an average particle size of 36.8 μm with purity of 99.97%, coral-like Al_2O_3 powder had an average particle size of 46.5 μm with purity of 99.95% and Al powder was of average particle size of 16.7 μm and purity of 99.8%. Epoxy resin powder had an average particle size of 14.1 μm (E12, 99% as the binder. The particles of the spherical Al_2O_3 powders had a smooth surface, while the coral-like Al_2O_3 powder displayed agglomeration of small sized platelet particle, with size of about 2 μm. Both the spherical and coral-like Al_2O_3 powders were mainly α-Al_2O_3. The Al powder consisted of irregularly shaped particles.

Three compositions, including Al_2O_3, coral-like Al_2O_3 and coral-like Al_2O_3/Al, were studied. They were mixed with 25 vol% E12, resulting in samples denoted as SA, CA and CA-Ax, respectively. In the CA-Ax group, the volume fraction of Al was in the range of 0–67.5 vol%, corresponding to $x = 0–67.5$. The mixtures were milled with a 3D mixer at rotational speed of 14 rpm for 6 h. All the composite powders exhibited sufficiently high flowability, thus meeting the requirement for powder the SLS printing experiment.

The SLS equipment used to print green samples was equipped with a CO_2 laser. The forming parameters, including laser power, layer thickness, scanning speed, hatch spacing and preheating temperature, were 7 W, 0.15 mm, 2000 mm · s^{-1}, 0.2 mm and 45 °C, respectively. The green bodied were first debinded at 600 °C for 2 h at a heating rate of 1 °C · min^{-1} and then heated at 1000 °C for 2 h at a heating rate of 2 °C · min^{-1}. Finally, Al_2O_3 ceramic foams were achieved after sintering at 1600 °C for 2 h, at a heating rate of 3 °C · min^{-1}.

The green samples from CA powder had higher porosity and lower bending strength than that from the SA powder. The high porosity was ascribed to the micro-porous structure on the rough surface of coral-like particles, which blocked melted E12 to diffuse in. The strength of the green samples from CA-A67.5 powder was about two times that of the one from CA powder, owing to the optimal gradation and accumulation of the particles. Specifically, the ceramic foams from the CA-A67.5 powder exhibited near-zero shrinkage of 0.91%, high porosity of 73.7% and high strength of 7.37 MPa, which were much better than those from of the samples derived from the SA and CA powders.

Figure 4.22 shows fractured surface SEM images of the sintered ceramic foams derived from different powders. Without the addition of Al, interspaces and sintering necks between the Al_2O_3 particles were clearly observed, as demonstrated in figures 4.22(a1)–(b2). In the sample from the CA powder, the micro-porous structure on the rough surface of the coral-like particles was retained, which was responsible for the high porosity, as revealed in figures 4.22(b1) and (b2). The micro-porous structure effectively separated the Al_2O_3 particles, while the rough surface resulted in loose particle accumulation [44].

However, the rough surface of the particles of CA powder led to weak sintering necks, so that the sample had accordingly low mechanical strength. To address this issue, metallic Al with suitable content was incorporated. As seen in figures 4.22(c1) and (c2), numerous hollow-spherical structures with shell thickness of about 1 μm were present, while coral-like structure was also visible occasionally. The hollow structure was originated from the Al particles after oxidation, owing to the Kirkendall effect. At the early stage of oxidation, core–shell structure was formed, with oxide layer to coat on the metallic core. Eventually, hollow structure was developed, as the metallic component entirely diffused out and completely oxidized. Massive molten Al between hollow spheres and coral-like particles are oxidized rapidly and twisted together to form the complex bond-bridge structure, which is beneficial for further enhancing the bonding force between particles.

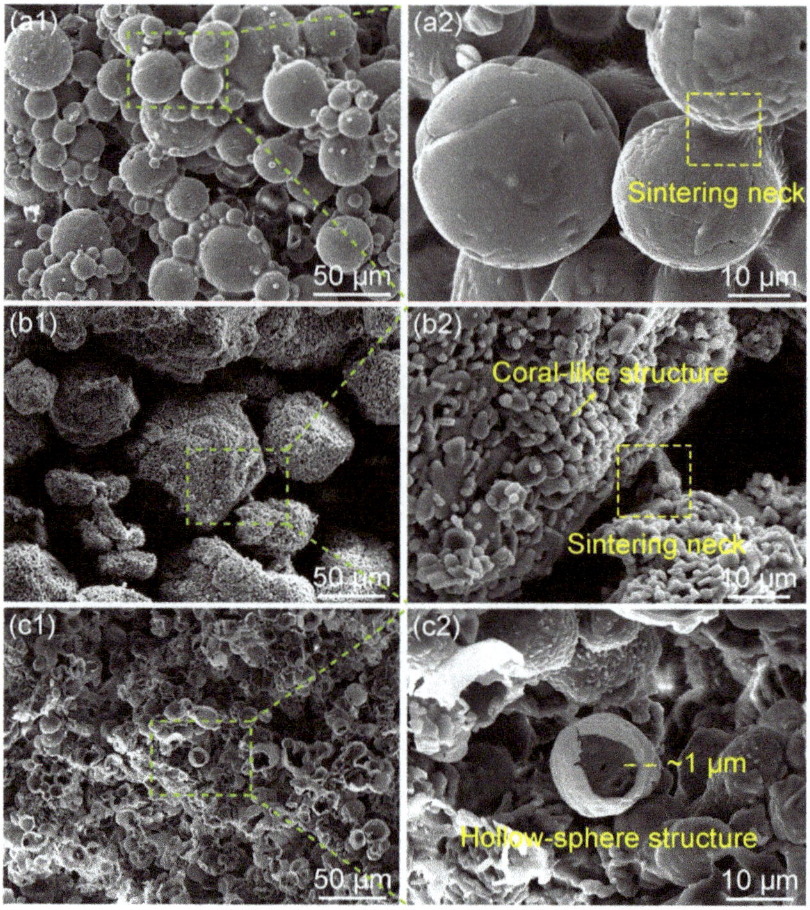

Figure 4.22. Fractured surface SEM images of the sintered samples from different powders: (a1, a2) spherical Al_2O_3 powders, (b1, b2) coral-like Al_2O_3 powders and (c1, c2) coral-like Al_2O_3/Al composites with 67.5 vol% Al. Reprinted from [43], Copyright (2021), with permission from Elsevier.

In the samples with Al content of >37.5 vol%, Al granules precipitated on the surface of the pre-sintered samples, due to the low wettability of the Al melts to Al_2O_3. Such precipitation occurred most likely at temperatures of <1000 °C. At high temperatures, the wettability was increased and oxidation took place quickly, so that voids were formed and accumulated, instead of the appearance of surface precipitation [45]. The sintered samples only had slight delamination, without the observation of serious defects. However, the content of Al had a strong effect on densification behaviors of the samples. The shrinkage in the length and height directions first increased till the content of Al to be 45.0 vol% and then decreased for the content to be increased from 45.0 to 67.5 vol%. The samples with the Al content of 15.0 and 67.5 vol% exhibited the smallest shrinkages.

As the coral-like Al_2O_3 was the main component, i.e., the Al content was <45.0 vol%, the expansion was induced due to the precipitation of molten Al and the out-diffusion of Al_2O_3. The expansion was accompanied by internal stress, which was increased with increasing content of Al. Once the internal stress was over the interlayer bonding strength, delamination would occur. When the content of Al was >45.0 vol%, Al was the main component. In this case, the Al melts preferentially formed clusters instead of precipitate, so that the expansion and internal stress were suppressed. As a result, the sintered ceramics were defect free with the CA-A67.5 composite powders, resulting in near-zero shrinkage.

According to XRD results, the sintered ceramic foams derived from the precursors with different content of Al all possessed single phase α-Al_2O_3, suggesting that the Al was completely oxidized at the sintering temperatures. Figure 4.23 shows SEM images of the ceramic foams with different contents of Al. With the addition of Al, the coral-like and hollow-sphere structures were well retained after sintering. With increasing content of Al, the number of coral-like structures was decreased, while that of hollow-sphere structures was increased.

The micro-porous structure of the coral-like particles was nearly unchanged, implying that they were not filled by the Al melts. Moreover, the grain size of the in-situ formed Al_2O_3 hollow structures was about 2 μm, which was smaller than those of the samples derived from the single spherical or coral-like Al_2O_3 powders, as depicted in figure 4.23(a). The hollow spheres and coral-like particles were bridged together, which was more and more pronounced, as the content of Al was increased, as seen in figure 4.23(c).

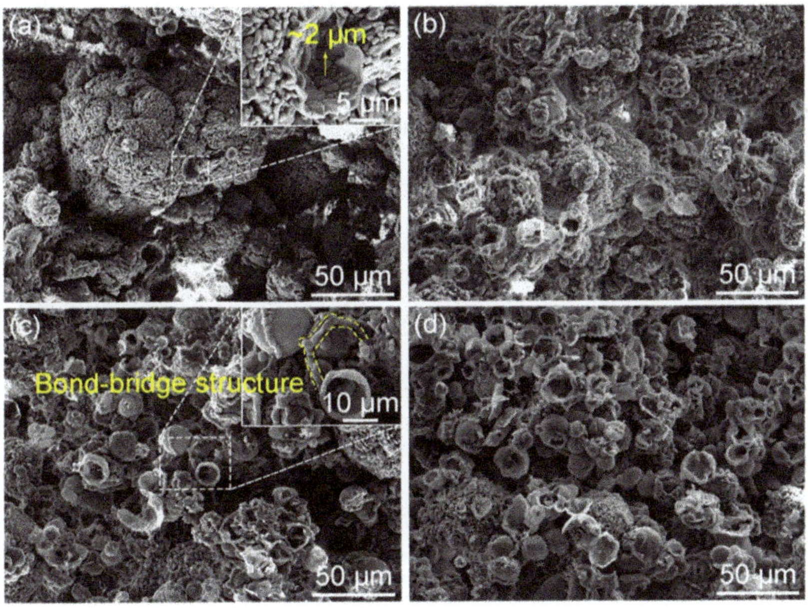

Figure 4.23. Fractured surface SEM images of the sintered samples prepared from coral-like Al_2O_3/Al composites with different contents of Al: (a) 15.0 vol%, (b) 22.5 vol%, (c) 45.0 vol% and (d) 67.5 vol%. Reprinted from [43], Copyright (2021), with permission from Elsevier.

Total porosity of the ceramic foams was all about 73.3%, as the content of Al was in the range of 0–67.5 vol%. This was because increase in porosity related to the hollow-sphere structure was compensated by the decrease in that due to the bond-bridged structure. The porous structure was characterized by large fraction of open pores and small fraction of closed pores. The porosity of open pores was first decreased and then increased, with increasing content of Al. The variation in closed pores followed an opposite trend. The increase in closed porosity was directly caused by the sealing effect of the Al melts.

The sintered ceramics without Al had large pores with an average diameter of 30 μm, close to the size of the voids due to the particle stacking. The small sized pores had an average size of 2 μm, corresponding to those inside the coral-like particles. With the addition of Al, both the number and size of the large pores were reduced, because of the filling of the small Al particles into the spaces of the large coral-like Al_2O_3 particles and the generation of the bond-bridged structures. Meanwhile, the number of small pores with diameters of <5 μm was increased, owing to the formation of the hollow-sphere structures.

Bending strength of the ceramic foams versus the content of Al is depicted in figure 4.24(a). With increasing content of Al powder, the bending strength was increased, due to the formation of more complex and robust bond-bridges structures and the presence of finer Al_2O_3 grains. Over the range of 30.0–45.0 vol%, this increase in bending strength was not very significant, probably because of the excessive expansion and delamination effects. Nevertheless, defect-free ceramic components with different structures, such as honeycomb structure and lattice structure, with very low shrinkages of 1.96% and 1.80%, are demonstrated in figure 4.24(b).

Highly dense (>95%) $Al_2O_3/GdAlO_3$(GAP) eutectic composite ceramics with large smooth surfaces were fabricated by using SLM process, from ceramic powders through one-step melting and solidification [46]. Near net-shaped plates with dimension of $73 \times 24 \times 5$ mm^3 were achieved by using laser pre-heating and multi-track deposition, without the use of binders. The stress between the substrate

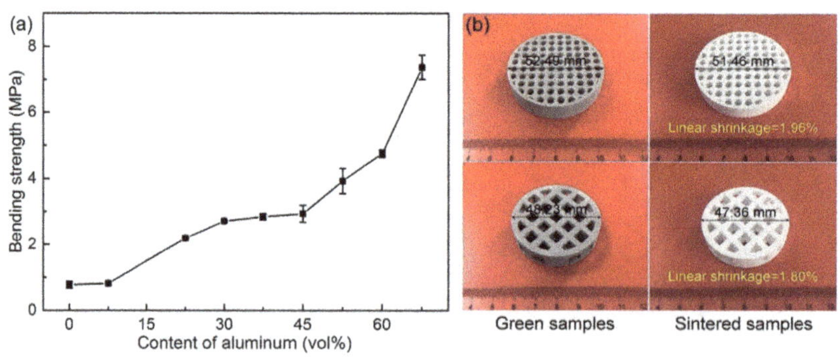

Figure 4.24. (a) Bending strength of the sintered samples versus the content of Al. (b) Photographs of the green samples (left) and sintered samples (right) with honeycomb structure and lattice structure from the CA-A67.5 powders. Reprinted from [43], Copyright (2021), with permission from Elsevier.

and deposited layers during the SLM printing could be minimized by using step-up preheating. With increasing scanning rate, microstructure of materials was transformed from ultra-fine irregular eutectic to complex regular eutectic. The ceramics exhibited microhardness 17.1 ± 0.2 GPa and fracture toughness of 4.5 ± 0.1 MPa · $m^{1/2}$.

Commercially available high purity ceramic powders, Al_2O_3 (>4 N) and Gd_2O_3 (>4 N), were mixed at the eutectic composition, with a mole ratio of 77:23. The Al_2O_3 and Gd_2O_3 had particle sizes of 1–2 and 2–5 μm, respectively. The mixture was ball milled at 400 rpm for 4 h, with 5 wt% polyvinyl alcohol (PVA) solution and 10 wt% ethyl alcohols. Substrates and spherical ceramic powders were then made from the mixture. The substrates were prepared by pressing the mixture at 80 MPa to form green bodies, followed by sintering at 1500 °C for 2 h.

Spherical ceramic powders, with particle sizes of 10–50 μm, were prepared by using the high speed centrifugal spray granulation method. The milled mixture powder was dispersed in DI water, followed by stirring to form ceramic slurry, with desired solid loading levels. The ceramic slurry was pumped to rotating nozzle and spraying dried, forming granulated ceramic powder. The outlet air temperatures were 120 °C–150 °C. The nozzle rotation frequency was 20–25 Hz, while the slurry feeding rate was 5–30 rpm. The powder was sieved and heated at 500 °C for 2 h to bourn out the polymer binder.

SLM printing experiment was conducted in Ar environment, with A 1500 W continuous-wave CO_2 laser (ROFIN DC015) as for scanning. Computer numerical control (CNC) process was developed to control the laser beam through programming, with precise movement in x, y and z-axis directions, as predesigned scanning path of the laser. Processing parameters included laser power of 100–450 W, laser scanning rate of 4–60 mm · min^{-1}, laser beam size of 10 mm, powder bed thickness of 0.5 mm and overlap ratio of 50%.

The powder material consisted of Al_2O_3 and Gd_2O_3 powders, with theoretical mole ratio of Al_2O_3:Gd_2O_3 = 77:23, corresponding to theoretical volume ratio of Al_2O_3: Gd_2O_3 = 63.65:36.35. The sintered ceramic substrate was composed of Al_2O_3 and Gd_2O_3 with theoretical volume ratio of Al_2O_3:Gd_2O_3 = 63.65:36.35. The solidified eutectic ceramics had theoretical volume ratio of Al_2O_3:$GdAlO_3$ = 49.7:50.3.

Macro-morphologies of the SLM processed Al_2O_3/GAP eutectic ceramics with two layers, at different laser powers, at the same scanning rate of 30 mm · min^{-1}, were characterized. The sample processed low laser power of 100 W experienced balling behavior on the surface, due to the agglomeration of the particles caused by the ununiform heating by the laser. This issue can be tackled the optimization of the processing parameters [47, 48]. For instance, as the laser power was >150 W, the balling phenomenon was absent.

The Al_2O_3/GAP eutectic sample, with size of 10 mm × 7 mm × 2 mm, made at laser power of 200 W and scanning rate of 8 mm · min^{-1}, possessed promising quality. The formation process of eutectic ceramics has three areas, corresponding to different morphologies, i.e., initial area, stable area and quenching area. Cracks could be formed in the initial area, due to the large thermal stress between the laser beam with high energy and the cold powder bed. The surface became smooth in the stable area. A large melting pool was formed in the quenching area, where the laser was suddenly stopped scanning. As the scanning rate was increased to 16 mm · min^{-1}, a waved surface was present in the stable area of the sample.

The bend of the periodical striations was along the scanning direction of laser, owing to the periodical melting-solidification process. In this case, heat convection was driven by the surface tension in the molten pool, while the gradients in pressure over the melt were related to the evaporation, bubble formations, mechanical or power fluctuations [49].

With a Gaussian distribution in laser energy, the temperature at the center of the molten pool was higher than that in the other areas. As a consequence, the surface tension was increased, so that the center part of the molten pool was pushed to the edge. Therefore, ordered stripes were formed, after the molten pool was rapidly solidified. Furthermore, striations were present as an arc shape, owing to the ellipsoidal shape of the molten pool. As the scanning rate was properly decreased, the molten pool would be more stationary, while the formation of the stripes was suppressed or even prevented.

The sample prepared with a pre-heating method, at a laser power of 250 W and scanning rate of 1000 mm · min^{-1} for one time, followed by a single layer at laser power of 250 W and scanning rate of 30 mm · min^{-1}, exhibited cracks and serious warping deformations on the surface. Even when the pre-heating powder bed was repeated three times, at the same laser power of 250 W and scanning rate of 1000 mm · min^{-1}, cracks were present at the edge of the ceramics.

When a step-up pre-heating method was adopted, the surface was flat and deformations was suppressed. In thc cxperiment, the powder bed was preheated one time, at a laser power of 150 W and scanning rate of 1000 mm · min^{-1}. After that, the powder bed was preheated for the second time, at 200 W and 1000 mm · min^{-1}, followed by a third pre-heating at 250 W and 1000 mm · min^{-1}. Then, the single layer was printed, at 250 W and 30 mm · min^{-1}.

As the laser power was decreased to 200 W, the same step-up preheating process remained, and the surface became smoother and more uniform. Meanwhile, the surface of the sample bended to the substrate, due to the Gaussian distribution of the laser energy. Cracks were present in the overlap areas. The samples printed with the same preheating process, i.e., 150–200–250 W and 1000 mm · min^{-1}, demonstrated higher surface quality, with low surface roughness of 3.28 and 2.09 μm. Al_2O_3 and GAP were the only phases in the as-solidified Al_2O_3/GAP eutectic ceramic sample. $Gd_3Al_5O_{12}$ and $Gd_4A_{12}O_9$ in the Al_2O_3–Gd_2O_3 binary phase system were not formed.

Figure 4.25 shows the SEM images of the as-solidified Al_2O_3/GAP eutectic ceramic samples, processed at a scanning rate of 16 mm · min^{-1} and laser power of 250 W. As seen in figure 4.25(a), high quality interlayer metallurgical bonding was observed in the as-solidified samples, without gapping or delaminating from the substrates. However, the layer-to-layer interface had microstructure that was rougher than that of the interior of the structures, simply because the heat-affected zone (HAZ) was reserved from the previous melted layer [50]. There were pores and cracks in the coarse area of the interface. The pores were formed either due to the evaporation of organic components or from the substrate or powders, because of the high-temperature laser irradiation or high scanning rate [51, 52].

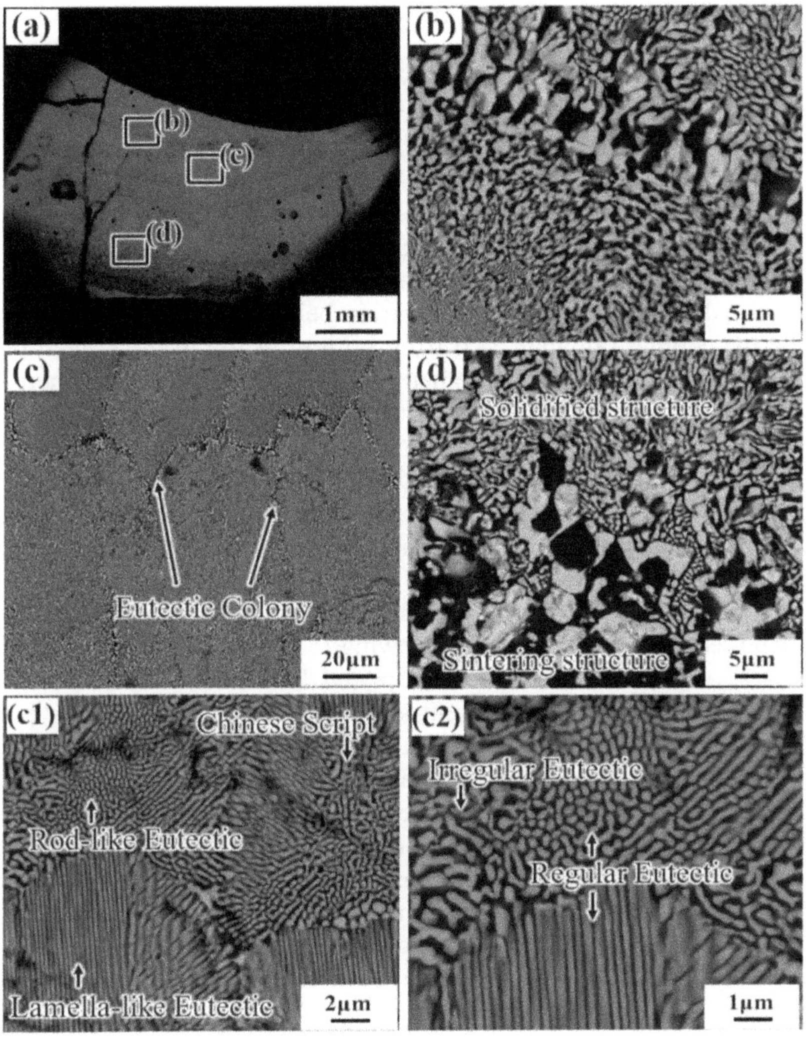

Figure 4.25. SEM images of the samples in different regions of the longitudinal section along the depositional direction: (a) the macroscopic morphology, (b) layer interface morphology, (c) intra-layer microstructure, (c1, c2) magnified morphologies and (d) bottom section of the sample. Reprinted from [46], Copyright (2022), with permission from Elsevier.

A high-magnification SEM image at the interface between two adjacent layers is illustrated in figure 4.25(b). The Al_2O_3 phase and GAP phase were intertwined around one another. The Al_2O_3 tended to form a faceted microstructure, whereas the faceted microstructure was absent for the GAP phase [53]. The concurrent growth behaviors of the two phases led to the formation of the irregular structure. As presented in figure 4.25(c), the colony structures were produced, owing to the interface perturbations and the instability of the planar solid–liquid interface, which was responsible by the high solidification rate [54–56].

High magnification image of the eutectic colonies is shown in figure 4.25(c1), revealing their complex microstructures, in which there were script, lamellar and rod-like patterns exist, with the script patterns dominating the boundary of the colonies. The lamellar and rod-like patterns belonged to regular eutectic structures [57]. The lamellar patterns had structure that was parallel to the direction of the heat flow.

High magnification SEM image at the bottom of the sample in figure 4.25(a) is illustrated in figure 4.25(d). The interface area between the substrate and the printed layer possessed coarser microstructure than the interior of the ceramic layer, owing to the restrained nucleation and crystallization behavior of the materials. As stated previously, the cooling rate increased from bottom to top in the molten pool, thus posing an influence on the development of the microstructures. Therefore, the bottom of the molten pool had microstructures much coarser than that the top of the molten pool.

SEM images of the samples solidified with different preheating processes are shown in figure 4.26. The variation in microstructures of the samples was neglected. Because the scanning rate in the preheating process was 1000 mm · min^{-1}, which was much higher than that used for the solidifying process, the effect on microstructure of the samples was very weak. In comparison, the eutectic spacings were slightly increased from 0.20 to 0.25 μm, owing to the remelting of the samples.

Figure 4.27 shows SEM images of top and bottom parts of the Al_2O_3/GAP eutectic ceramics processed, at different scanning rates, while the laser power was 200 W. Refinement in microstructures at top of the samples was observed with increasing scanning rate. Meanwhile, the average eutectic spacing was decreased, while the number of the eutectic colonies was increased. In contrast, the microstructure of the bottom region kept almost unchanged, with irregular script patterns to be slightly refined with increasing scanning rate.

The microstructure at the top was changed from the script patterns to colonies, regular rod-like and lamella-like eutectic ones, with increasing scanning rate, as observed in figure 4.27(c1). At the same time, the structures became to be from irregular to regular, when the scanning rate was relatively high [58]. According to the results of finite-element numerical simulation, both the temperature gradients and the solidification rates were decreased from top part to bottom part. The closer to the top part, the more regular of the microstructure would be.

At the scanning rates of $\geqslant$30 mm · min^{-1}, the microstructure of the top part became nearly entirely regular. With a high-temperature gradient and the rapid solidification process, the faceted irregular eutectic structures usually have no sufficiently high stability. Therefore, the faced growth behavior was weakened or transient to non-faceted growth mode at high solidification rates [59–61]. This explains the transition to complex regular eutectic structures in the Al_2O_3/GAP eutectic growth processes.

Eutectic spacing is directly dependent on scanning rate, which has strong influence on mechanical properties of eutectic ceramics. Microstructures of the top, middle and bottom parts of the solidified eutectic samples processed at the scanning rates of 4 and 30 mm · min^{-1} were compared. The eutectic spacing was

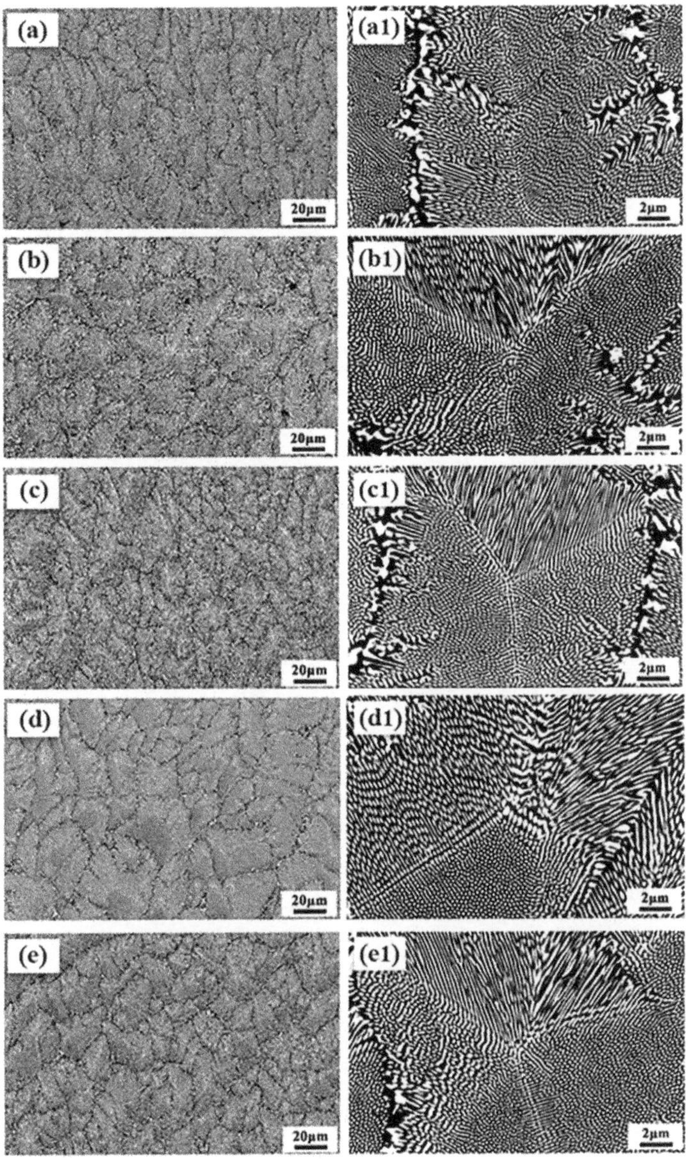

Figure 4.26. Cross-sectional SEM images of the multi-track Al_2O_3/GAP eutectic ceramics processed by using different pre-heating strategies with low (a–e) and high (a1–e1) magnifications: (a) pre-heating at laser power of 250 W and scanning rate of 1000 mm · min^{-1} for one time and then 250 W + 30 mm · min^{-1} for SLM single layer, (b) pre-heating at 250 W + 1000 mm · min^{-1} for three times and then 250 W + 30 mm · min^{-1} for one SLM layer, (c) step-up pre-heating at 150 W + 1000 mm · min^{-1}, 200 W + 1000 mm · min^{-1} and 250 W + 1000 mm · min^{-1}, gradually and then 250 W + 30 mm · min^{-1} for SLM single layer, (d) surface remelting after the scanning strategies in panel (c) and (e) step-up pre-heating at 150 W + 1000 mm · min^{-1}, 200 W + 1000 mm · min^{-1} and 250 W + 1000 mm · min^{-1}, gradually and then 200 W + 30 mm · min^{-1} for SLM single layer. Reprinted from [46], Copyright (2022), with permission from Elsevier.

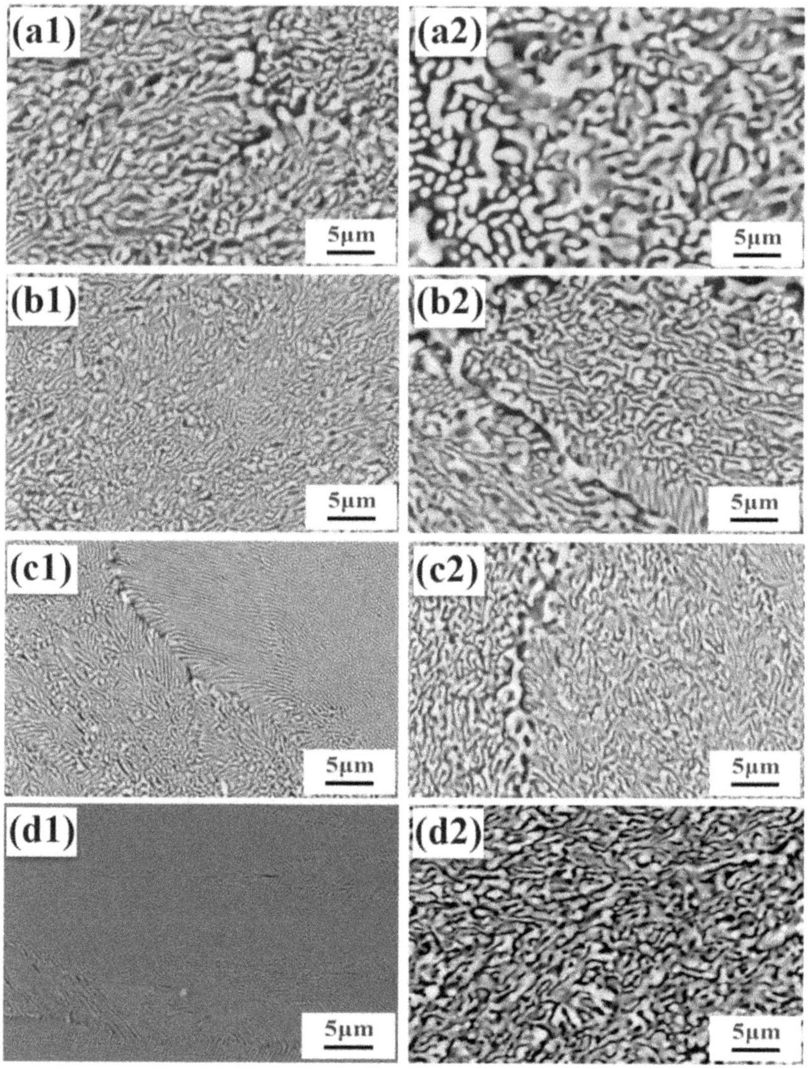

Figure 4.27. SEM images at the top (a1–d1) and bottom (a2–d2) parts of the Al_2O_3/GAP eutectic ceramics processed at different scanning rates: (a1, a2) 4 mm · min^{-1}, (b1, b2) 8 mm · min^{-1}, (c1, c2) 16 mm · min^{-1} and (d1, d2) 30 mm · min^{-1}. Reprinted from [46], Copyright (2022), with permission from Elsevier.

gradually increased from top part to the bottom part. The growth behavior of the phases was directly related to the thermal undercooling. The higher the undercooling, the finer the grains of the samples would be. Nevertheless, the solidification rate positively affected the thermal undercooling.

Maximum average hardness and fracture toughness of the SLM processed Al_2O_3/GAP binary eutectic ceramics were 17.1 ± 0.2 GPa and 4.5 ± 0.1 MPa · $m^{1/2}$, respectively, as the laser power was 200 W and scanning rate was 10 mm · min^{-1}. When the scanning rate was increased from 4 to 60 mm · min^{-1}, the hardness was slightly increased. The fracture toughness was first increased and then decreased. Mechanical properties of the SLM processed Al_2O_3/GAP eutectic *in-situ* composite ceramics were higher than that fabricated by using other methods, such as floating zone melting method [62].

The same authors reported the fabrication of highly dense $Al_2O_3/GdAlO_3/ZrO_2$ ternary eutectic ceramics, by using SLM processes, with mixed ceramic powders without post-sintering treatment [63]. The as-processed eutectic ceramics were composed of the three phases of Al_2O_3, $GdAlO_3$, with homogeneous three-dimensional network structure and sub-micron sized grains. Three transitions in microstructure were observed, i.e., from 'Chinese script' structure to rod-like eutectic, 'Chinese script' structure to lamellar eutectic and lamellar to rod-like eutectic ones, in the single melted pool, owing to the variation in the eutectic growth behavior and the volume fraction of three phases.

The interphase spacing was decreased from 0.92 to 0.48μm, with increasing solidification rate, following the relationship of $\lambda = 4.82 \times V_s^{-1/2}$. Hardness values were 14.3 and 15.3 GPa, while the fracture toughness levels were 6.1 to 7.8 $MPa \cdot m^{1/2}$, for the top to bottom parts of the molten pool. The difference in mechanical properties in the molten pool was ascribed to the factor that the cracks propagated along a straight line in the top zone, while crack bridging and arrest were present in the GAP phase near the zone of bottom.

Commercial powders (purity >4 N) of Al_2O_3, Gd_2O_3 and ZrO_2 were used to prepare the ternary samples. The three powders, with the eutectic composition of $Al_2O_3:Gd_2O_3:ZrO_2$ = 58:19:23 mol%, together with 10 wt% PVA, were mixed through ball milling for 5 h at 550 rpm. Half of the mixture was spray dried to form spherical powders. After calcination at 400 °C for 0.5 h, the PVA was removed, leading to powder with particle sizes of 10–50 μm. The other half was pressed into pellet green bodies and then sintered at 1500 °C for 2 h, which were used as substrates.

A lab-made SLM facility, equipped with a continuous wave CO_2 laser (Rofin DC-015), having maximum power of 1500 W, was used to fabricate the eutectic ceramic samples. The spherical ceramic powders were coated on the substrate, with a thickness of 0.5 mm, by using a roller. The thin layer of the powders was preheated for several times with laser scanning, at stepped powers of 50–200 W and the scanning rate of 1000 $mm \cdot min^{-1}$, preventing it from cracking. The practical experiment parameters included a laser power of 200 W, scanning speed of 6 $mm \cdot min^{-1}$ and laser beam size of 8 mm. The printing process was conducted in Ar environment to minimize porosity of the printed samples.

The as-printed sample macroscopically exhibited a concave shape, which was 11 mm in width and 60 mm in length. The width was up to ten times larger than that of the eutectic samples obtained by using laser floating zone melting method [62, 64, 65]. To ensure the quality of eutectic ceramics, it is necessary to prevent the gas bubble induced porosity and the thermal stress induced cracks. With optimal processing parameters, the SLM eutectic ceramics possessed glossy surface, with relative density of >98%. The solidified layers and the substrate were strongly bonded, without visible space. Except the quenching zone, almost no pores or cracks were observed in the samples.

The cross section of the molten pool had an arc-shaped profile, which was closely related to the Gaussian distribution in energy of the laser beam. The energy is highest at the center of the laser beam, so that the center area of the molten pool was melted most intensively, as compared with the edge areas. As a result, the solid–liquid interface macroscopically displayed a non-planar profile. In this regard, the solidification direction and solidification rate could be different between the bottom and surface parts of the molten pool, thus influencing microstructure of the eutectic ceramics.

Figure 4.28 shows SEM images of the molten pool in the direction of the edge and vertical center. The microstructure was developed parallelly in the direction that was perpendicular to the boundary of the molten pool, as depicted in figures 4.28(a)–(c). There was an angle between the growth direction of the microstructure and the scanning direction of the laser beam, while the value of the angle was reduced from the bottom part to the top part of the molten pool, as observed in figures 4.28(d)–(f). The growth direction of the microstructure is strongly dependent on the solidification process [66]. Usually, the heat dissipation is mainly through the substrate in the process of SLM [67]. Therefore, the heat flux is released most quickly in the direction perpendicular to the solid–liquid interface, so that the microstructure is developed in the direction opposite to that of the heat flux.

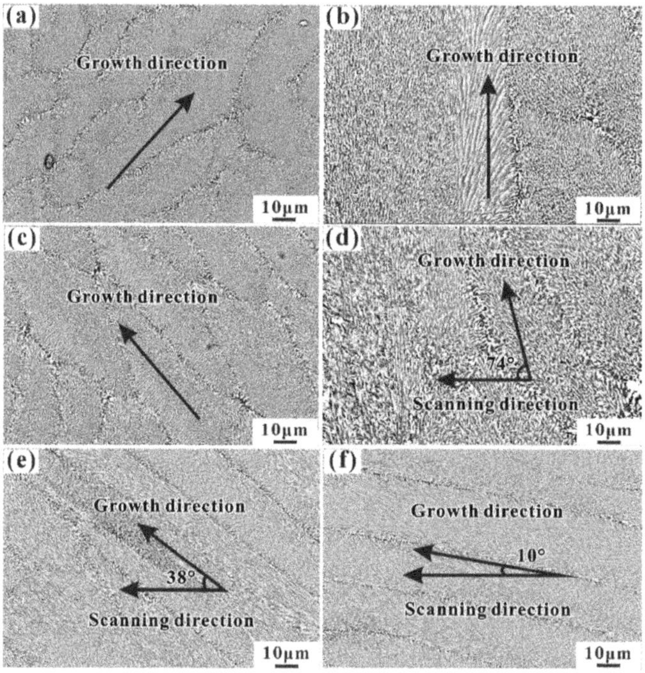

Figure 4.28. SEM images showing the directions of microstructure development in different regions of the molten pool: (a) left boundary, (b) bottom boundary, (c) right boundary, (d) bottom zone, (e) middle zone and (f) top zone. Reprinted from [63], Copyright (2018), with permission from Elsevier.

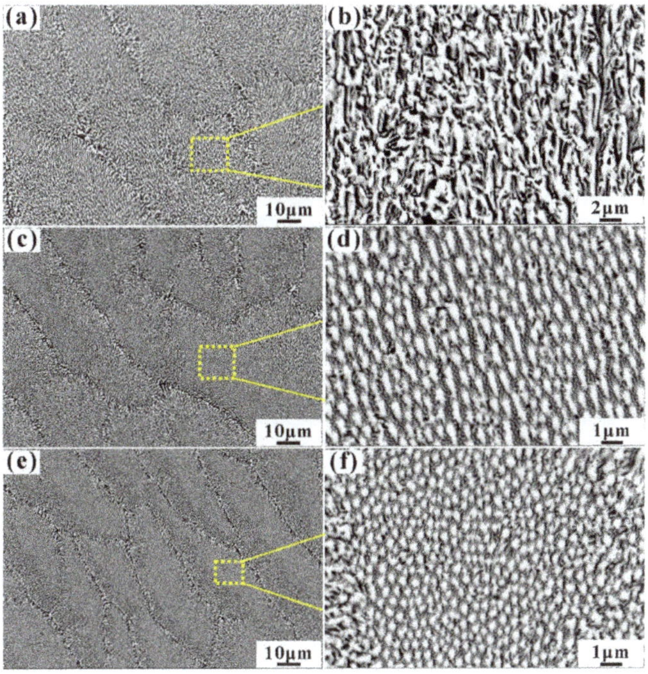

Figure 4.29. SEM images of different regions of the molten pool: (a, b) bottom zone, (c, d) middle zone and (e, f) top zone. Reprinted from [63], Copyright (2018), with permission from Elsevier.

Figure 4.29 shows SEM images of different regions of the molten pool. According to the high magnification images, three phases could be identified, which were Al_2O_3, GAP and stabilized tetragonal ZrO_2. The Al_2O_3 and GAP were intergrown, while finer ZrO_2 phase was dispersed at the Al_2O_3/GAP interface of, as presented in figures 4.29(b), (d), and (f). Because of their high fusion entropies, they exhibited typical irregular and complex-regular eutectic microstructures [68]. Irregular microstructures are generally formed, as the faceted phase growth takes place [69].

The eutectic ceramics exhibited different microstructural morphologies, including scrip structure, lamella structure and rod-like structures, as illustrated in figure 4.30 (a). There were structural transitions in the eutectic ceramics, including script-rod transition, script-lamella transition and lamella-rod transition, from the bottom part to the top part of the molten pool, as demonstrated in figures 4.30(b)–(d). The eutectic structural transitions were ascribed to the changes in the growth behavior and volumetric content of the component phases.

The growth behavior of compounds is also dependent on the degree of undercooling [70]. As the undercooling was sufficiently high, the compound with a high entropy of fusion would grow into a non-faceted structure. Given the relationship between undercooling and solidification rate, the degree of undercooling was increased from the bottom part to the surface part of the molten pool. As a consequence, Al_2O_3 that should be grown into a faceted structure actually was

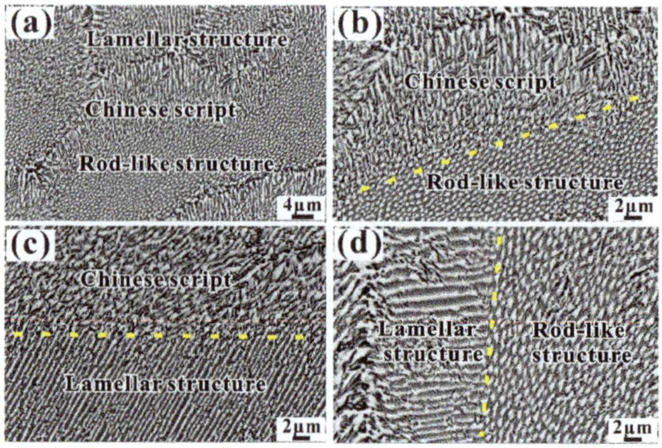

Figure 4.30. SEM images illustrating microstructure and morphology development of the SLM processed Al_2O_3/GAP/ZrO_2 eutectic ceramics: (a) presence of three structures, (b) transition from script to rod-like eutectic structure, (c) transition from script to lamellar eutectic structure and (d) transition from lamellar eutectic structure to rod-like eutectic structure. Reprinted from [63], Copyright (2018), with permission from Elsevier.

present as a non-faceted one in some parts of the molten pool, thus resulting in the faceted/non-faceted eutectic growth pattern.

When the eutectic growth follows the faceted/non-faceted mode, a lamellar or rod-like structure will be developed, depending on the volumetric ratio of the phases in the systems [69]. When isotropic surface energy is dominant, a rod-like structure is present as the volumetric content of the minority phase is $<1/\pi$, whereas a lamellar structure is formed as the content is $>1/\pi$. For the laser melted Al_2O_3/GAP/ZrO_2 eutectic ceramics, volume fractions of the phases in the eutectic colonies in different parts of the molten pool were experimentally measured.

The volume content of GAP (F_G) was 44%, which was $>1/\pi$ and nearly remained constant. The values of Al_2O_3 (F_A) were 37.2% ($>1/\pi$) and 29.6% ($<1/\pi$), while those of ZrO_2 (F_Z) were 18.3% ($<1/\pi$) and 27.2% ($<1/\pi$), in the bottom and top parts of the molten pool. The variation in volumetric content of the phases in the eutectic ceramics was related to the compositional segregation from colonies to intercellular regions, thus leading to the formation of the rod-like eutectic microstructure in the top part of the molten pool, as observed in figures 4.29(e) and (f). Therefore, the proportion of rod-like structures was increased with increasing temperature gradient and/or growth rate or the introduction of other components.

The decrease in hardness of the samples was associated with the variation in the volumetric contents of the phases. The hardness of Al_2O_3 is the range of 14.7–16.2 GPa, while that of $ZrO_2 < 13$ GPa [71]. Therefore, the hardness was decreased in the height direction of the molten pool, because the content of Al_2O_3 was decreased and that of ZrO_2 was increased. Additionally, the number of colonies was increased with increasing degree in refinement of the microstructure, thus causing an increase in the concentration of defects, such as cavities and micro-cracks. The presence of more defects would bring about a decrease in hardness.

The decrease in fracture toughness was related to the effect of thermal residual stress. In terms of crack propagation, the regions with tensile stress are easily penetrated by the front of the cracks, while deflection or arrest occur in the regions with compressive stress, so that fracture toughness in tensile stress regions would be decreased, whereas that would be increased in regions with compressive stress. During the cooling process of the SLM processed Al_2O_3/GAP/ZrO_2 eutectic ceramics, thermo-elastic stresses were developed, owing to the difference in the coefficient of thermal expansion of different phases [71]. In the Al_2O_3/GAP/ZrO_2 system, coefficients of thermal expansion of the three phases are different. The GAP phase and Al_2O_3 phase are in compressive stress, while the ZrO_2 phase experiences tensile stress.

4.3.3 Zirconia

Parts of this section have been reprinted from [103], Copyright (2024), with permission from Elsevier.

Zirconia, ZrO_2, has been widely used for various industries, such as structural ceramics, solid oxide fuel cells, sensors, electronics and biomaterials [72–82]. It has an extremely high melting temperature of 2690 °C. Pure ZrO_2 has three phases, monoclinic (m), tetragonal (t) and cubic (c), at different temperatures at ambient pressure [83–85]. The monoclinic phase is stable from room temperature to 1170 °C. In the temperature range of 1170 °C–2370 °C, the stable phase is tetragonal, while the cubic phase is present above 2370 °C.

The three phases are reversibly transformed during the heating and cooling processes, which are accompanied by a variation in volume. Such variations in volume are 0.5% for c $\rightarrow$ t and 4% for t $\rightarrow$ m, whose consequence is the occurrence of structural catastrophic failure. To address this problem, zirconia ceramics should be partially or fully stabilized. The stabilized zirconia exhibits outstanding mechanical properties, such as extremely high strength and fracture toughness. Moreover, the related materials have high chemical stability, special electrical properties and high biocompatibility.

Zirconia with multi-phase compositions is usually called partially stabilized zirconia (PSZ), which possesses special properties for applications [86–88]. For instance, tetragonal grains are distributed in a cubic matrix, through the incorporation of CaO. As the tetragonal phase, which is metastable, is transformed to the monoclinic phase, the fracture toughness and strength are significantly improved. The 3.5 wt% CaO doped zirconia is fully stabilized to become the cubic phase.

Various oxides, including MgO, Y_2O_3, La_2O_3, CeO_2, have been employed to stabilize the phase structures of zirconia. Y_2O_3 doped zirconia, known as yttria stabilized zirconia (YSZ), is the most widely and extensively studied [89–91]. It is also denoted as *Y*-TZP, according to the content of molar ratio of yttria. For example, 3 mol% yttria stabilized zirconia is named 3YSZ or 3Y-TZP.

Spherical zirconia-polypropylene (ZrO_2-PP) composite powder was prepared through a thermally induced phase separation (TIPS) approach for SLS processing to fabricate zirconia ceramics [92]. At optimal conditions, the SLS-printed green bodies had a relative density of up to 36%. Then, pressure infiltration (PI) and warm isostatic pressing (WIPing) were utilized to further increase the density of the ZrO_2-PP green bodies. After infiltrating in 30 vol% ZrO_2 aqueous suspension, the density was increased from 32% to 54%. With WIPing at 135 °C and 64 MPa), the density of the final 3YSZ ceramics reached 92%.

The 3 mol% Y_2O_3–ZrO_2 (3YSZ, Tosoh, Japan) powder, with $D_{50} = 30$ nm, prepared through chemical co-precipitation, was used as raw material, while isotactic polypropylene (PP) with an average molecular weight (M_w) of 12 000 ($M_w/M_n = 2.4$, Sigma-Aldrich) was used as a binder. Xylene (p-xylene, reagent grade, Sigma-Aldrich) was utilized as a solvent for thermally induced phase separation. Ethanol (Chem-lab NV, Belgium) was employed to wash and clean the printed structures.

The 3YSZ powder was dispersed in xylene through ball milling for 24 h. The suspension was diluted and stirred in N_2. After heating at 133 °C, a 9 wt% PP xylene solution derived suspension was obtained. The suspension was cooled down to room temperature, so that precipitation occurred. The precipitate was washed with ethanol several times to eliminate xylene, followed by drying at 65 °C, leading to composite powders with 30 and 40 vol% ZrO_2.

A Sinterstation 2000 (DTM Corporation/3D Systems) used for SLS printing was equipped with a 100 W CO_2 laser (f100, Synrad, USA), while the laser beam diameter was 400 μm. SLS parameters included powder bed preheating temperature and thickness of single printed layer, whereas laser parameters included power, scan spacing and scan rate. The SLS printing experiment was carried out in N_2 protection. After printing, the non-sintered composite powder was collected for recycling.

30 vol% 3YSZ aqueous suspension was used for slurry pressure infiltration. The suspension was prepared through multidirectional mixing (Turbula type mixer) for 24 h and electrostatically stabilized by the addition of HNO_3. Both the ZrO_2 suspension and SLS processed green bodies were put into stainless steel container, followed by applying pressure for 5 min at levels of 3, 16 and 32 MPa. After infiltration, the samples were dried at 60 °C for 2 h.

For WIPing, the device was combined with a heating jacket. The SLSed and SLSed/PI samples were all subject to WIPing. All the green bodies were calcined at 600 °C in air at a heating rate of 0.1 $°C \cdot min^{-1}$ to burn out the binder and then sintered at 1450 °C for 2 h in air. The sintered samples were polished and then thermally etched in air at 1350 °C for 20 min for SEM observation.

In the TIPS process, the polymer is dissolved in solvent through heating, after which phase separation is induced as the solution is cooled down. The concentration of polymer in the suspensions has a strong effect on morphology and structure of the precipitates. In a polymer-solvent diagram, the point at which the binodal and spinodal curves meet is known as critical composition for a homogeneous polymer solution [93]. When the solutions with critical composition are cooled from a temperature that is above the binodal line, liquid-liquid phase separation will spontaneously occur in the stable region, resulting in polymer-rich and polymer-scarce phases. With further cooling, the system will experience crystallization, forming morphologies of bi-continuous membranes.

For the solutions with polymer concentrations not on the critical point, the liquid-liquid phase separation is merely present, as the fluctuation in concentration is sufficiently high, so that the potential barrier is overcome in the metastable region. In this region, the liquid-liquid phase separation is governed by a nucleation-growth mechanism. As the concentration of polymer is above the critical point and below the monotectic point, the polymer-rich phase would become the matrix, while the phase with less polymer would be expelled. As a result, cellular membrane structures are formed after solidification [94]. In the solutions with polymer concentration below the critical point, droplets of polymer-rich phase are generated in the polymer-poor matrix.

The PP-zirconia suspensions, with 30 vol% ZrO_2 (D_{50} = 27 μm) and 40 vol% (D_{50} = 49 μm) were derived from the PP xylene solution with concentration of 9 wt%. The composite particles with spherical morphology had size distribution of 5–150 μm. To indirectly SLS print the PP-zirconia composite powder, the powder bed was preheated to 145 °C, which was slightly lower than the melting temperature of the polymer. The thickness of the powder layer to be printed was set to be 0.13 mm.

The processing parameters were optimized to ensure the mechanical strength of the green bodies, including laser powers of 3–7 W, scanning rates of 500–1250 mm · s^{-1}, scanning spacings of 0.1–0.2 mm, with dimension of structures to be 15 mm × 15 mm × 10 mm. Laser energy density was in the range of 0.09–1.07 J · mm^{-3}. Too high a laser energy density led to degradation of the polymer, whereas too low an energy density resulted in delamination due to the incomplete molting of the polymer. Therefore, the laser energy density should also be optimized.

The green bodies of the PP-40 vol% ZrO_2 samples however were too weak in mechanical strength for further treatment, while those derived from the PP-30 vol% ZrO_2 composite powder had sufficiently high strength. The density of the green bodies showed a decreasing trend with increasing laser energy density. The samples printed with laser energy densities below 0.2 J · mm^{-3} possessed relative density of 36%, while those printed with higher laser energy densities were highly porous and low densities, owing to the degradation of PP.

Figure 4.31 shows fracture surface SEM images of the as-SLS and WIPed SLS processed samples, presenting inhomogeneous microstructure, with the presence of connected microspheres of the composite powder. The inter-agglomerated voids were retained, owing to the low plastic flowability during the SLS printing, as observed in figure 4.31(a). After the WIPing process, the microstructure became more homogeneous, as illustrated in figures 4.31(b) and (c), suggesting that plastic deformation was induced by the isostatic pressing and hence leading to increased density.

Figure 4.32 shows SEM images of representative zirconia ceramics after sintering. The sample processed with just SLS printing had a relatively low density of 32%, while the layered structure was visible, as seen in figure 4.32(a). The pores were mainly from the intergranular spaces correlated the microsphere layers, as demonstrated in figure 4.32(b). The ceramics processed with pressure infiltration had higher density and more homogeneous microstructure, as revealed in figure 4.32(c). The large intergranular pores were absent, while the pore size was reduced to about 5 μm,

Figure 4.31. Fractured surface SEM images of the green bodies processed at different steps: (a) SLS and (b, c) SLS + WIP. Reprinted from [92], Copyright (2014), with permission from Elsevier.

as illustrated in figures 4.31(e) and (f). This sample had large cracks in the center area, due to the large shrinkage caused by WIPing process. The linear shrinkages were 48% in the *X*-/*Y*-direction and 52% in the printing direction.

As the PI and WIPing processes were combined, the large cracks were prevented, although the microstructure was observed, as presented in figures 4.31(g) and (h). The linear shrinkages of the ceramics derived from the SLS processed green bodies after PI and WIPing were 38% in the *x*-/*y*-direction and 42% in the printing direction. The PI/WIPed ZrO_2 ceramics exhibited a certain porosity. The grain size in the dense areas was at the submicrometer scale, implying that sintering parameters were suitable for the experiment.

7YSZ zirconia ceramics were prepared by using the SLM process with a 1 μm wavelength fiber laser and high-temperature preheating at 1500 °C–2500 °C [95]. An Nd-YAG laser could be used to preheat over an range of 10 mm diameter, with which orderly cracks were transformed into disordered small cracks. With preheating at 1500 °C–2500 °C, the relative density of the zirconia ceramics, made from the mixture of 20 wt% fine powder (9–22.5 μm) and 80 wt% coarse powder (22.5–45 μm), was increased to 90%–91%, from 84% for the counterpart without preheating. The preheating was beneficial to having a tetragonal phase. The SLM machine was equipped with a high power Nd-YAG laser preheating system and a 120 W power Gaussian profile fiber scanning laser. The printed samples were 25 mm^3 cubic.

The profiles of surface temperature versus time for different preheating temperatures were examined, consisting of three sections. In the preheating process, the real temperature reached the predetermined value after 17–27 s, depending on the parameters of the preheating laser. The second was the powder melting process due to the irradiation of the SLM laser. The rest was related to the new layer deposition and the cooling. After printing, surface temperature of the samples dropped to 600 °C in 20–30 s.

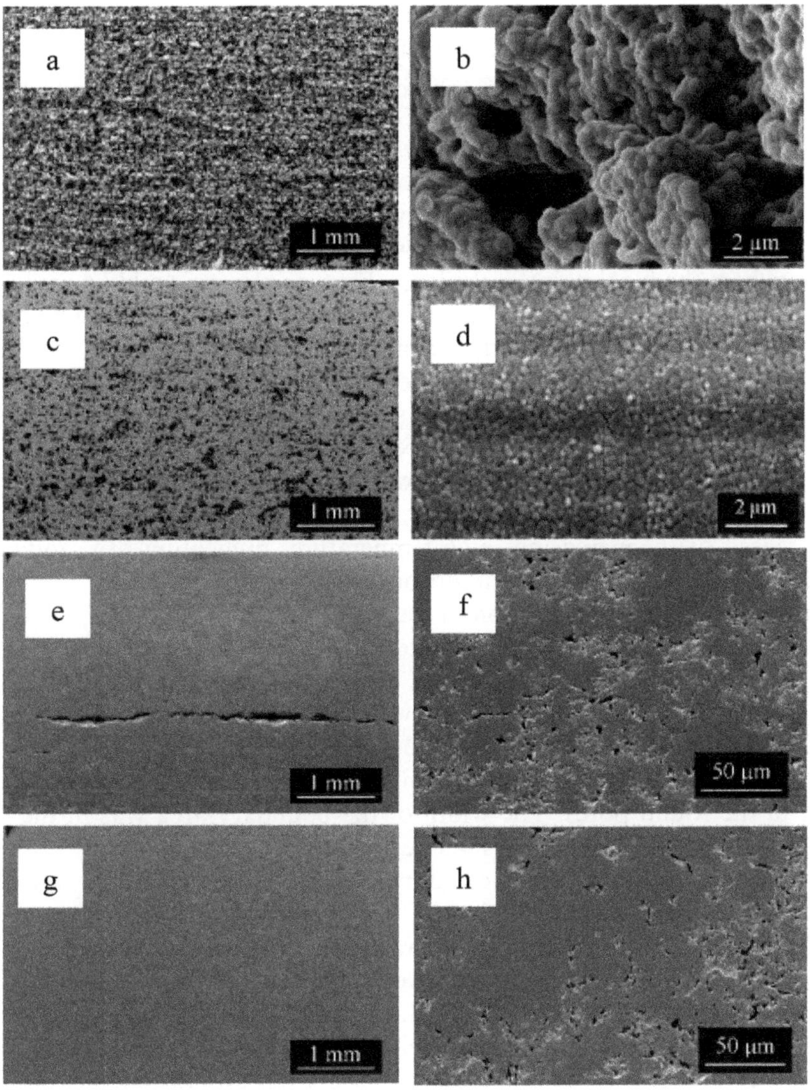

Figure 4.32. Cross-sectional SEM images of the sintered samples produced with different processing steps: (a, b) SLS, (c) SLS + PI, (e, f) SLS + WIP and (d, g, h) SLS + PI + WIP. Reprinted from [92], Copyright (2014), with permission from Elsevier.

As the preheating temperature was set to be 1500 °C, the average surface temperature of the sample was raised rapidly to 1800 °C, with an increase magnitude of 300 °C. However, for preheating at 2000 °C and 2500 °C, the laser irradiation-induced surface temperature was increased by 100 °C, because too high a preheating temperature reduced the input energy of the scanning laser.

The center of the preheating area of the samples printed at 60 W at different preheating temperatures were evaluated. Upon preheating at 1500 °C, the powder in the preheating area was slightly pre-sintered during the printing process. After the printing process was finished, the powder around the scanned area could be easily taken away. However, if the preheating temperature was 2500 °C, the powder in the preheating area was adhered to the borders of the samples. The powder within the SLM area was entirely melted, while the samples had quite a smooth surface. No spherical particles were present on the surface of the printed samples, so that a new layer of the powder could be readily deposited.

During the high-temperature preheating process, the ceramic powder had been sintered to a certain degree, so that the powder was prevented from blowing away by the plasma from the focal point of the laser, because plasma could be generated due to the high temperature at the SLM laser focal point. Meanwhile, the energy input of the SLM laser was reduced, thus decreasing the formation tendency of plasma.

Figure 4.33 shows SEM images of the samples at different preheating temperatures, with laser power of 60 W. Cracks with relatively high concentration were present. When increasing preheating temperature, the microstructure was improved, in terms of the distribution and morphology of the cracks. Vertical cracks become disordered and the length was reduced, because the SLM laser input energy density was decreased with increasing high-temperature preheating. In addition, the preheating process enabled more uniform distribution of laser energy, which was in turn responsible for the variation in the vertical cracks.

SEM images of the powder preheated to 2000 °C are shown in figures 4.34(a) and (b). The particles were partly connected due to the sintering effect, although the shape of the particles was retained. There were circular pores in the border area, with sizes of 20–100 μm. High viscosity melt pool was formed at the boundary, due to the laser irradiation, which penetrated the unscanned powder and then solidified, thus resulting in the round pores. The width of the boundary area was in the range of 100–200 μm, as observed in figures 4.34(c) and (d).

The roughness of ceramics fabricated by using the SLM printing technique could be as large as 150 μm, while the values of the ceramics processed with conventional ceramic processes is $\leqslant$50 μm [96]. In this work, the roughness values of top and side surfaces were 55 and 154 μm, respectively. The top surface of the sample was irradiated by the laser beam, on which the powder was melted. As a consequence, the viscosity of the melt pool was relatively low, so that the surface of the solidified melt pool was smooth owing to the effect of surface tension. In comparison, since the powder at the boundary was not completely melted, the melt pool had higher viscosity, so that the powder that was not scanned by the laser was agglomerated, thus leading to higher surface roughness.

Relative density of the samples preheated at 1500 °C, 2000 °C and 2500 °C was higher than that of the one preheated at 250 °C. The relative density reached a maximum value of 91%. Moreover, with preheating at high temperatures, the surface density of cracks was reduced, while the microstructure was improved. When the laser power is relatively low, the samples processed at relatively low power of laser had high porosity. With increasing preheating temperature, the porosity was also decreased.

According to XRD characterization results, the samples were mainly of tetragonal phase, irrespective of preheating temperature, although the peak intensity was

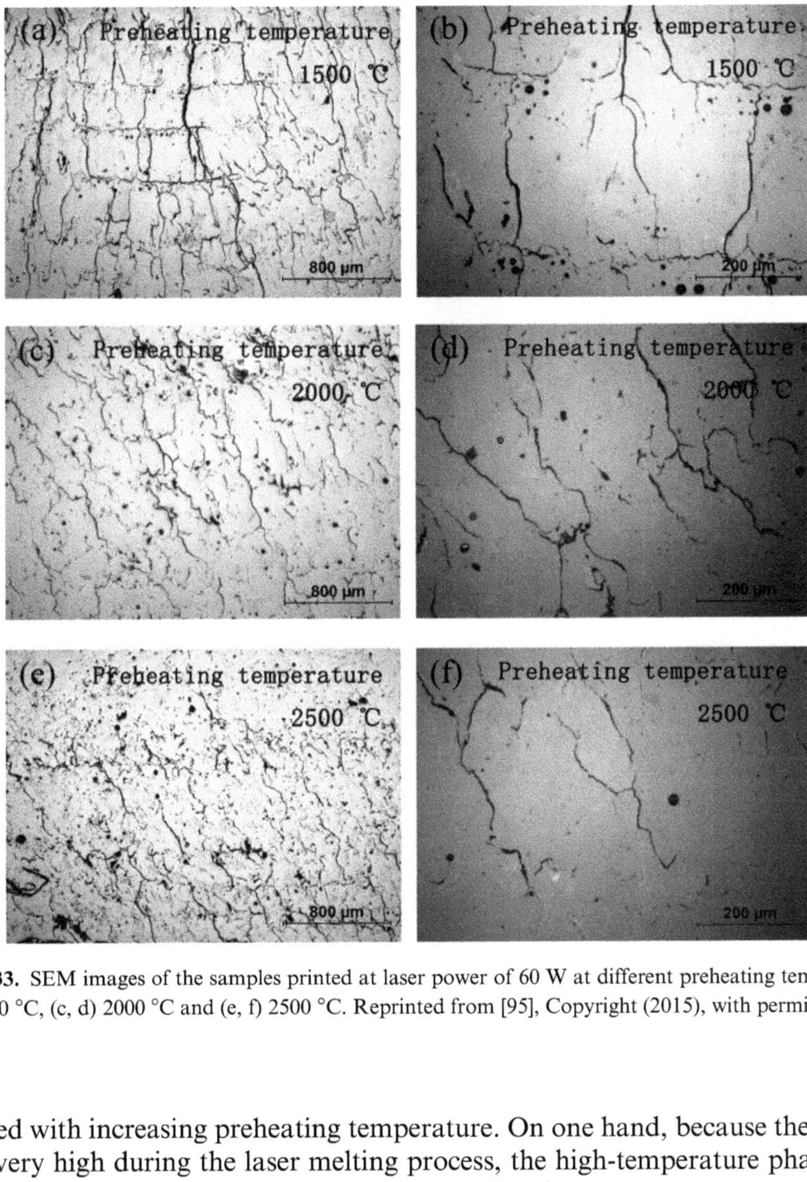

Figure 4.33. SEM images of the samples printed at laser power of 60 W at different preheating temperatures: (a, b) 1500 °C, (c, d) 2000 °C and (e, f) 2500 °C. Reprinted from [95], Copyright (2015), with permission from Elsevier.

increased with increasing preheating temperature. On one hand, because the cooling rate is very high during the laser melting process, the high-temperature phases, i.e., tetragonal and cubic, could be retained after cooling down to room temperature. On the other hand, the diffusion of yttria was enhanced due to the laser irradiation.

Figure 4.35 shows SEM images of the single line scanning tracks printed at different preheating temperatures and different scanning rates, while the laser power was 60 W and other process parameters were the same. Width of the single line scanning track was in the range of 90–300 μm, showing an increase trend with decreasing laser scanning rate. For the sample processed at preheating temperature of 25 °C and scanning rate of 0.400 $m \cdot s^{-1}$, the particles were incompletely melted, while there were particles to be bonded to the scanning track of the single line.

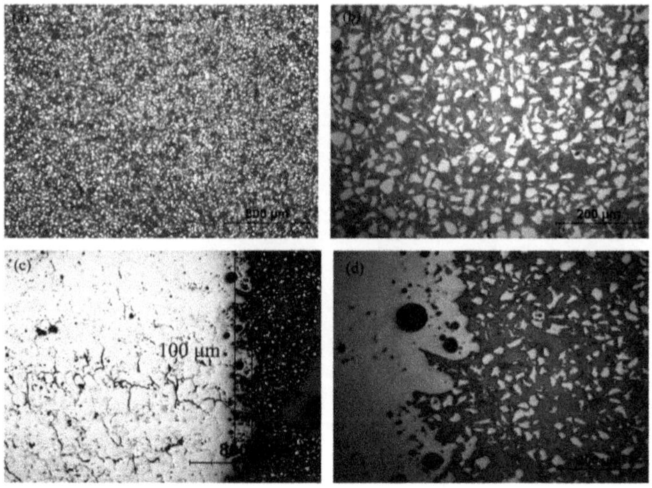

Figure 4.34. SEM images of representative samples preheated at 2000 °C: (a, b) powder and (c, d) border. Reprinted from [95], Copyright (2015), with permission from Elsevier.

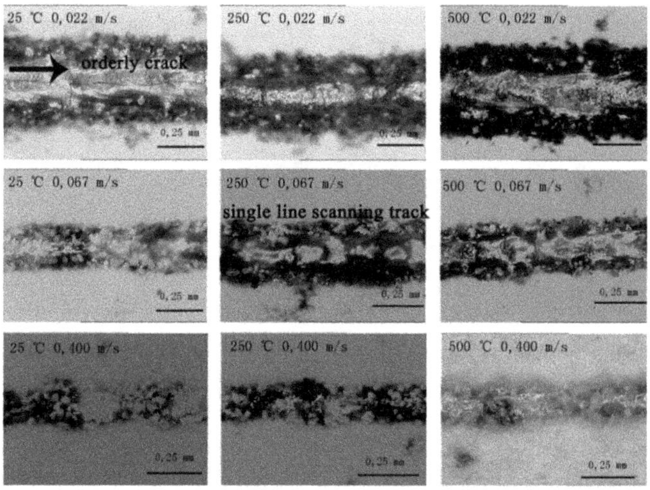

Figure 4.35. SEM images of the single line scanning tracks printed at different preheating temperatures (25 °C, 250 °C and 500 °C) and scanning speeds (0.022, 0.067 and 0.400 $m \cdot s^{-1}$). Reprinted from [95], Copyright (2015), with permission from Elsevier.

The bonded particles could be easily taken away, whereas the scanning track was not continuous.

As the scanning rate was reduced to 0.067 $m \cdot s^{-1}$, the powder could be completely melted, so that the single line scanning track was continuous in the center. When the scanning was further decreased to 0.022 $m \cdot s^{-1}$, the cross section of the single line scanning track became an arc, which was the consequence of surface tension. When the preheating temperatures were set to be 250 °C and 500 °C, the effect of surface tension was suppressed. As a result, a wider scanning track was available.

For the single line scanning tracks at 25 °C, as the scanning speed was higher than 0.067 m · s^{-1}, the sample possessed very low density, because there are many unbelted particles on both sides of the single line scanning track. Therefore, the center of sample was highly porous, resulting in a rough surface. As the scanning rate was below 0.033 m · s^{-1}, the width of the single line scanning track was >200 μm. The sample more seriously deformed. Optimal scanning rate was within the range of 0.100–0.033 m · s^{-1}, so that the particles were completely fused inside the scanning track and the deformation was prevented.

The presence of the pores inside the tracks originated from the air that could not escape during the printing process. With increasing scanning rate, the pores were destroyed, so that dense single line scanning tracks were gradually formed. Owing to the Gaussian distribution in energy of the laser beam, the central area had a higher energy density than the border. During the cooling process, the cooling rate was different in the different areas. For instance, the center was cooled more slowly than the edges of the tracks. In other words, the edges were solidified earlier than the central area. Therefore, the central area was under tensile stress. In this case, if the central region was not sufficiently strong to withstand the stress, a vertical crack would be produced in the central area.

As the preheating temperature was much lower than the melting point of YSZ, the energy for the melting of YSZ was only supplied by the laser. Therefore, the difference in temperature between the melted area and the surrounding powder was increased. The cooling rates in different areas would be different, thus generating cracks. In addition, the increase in laser power and irradiation time would also result in inhomogeneity of energy distribution in the laser focal spot.

Because the scan paths were at the same position in the adjacent odd/even layers, short ordered cracks tended to be connected to produce long ordered cracks, so that vertical through cracks would be in the vertical section of the sample. As the preheating temperature was very close to the melting point of YSZ, the energy input to melt the powder could be relatively lower. As a result, the energy of the laser would be absorbed by the surface of the ceramics, while the solidified layer would not be reheated. In this case, the use of high-temperature preheating effectively suppressed the formation of vertical through cracks.

Dense 8YSZ ceramics were prepared by using SLS printing technology [97]. To enhance laser–matter interaction for YSZ, graphite was introduced into the YSZ powder, resulting in an increase in laser energy absorption efficiency from 2% to 60%. At optimal conditions, the YSZ ceramics could reach high relative density of 96.5%. A commercial SLS machine, ProX 200 from 3D Systems, was utilized for the experiment, with Nd:YAG laser having maximum power of 300 W. The building platform had dimensions of 140 × 140 mm^2, while the maximal height was 100 mm. The printing process was conducted at ambient atmosphere.

A representative cross-sectional microscopic image, which was parallel to the building direction, is shown in figure 4.36(a), demonstrating that the consolidated YSZ layer was dense and homogeneous, with very few cracks. This suggests that the

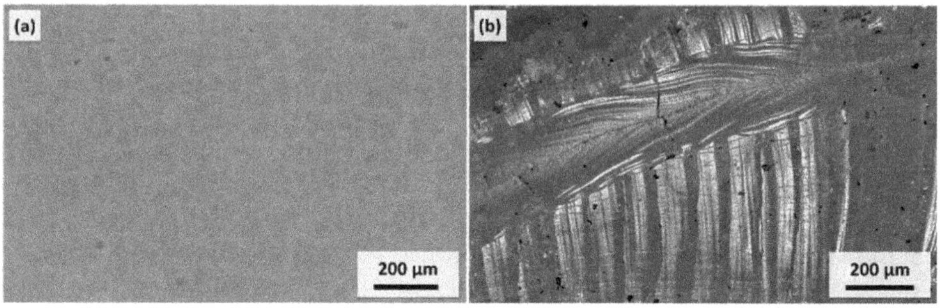

Figure 4.36. Optical microscopic images of the YSZ sample prepared by using the SLS process: (a) cross-section and (b) surface. Reprinted from [97], Copyright (2018), with permission from Elsevier.

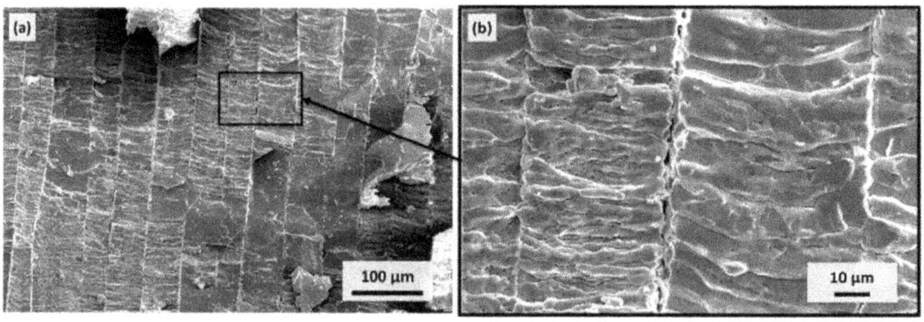

Figure 4.37. Fractured surface SEM images of a representative laser-sintered YSZ sample: (a) low magnification and (b) high magnification. Reprinted from [97], Copyright (2018), with permission from Elsevier.

printed items were not obviously laminated. The samples had isotropic mechanical properties with a smooth surface, as illustrated in figure 4.36(b). The starting powder was gray in color due to the addition of graphite, while the final ceramics were white, indicating that the graphite had been completely eliminated during the printing process.

The printed samples exhibited a laminated microstructure, as seen in figure 4.37. The printed layer was characterized by the stacking of grains with an elongated shape perpendicular to the scanning direction. Such microstructures were different from that of the same starting powder processed by using the conventional process, after sintering at 1600 °C for 2 h, as revealed in figure 4.38. The layer thickness was 50 ± 8 μm, while the grain size 10 ± 2 μm. The formation of the characteristic microstructures was a result of the combination of the processing parameters.

8YSZ ceramics were developed by using SLS/SLM process, with Nd:YAG laser [98]. Particularly, titanium carbide (TiC) was added to the starting powder as absorbance enhancer to increase the laser energy absorbing efficiency. TiC was selected because of its for relatively low weight/absorbance ratio, as compared with silicon carbide, carbon black and graphite.

The 8YSZ powder was prepared by using combustion method. Zirconium nitrate ($Zr(NO_3)_4 \cdot 5H_2O$) and urea (CH_4N_2O) were dissolved in DI water, while Y_2O_3 was

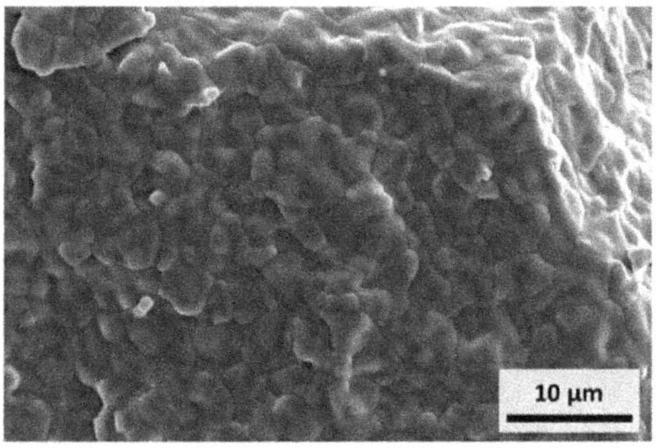

Figure 4.38. Fractured surface SEM image of a representative YSZ ceramics prepared by using the conventional process. Reprinted from [97], Copyright (2018), with permission from Elsevier.

dissolved nitric acid (HNO_3). The three solutions were mixed and then heated to combust, obtaining 8YSZ powder. The 8YSZ powder was dry milled for 1 h with zirconia balls of Ø = 2 cm and then wet milled for 30 h in ethanol with zirconia grinding balls of Ø = 1.1 μm. The wet milled 8YSZ suspension was dried at 100 °C for 12 h. Wet dried 8YSZ powder was dispersed in DI water, together with binder. This suspension was then spray dried, debinded at 500 °C to remove the residual binder and then calcined at 1200 °C for 2 h to increase cohesion and prevent the spheres from breaking.

Absorbance enhancing agents included SiC, TiC, carbon black (CB) and graphite, with average particle sizes (D_{50}) of 0.55, 0.85, 1.7 and 1.5 μm, respectively. They were mixed respectively with the 8YSZ powder through ball milling for 24 h. A 3D Systems ProX 200 was used for the SLS printing experiment. Laser parameters included power of 30–96 W, scanning rate of 8–80 mm · s^{-1}, hatch distance of 35–85 μm and laser focus range of −0.5–1.5 mm. Layer thickness, spot size and compaction rate were kept as 100 μm, 75 μm and 300%, respectively. The printing platforms were Al plates with dimensions of 14 cm × 14 cm. The samples to be printed were cuboids with dimensions of 0.8 cm × 0.8 cm × 0.3 cm spaced by 0.4 cm and 1.4 cm × 1.4 cm × 0.4 cm cuboids spaced by 2 cm.

The composition of the 8YSZ power was $Zr_{0,84}Y_{0,16}O_{1,92}$, with cubic structure. XRD patterns and SEM images of the powders, after as-synthesis, wet milling and spray drying, are shown in figures 4.39 and 4.40, respectively. No phase transition occurred for the 8YSZ powder during the processing. After the combustion reaction, the powder exhibited agglomeration with irregular porous morphology, as presented in figure 4.40(a). A finer morphology was obtained after wet milling, as depicted in figure 4.40(b), had average particle size of D_{50} = 1.1 ± 0.05 μm. The spray dried powder consisted of spherical particles, with uniform particle size distribution and average size of D_{50} = 34.7 ± 0.14 μm, as observed in figure 4.40(c).

The absorbance levels of TiC, SiC, CB and graphite to the laser were 82%, 78%, 80% and 90%, respectively. The mixture with 0.25 wt% TiC displayed an increase in laser absorption by 60%, having the minimal weight/absorbance ratio. Samples were prepared with laser power in the range of 30–57 W at scanning rates of 8–80 mm · s^{-1}.

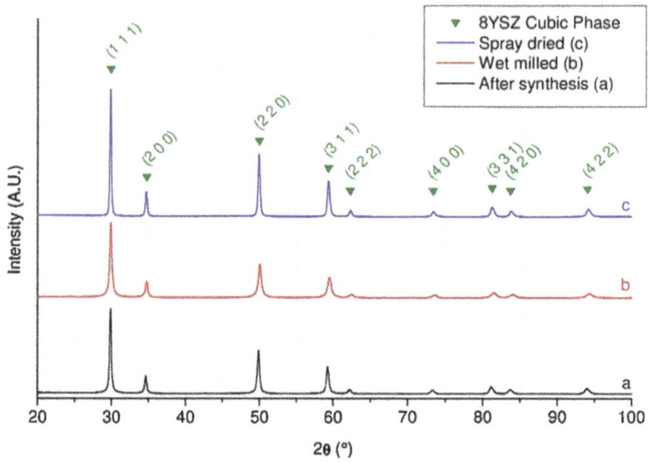

Figure 4.39. XRD patterns of the 8YSZ powders with cubic crystal structure: (a) as-synthesized, (b) after wet milling and (c) after spray drying. Reproduced from [98]. CC BY 4.0.

Figure 4.40. SEM images of the 8YSZ powders: (a) as-synthesized, (b) after wet milling and (c) after spray drying. Reproduced from [98]. CC BY 4.0.

It is desired that the 8YSZ + 0.25 wt% TiC powder was consolidated after attaching the substrate, through metal-ceramic interaction. In terms of formation of consolidated structures, the optimal processing parameters were laser power of 57 W and scanning rate of 32 mm · s^{-1}. Too high a laser power would induce degradation of the platform and consolidation of the powder outside the platform.

Figure 4.41 shows the outcomes printed with laser power ranging from 72 to 96 W, while the scanning rates were in the range of 60–70 mm · s^{-1}, with hatch distance of 50 μm. With increasing laser power, density of the printed samples was gradually increased. Comparatively, density of the sample processed at a scanning

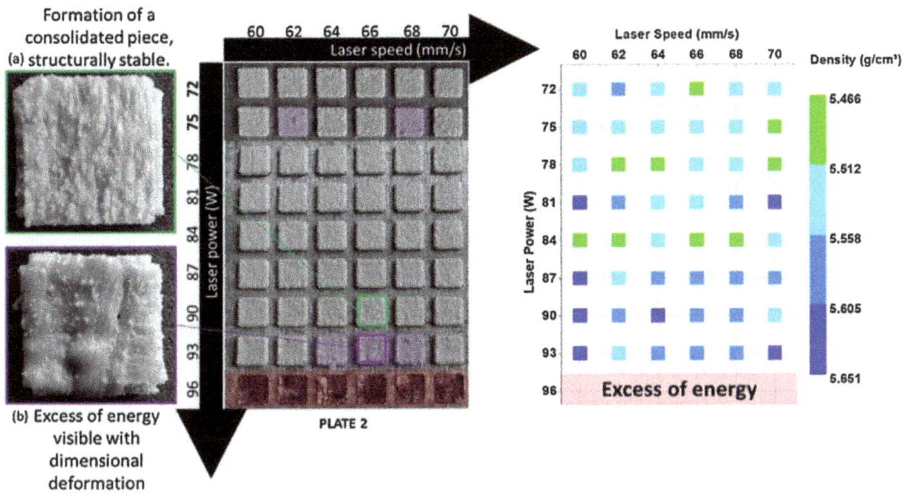

Figure 4.41. Details of the manufactured plate number 2. 54 cuboids with dimensions of 0.8 cm × 0.8 cm × 0.3 cm, printed at laser scanning rates of 60–70 mm · s^{-1} and laser powers of 72–96 W. In the red zone highlighted, excess energy caused degradation of the platform. The chart is to display the density of the corresponding sample. (a) A representative sample without dimensional deformation and (b) an example with dimensional degradation. Reproduced from [98]. CC BY 4.0.

rate of 60 mm · s^{-1} was higher than that of the one printed at 70 mm · s^{-1}. At a high laser power of 93 W, dimensional defects were present on the surface of samples.

With a laser speed of 60 mm · s^{-1}, laser power of 75 W and laser strategy at N90, but varying the hatch distance from 35 to 85 μm, the variation in relative density was just 1.6%. The sample processed at a hatch distance of 35 μm presented excess energy, without consolidation. Over the range of 40–85 μm, the density was dependent linearly on the hatch distance. With a laser speed of 60 mm · s^{-1}, laser power of 75 W and hatch distance of 50 μm, the variation in density was 1.48%, as a function of laser strategy. The 8YSZ ceramics had Vickers micro-hardness (HV0.1) in the range of 1550–1930 Hv.

Figure 4.42 shows a representative microscopic image of the samples. An example of a transition zone between a layer at 0° and a subsequent layer at 45° is schematically presented in figures 4.43 and 4.44, and depicts a photograph of a printed item, evidencing that the surrounding areas had relatively higher density than the center area, thus leading to different shrinkage rates between the two areas. As a consequence, the sample experienced dimensional deformation.

Figure 4.45 shows an example of the dense part processed using some of the optimal parameters presented in this study. It was processed using a N90 laser strategy, a laser power of 75 W, a laser speed of 60 mm · s^{-1}, and a hatch distance of 70 μm. It presented 97% of relative density. Figure 4.46 shows a fractured surface SEM image of a representative sample printed with a N90. Bother intralayer and interlayer cracks were present, together with localized cleavage planes caused by the rapid rupture of the brittle sample.

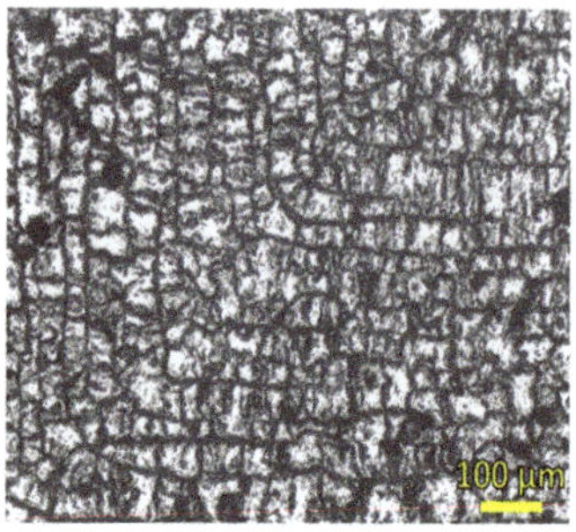

Figure 4.42. Representative optical microscopic image of the 8YSZ + 0.25 wt% TiC sample processed. Reproduced from [98]. CC BY 4.0.

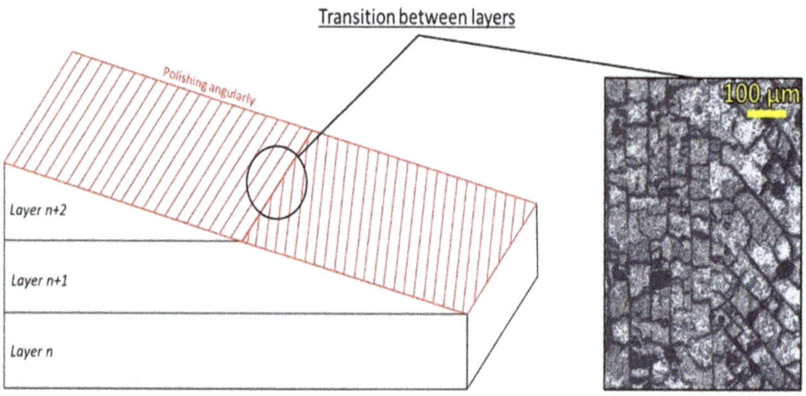

Figure 4.43. Schematic diagram and optical microscopic image microstructure of the 8YSZ + 0.25wt% TiC sample processed with a linear laser strategy with rotation of 45°. Reproduced from [98]. CC BY 4.0.

Figure 4.44. Photograph of a cuboid of 1.4 cm × 1.4 cm × 0.4 cm displaying mushroom-like deformation, duo to overheating of the upper surrounding regions. Reproduced from [98]. CC BY 4.0.

Figure 4.45. Representative photographs of 0.8 cm × 0.8 cm × 0.7 cm cuboid processed with the parameters within the optimal window: (a) as-printed and (b) lateral view with polished top surface. Reproduced from [98]. CC BY 4.0.

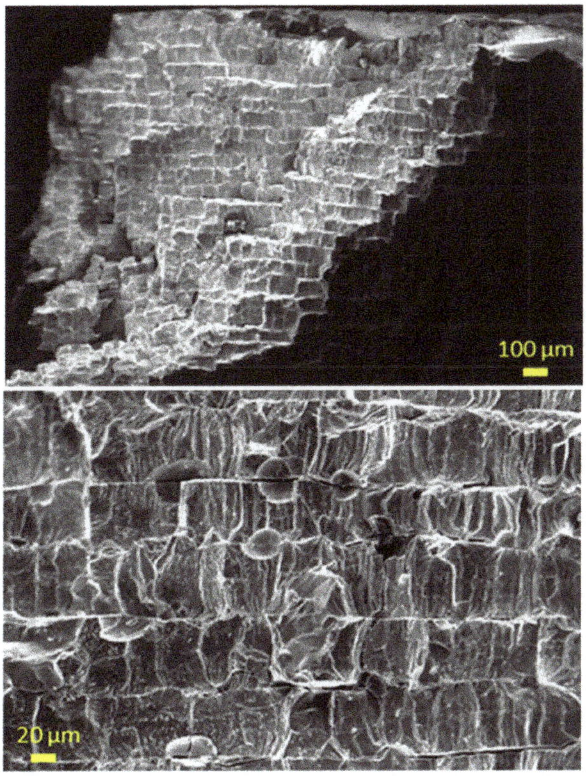

Figure 4.46. Fractured surface SEM image of a representative sample processed with N90 laser strategy, showing the presence of lamination and cracks. Reproduced from [98]. CC BY 4.0.

Alumina-toughened zirconia (ATZ) ceramics were fabricated by using direct powder bed selective laser processing (PBSLP), with a commercial ProX 200 (*3D Systems*) machine, equipped with a Nd:YAG laser [99]. With the addition of 0.75 wt% graphite powder into the ATZ powder, the laser absorbency was increased from 2% to

60%. Dense and geometrically accurate ATZ structures could be readily obtained by optimize the combination scanning rate and laser power, with a high relative density of 96.7%.

Tosoh alumina-toughened zirconia (ATZ) powder was used as the starting material, while graphite powder was supplied by TIMREX KS44, with $D_{50} = 19$ μm. It was reported that ATZ-graphite (ATZ-G) mixture with 0.75 wt% graphite exhibited the highest laser absorptance of 81% [100]. The ATZ-G mixture was milled for 1 h with a three-dimensional mechanical shaker mill. A commercial PBSLP machine, ProX 200 (3D SYSTEMS), was used in the printing experiment. It was equipped with a Nd:YAG laser, with laser powers of 30–300 W. The scanning rates were in the range of 50–3200 mm $\cdot$ s^{-1}. The dimensions of the building platform (AlSi12-based alloy) were 14 × 14 cm^2, with a maximum height of 10 cm.

Samples with dimensions of 1.5 cm × 1.5 cm and 30 layers in total were printed with different combinations of laser power and scanning rate. There were *A* and *B* groups of samples, with decrease in the laser power at an interval of 9 W (except B8) and increase in the scanning rate at an interval of 50 mm $\cdot$ s^{-1} (except B1). Therefore, the energy density was decreased with magnitudes from 11% to 83% (12 samples). There were also X and Y groups of samples, with decrease in the scanning rate at an interval of 10 mm $\cdot$ s^{-1} and increase in the laser power at an interval of 6 W (except Y1 and Y8). In this case, the energy density was increased with magnitudes from 7% to 133% (12 samples). Cuboids with dimensions of 15 × 15 × 5 mm^3 were printed for evaluation.

Particle sizes of the ATZ powder were measured to be $D_{10} = 9$ μm, $D_{50} = 22$μm and $D_{90} = 54$ μm. The ATZ-G mixture powder had a laser absorptance of 66% ± 3%. Tapped density and loose packing density of the ATZ-G powder were 1.85 and 1.56 g $\cdot$ cm^{-3}, respectively, with HR of 1.18. The HR value was <1.25, thus having suitable flowability for the printing experiment. The ATZ powder contained tetragonal and monoclinic zirconia phases, together with alumina. Mean crystallite sizes calculated from XRD patterns were 28.8, 30.1 and 38.7 nm, for monoclinic zirconia, tetragonal zirconia and α-alumina, respectively.

Figure 4.47 shows photographs of representative ATZ samples with complex shapes and different sizes, printed with parameters of B2. The ATZ samples had a density of 5.42 g $\cdot$ cm^{-3}, corresponding to a relative density of 97.4%. Figure 4.48 depicts a representative cross-sectional SEM image of the ATZ printed items. The ceramic layer was coherently attached on the metallic platform (AlSi12-based alloy), without separation. Meanwhile, the printed ceramic layers could not be identified, suggesting that the energy transferred to the powder bed was sufficiently high to melt the powder during the printing.

However, due to the high thermal shock, the ATZ ceramics possessed internal cracks, while the porosity was about 4.1%. The cracks extended in both directions, with lengths from hundreds of microns to millimeter scale. Two types of cracking patterns were observed from cross-sectional microstructure, labelled as box 1 and box 2 in figure 4.48. Box 1 is near the top surface of the ceramic item, with thickness of 480 μm and porosity of about 3.9%. The cracks in this zone were mainly parallel

Figure 4.47. Complex and simple shaped ATZ ceramics printed with optimal PBLSP parameter combinations (B2). Reproduced from [99], with permission from Springer Nature.

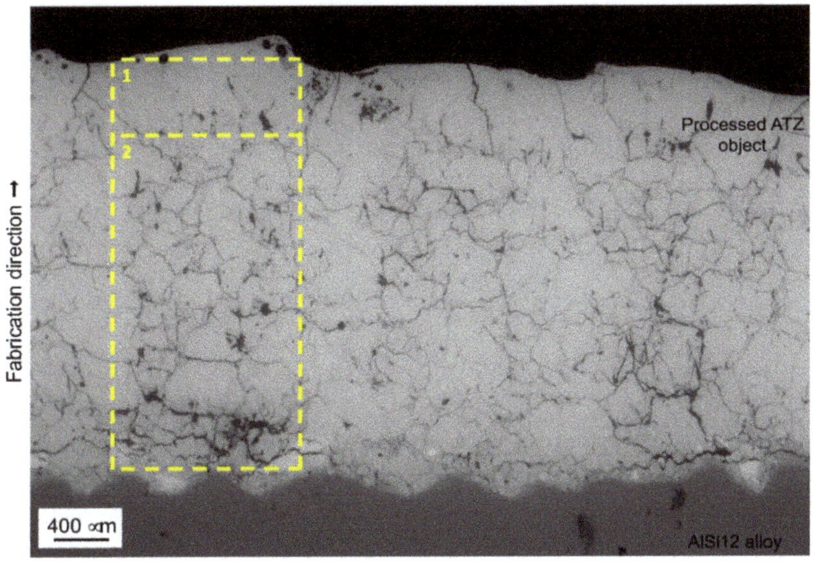

Figure 4.48. Polished cross-sectional SEM image of a presentative sample with two zones (1 and 2) of cracking modes. Reproduced from [99], with permission from Springer Nature.

to the printing direction, with widths of 5–13 μm. Round pores were occasionally present, with diameter of up to 80 μm, which were close to the top surface.

Box 2 had thickness of 2.5 mm, much larger than box 1. It presented a high porosity of 9.6%. More than 40% of the cracks in this zone were perpendicular to the printing direction. The profile of the cracks could be closely related to the difference in thermal diffusion rates inside the ceramics. There would be higher thermal flow at the bottom of the printed bodies, owing to the higher thermal conductivity of the AlSi12 alloy, as compared with the ATZ ceramics [101].

For box 1, the previously deposited ceramic layer acted as a thermal insulator, so that this area experienced high temperature. Therefore, a longer time was required for it to cool down, thus suppressing the formation of cracks, which was confirmed when monitoring the printing platform. When the first layer was printed, the interface was glowing, whereas only embers were observed for the rest layers. Slightly more than 30% of cracks were parallel to the printing direction, while 27.5%

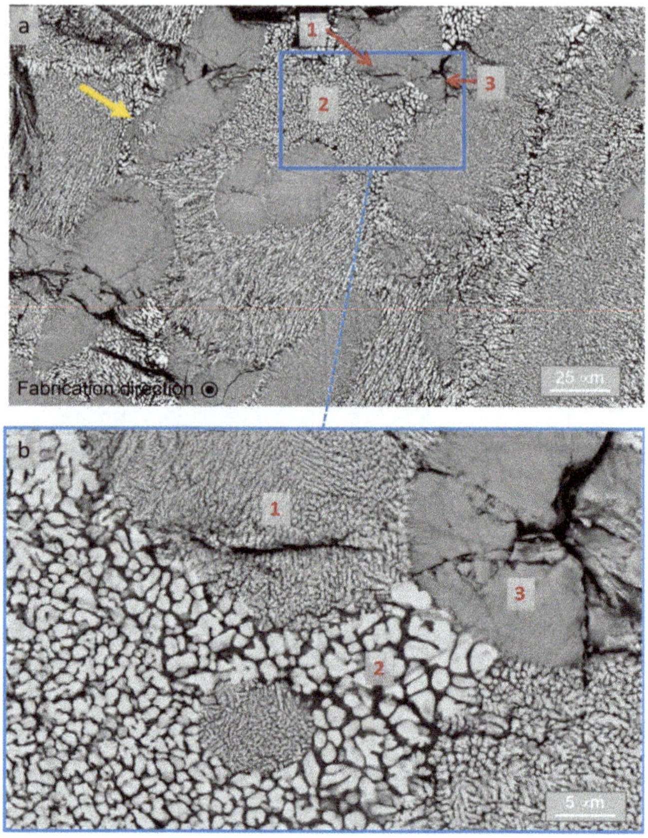

Figure 4.49. High magnification vertical cross-sectional SEM images of the ATZ ceramics with three structural forms: (a) global view and (b) detailed view. Reprinted from [103], Copyright (2024), with permission from Elsevier.

of the cracks were randomly distributed without orientation, similar to those reported in the open literature [102].

Figure 4.49 shows higher magnification SEM images of the ATZ ceramics. A dendritic-like microstructure was observed in core of the sample, which were present in three different forms. The rounded-elliptical types were dendritic eutectic structures, marked with No. 1 and No. 3. The lengths of the major axis of the ellipses were in the range of 38.2–95.3 μm. The lamellar spacing or the average distance between adjacent two layer of No. 3 was smaller than that of No. 1. In some locations, the No. 3 form was enveloped inside the No. 1 form, as seen in figure 4.49 (a). The form indicated with No. 2 was of a cellular eutectic structure, with widths of 0.2–0.8 μm.

A transient selective laser (TSL) technology was employed for microprocessing of ZrO_2 ceramics, where ultrashort-pulse lasers (USPL) and a continuous-wave (CW) laser were combined to realize ultrafast drilling of opaque ZrO_2 ceramics [103]. ZrO_2 could be drilled ultra-highly efficiently, with efficiency being more than three orders of magnitude higher than that of USPL drilling. The mechanism of material removal, process hole formation and effects of processing parameter dependence for the TSL drilling were studied, through direct observation of the internal occurrence at the scale of microseconds.

The variations in surface characteristics of processed samples were examined, including surface morphologies, chemical compositions and crystalline phases. The effect of different excited electron regions, at picosecond time scale induced by the USPL, on the performance of TSL drilling was evaluated. The productivity and stability of the TSL processing were confirmed through the fabrication of ZrO_2 microarrays with different sizes.

The setup for the TSL printing experiment is schematically depicted in figure 4.50 (a). Two types of lasers were used in the experiment. The USPL (PHAROS, LIGHT CONVERSION) had central wavelength of 1030 nm, pulse durations of 5 ps and 180 fs and pulse energy of 100 μJ. The CW laser (RED POWER, SPI Laser) had wavelength of 1070 nm and power of up to 250 W. A ZrO_2 ceramic sample (Kyocera) with dimensions of 20 mm × 20 mm × 0.4 mm, was used for the microfabrication experiment, whereas the side surface (20 mm × 0.4 mm) of the sample was subject to TSL processing.

The USPL and CW lasers were focused on the sample through an objective lens (OL1) with a numerical aperture (NA) of 0.4 (M Plan Apo NIR 10×, Mitutoyo), where the spot diameters were 5.7 and 8.6 μm, respectively. The focal plane distance of the two lasers was measured to be about 120 μm, as demonstrated in figure 4.50 (c), which was unchanged throughout the experiment. In fact, the small distance between the two laser focal planes had almost no effect on the experimental results,

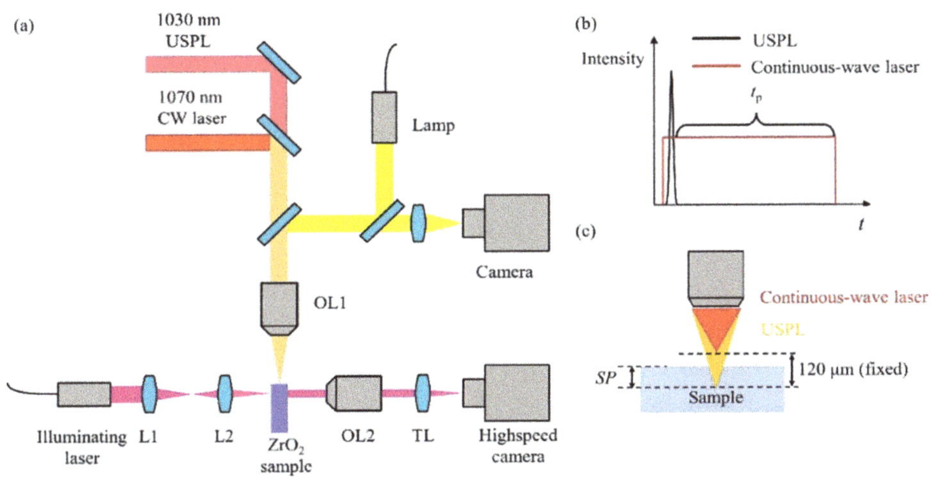

Figure 4.50. (a) Setup of the TSL processing experiment. (b) Timing profile of the laser irradiation. (c) Focal position of the CW laser and USPL: (L1) lens with a focal length of 150 mm, (L2) lens with a focal length of 30 mm, (OL) objective lens and (TL) tube lens. Reprinted from [103], Copyright (2024), with permission from Elsevier.

since the USPL was merely used to induce the electron excitation region inside the ZrO_2 ceramic sample instead of processing.

The timing profile of the irradiation of the two lasers is illustrated in figure 4.50 (b), which was controlled by using an electronic circuit and a delay generator (DG645, Stanford Research Systems), with precision at the scale of nanoseconds. The irradiation time of the CW laser after USPL was in the range of 24–134 s. The distance between the surface of the sample and the focal plane of the USPL was used to represent the sample position. SP was adjusted by sliding the sample along the optical axis of the lasers.

The processing was conducted at quite high speed, which was up to 2 million frames per second. It was monitored with a high-speed camera (Hyper Vision HPV-X2, Shimadzu) and illumination laser (CAVILUX HF, Cavitar) at the wavelength of 640 nm with pulse duration of 50 ns. Two lenses, with focal lengths of 150 mm (L1, AC254-150-B, Thorlabs) and 30 mm (L2, AC254-030-B, Thorlabs), were used to strengthen the probe beam. After experiments, the background was subtracted with an image that was captured before the experiments, so as to minimize noise in the images.

The processed samples were cleaned with an ultrasonic cleaner for 5 min. After that, diameter, topography and cross-sectional image of the drilled holes were recorded with a laser microscope (LEXT OLS4100, Olympus). Then, the depth of the holes was estimated from the images that were captured with a high-speed camera. Cross-sectional images of the drilled holes were obtained with a laser microscope. The samples were cut along the center of the holes with a low-speed cutter, for side view observation.

The excited electron regions were induced by USPL, which influenced the TSL processing of ZrO_2. To reveal this effect, different excited electron regions were generated, with the processes to be monitored with a modified pump–probe method [104]. In the pump–probe experiment, the USPL parameters included a pulse energy of 100 μJ, SP value of 60 μm and pulse durations of 5 ps and 180 fs. Side-view images of the excited electron region were obtained at 75 ps, which was the time for the excited electron region expanded to the maximum length after USPL irradiation.

Figure 4.51 shows the process of material removal during the TSL drilling of the ZrO_2 ceramics, with P of 100 W, t_p of 24 μs and SP of 60 μm. A scattered CW laser was present before the irradiation of USPL, as demonstrated in figure 4.51(a), but no material was removed. Upon the irradiation of USPL on the material, a bright area was present on the sample surface, as seen in figure 4.51(b). This observation was ascribed to the strong absorption of CW laser energy in the high electron density area induced by the USPL [105, 106]. The absorption of the CW laser energy meant that the material would be removed.

Once the material removal began, the energy absorption became stronger at the bottom of the drilled hole, whereas high-temperature plasma was produced, which was accompanied by intense luminescence, as revealed in figures 4.51(b)–(k). The ejection of material at the surface of the sample was mainly related to the evaporation of the material caused by the CW laser irradiation [107]. There was also liquid material ejected from the drilled hole near the surface of the sample. As

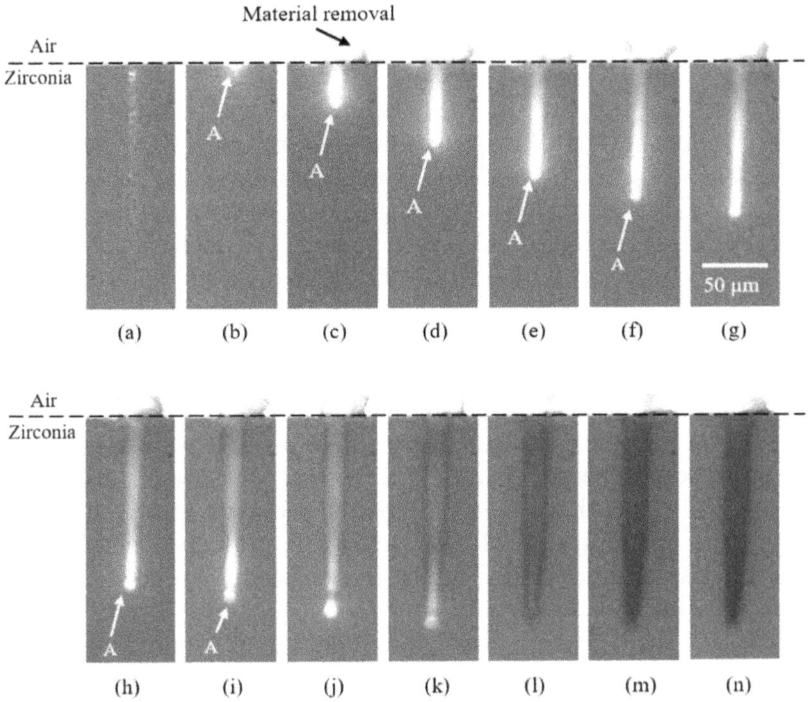

Figure 4.51. Drilling process with the TSL processing technology, with P = 100 W, t_p = 24 μs, SP = 60 μm. The images are the side views at different times after the the start of the ultrashort pulse laser irradiation: (a) −0.5 μs, (b) 1 μs, (c) 3.5 μs, (d) 6 μs, (e) 8.5 μs, (f) 11 μs, (g) 13.5 μs, (h) 16 μs, (i) 18.5 μs, (j) 21 μs, (k) 23.5 μs, (l) 26 μs, (m) 28.5 μs, (n) 31 μs. The mark A shows the absorption position of the CW laser. Reprinted from [103], Copyright (2024), with permission from Elsevier.

the laser irradiation was stopped, as presented in figure 4.51(l)–(n), the material removal ceased, so that whole area of hole was darkened.

The depth of the hole versus processing time was studied, while the processing speed was calculated through polynomial fitting. The processing speed reached the maximum value of 15 $m \cdot s^{-1}$ very quickly upon beginning the experiment and gradually declined to 3.9 $m \cdot s^{-1}$ after processing for 24 μs. This observation could be attributed to the decrease in effective CW laser energy absorbed in the area inside the material. The decrease in the effective CW laser energy was caused by two factors. On one hand, high temperature plasma produced during the TSL drilling process in the hole partly blocked the CW laser energy [108, 109]. On the other hand, multiple reflections of the CW laser due to the inner wall of the processed hole would cause energy loss, before approaching the absorption area at the bottom of the sample.

The conventional USPL process was compared with the TSL process. The hole depth and aspect ratio of the drilled holes were almost the same when using the two methods, including a hole depth of about 150 μm and aspect ratio of about 8.5. However, the TSL process had much higher processing efficiency. The average processing speed of TSL was 169.2 μm/23.5 μs = 7.2×10^6 $\mu m \cdot s^{-1}$, while that of

USPL was 138.78 μm/35 ms = 3.97×10^3 μm · s^{-1}, corresponding to a ratio of >1800, at the repetition rate of 10 kHz. Therefore, the TSL process ultrahigh speed drilling of ZrO_2 could be realized by using.

Depth of holes was dependent on CW laser power, processing time of CW laser and sample position. As the CW laser power was increased from 50 to 250 W, the depth of the hole was raised from 67 to 215 μm. Meanwhile, the stability of the depth value was decreased. When the processing time of CW laser was prolonged from 24 to 134 μs, the depth of the hole depth was increased. The depth of the hole was at the maximum, in the sample position of 100–120 μm, because the focal plane of the CW laser was closest to surface of the sample, thus enabling effective removal of the material [110]. In addition, the focal plane of the CW laser also had an effect on the shape of the hole, owing to the dependence of the beam shape and size of the CW laser near the surface of the sample [111].

For SP in the range of 100–120 μm, the depth of the hole was maximized, because of the higher processing speed. As the SP level was >150 μm or <60 μm, the holes had relatively shallow depth, due to the decreased processing speed. In the case of SP being 80 μm, the maximum speed was almost the same as the initial processing speed. Nevertheless, when reaching 30 μs, the processing speed started to decline, resulting in reduction in hole depth.

The diameter of the hole was strongly dependent on three processing parameters. As the laser power was increased from 50 to 100 W, the diameter of the hole was increased. However, when the value of P was further increased from 100 to 250 W, the increase in the hole diameter stopped, while it remained 41 μm. When using the TSL process to drill holes on ZrO_2 ceramics, the starting diameter of the hole was dependent on two parameters, i.e., the excited electron region (filament) and the spot size of the CW laser. Since the excited electron region had relatively large sizes of 70–100 μm [112], the limit of while the beam spot size of the CW laser was 43.4 μm for SP = 60 μm, the diameter of the hole was not increased.

When the value of t_p was increased from 24 to 84 μs and then to 134 μs, the diameter of the drilled holes was only very slightly changed. Prolonged t_p only caused the extension of the inner walls. In the SP of 60–150 μm, the variation in hole diameter was not significant, because the focal plane of the CW laser was very close to the surface of the sample, so that the shapes of the beams at the surface were similar. However, for SP values of 200 and 250 μm, the hole diameters were increased to 65.3 and 77.4 μm, respectively, owing to the increase in the beam diameter near the surface of the sample.

The variation scenario in morphology, in the processing with power of 100 W, t_p of 134 μs and SP of 60 μm, is demonstrated in figure 4.52(a). As the processing time was 5 μs, a sharp tapered morphology was formed near the surface of the sample. With increasing processing time, the diameter of the hole below the tapered area was greatly increased, while the diameter near the surface of the sample was unchanged. The extension process of the inner wall is schematically depicted in figure 4.52(b). As the CW laser beam reached the bottom of the hole, it experienced multiple reflections by the inner wall of the hole. Meanwhile, the material removal was triggered owing to the weak absorption of the laser energy, thus widening the holes

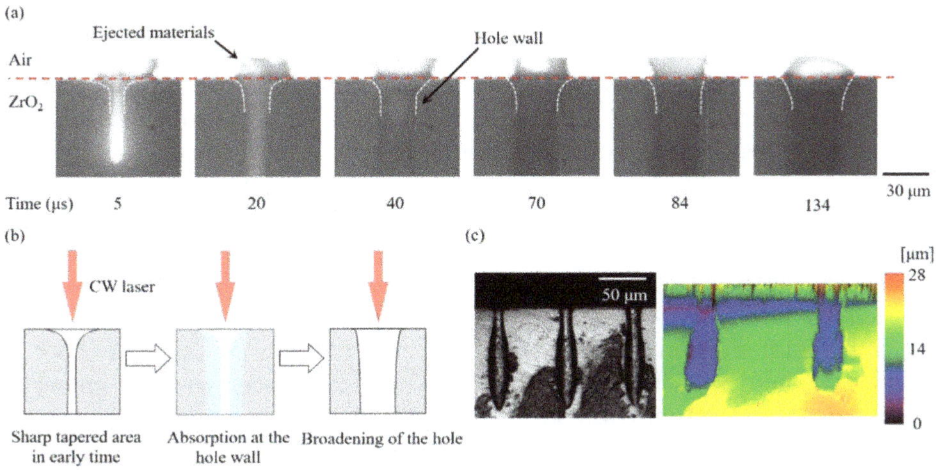

Figure 4.52. (a) Processing phenomenon near the surface, with $P = 100$ W, $t_p = 134$ μs, SP = 60 μm. (b) Conceptional broadening process of the hole. (c) Cross-section images of the holes processed with $P = 100$ W, $t_p = 24$ μs, SP = 60 μm. Reprinted from [124], Copyright (2013), with permission from Elsevier.

[113]. Cross-sectional images of the drilled holes, with processing parameters of P = 100 W, $t_p = 24$ μs and SP = 60 μm, are illustrated in figure 4.52(c). In this case, the drilled hole was not blocked by the removed material, while the area near the holes was not cracked, thus confirming the capability of the TSL technology in processing ZrO_2 ceramics with high-precision and high-quality.

Compared with the ZrO_2 samples processed by using a femtosecond laser, those processed with the ultrahigh TSL technology had smoother surfaces and much less residue. This was simply because the TSL processing technique possessed much higher efficiency in terms of material vaporization and removal processes, which took place within the microsecond scale through the absorption of the CW laser energy, so as to ensure intense mass transportation. The morphologies of the TSL processed samples were also dependent on sample positions. It was found that high sample positions resulted in rougher surface morphology.

The sample before processing consisted of mainly monoclinic phase, along with minor cubic and tetragonal phases. After TSL processing with $t_p = 24$ μs, the content of the monoclinic phase was significantly decreased, while that of the cubic phase was tremendously increased, suggesting that the TSL processing triggered the phase transition from monoclinic to cubic. The phase transition in the ZrO_2 ceramics was ascribed to the high temperature induced due to the absorption of the CW laser energy in h-electron-density area. In addition, the contents of the cubic and tetragonal phases were further increased as t_p was prolonged to 84 μs. Therefore, a longer processing time was beneficial to have stable cubic phase for ZrO_2 ceramics.

The mechanism of the material-removal for the TSL drilling of ZrO_2 ceramics was proposed. Initially, as the material was irradiated by the USPL pulse, single filaments in the excited electron region were formed near the irradiated areas. Then, the top of the filament areas would absorb the CW laser energy. Such absorption was known as the initial absorption, which occurred due to the USPL pulse irradiation. Consequently, plasma was produced in the high temperature area owing to the laser energy absorption. Meanwhile, material removal was initiated. The continuous irradiation of the CW laser would expand the high temperature area, so that the sample was heated from top to bottom.

This phenomenon was known as follow-up absorption, which was caused by not only the direct absorption of CW laser energy into the high-temperature areas but also the energy transfer from the absorption site to the interior of the sample. This long-standing absorption that occurred at the microsecond time scale might be supported by thermal ionization and thermally induced defects, whereas the effect of the USPL irradiation was absent. When the irradiation of the CW laser was suspended or the depth of the drilled hole was increased such that the intensity of the CW laser at the bottom was not sufficiently high to maintain the process, the hole-drilling process was finished.

The effect of the excited electron area induced by a single USPL pulse on the TSL drilling efficiency was examined. Obvious differences in the excited electron region, from the viewpoint of size and distribution induced by different pulse durations of the USPL were observed, whereas the overall TSL drilling process and processing speed exhibited no significant variation. The time required to drill a hole was several tens of microseconds, while the electron excitation area induced by an ultrashort-pulse laser was formed in just several nanoseconds. Throughout the lifetime of the electron excitation, the CW laser energy absorbed by the excited electron region was all on the surface of the sample. Therefore, the internal excited electron area induced by the USPL had a very weak effect on the drilling efficiency. In this regard, the subsequent extension of the processed area was mainly ascribed to the CW laser absorption in the high temperature areas.

The application of the TSL processing technique was verified by the fabrication of microarrays, consisting of blind and through holes. Blind hole microarrays with different diameters and depths could be readily fabricated by controlling the processing time. The through hole microarrays were achieved by drilling for 600 μs. Therefore, the TSL processing technique could be an effective strategy for ultrafast and robust machining of ZrO_2 ceramics, with the possibility to extend to other functional microstructures for different applications.

4.3.4 Bioceramic materials

Calcium phosphate (CaP) is among the most important biomaterials, with various applications in the field of biomedicine [114–117]. Biomaterials made with 3D printing could have patient matching characteristics. It is also very convenient to fabricate scaffolds with desired porous networks and interconnectivity [118, 119]. Besides biological properties, biomaterials should have sufficiently high mechanical strength in some specific applications, such as bone implants [120].

Ceramic implants exhibit superior chemical resistance than metal implants. The human body contains water, salt, dissolved oxygen, bacteria, proteins and various ions, which is a very harsh environment for metals. Hydroxyapatite (HAP) coatings have outstanding chemical stability, so it has been extensively employed as coatings on surfaces of base metals to increase their biocompatibility. In addition, HAP has a piezoelectric effect, thus being able to accelerate the healing of bone fractures [121–123].

One issue of SLS/SLM to process CaP ceramics is the low laser energy absorption. Solutions to solve this problem include intensifying the power of the lasers, reducing the scanning speed, using alternative sources of energy and so on. Meanwhile, it is necessary to take into account productivity, final quality of the scaffolds and cost-effectiveness. In this regard, the most effective way is to use an absorption additive, which is better for making the materials biocompatible, biodegradable, soluble or thermal degradable.

Porous scaffolds based on calcium phosphate were developed by using the SLS process, with various TCP/HAP weight ratios [124]. Fracture toughness and compressive strength of the samples both were increased, with increasing content of TCP in the range of 0–30 wt% and then declined with further increase in the content of TCP. The scaffold with TCP/HAP ratio of 30/70 displayed optimal mechanical performances, with fracture toughness of 1.33 $MPa \cdot m^{1/2}$ and compressive strength of 18.35 MPa. After soaking in SBF for one week, apatite was crystallized on surface of the scaffolds. Meanwhile, the dissolution rate of the scaffolds was increased with increasing content of TCP. For cell adherence and proliferation, biological stability and biodegradation rate should be balanced.

The HAP powder had a Ca/P ratio of 1.67, with purity of 99% and irregular shaped particles having sizes of 0.06–0.1 μm. The β-tricalcium phosphate (β-TCP) powder had particle sizes of 0.1–0.3 μm, Ca/P ratio of 1.5 and purity of 98%. The BCP powders were prepared by mechanically mixing the β-TCP and HAP powders, with weight ratios of 0/100, 10/90, 30/70, 50/50, 70/30 and 100/0.

A lab-built SLS system was used to print the porous ceramic scaffolds. During the printing process, the particles of the powders were bonded through the thermal effect of the laser energy, thus forming three-dimensional ceramic scaffolds. The printing parameters included a powder layer thickness of 0.2 mm, laser power of 12 W, spot diameter of 800 μm and laser scanning rate of 100 $mm \cdot min^{-1}$.

Figure 4.53 shows a representative porous biphasic calcium phosphate (BCP) ceramic scaffold. The scaffold had 3D orthogonal porous square channels, with width, height and space of 13, 7 and 2 mm, respectively. The scaffold had a microporosity of 61%, while its interconnected macroporous structure had rectangular pores with sizes of 0.8–1.2 mm.

According to XRD results, it was found that there was α-TCP phase, owing to the phase transformation of β-TCP at temperatures of >1180 °C [125]. Therefore, the content of α-TCP peak continuously increased with increasing composition of β-TCP. Meanwhile, HAP experienced decomposition if the sintering temperature was too high. The weight content of HAP in a sample of TCP/HAP (50/50) after SLS processing was calculated based on the XRD results, which was reduced to 40.2 wt %, confirming that HAP decomposed partly during the SLS processing.

The fracture toughness of HAP was 0.83 $MPa \cdot m^{1/2}$, whereas that of β-TCP 0.98 $MPa \cdot m^{1/2}$. The BCP ceramic scaffolds displayed higher fracture toughness and

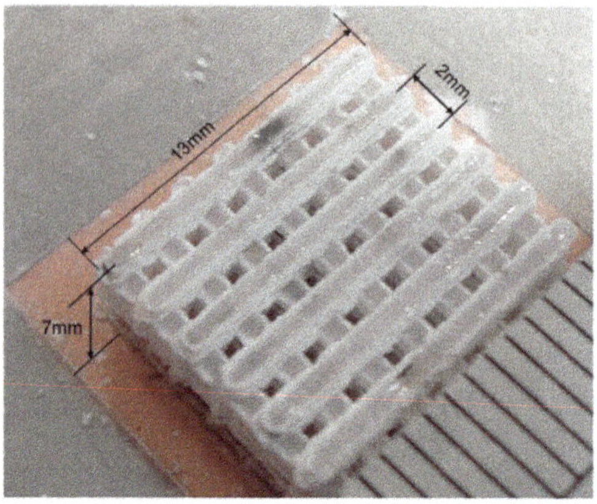

Figure 4.53. Representative photograph of the porous BCP scaffolds fabricated by using SLS processing technology. Reprinted from [124], Copyright (2013), with permission from Elsevier.

compressive strength than the two components. A low content of β-TCP could enhance the mechanical strength of HAP. The fracture toughness was increased with increasing content of β-TCP, reached maximum value of 1.33 MPa · $m^{1/2}$ for the sample with the TCP/HAP ratio of 30/70 and then started to decline. Similarly, the scaffold had the maximum compressive strength of 18.35 MPa, at the TCP/HAP ratio of 30/70.

Surface microstructures of the samples after SLS printing were examined by using SEM. All the scaffolds exhibited a full dense microstructure. Figure 4.54 shows SEM images of the ceramic scaffolds from TCP/HAP(30/70) and TCP/HAP(70/30). At low concentrations, the TCP phase served as a reinforcement agent in the HAP matrix, thus leading to improvement in mechanical strength. As the content was increased to the threshold value, micro-cracks would be formed, as observed in figure 4.54(b), owing to the mismatch in thermal expansion of the TCP and HAP phases and the volume variation originated from the phase transformation of β-TCP to α-TCP [126]. Therefore, mechanical properties of the ceramic scaffolds would be decreased. The threshold value about 30 wt% TCP for the TCP/HAP samples.

Figure 4.54 shows SEM images of the scaffolds after they were soaked in SBF for one week. The surface of the HAP scaffolds was almost flat without variation in morphology after immersion in SBF, as seen in figure 4.54(a). Bone-like apatite crystals were formed on the surface of the scaffolds. Meanwhile, the precipitates of agglomerated apatite were present on surfaces of the TCP/BCP scaffolds, as illustrated in figures 4.54(b)–(f). Surface roughness of the TCP/HAP(10/90) scaffold was increased, on which apatite particles were randomly distributed.

The TCP/HAP(30/70) scaffold had cavities uniformly distributed on the surface, as revealed in figure 4.54(c). The TCP/HAP(50/50) scaffold exhibited similar surface morphology, with larger sized cavities, as demonstrated in figure 4.54(d). This is because TCP was more preferentially dissolvable in SBF than HAP. For the TCP/HAP(70/30) scaffold, the surface morphology was significantly different. As observed in figure 4.54(e), the surface was almost fully covered by a flake-like

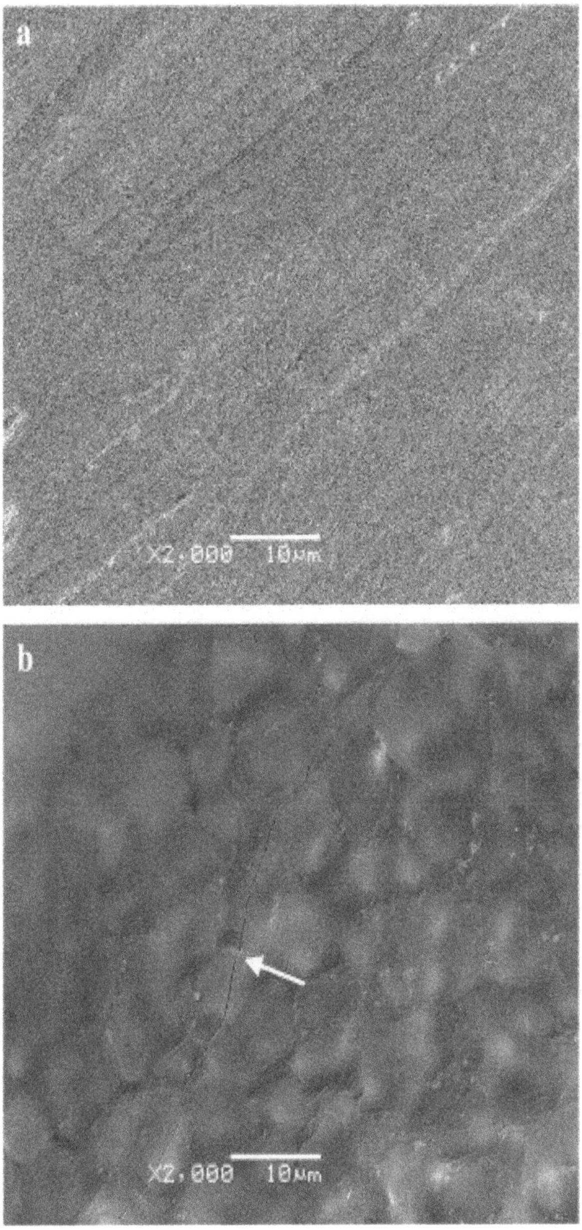

Figure 4.54. Surface SEM images of representative scaffolds: (a) TCP/HAP(30/70) and (b) TCP/HAP(70/30). Reprinted from [124], Copyright (2013), with permission from Elsevier.

apatite layer. In contrast, the surface of the β-TCP scaffold had large-sized flake-like morphology, where the apatite particles were more like bone, as demonstrated in figure 4.54(f), suggesting that β-TCP is more suitable than TCP/HAP(30/70) for the growth of apatite [127] figure 4.55.

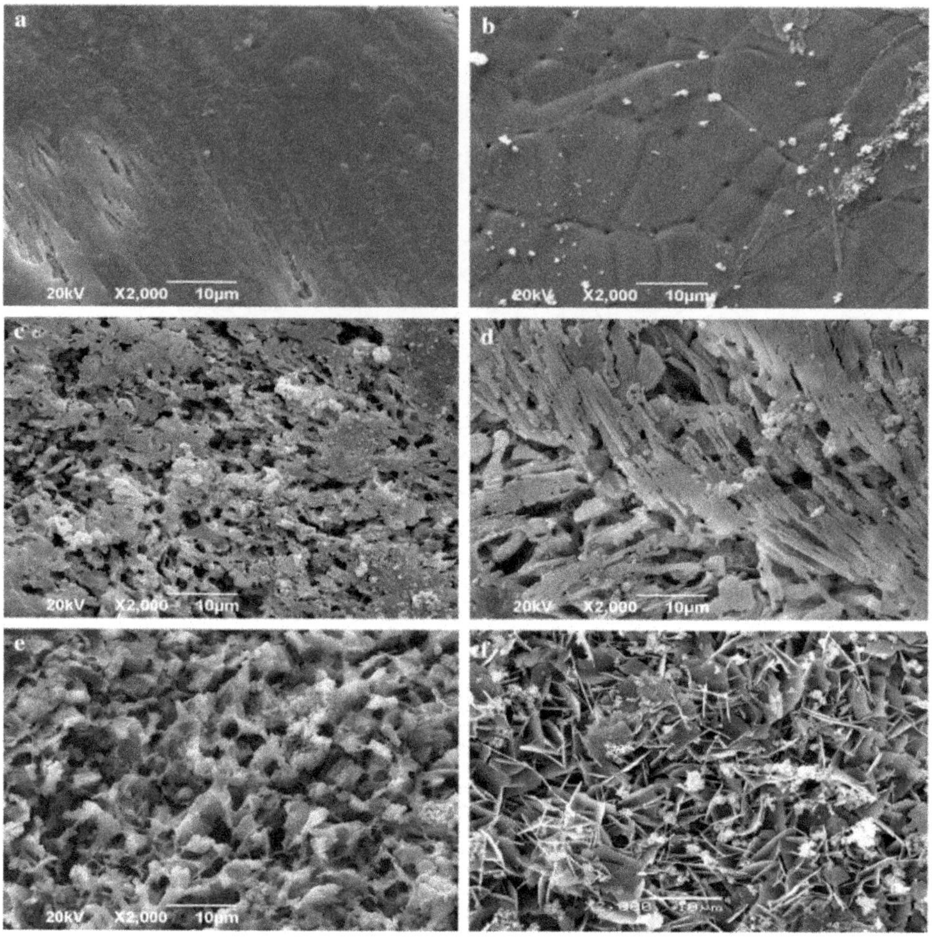

Figure 4.55. Surface SEM images of the scaffolds after soaking in SBF for one week: (a) HAP, (b) TCP/HAP (10/90), (c) TCP/HAP(30/70), (d) TCP/HAP(50/50), (e) TCP/HAP(70/30) and (f) TCP. Reprinted from [124], Copyright (2013), with permission from Elsevier.

Calcium phosphate powders were tailored SLS/SLM applications [128]. Hydroxyapatite and chlorapatite were used as the raw materials, while graphite powder was utilized to enhance laser absorptance. The incorporation of graphite powder widened the process window of the powder blends. Without the addition of graphite, the powders were not processable for the SLS/SLM printing. Hydroxyapatite experienced a phase transition to other calcium phosphate compounds, whereas chlorapatite was highly stable during the process. In comparison, the sample with hydroxyapatite and graphite required longer times to be printed. The production of the printing could be optimized by adjusting the processing parameters. The samples exhibited highly interconnected pores, which would be beneficial in terms of bioactivity if they were used as bioceramic scaffolds. Meanwhile, a suitable post-treatment increased the content of the hydroxyapatite phase and mechanical strength as well.

HAP microsphere powder was supplied by Urodelia (SA Company, Saint–Lys, France), synthesized with chemical coprecipitation method, combined with a spray-drying process (product Reference 300.08.2) [129, 130]. The spray drying was to ensure flowability of the HAP powder. Chlorapatite (ClA) powder was obtained by using a solid-gas reaction (chlorination) process, from HAP powder [131]. The HAP and ClA powders were mixed with graphite powder, with a total quantity of 0.5 kg, followed by mixing with a Turbula shaker mixer at 42 rpm for 1 h.

TA6V plates, with sizes of 10 cm in diameter and 2 mm in thickness, were coated with 0.1 mm HAP through plasma spray, which were used as the substrate for the subsequent 3D printing. SLS/SLM printing was conducted with a 3D Systems ProX DMP 200. Figure 4.56 shows a schematic diagram of the main components in the printing chamber. The roller was 110 mm in diameter, being able to move in a straight line driven by a carriage at speed of 400 mm $\cdot$ s^{-1} and accuracy of <1 mm and rotate at speed 3.7 turns per second and accuracy of $<5°$. Powder thickness of the layer was dependent on the powder properties, such as particle size, size distribution, flowability and so on. A scraper was employed to transfer the ceramic powder from the feeding piston to the sintering piston.

Structures were printed on the HAP coated TA6V substrates. The un-sintered powders were collected with a collecting tank for recycling. Since no material is removed, waste is reduced significantly. In the Class I Laser System, the main optical devices include a fiber laser source with continuous operation, single-mode and maximum power of 300 W, at wavelength of 1070 nm, a collimator with outlet beam diameter at $1/e^2$ 15 mm, a scanner head and a flat field F-Theta lens with focus length of 420 mm. In this case, the beam diameter of the focal spot was about 38 μm. Three groups of samples were printed to identify the printing parameters and feedstock formulations. In each group, the samples were printed with different 3D models, by using two laser scanning modes, as schematically presented in figure 4.57.

The first group of samples were circular patterns. The 3D model was composed of a short cylinder, with a diameter and height of 8 and 0.8 mm, respectively. Accordingly, the laser was scanned to form a hexagonal island, as demonstrated in figure 4.57(a). The hexagons had a diameter of 0.2 mm, with an overlap of

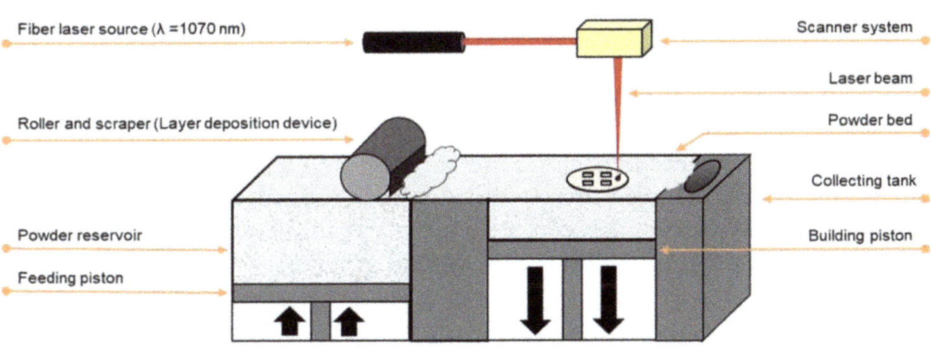

Figure 4.56. Schematic illustration of the main components in the printing chamber of a 3D Systems ProX® DMP 200 printer. Reproduced from []. CC BY 4.0.

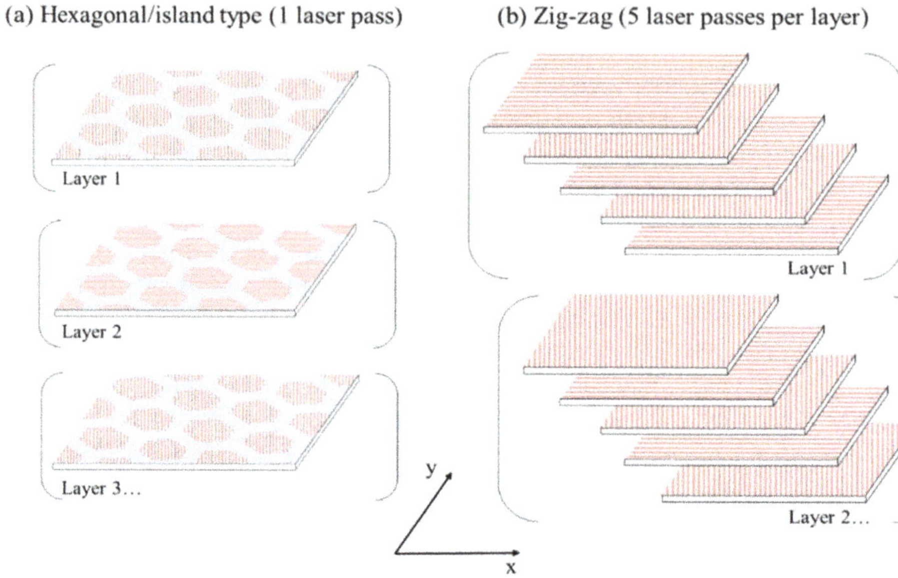

Figure 4.57. Schematic diagrams of the two laser scanning strategies for the three approaches: (a) hexagonal island type scanning (1 laser pass per layer) and (b) zig-zag scanning (5 laser passes per layer). In red color areas, the patterns were formed by the laser scanning, while the blue regions refer to the overlap between the hexagons. Reproduced from [128]. CC BY 4.0.

0.1 mm. Meanwhile, they were rotated by 90° between every two adjacent layers. By doing this, the printing parameters and the process windows for different powder mixtures could be identified. By the way, short time printing allowed for quick characterization, without removing from the substrate.

The second group was of complex shape, with two 3D models. One model was a cylindrical scaffold, with dimensions of 20 mm × 17 mm and designated internal porosity, whereas the second one consisted of a trapezoid, with a toothed bottom surface of 8 mm × 25 mm × 13 mm. The hexagon shape was the same as that in the first group. In this group, the laser parameters for the experiment included laser power of 30 W, scanning speed of 25 mm $\cdot$ s^{-1}, energy density E_d of 240 J $\cdot$ cm^{-3}. The printed scaffold samples were detached from the substrate with an endless diamond wire saw.

The third group was HAP cylinders. The 3D model was a cylinder with diameter and height of 10 and 15 mm, respectively, so that the samples could be used for mechanical property measurements, as required in ISO 13175-3:2012. The printing parameters were selected according to the experiments of the first group, including a laser power of 36 W, scanning speed of 75 mm $\cdot$ s^{-1}, and energy density E_d of 96 J $\cdot$ cm^{-3}. Noting that the printing rate in the second group was relatively low, owing to the hexagonal-based laser scanning mode, a zig-zag scanning mode was used to increase the printing rate, as illustrated in figure 4.57(b). An overlapping of 5 laser passes per layer were used over the same irradiated area. The orientation of the laser tracks was rotated between both passes and layer steps. The printed samples could be subject to compression strength tests and chemical and structural characterization. Post-treatment of the samples was conducted in air at 1250 °C for 2 h, at heating and cooling rates of 2 and 1 °C · min^{-1}.

The initial white HAP powder became yellowish after chlorination with ClA powder, with the laser absorptance to be increased from 2.4% to 12.5%. After the addition of graphite powder, the mixtures exhibited a greyish color, while the darkness was increased with increasing content of graphite powder. The samples with 5 and 10 wt% graphite powder possessed absorbances of 66.8% and 73.7%, respectively.

Figure 4.58 shows photographs of the printed circular patterns with different powder mixtures and laser energy density E_d. Figure 4.59 depicts the corresponding printing process windows. The samples with different powders, contents of graphite and E_d were denoted as HA_5G_96, HA_10G_57.6, HA_10G_96, HA_10G_288, ClA_5G_96 and ClA_5G_336. Three quality indicators were used to evaluate the printing effects at each assigning point, which were non-densification, sintered/melted powder and thermal cracking. When the laser energy was not sufficiently high, the powder was not densified. However, if the energy was too high thermal cracks would be formed, because of the high thermal gradients and stresses, so that the printed structures could not be maintained.

Pure HAP powder could not be densified, regardless to the processing parameters. For HA_5G and HA_10G, the parameter windows of powder sintering/melting occurred were not quite wide enough. With increasing content of graphite, the process window was widened, so that the selection of the laser parameters to sinter/

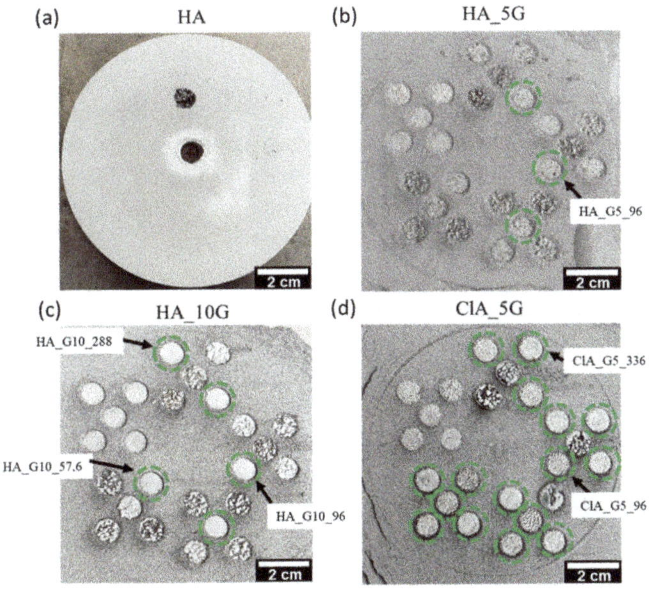

Figure 4.58. Photographs of the samples with the single circular patterns printed with different powder mixtures and printing parameters to figure out the processing windows: (a) HA, (b) HA_G5, (c) HA_10G and (d) ClA_5G. Reproduced from [128]. CC BY 4.0.

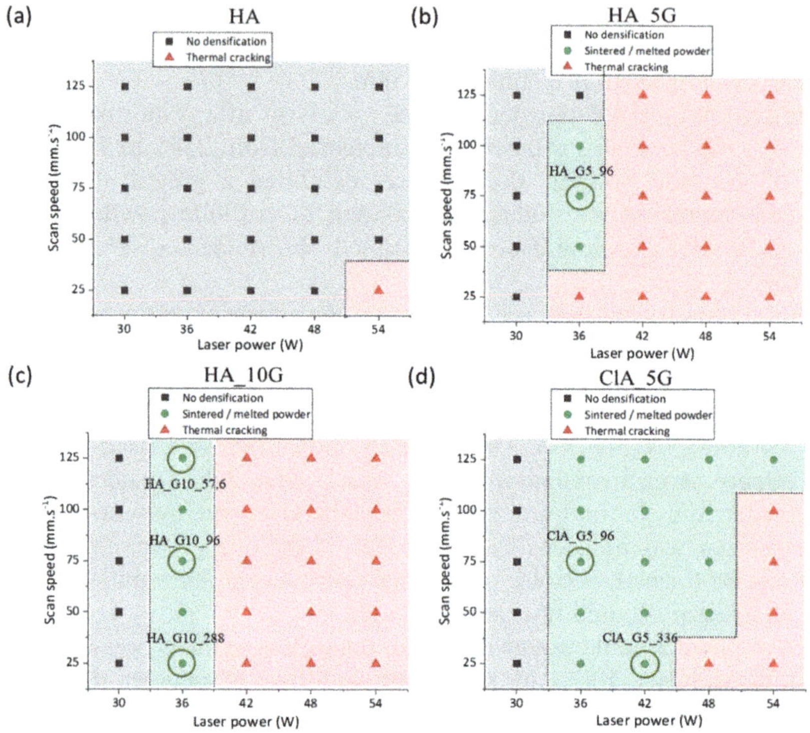

Figure 4.59. Processing windows for different powder mixtures: (a) HA, (b) HA_5G, (c) HA_10G and (d) ClA_5G. The circles indicate the characterized samples for comparison. Reproduced from [128]. CC BY 4.0.

melt the powder was more flexible. For HA_10G, densification of samples treated at a given energy density was dependent on scanning rate and laser power. For instance, when the processing parameters were 36 W and 75 mm $\cdot$ s^{-1} versus 48 W and 100 mm $\cdot$ s^{-1}, although they had the same energy density of 96 J $\cdot$ mm^{-3}, only the low scanning rate resulted acceptable densification.

In fact, as the energy density was kept unchanged, the effects of laser hatch distance, scanning rate and laser power were different [132]. They are in the order of laser scanning rate $<$ hatching distance $<$ laser power. Therefore, laser energy density could not be used as the key criterion in optimizing the process parameters for SLS/SLM printing. As the HAP was modified to be CIA, it was more difficult to widen the processing, simply because CIA has higher stability than HAP [133].

Figure 4.60 shows SEM images of the circular patterns printed at the same energy density, from different powder mixtures. A balling effect was observed in the samples. There are two types of balling phenomena [134]. The first type of balls was coarse in appearance, with interrupted dendritic structure and relatively weak strength. The rough surface was resulted from the limited formation of liquid and low degree of undercooling, due to the low energy input of the laser.

Figure 4.60. SEM images of the laser-induced circular patterns printed at $E_d = 96$ J · cm^{-3} with different powder mixtures at low and high magnifications. Reproduced from [128]. CC BY 4.0.

The second type characterized by numerous micrometric balls in the areas irradiated by the laser. These balls had larger sizes than the HAP and ClA microspheres. The smoother surface of the balls reflected that the samples were well melted during the printing process, owing to the high capillary instability of the melt [135]. The sample HA_10G_96 exhibited the second type of balling effect, as evidenced by the micrometric balls on the surface. Meanwhile, the surface was flat and smooth. In comparison, the ClA_5G_96 sample displayed the first type of balling effect, with a rough surface. As expected, the sample HA_5G_96 had both kinds of balls. The difference in surface finishing of the samples was closely related to the laser absorptance and powder stabilities. Close inspection indicated that pinholes with a diameter of 20 μm were visible on the sample surface, with random distribution, which could be attributed to the formation of micron gas bubbles of CO and CO_2 produced due to burnout of graphite.

Figure 4.61 shows SEM images of the samples from HA_10G, processed at different E_d. After processing at high values of $E_d = 288$ and 96 J · cm^{-3}, the second type of balls was observed in the samples. In contrast, as the value of E_d was lowered to 57.6 J · cm^{-3}, the number of balls was reduced, so that the surface finishing was relatively flat. Meanwhile, the three samples all had grain boundary cracks on the surface. With increasing E_d, the size of both the grain and crack was increased.

Figure 4.62 shows SEM images of the sample ClA_5G_336, after printing at a high level of E_d. In this case, the first type of balls was absent, leading to a flat surface. The sample surface was randomly distributed with large voids, which had diameters of 100 μm. Inside the voids, there were rod-like ClA particles grown in

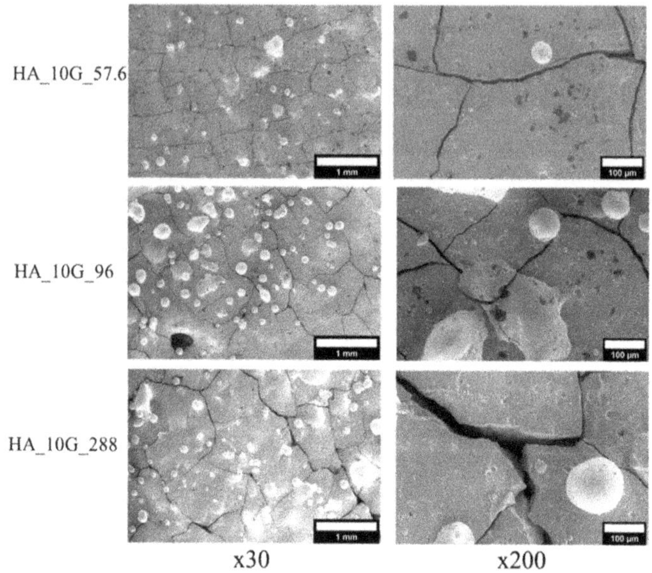

Figure 4.61. SEM images of the laser-induced circular patterns printed with same powder mixture (HA_10G) at different values of E_d, 288, 96 and 57.6 J · cm^{-3}. Reproduced from [128]. CC BY 4.0.

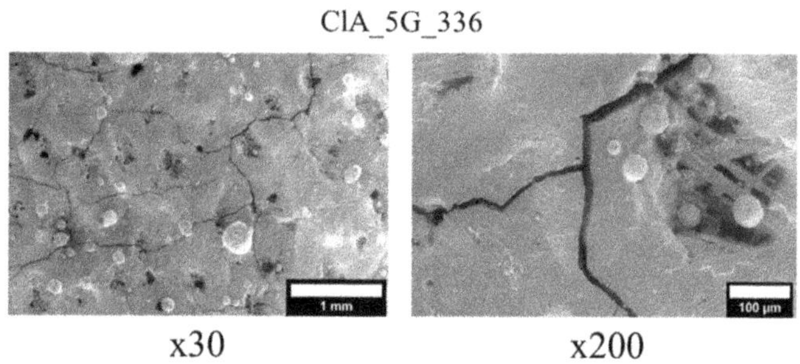

Figure 4.62. SEM images of laser-induced circular patterns printed with ClA_5G powder mixture at $E_d = 336$ J · cm^{-3}. Reproduced from [128]. CC BY 4.0.

parallel to the plane of the surface, which was typical morphology of CIA processed at high temperatures [136].

The second approach was only applied to the sample HA_5G, with a 3D model, processed at 240 J · cm^{-3}. Due to the change in thermal conductivity, it was possible to make three-dimensional structures, although the samples exhibited relatively high roughness. Two types of pores were present inside the structures, which were homogeneous interconnected pores in whole body and closed pores with smaller size. Two possible reasons were responsible for the presence of the small closed pores. On one hand, such pores could be related to microbubbles of CO and CO_2 gases, due to the burnout of the graphite. On the other hand, they could be caused by the decomposition of the carbonate in the starting HAP powder.

The printing induced distortion and cracks in the microstructure of the samples, due to the high temperature. Meanwhile, the samples were of relatively high porosity, which was up to 24%. The porous structures of the HA_Trapezoid and HA_Scaffold samples were of similar homogeneous architectures. Figure 4.63 shows SEM images of the HA_Trapezoid sample. The bottom surface of the sample was designed as tooth-like structure, to make it easy for the detachment of the printed items from the substrate. Interconnected porous microstructure was clearly observed, which was ascribed to the balling phenomenon as stated earlier. Cracks were also similarly present.

HA_10G powder mixture was subject to printing at an energy density of 96 J · cm^{-3} to prepare cylinder samples for mechanical property characterization. Figure 4.64 shows photographs of the cylinders with diameter and height of 10 and 15 mm, respectively. The color was changed from greyish powder to brown, owing mainly to the presence of the secondary phases in the samples, while no variation in color was observed after the thermal treatment.

Figure 4.65 shows three-dimensional reconstruction profiles of the samples before and after thermal treatment, by using x-ray micro-computed tomography (CT). Similarly, both closed and interconnected pores were present in the two samples. Porosities revealed by CT techniques were 48 and 50 vol%, before and after the thermal treatment, close to the porosity of common cancellous bone, in the range of 50%–90% [136]. Compressive strengths of the samples before and after thermal treatment were 0.010 and 0.049 MPa, respectively, which however were much lower than that reported for cancellous bone and hence needed to be further optimized.

Alternatively, calcium phosphate was blended with polymers to form composites for SLS/SLM printing [137]. The polymer components are melted during the printing process, acting as a binder in the green bodies, which are then removed

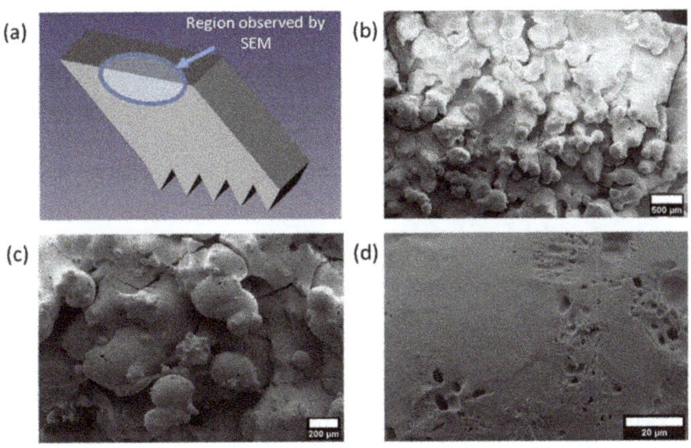

Figure 4.63. (a) Schematic diagram of the HAP_Trapezoid sample with the circle indicating the area for SEM observation. (b–d) Different magnification SEM images of the HAP_Trapezoid sample printed with HA_5G powder mixture at 240 J · cm^{-3} (30 W, 25 mm · s^{-1}) with hexagonal laser printing mode. Reproduced from [128]. CC BY 4.0.

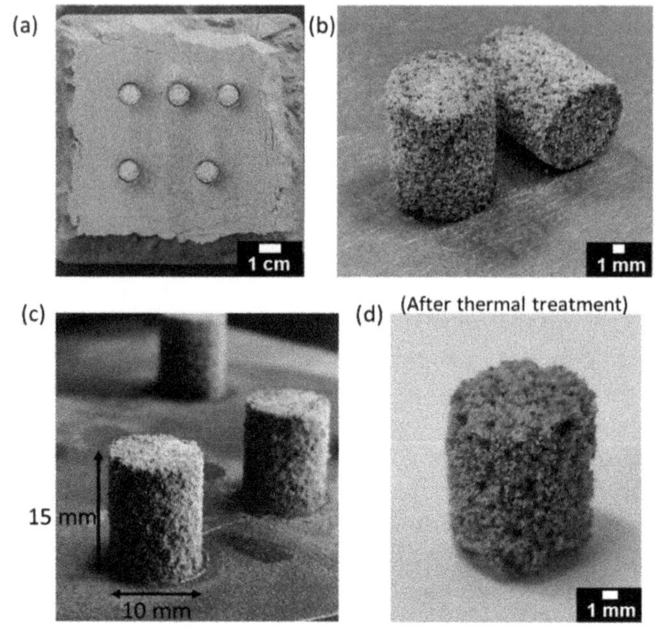

Figure 4.64. Photographs of the HAP cylinder samples printed at 96 J · cm^{-3} with zig-zag laser scanning mode (5 passes): (a) as-printed, (b) after detached from the substrate, (c) general view and (d) sample after thermal-treatment. Reproduced from [128]. CC BY 4.0.

through calcining at high temperatures. The resultant products would have certain porosities. Various polymers have been used for such applications.

Aliphatic-polycarbonate/hydroxyapatite (a-PC/HAP) composite scaffolds for medical applications were fabricated by using SLS [138]. The optimized composition of the a-PC/HAP composite was 10 wt% HAP, while the optimal processing parameters included laser power of 10 W, scanning rate of 2 m · s^{-1}, scanning space of 0.15 mm and layer thickness of 0.17 mm. Accordingly, the scaffold had a porosity of 77.4% and compressive module of 26 MPa. The samples had inter-connected pores, which was beneficial to cell ingrowth as a bone scaffold. The HAP phase with nanosized particles was well distributed in the a-PC matrix, without decomposition during the SLS processing, whereas bioactivity was unchanged.

Meanwhile, a complex scaffold with a smooth surface could be controlled by adjusting the content of HA and the processing parameters as well. A HRPS-IV-type SLS machine was equipped with a continuous wave CO_2 laser, which had a spot size of 0.2 mm, focusing distance of 640 mm, maximum power of 55 W and maximum scanning speed of 5 m · s^{-1}. A fixed one-way scanning mode was used for the printing experiment.

High-density polyethylene/hydroxyapatite (HDPE/HAP) composites were fabricated as functional graded scaffolds by using SLS processes [139]. The composite scaffolds had interconnected pores with diameters of 30–180 μm and porosity of 45%–48%. The exhibited flexural modulus of 36–161 MPa and ultimate strength of 4.5–33 MPa, with potential applications for bone and cartilage tissue engineering.

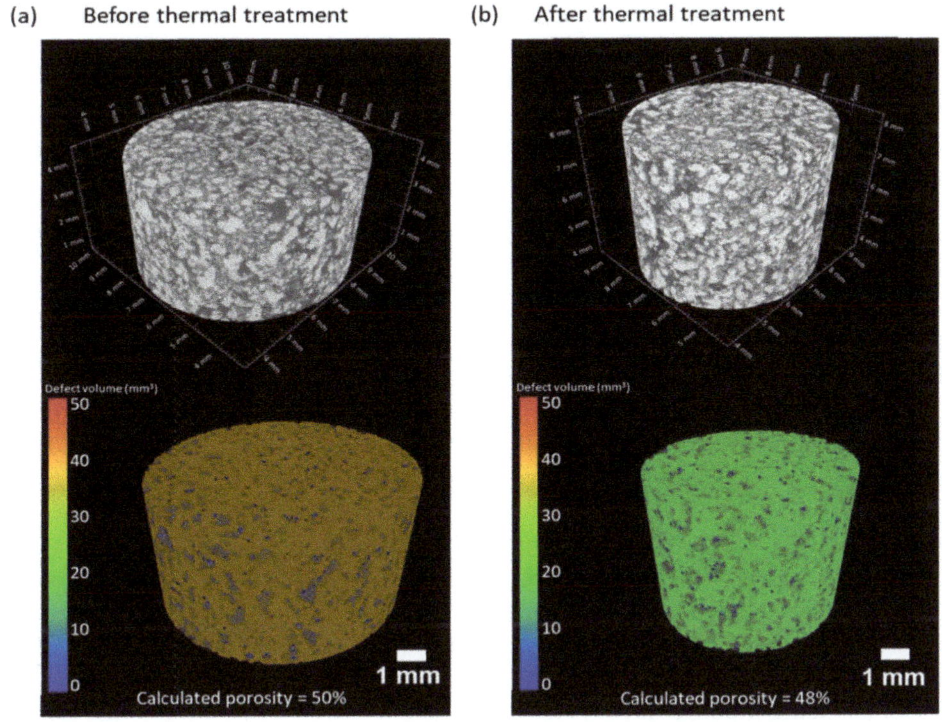

Figure 4.65. X-ray micro-computed tomography profiles of the sample HA_Cyl before (a) and after (b) the thermal treatment. Upper panel: volume of the samples. Bottom panel: porosity distribution in the region of interest (blue: closed porosity, orange and green: interconnected porosity). Reproduced from [128]. CC BY 4.0.

The HDPE powder (HD7555 Ipiranga) had a melting point of 127.7 °C, with particle sizes in the range of 150–212 μm. The HAP powder (Sigma-Aldrich) had particle sizes of 5–10 μm. The dimension of the scaffold printed was 35 mm × 5 mm × 1.4 mm. The HDPE and HAP powders were mixed mechanically by using a *Y* cylindrical mixer at 90 rpm for 70 min. Before SLS printing, the powder mixtures were preheated at 95 °C. The printing machine was equipped with a CO_2 laser at a power of 5 W, with a beam size of 250 μm and scan speed of 57 mm · s^{-1}. The chamber temperature was set to be 80 °C. The layer thickness was 200 μm, while the spacing between the laser scans was 125 μm.

Porous scaffolds based on polyamide (PA) and HAP were fabricated with SLS printing, with different configurations and pore geometries [140]. The minimum pore size was 400 μm, while the porosity was in the range of 40%–70%. Configurations of the scaffolds were designed according to finite-element analysis results, with different PA/HAP ratios. The printed scaffold samples were subject to the evaluation of mechanical properties, while in-vitro studies were conducted with a human osteosarcoma cell line for cell growth.

Three-dimensional (3D) bone scaffolds with multi-scaled porosity, acceptable mechanical strength and promising biocompatibility were prepared by using SLS printing processes, with polycaprolactone-HAP (PCL-HAP) composite microspheres [141]. The PCL-HAP composite microspheres were obtained from PCL and HAP powders by using a modified solvent evaporation method, which had uniform size and narrow size distribution. The 3D scaffolds exhibited both *in vitro* and *in vivo* manipulate multiple stem cell behaviors, along with outstanding histocompatibility.

The HAP powder had rod-like nanosized morphology, with diameters of 20–30 nm and lengths of 80–100 nm. The PCL had a molecular weight of 50 000. To prepare the composite microspheres, 1 g PCL was dissolved in 15 ml dichloromethane (DCM), while HAP powder was dispersed in 5 ml ethanol with ultrasonication. The two systems were mixed to form S/O nanosuspensions, which were subsequently added dropwise into polyvinylalcohol (PVA) solution (1.5 $w \cdot v^{-1}\%$). After stirring for 5 h to evaporate the organic solvent, the microspheres were collected through centrifugation, followed by washing with DI water and then freeze-drying.

Cuboid porous scaffolds had 3D orthogonal periodic porous structures, with length and width of 8.8 mm, height of 18.4 mm and pore size of 800 μm, as illustrated in figure 4.66(a). The structures were designed with Pro/Engineer 3D modeling software, which was exported to a STL file format that could be identified by the SLS system that was equipped with a 50 W CO_2 laser (HRPS-IV, BinHu Mechanical & Electrical Co., Ltd, Wuhan, China). Three groups of samples were printed, including pure PCL, 10% HA/PCL and 20% HA/PCL. Owing to the presence of HAP, the composite powders were processed at a lower scanning speed and higher laser power. The printing process is schematically presented in figure 4.66(b).

SEM images of the pure PCL and 10% HA/PCL, 20% HA/PCL composite microspheres are depicted in figure 4.67(a), revealing their regular spherical shape with an average diameter 100 μm. Because there were HA nanoparticles attached on the surface of the composite microspheres, they exhibited a rough surface as compared with the pure PCL one. The internal profiles of the microspheres were demonstrated by the cross-sectional fluorescent images. PCL and HAP were

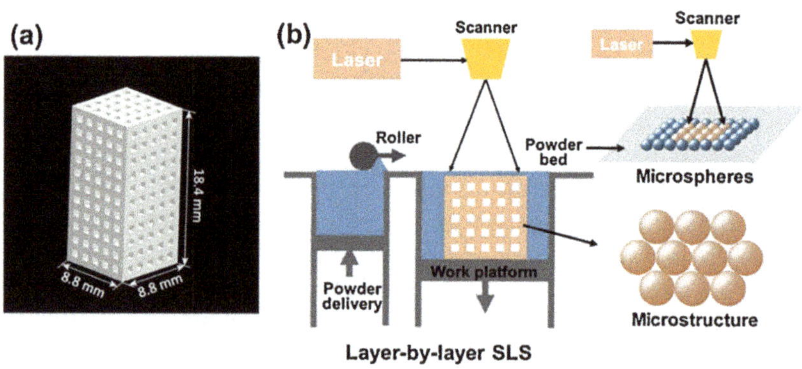

Figure 4.66. Design structure (a) and fabrication process (b) of the microsphere-based 3D scaffolds by using SLS processing technique. Reprinted from [141], Copyright (2015), with permission from Elsevier.

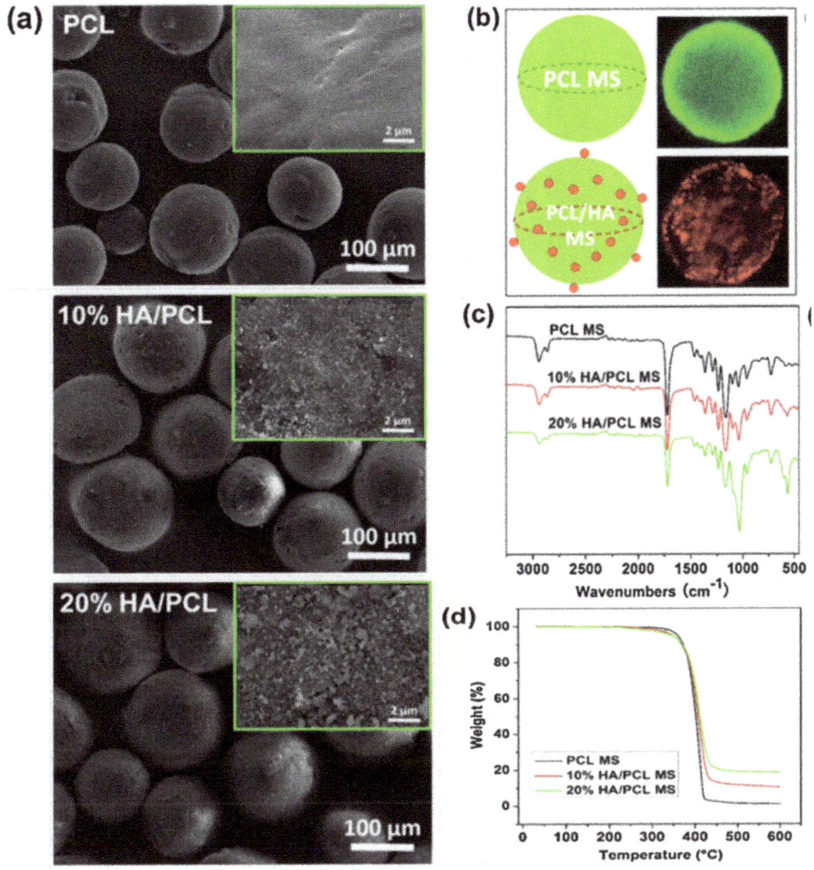

Figure 4.67. Characterization results of the microspheres: (a) SEM images of PCL, 10% HA/PCL and 20% HA/PCL with the inserts to be the corresponding surface SEM images, (b) confocal images of the samples indicating the distribution of in the composite microspheres of PCL/HA, (c) FTIR curves and (d) TGA curves of the microspheres with different contents of HA (MS: microspheres). Reprinted from [141], Copyright (2015), with permission from Elsevier.

represented by green and red fluorescence, indicating that the HAP nanoparticles were uniformly distributed in the composite microspheres, as seen in figure 4.67(b).

FTIR spectroscopy was used to characterize the PCL and HAP. Compared with the pure PCL, the 10% HA/PCL possessed characteristic peaks of PO_4 at 560 and 1033 cm^{-1}, with the peak intensity to be increased with increasing content of HAP. Accordingly, the peak intensity of the alkyl group CH at 2940 cm^{-1} and the carbonyl group CO at 1726 cm^{-1} in PCL was decreased, as observed in figure 4.67 (c). The TGA curves are shown in figure 4.67(d). The thermal decomposition began at about 300 °C, while the PCL completely decompose before about 500 °C. The HAP contents were estimated to be 9.4% and 17.6%, in rough agreement with the designed compositions of 10% HA/PCL and 20% HA/PCL.

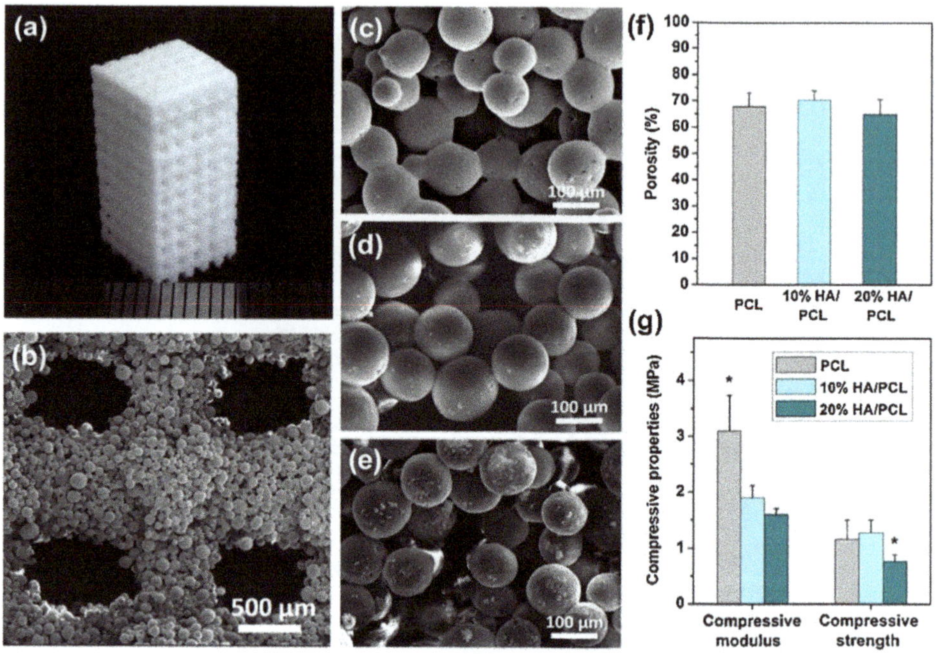

Figure 4.68. Characterization results of the SLS printed 3D scaffolds: (a) Photographs of a scaffold with dimension of 8.8 mm × 8.8 mm × 18.4 mm and highly ordered porous cuboid structure, (b) SEM image of the 10% HA/PCL scaffold, (c–e) SEM images of the PCL, 10% HA/PCL and 20% HA/PCL scaffolds, (f) porosity and (g) mechanical properties. Reprinted from [141], Copyright (2015), with permission from Elsevier.

The scaffold had precise size and structural integrity, as illustrated in figure 4.68 (a), whereas the pore sizes of the 3D interconnected pores were in the range of 600–800 μm, as presented in figure 4.68(b). All the samples that were printed with optimized parameters had been well densified, with typical sintered morphology. The microspheres were strongly connected to one another, without damage. Meanwhile, coalescence of the microspheres was clearly visible, with the presence of sintering necks, thus forming an interconnected porous network, with micropore sizes of 30–100 μm.

Comparatively, the PCL scaffold exhibited stronger sintering necks than that of the composite HA/PCL scaffolds, because of the hindering effect of the HAP particles, as demonstrated in figure 4.68(c)–(e). The porosities of the printed scaffolds were in the range of 65%–70%, without a great difference among the three groups of samples, as depicted in figure 4.68(f). Therefore, the incorporation of HAP had a very weak influence on porosity of the scaffolds. The compressive modulus was decreased from 3.1 to 1.6 MPa, as the content of HAP was increased by 20%. At the same time, the compressive strength of the 20% HA/PCL scaffolds was much lower than the pure PCL and 10% HA/PCL scaffolds, as seen in figure 4.68(g).

Bioresorbable scaffolds of PCL-HA with mechanical strengths meeting the requirements of bone tissue engineering were fabricated by using SLS printing process, with the models to be designed through finite-element analysis (FEA) [142].

Solid gage parts and scaffolds, with 1D, 2D and 3D orthogonal periodic porous architectures, were printed from PCL-HAP powder mixtures with HAP of up to 30 vol%. The PCL:HAP scaffolds exhibited a relative density of 99%, with outstanding geometric and dimensional integrity. The gage samples processed at optimal SLS conditions displayed compressive moduli of 299.3, 311.2, 415.5 and 498.3 MPa from powder mixtures with PCL:HAP ratios of 100:0, 90:10, 80:20 and 70:30, respectively. The compressive effective stiffness was increased and the porosity was decreased, with increasing content of HAP. For the 3D porous scaffolds, the compressive modulus was raised from 14.9 to 36.2 MPa. The compressive modulus was analyzed by using a micromechanical FEA model, revealing the reinforcement effect of HAP. A first-principles based approach was employed to describe the random distribution of the HAP particles in the PCL matrix. The experimental and the FEA modeled results were in good agreement each other.

The HAP powder (Astaris, St Louis, MO) had an average particle size of 45 μm, while PCL powder (Solvay Caprolactones, Warrington, UK) was used as the polymer matrix, which contained semicrystalline (56%) aliphatic thermoplastic, with a melting point of 58 °C–60 °C and glass transition temperature of about −60 °C. PCL:HAP powders were mixed with a rotary tumbler (784 AVM, US Stoneware, OH) for 24 h. The printing experiment was conducted with a Sinterstation® 2000 commercial SLS machine (3D Systems Inc., Valencia, CA). Preheated layers from the powder mixtures were sequentially deposited through scanning with a low-power continuous wave CO_2 laser, which had a wavelength of $\lambda = 10.6$ μm, power of <10 W and spot size of 450 μm.

Scaffolds were designed to have 1D, 2D and 3D orthogonal porous square channels, with a width of 2 mm and space of 0.7 mm, corresponding to porosities of 51.1%, 68.5% and 80.9%, respectively. Figure 4.69 shows a modeling strategy to predict mechanical properties of the composite scaffolds with different contents of HAP, using a three-stage process based on first-principles modeling. The distribution profile of the HAP particles in the PCL matrix was treated with a random sphere-packing model. Mechanical properties of the composites were predicted, as the location data from the packing model were exported to a micromechanical model.

The elastic modulus of PCL was 299.3 MPa, while that of HAP was 3.69 GPa. The Poisson's ratio was assumed to be 0.3. Nearly full densification was realized for the SLS printed scaffolds. Figure 4.70 shows representative cross-sectional SEM images of the scaffold cross-sections, further confirming the high density of the samples. No delamination was observed even at 30% HAP loading, indicating good bonding between the polymer matrix and HAP particles.

Compressive strengths of the printed solids and the designed porous PCL:HAP scaffolds were systematically characterized. The addition of HAP resulted in a general increment trend in compressive modulus (E). A significant difference ($n = 5$, $P < 0.05$) was not observed in the compressive modulus of solid gages from pure PCL and the 10% PCL:HAP composite samples were nearly the same. For the 1D, 2D and 3D scaffolds, there was a large difference in compressive modulus between the pure PCL (100:0) scaffolds and PCL:HAP composite ones.

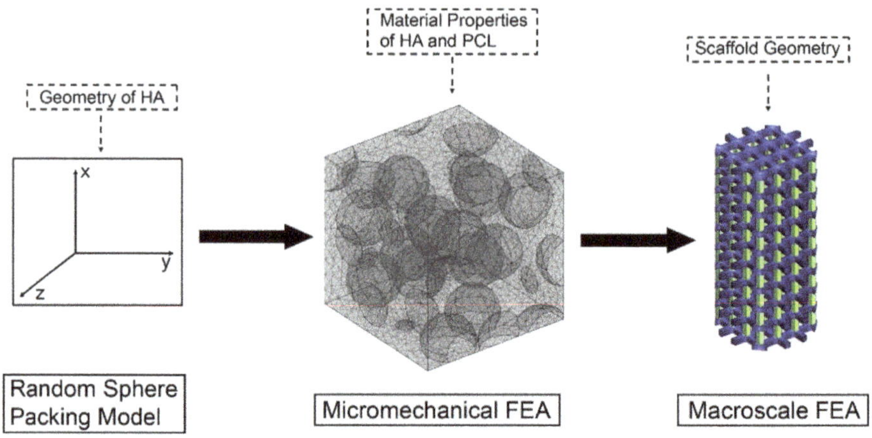

Figure 4.69. Schematic diagram of the three-stage modeling process developed to predict mechanical properties of the PCL:HAP composite scaffolds at different loading levels of filler. Reprinted from [142], Copyright (2012), with permission from Elsevier.

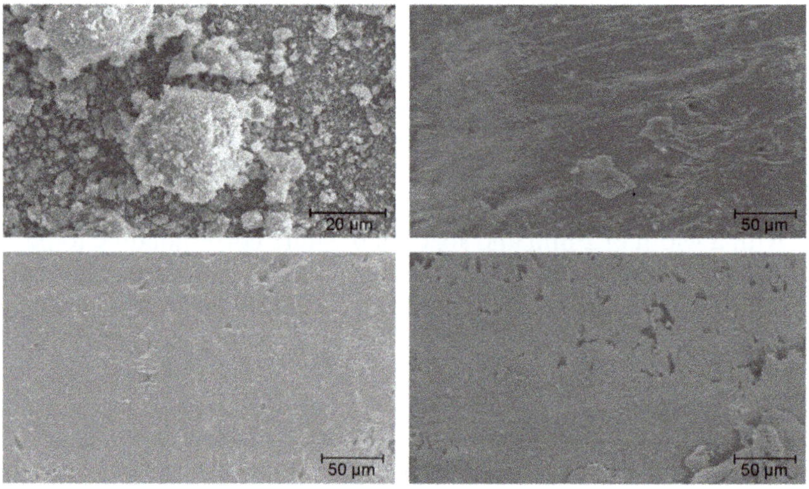

Figure 4.70. SEM images of the pure HAP powder (top left), SLS-printed 90:10 (top right), 80:20 (bottom left) and 70:30 (bottom right) composites. Reprinted from [142], Copyright (2012), with permission from Elsevier.

Solid gage samples showed an increase from 299.3 to 498.3 MPa when increasing the loading of HAP from 0% to 30%, and in the scaffold with the highest porosity (80.9%), the compressive modulus increased more than 100% (from 14.9 to 36.2 MPa). The improvement in compressive moduli for the scaffolds can be attributed to particulate reinforcement by the HA as a hard phase filler. The ultimate compressive strength (σ_{uc}) and strain at yield (ε_y) of the solid and scaffold specimens did not improve with increased loading of HAP.

Compressive Young's moduli were calculated for the samples with HAP content at 10, 20 and 30 vol%, by using a micromechanical model, which were well consistent with experimental results. Compressive effective moduli of the porous scaffolds predicted by FEA were also similar to the results from mechanical characterization results. Figure 4.71 shows plots of von Mises stress distributions of the porous scaffolds. For 1D scaffold, the distribution of the stress was highly uniform. However, for 2D and 3D scaffolds, the stress was concentrated at the corners of the columns. In the areas with concentration of stress, the von Mises stress was not over the yield stress for the bulk samples.

HAP was in-situ combined with poly(l-lactic acid) (PLLA) particles, forming composites used to prepare scaffolds by using SLS printing technology [143]. PLLA particles were first functionalized through the polymerization of dopamine oxide, thus introducing rich active catechol groups on the surface of the polymer particles. As a consequence, Ca^{2+} ions would be concentrated by the catechol groups SBF solution, while the Ca^{2+} ions absorbed PO_4^{3-} ions through electrostatic interaction, thus facilitating in-situ nucleation of HAP.

HAP was homogeneously grown on the surface of the PLLA particles, resulting in HAP/PLLA scaffolds with strong interfacial bonding between the two components. At the same time, the scaffolds exhibited outstanding bioactivity, in terms of inducing the growth of apatite, thus making it possible for human bone mesenchymal stem cell attachment, proliferation and osteogenic differentiation. Furthermore, *in vivo* experimental results indicated that the HAP/PLLA scaffolds could stimulate ingrowth of blood vessels and formation of new bone. Correspondingly, the bone volume fraction and bone mineral density were increased by 44.44% and 41.73%, respectively.

PLLA powder, DA hydrochloride, tris–HCl solution, fetal bovine serum (FBS), calcein-acetoxymethylester (calcein-AM), Dulbecco's modified Eagle's medium

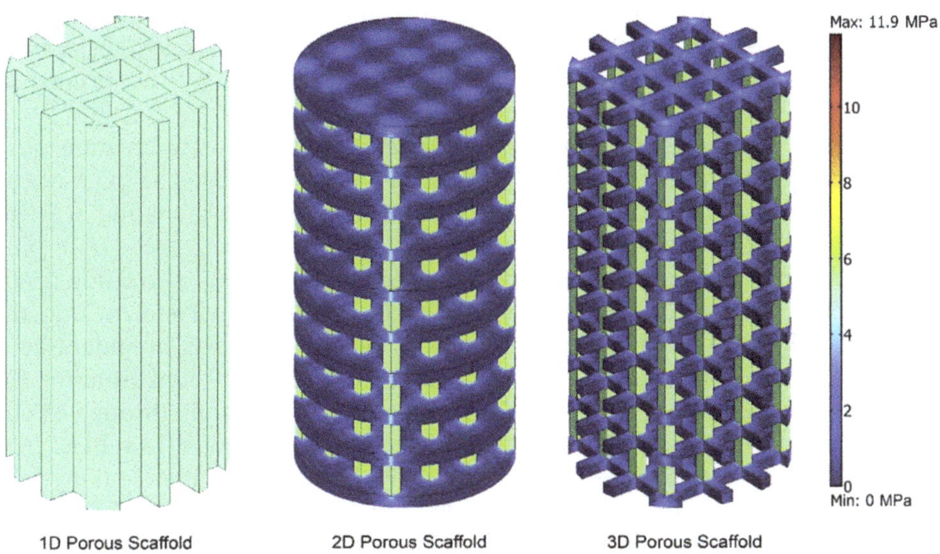

Figure 4.71. Von Mises stress plots of the porous compressive scaffold geometries for the scaffolds made from PCL:HA 70:30 powder mixture. Reprinted from [142], Copyright (2012), with permission from Elsevier.

(DMEM), 4,6-diamidino-2-phenylindole (DAPI), propidium iodide (PI), penicillin, streptomycin and PBS were all commercially available from Sigma-Aldrich (Shanghai, China). To graft DA onto the surface of the PLLA particles, 60 mg DA hydrochloride was added into 30 ml Tris–HCl solution, forming pDA solution. Then, 1 g PLLA powder was dispersed into the pDA solution, with aid of magnetic stirring for 24 h at room temperature. After that, the suspension was centrifuged for 10 min at 8000 rpm. The unattached pDA was washed with DI water, followed by drying at 40 °C for 24 h, leading to a powder sample.

To obtain HAP/PLLA composite powder, the pDA/PLLA powder was soaked in a 1.5 × SBF solution in a water bath at 37 °C for 1–5 days in an incubator. The SBF solution was replaced with a new one every 24 hours, in order to ensure sufficient Ca^{2+} and PO_4^{3-} ions for the nucleation of HAP. The samples soaked for different time durations were collected, washed and dried at 40 °C, which were named as HAP/PLLA-1, HAP/PLLA-3 and HAP/PLLA-5 powders, for soaking times of 1 d, 3 d and 5 d, respectively. Meanwhile, PLLA powder without pDA coating was also prepared as a comparison.

For SLS printing, the powder was uniformly spread on a platform, which was then scanned with a laser beam following a designed pattern. Once the first layer was printed, the platform was lowered by a distance equivalent to the thickness of one layer, where the powder was spread again with the same thickness, thus forming the second layer after printing. A scaffold was developed after repeating this process. After that, the powder bed was cooled down, so that the scaffold could be taken out.

The optimized processing parameters included a laser power of 2.6 W, scanning rate of 120 mm · min^{-1}, laser beam spot size of 580 μm, scanning line space of 1 mm, powder bed temperature of 80 °C and powder layer thickness of 0.1–0.2 mm. The corresponding laser energy density E was 1.3 J · mm^{-2}. The composite samples derived from the HAP/PLLA-1, HAP/PLLA-3 and HAP/PLLA-5 powders were denoted as HAP/PLLA-1, HAP/PLLA-3 and HAP/PLLA-5 scaffolds, respectively. A pure PLLA scaffold without the modification with pDA was printed for comparison.

Mechanisms for the formation of the pDA coating and in-situ creation of the HAP particles on surfaces of the PLLA particles were examined. The pDA coating had a thickness of about 1 μm. The formation of the pDA coating was also confirmed by XPS results. After soaking in SBF solution, the HAP phase precipitated on the surface of PLLA, as supported by XRD characterization results. DA was self-polymerized into pDA in the slight alkaline environment and then anchored on the surface of the PLLA particles [144]. Therefore, abundant active catechol groups were produced as a result, which attached Ca^{2+} ions from the SBF solution through chelation, while Ca^{2+} ions attracted PO_4^{3-} ions through electrostatic interaction, thus facilitating the nucleation of HAP crystals [145–147].

Surface SEM images of the PLLA and pDA/PLLA powders after soaking in 1.5 × SBF solution for different time durations are depicted in figure 4.72(a). The particle surfaces of the HAP/PLLA-1, HAP/PLLA-3 and HAP/PLLA-5 powders were decorated with particle clusters, which were not present for the PLLA powder. The HAP/PLLA-3 powder particles were entirely covered by the particle clusters,

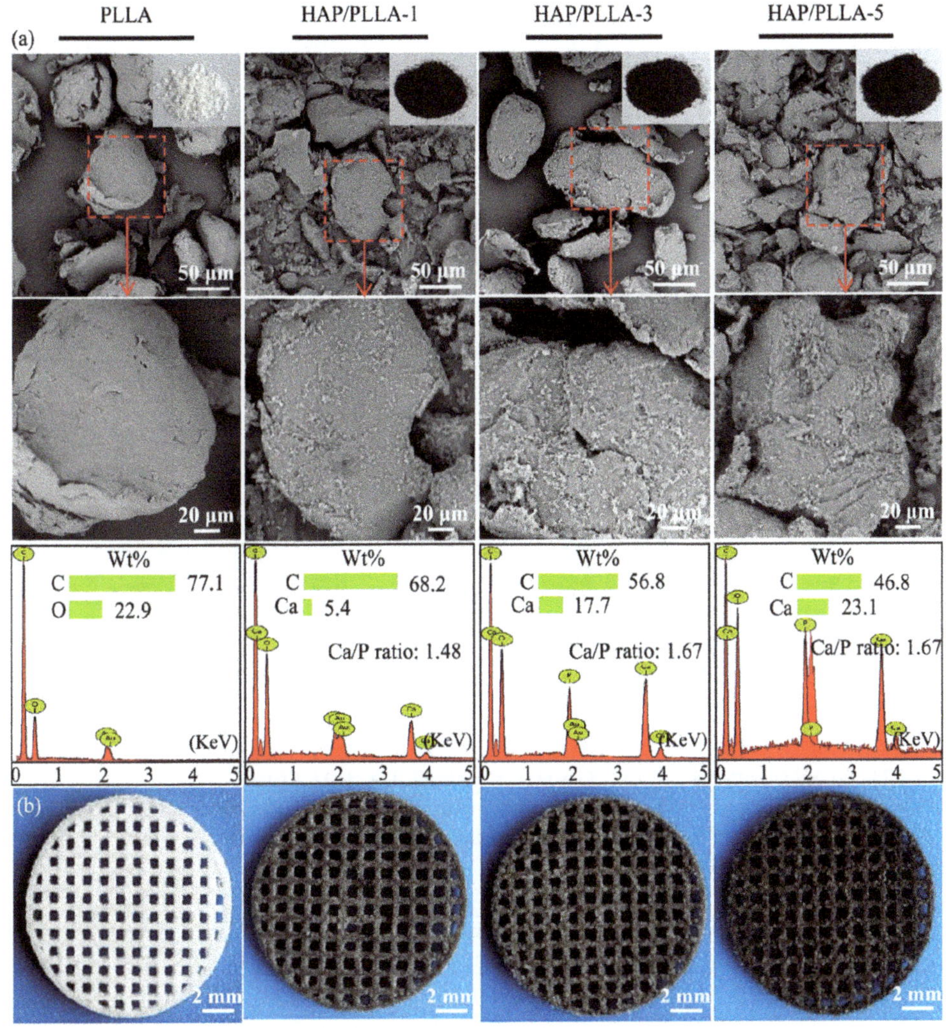

Figure 4.72. (a) SEM images and the corresponding EDS mapping spectra of the PLLA, HAP/PLLA-1, HAP/PLLA-3 and HAP/PLLA-5 powders. The insets in SEM images are the photographs of the PLLA, HAP/PLLA-1, HAP/PLLA-3 and HAP/PLLA-5 powders. (b) Photographs of the printed PLLA, HAP/PLLA-1, HAP/PLLA-3, and HAP/PLLA-5 scaffolds. Reprinted with permission from [143]. Copyright (2020) American Chemical Society.

while there was a layer of the particle clusters on the surface of the HAP/PLLA-5 powder. The EDS spectrum of the PLLA powder consisted of only C and O peaks. In comparison, for the three HAP/PLLA powders, the characteristic peaks of Ca and P were detected, where the Ca/P ratio was 1.67 for the HAP/PLLA-3 and HAP/PLLA-5 powders, confirming that the particle clusters were HAP.

PLLA is biologically inactive, so no mineralization of HAP was induced [148, 149]. With increasing soaking time duration, the content of Ca was increased from 5.4 to 23.1 wt%, from the HAP/PLLA-1 to the HAP/PLLA-5 powders. In the HAP/

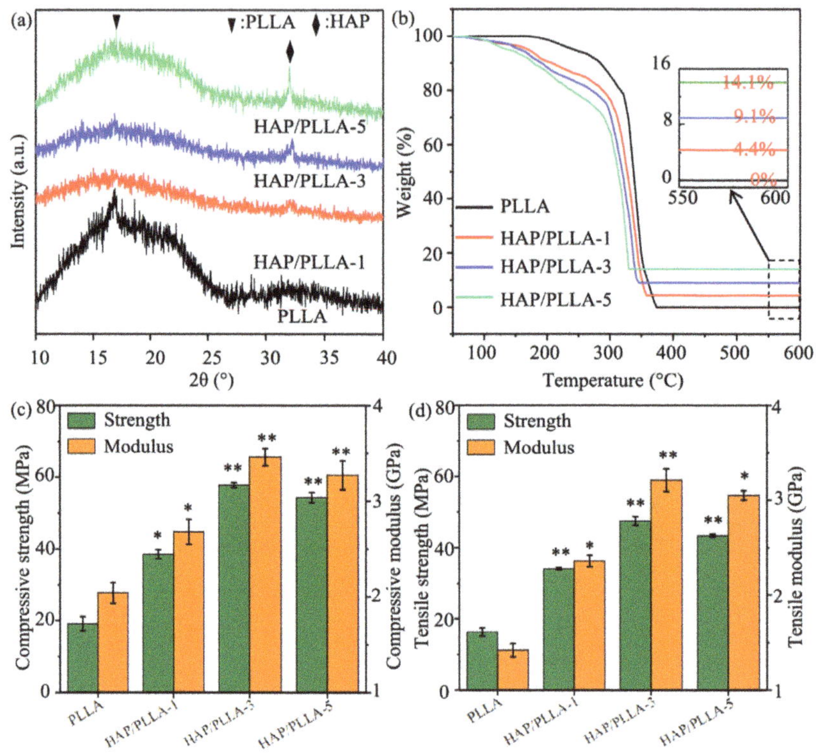

Figure 4.73. (a) XRD patterns and (b) TGA curves of the PLLA, HAP/PLLA-1, HAP/PLLA-3 and HAP/PLLA-5 scaffolds. Compressive (c) and tensile (d) strengths of the scaffolds (*represents statistical difference: $*P < 0.05$, $**P < 0.01$, as compared with the PLLA scaffold). Reprinted with permission from [143]. Copyright (2020) American Chemical Society.

PLLA composite powders, mineralization of HAP occurred on the surface of the PLLA particles, owing to the aid of pDA. Photographs of the SLS printed PLLA, HAP/PLLA-1, HAP/PLLA-3 and HAP/PLLA-5 scaffolds are depicted in figure 4.72 (b). All the samples had high porosity, while the pores were well-interconnected with an average size of 500 μm, suitable for bone formation and vascularization [150].

XRD patterns of the PLLA, HAP/PLLA-1, HAP/PLLA-3 and HAP/PLLA-5 scaffolds are presented in figure 4.73(a). Since semi-crystalline PLLA has a peak at $2\theta = 16.5°$, corresponding to its (110)/(200) plane, implying that amorphization of PLLA was induced by laser irradiation during the printing process. A small characteristic peak occurring at about $2\theta = 31.8°$ was derived from the (211) plane of HAP, and the intensity of this peak increased as the HAP content in the PLLA matrix was increased.

The content of HAP in the HAP/PLLA composite scaffolds was estimated according to their degradation behaviors, with TGA curves up to 600 °C as demonstrated in figure 4.73(b). The PLLA scaffold started to lose weight at about 350 °C, while the decomposition was completed at about 430 °C. Because HAP was thermally stable, the weight loss was fully responsible by the degradation of PLLA.

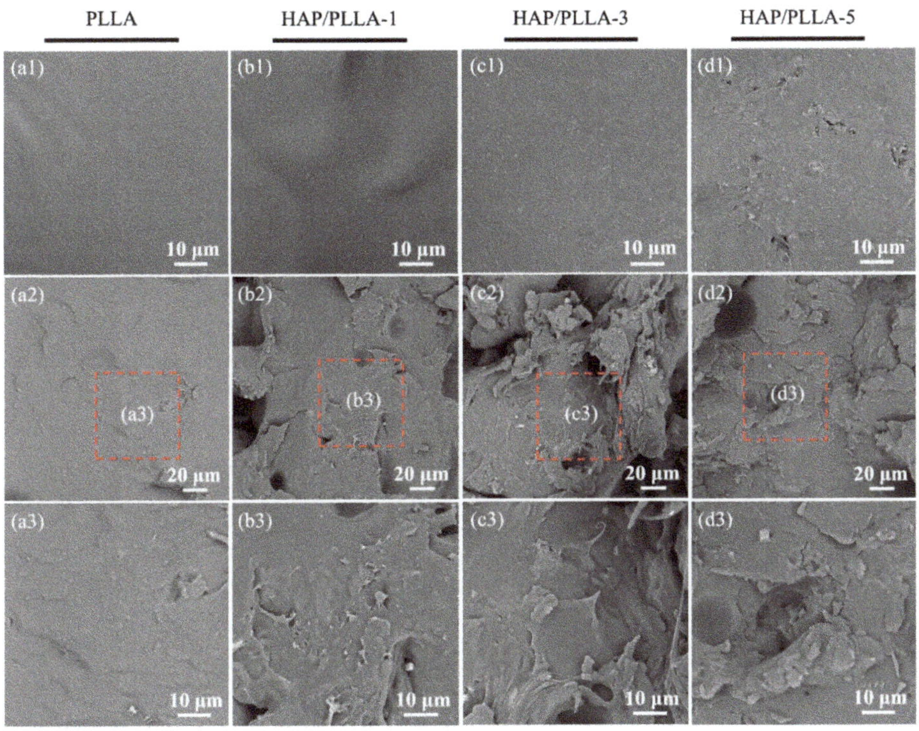

Figure 4.74. SEM images of the PLLA, HAP/PLLA-1, HAP/PLLA-3 and HAP/PLLA-5 scaffolds: (a1–d1) natural surface and (a2–d2 and a3–d3) fractured surface. Reprinted with permission from [143]. Copyright (2020) American Chemical Society.

In this regard, the weight contents of HAP in the HAP/PLLA composite scaffolds were 4.4%, 9.1%, and 14.1% for the HAP/PLLA-1, HAP/PLLA-3 and HAP/PLLA-5 scaffolds, respectively.

Mechanical properties of the scaffolds, i.e., compressive and tensile properties, are plotted in figures 4.73(c) and (d). Both the compressive strength and compressive modulus were increased from 19.14 MPa and 2.04 GPa for the pure PLLA scaffold to 57.83 MPa and 3.46 GPa for the HAP/PLLA-3 scaffold and then decreased to 54.34 MPa and 3.27 GPa for the HAP/PLLA-5 scaffold. The mechanical properties of the HAP/PLLA-3 scaffold were within the scopes of cancellous bones and cortical bones. The tensile properties exhibited a similar variation trend, with improvements by 192% and 126%, respectively.

Figure 4.74 shows natural surface and fractured surface SEM images of the PLLA, HAP/PLLA-1, HAP/PLLA-3 and HAP/PLLA-5 scaffolds. Surfaces of the HAP/PLLA composite scaffolds were characterized by the presence of bright spots, which were caused by the mineralized HAP particles, as observed in figures 4.74 (b1)–(d1). In comparison, the surface of the PLLA scaffold was highly smooth and dense, as seen in figure 4.74(a1). HAP particles were evenly distributed inside the PLLA matrix, without the presence of aggregation. Nevertheless, pores were visible as defects near the HAP particles, which could be related to their high concentrations in the PLLA matrix.

The fractured surface of the PLLA scaffold was represented by the brittle fracture mode, without plastic deformation, as revealed in figures 4.74(a2) and (a3). In contrast, the fractured surfaces of the HAP/PLLA composite scaffolds were rough and irregular, with the observation of ductile deformation for the matrix, as seen in figures 4.74(b2–d2) and (b3–d3). In the HAP/PLLA-3 scaffold, elongated polymer filaments protruded from the matrix, due mainly to the formation of strong interface bonding between HAP and PLLA phases. Therefore, the PLLA molecules near the HAP particles were stretched and elongated, when the fractured surface approached the HAP particles. However, they would not be broken, before a certain length was reached owing to the stretching and elongating. Because of the strong interfacial bonding, the fracture energy was effectively consumed, thus alternating the fracture mode from brittle to ductile fractures [144].

Polymethyl methacrylate (PMMA) and β-tricalcium phosphate (β-TCP) were combined to form composite powders, which were used to fabricate biomaterials with SLS printing technology [151]. PMMA powder had an average particle size of 75 μm, with melting and the glass transition temperatures of 160 °C and 106 °C. The β-TCP powder had a density of 3.18 $g \cdot cm^{-3}$. Two compositions were selected for study, i.e., 10% and 20% β-TCP.

In SLS printing, the critical processing parameters included laser power (*P*), scan speed (v) and laser beam size (*D*). A makeshift SLS system was used for the experiment, equipped with a 60 W CO_2 laser, with a power output in the range of 0.6–60 W and scan speeds of up to 5000 $mm \cdot s^{-1}$. The spot size and focal length of the laser were 540 μm and 370 mm, respectively, while a computer-controlled optical deflection system was employed to realize the desired raster scan patterns. The items were printed on aluminum substrate, whereas the powder bed was maintained at about 100 °C, which was slightly lower than the glass transition temperature of PMMA.

Laser power and scanning rate were tested with single layer samples, in the ranges of 13–24 W and 250–500 $mm \cdot s^{-1}$, respectively. The samples were then characterized by using SEM to examine the effects of processing parameters on the interparticle bonding and melt-flow behavior. Single-layer samples were printed with 10% TCP + PMMA composite powder at the laser power at 22.8 W, while different scanning speeds were used. The higher the scanning speed, the coarser the structure would be. At relatively low scanning speeds, e.g., 381 and 254 $mm \cdot s^{-1}$, interparticle coalescence was present in the base matrix, which could be understood in terms of sintering time during the printing process.

At a given scanning speed, with increasing laser power, the microstructure was gradually improved, as reflected by the increase in the area having coalesced PMMA to form continuous matrix networks to hold the β-TCP particles. However, as the laser power was over a certain limit, the polymer matrix started to degrade, so that the printed structure was disintegrated. Comparatively, the sample printed with the lowest speed of 254 $mm \cdot s^{-1}$ exhibited the optimal microstructure.

The laser power and scanning speed also had a strong effect on thickness of the printed layers. At a scanning speed of 381 $mm \cdot s^{-1}$, thickness of the printed layer was increased with increasing laser power, owing to the increased energy absorption. As a result, the sintered area was enlarged. The coalescence of the particles was undesired, as the laser power was lower than 13 W, while uniform layers could be achieved for the laser powers in the range of 13–35 W. Once the laser power was over 35 W, charring, discoloring and degradation of the materials took place, accompanied even by smoking.

The effect of laser scanning speed on thickness of printed layers was also related to the effective energy absorption, i.e., the thickness was decreased with increasing scanning speed at given powers. The thickness of the layers was stabilized at a value of about 500 μm, as the scanning speed was over a given level. Although the thickness of the layers was proportional to the energy flow rate into the substrate, there was a limit for scanning speed, beyond which the thickness was kept unchanged. In fact, too high a scanning speed led to insufficient densification of the materials.

According to the SEM images of the single layers printed at the laser power of 18 W, it was observed that the densification behaviors were significantly different for the samples with different contents of β-TCP. For the sample from the 80% PMMA + 20% β-TCP powder, degradation and decolorization were observed, as the laser power was above 18 W, which could be attributed to the variation in rheological properties of the materials with increase in the content of β-TCP. In other words, the higher content of β-TCP could result in an alternation in the thermodynamics of the process, so that overheating and degrading occurs in the matrix near the laser irradiation spot.

Experiments were conducted to further optimize the laser power and scanning speed, for the sample with 10% β-TCP. Figure 4.75 shows surface SEM images of representative samples. Although they exhibited a similar degree of coalescence and densification behavior, a gradual improvement occurred with reducing scanning speed, at given laser powers. The optimal range of the ED levels was 0.14–0.16 $J \cdot mm^{-2}$, in terms of densification of the materials. 18 W was a critical laser power, below which only partial melting and partial neck formation were observed, whereas effective densification was present at laser powers of beyond 18 W, irrespective of level of ED. Porosity was decreased with increasing coalescence and densification of the layers.

At an ED of 0.16 $J \cdot mm^{-2}$, strong fusion among the particles was achieved, while the pore size was obviously decreased, suggesting the occurrence of melting, interparticle coalescence and bonding. Large pores are occasionally present, owing mainly to the presence of the β-TCP particles that raised viscosity of the melt pool and increased flowing resistance of the melts. Figure 4.76 shows SEM images of the single-layer samples printed with the modified conditions, with laser powers in the range of 37–42 W. Although ED of 0.1 $J \cdot mm^{-2}$ could be used to realize promising densification, a higher ED was beneficial to further enhance the sintering behavior. At a laser power of 42 W, the optimal scanning speeds were in the midrange. However, the powder bed was overheated and degraded, as the laser power was over 37 and 42 W, for ED of 0.16 and 0.14 $J \cdot mm^{-2}$, respectively.

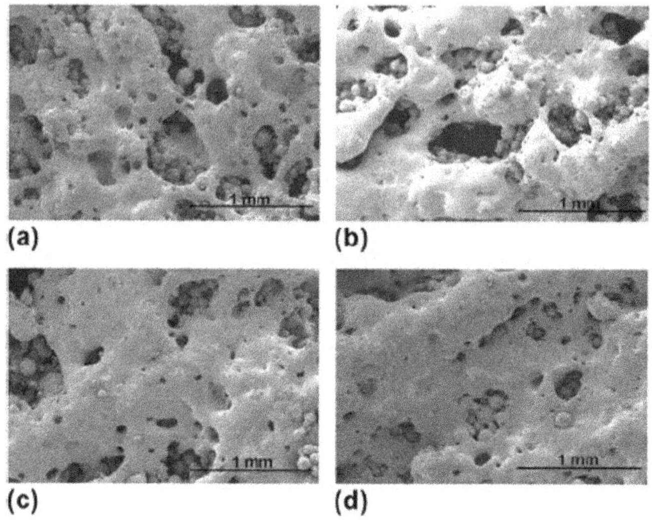

Figure 4.75. SEM images of the PMMA + β-TCP composite single layers printed with different values of ED, laser power (*P*) and scanning speed (*s*): (a) ED = 0.1 J · mm^{-2}, $P = 22.8$ W, $s = 422.17$ mm · s^{-1}, (b) ED = 0.12 J · mm^{-2}, $P^{-1} = 22.8$ W, $s = 352$ mm · s^{-1}, (c) ED = 0.14 J · mm^{-2}, $P^{-1} = 22.8$ W, $s = 300$ mm · s^{-1}, (d) ED = 0.16 J · mm^{-2}, $P^{-1} = 22.8$ W, $s = 263$ mm · s^{-1}. Reproduced from [151], with permission from Springer Nature.

It was found that the porosity of the scaffolds might not be fully controlled in some cases. Insufficiently high energy levels, caused by too low a laser power or too high a scanning speed, could result in weak interparticle coalescence, thus leading to poor interparticle connection, as evidenced in figure 4.77(a). Meanwhile, excessive laser energy could induce too-high temperatures, thus resulting in pyrolysis of the polymer matrix and hence increment porosity of the printed structures [152, 153]. In addition, if the energy input was too high, gases could be trapped inside the polymer matrix, as illustrated in figure 4.77(b).

The higher the temperature, the large the gas bubbles would be, because of the variations in pressure and viscosity [154]. Consequently, pickup of gas phases, formation and growth of bubbles and rupture of the matrix, could all be responsible for the presence of uncontrolled pores, which were closely linked to the laser power and scanning speed. In these cases, the β-TCP particles would increase the viscosity and hence retard the freed movement of the PMMA matrix. Therefore, the uncontrolled porosity was not pronounced in the samples from the PMMA + β-TCP powders.

In these PMMA + β-TCP composite scaffolds, the PMMA acted as the matrix, in which the β-TCP particles were dispersed, serving as an active component with bioactivity. Therefore, it is expected for the β-TCP particles to be distributed on the surface of the printed scaffolds, so as to more effectively facilitate the biological tissue interaction. For the 10% β-TCP sample, the formation of the PMMA matrix layer was optimized at a laser power of 42 W. In this case, the PMMA phase was most likely fully densified, with a smooth surface, on which the relatively light

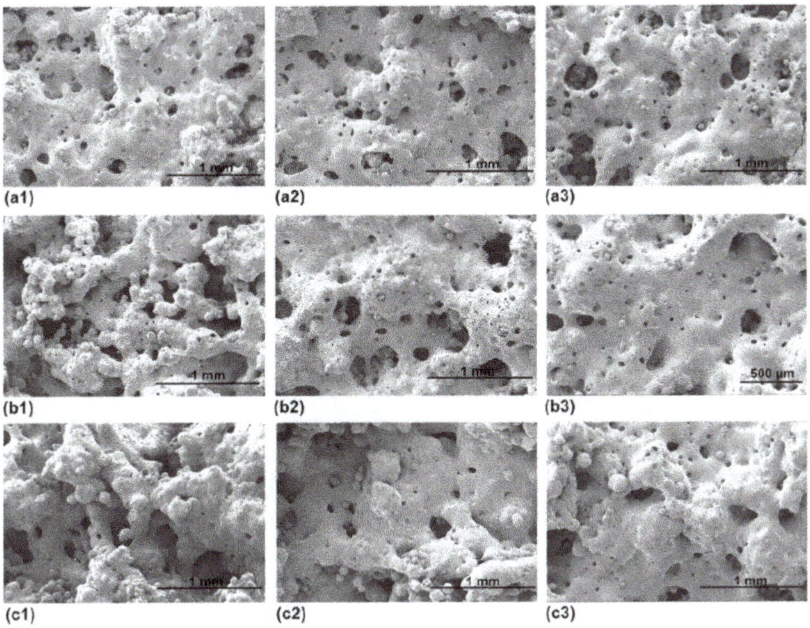

Figure 4.76. SEM images of the PMMA + β-TCP composite samples printed with different process parameters at high powers: (a1) $P = 37.2$ W, $s = 711$ mm · s^{-1}, (a2) $P = 39.6$ W, $s = 762$ mm · s^{-1}, (a3) $P = 42$ W, $s = 812$ mm · s^{-1}, ED = 0.1 J · mm^{-2}, (b1) $P = 37.2$ W, $s = 635$ mm · s^{-1}, (b2) $P = 39.6$ W, $s = 655$ mm · s^{-1}, (b3) $P = 42$ W, $s = 736$ mm · s^{-1}, ED = 0.12 J · mm^{-2}, (c1) $P = 37.2$ W, $s = 457$ mm · s^{-1}, (c2) $P = 39.6$ W, $s = 482$ mm · s^{-1}, (c3) $P = 42$ W, $s = 508$ mm · s^{-1}, ED = 0.14 J · mm^{-2}. Reproduced from [151], with permission from Springer Nature.

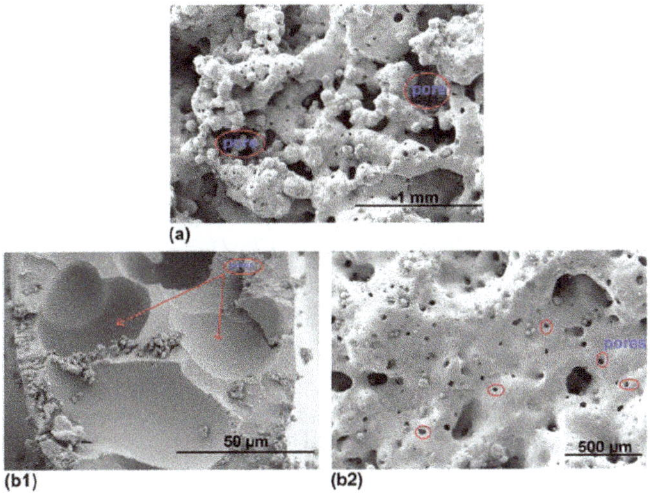

Figure 4.77. SEM images of the samples with uncontrolled porosities in printed samples caused different reasons: (a) insufficient melting and fusion and (b1, b2) gas pickups. Reproduced from [151], with permission from Springer Nature.

β-TCP particles were floated and segregated. Eventually, as multilayers were printed, the β-TCP particles would be present in between the interlayer spaces. However, as the laser power was increased to 45 W, the PMMA layer was bubbled and undulated, while the β-TCP particles were still over the surface of the printed layer.

For the sample with 20% β-TCP composite, the β-TCP particles were wrapped by the PMMA matrix, leading to the structures with irregular morphology. This observation implied that the thermal behavior of the materials was altered, as the content β-TCP was too high, which could be attributed to the factor that the laser energy absorption was enhanced, thus causing overheating of the PMMA component. Then, agglomeration of the β-TCP particles became more serious.

4.3.5 Other ceramic materials

Silicon carbide (SiC) is an important engineering material, with applications as both structural and functional advantages, including high mechanical stiffness, low density, wide bandgap, low coefficient of thermal expansion (ETC), high thermal stability, strong corrosion resistance and so on [155–159]. SiC ceramics have been made by using SLS/SLM, while there are issues to be addressed, such as poor densification and hence low mechanical strength. Therefore, indirect SLS is usually utilized to fabricate SiC ceramic structures, from composite powders with the incorporation of polymers.

SLS was combined with reaction-bonded (RB) technology to prepare SiC/Si composites with superior performances [160]. Epoxy resin was used as a binder, whose effects on properties and microstructures of the printed structures were evaluated. Graphite with low reactivity was utilized as a slow-release carbon source to increase carbon density of the preforms and densification of samples during the printing process. With increasing content of graphite, clusters of nanometer-sized SiC grains in the form of clusters were developed, while the content and particle size of silicon were reduced. Moreover, by applying, the content of residual silicon content in the SiC/Si composites was further decreased after being subject to a two-step sintering process, so that high temperature flexural strength of the final ceramics was enhanced.

The raw materials included α-SiC powder with purity of 98.6% and $D_{50} = 4$ μm, spheroidal graphite with $D_{50} = 10$ μm and 25 μm and coarse Si powder with purity of 97.8% and $D_{50} = 4$ mm. Epoxy resin (E12, $D_{50} = 5$ μm) and dicyandiamide (Q-curing agent SH-500A), with mass ratio of 10:1, were mixed and cured at 180 °C. The powders were mixed at 150 rpm for 24 h. A commercial 3D printer (Farsoon SLS eform, China), with a chamber of 300 mm × 300 mm × 400 mm, was used to prepare SiC/C preforms with complex shapes. A CO_2 laser with a maximum power of 18 W and the beam size of 0.3 mm was employed for the printing experiment. The scanning speed was 3500 mm · s^{-1}, while the thickness of each printed layer was 0.1 mm.

According to the TG analysis results, the carbon content of E12 was 12.4%. The preforms were put on the Si particles, with the mass being twice the total mass of the preforms. Therefore, the molten Si was infiltrated into the preforms through the

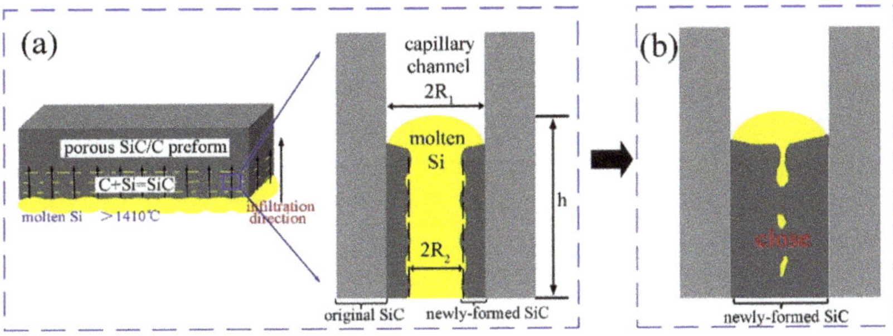

Figure 4.78. Schematic diagram showing the infiltration of molten Si in the preforms. Reprinted from [160], Copyright (2020), with permission from Elsevier.

capillary effect of the liquid Si at 1600 °C. The SiC/C preforms were penetrated by the Si melts and then the reaction between C and Si was triggered as the temperature reached 1414 °C, as schematically demonstrated in figure 4.78(a). Once Si was molten at the melting temperature, the liquid would infiltrate into the SiC/C preforms due to the capillary effect [161]. The capillary channel was gradually narrowed after the reaction occurred. As the capillary channel was completely enclosed, the infiltration of the Si melt was stopped, as illustrated in figure 4.78(b). Consequently, close pores would be formed, while residual carbon would be left in the matrix. The reactivity and content of carbon had a direct effect on the reaction rate and productivity of SiC.

Carbon black is usually amorphous with relatively high specific surface area, thus possessing high reactivity. To obtain silicon carbide ceramics through reaction bonding, carbon black was mixed with silicon carbide powder to form porous SiC/C preforms. At sufficiently high temperatures, the reaction between carbon black and molten Si was rapidly boosted. Therefore, massive SiC was formed in a very short time duration. The SiC from the reaction and the initial component were strongly bonded together. In this case, blockage of the capillary channels tended to happen [162].

At the same time, since the SiC and carbon black had different particle sizes, their mixture would have relatively poor fluidity and stability. To this end, spheroidal graphite with large particle sizes and ordered structure was used as the carbon source. Because of its relatively low activity and large particle size, the presence of spheroidal graphite would slow down the release of the carbon source, thus delaying the blockage of the capillary channels.

With laser irradiation, E12 was cross-linked to form network structures to bond with SiC particles, due to its thermosetting nature. As the content of E12 was 5 wt%, the preform had sufficiently high strength, but powder was loosely attached on the surface of the sample. When the content of E12 was increased to 10 wt%, the bonding strength was largely enhanced. The sample was sufficiently strong for handling, without cracking or collapsing. Therefore, 10 wt% with the optimal content of E12, in terms of printing quality and low linear shrinkage during densification.

It was observed that porosity and carbon density of preforms were decreased with increasing content of E12. Actually, E12 would fill the pores in the preforms, leading to reduction in porosity, while the carbon yield was reduced, so that the density of carbon was decreased. Meanwhile, the bulk density and flexural strength of the printed samples were both slightly decreased with increasing level of E12. Therefore, the content of E12 had only a limited effect on properties of printed ceramic samples. In order to reduce the content of carbon and improve mechanical strength of the SiC/Si composites, extra carbon should be incorporated into the preforms.

Photographs of representative SLS printed items with different structures are shown in figure 4.79(a). During the SLS printing process, E12 was melted as a binder to bond the SiC particles together. A typical fractured surface SEM image of the sample of figure 4.79(a) is depicted in figure 4.79(b). The SiC particles were decorated with E12 melt. As the carbonization temperature was increased, the incompletely melted E12 was softened, so that the SiC particles began to fell off, resulting in volume shrinkage. The largest linear shrinkage was observed in the z-direction, and more incompletely melted E12 was present in the z-direction. After finishing carbonization, the E12 was completely converted to carbon, leading to further shrinkage in volume of the preforms.

Owing to the infiltration of the Si melt, the printed items had nearly full densification bodies. The variation in dimension before and after the reaction bonding was <3%. Figure 4.79(c) shows the surface SEM image and XRD pattern (inset) of the printed ceramics. It was observed from the surface morphology that the printed ceramics consisted of two phases, with white and gray colors. SiC grains were gray, while Si grains were white. As revealed in the XRD pattern, the gray SiC are present as α-SiC and β-SiC. According to SEM images, the newly formed SiC and the original SiC grains were bonded together, forming a core–shell structure, as demonstrated in figure 4.79(d) [160, 162]. Due to the relatively low content of carbon, the newly formed SiC as the shell was thin as a result.

Figure 4.79(e) shows a fractured surface SEM image of the printed sample, revealing a typical brittle fracture mode. The Si phase was removed by using a HNO_3 and HF solution, with a volume ratio of 7:1, for clear observation. The two newly formed SiC grains were clearly visible, as demonstrated in figure 4.79(f). They were nanosized SiC grains in between the original SiC grains and the layer on the surface of the SiC. Owing to the high content of Si, a continuous Si framework was formed, which was responsible for the low performances of the printed ceramics. Therefore, the carbon density of the preforms should properly increase [163].

Two spheroidal graphite powders, with particle sizes of 10 and 25 μm, were adopted as slow-release carbon sources to increase the carbon density of the preforms. With the addition of the graphite powder with particle size of 25 μm, the porosity was increased, so that the increase in carbon density of the preforms was not pronounced. This implied that the graphite with large particle size was promising, while it tended to result in residual carbon. In contrast, when the particle size of the graphite was 10 μm, the porosity was decreased and the carbon density was increased. In this regard, spheroidal graphite with a smaller size was recommended.

Figure 4.79. Morphologies of the SiC/E12 preforms and printed samples: (a) photographs of representative green bodies with complicated structures, (b) fractured surface SEM image of the green body, (c) optical microscopic image of the printed ceramics, (d) surface SEM image of the printed sample, (e) fractured surface SEM image of printed ceramics and (f) fractured surface SEM image of the printed sample after acid etching. Reprinted from [160], Copyright (2020), with permission from Elsevier.

Figure 4.80(a) depicts surface microstructure of the printed sample with 5 wt% graphite, revealing two types of phases, i.e., SiC and Si, which had dark gray and grayish colors, respectively. The original SiC and newly generated SiC were combined, thus forming a core–shell structure. The shell was relatively thin, owing to the low content of the spheroidal graphite. At the same time, nanosized SiC particles were formed in the silicon phase. As the content of the spheroidal graphite was increased to 20 wt%, significant alternation in the structure of the printed structure was observed, as revealed in figure 4.80(b).

The Si in between the SiC particles was wrapped by a 'thick fog', with the thickness of the coating layer being increased. As the content of graphite was raised, the quantity of the new SiC was increased and hence the thickness of the shell layer

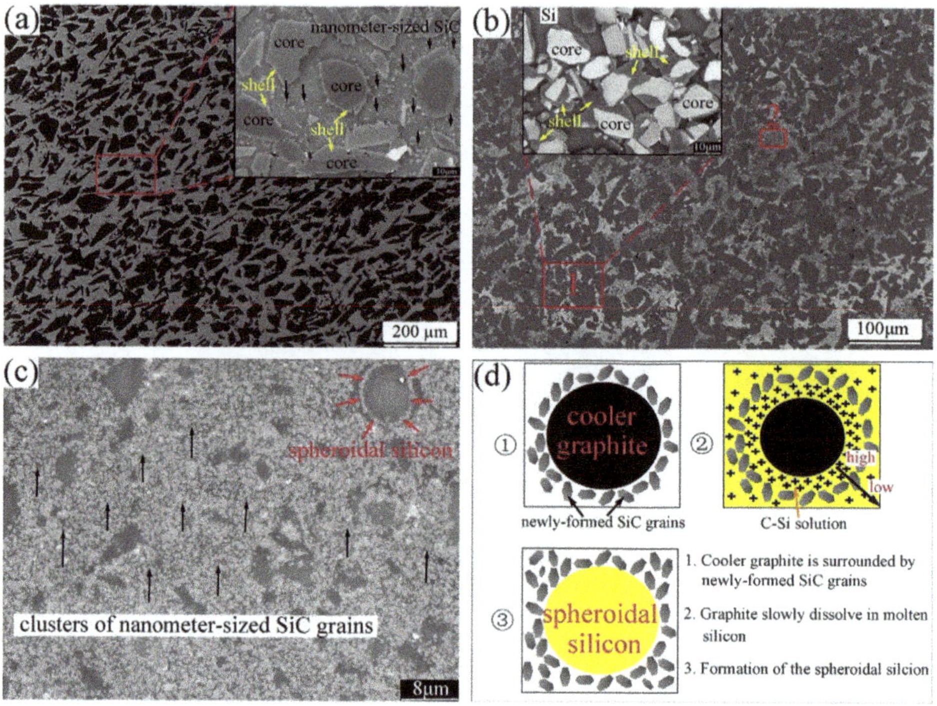

Figure 4.80. SEM images and schematic diagram of formation mechanism: (a) surface topography of the printed sample with 5 wt% graphite, (b) surface topography of the printed body sample 20 wt% graphite, (c) clusters of nanometer-sized SiC grains and (d) formation mechanism of spheroidal Si particle. Reprinted from [160], Copyright (2020), with permission from Elsevier.

was also increased. The 'thick fog' region depicted in figure 4.80(b) is enlarged in figure 4.80(c). The nanosized SiC grains were agglomerated as clusters, which were uniformly embedded in the Si phase. Therefore, the size and content of Si were tremendously reduced, which was responsible for the increment in mechanical strength of the printed samples. Schematic diagrams describing the formation mechanisms of the 'fog obscured' region and the spherical Si particles are illustrated in figure 4.80(d).

When the content of spheroidal graphite was raised to 25 wt%, numerous spherical Si particles and unreacted spheroidal graphite particles were present in the as-printed samples. After the reaction-bonding process, blockage of the capillary channel occurred, so that the residual carbon was unable to escape, owing to the infiltration of molten Si. To effectively eliminate the residual Si and carbon, the printed samples were subjected to subsequent calcination. Consequently, the residual Si reacted with the residual carbon through diffusion. After being calcined in Ar at 1750 °C for 2 h, the unreacted spheroidal graphite was completely absent. Furthermore, since the reaction between Si and carbon was accompanied by a large volume shrinkage, pores were easily produced in the printed samples.

Because Si has a melting point of 1414 °C, the effect of Si on high temperature properties was negative. As the content of spheroidal graphite was increased, the flexural strength was increased first and then started to decrease. Therefore, the sample with 20 wt% spheroidal graphite was optimal in terms of achieving high mechanical properties. Flexural strength of the printed sample from the powder with 25 wt% graphite after calcination was significantly enhanced. For instance, the flexural strength of the 25 wt% at 1350 °C was higher than that of the one with 20 wt% graphite. This was because the content of Si was higher in the sample with 20 wt% graphite than in one with 30 wt% graphite. When the temperature was 1350 °C, Si started to soften, which would negatively contribute to flexural strength.

SLS printing process has been employed to fabricate SiC ceramic composites, with carbon fibers as the enforcement agent combined with liquid Si infiltration [164]. Chopped carbon fibers (C_f) and phenolic resin (PR) were mixed to form powders for SLS printing experiments. The C_f/PR green bodies had a relatively low PR content of 25 vol%, exhibiting suitable printing stability and promising mechanical strength. SiC ceramics composites were eventually obtained through liquid Si infiltration, with optimal performances, including density of $2.72\ g \cdot cm^{-3}$, flexural strength of 266 MPa, elastic modulus of 248 GPa and fracture toughness of $2.87\ MPa \cdot m^{1/2}$.

Figure 4.81 shows a schematic diagram for preparation of the SiC ceramic composites, consisting of three steps, including preparation of printing powders, the fabrication of C_f/PR green bodies through SLS printing and reaction-formed SiC (RFSC) to form SiC composites. A sufficiently low content of PR ensured high solid content and strong mechanical properties of the final SiC ceramic composites. Owing to the formation of near-net-shape RFSC, the SiC ceramic composites

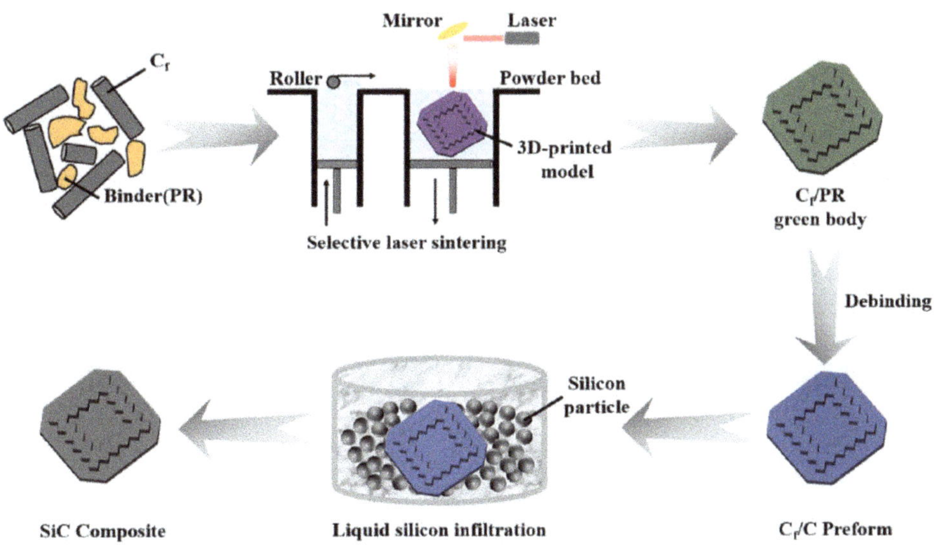

Figure 4.81. Schematic diagram of the process to fabricate the SiC ceramic composites. Reprinted from [164], Copyright (2022), with permission from Elsevier.

experienced no shrinkage and deformation, without the requirement of post-processing steps.

Commercial chopped carbon fiber powder had a density of 1.76 g · cm^{-3}, D_{50} = 75 μm and an average diameter of 7 μm. Thermoplastic PR had a density of 1.22 g · cm^{-3} and D_{50} = 20 μm, which was mixed with 10 wt% methenamine as a hardener. The C_f powder and 25 vol% PR were mixed through ball milling at 70 rpm for 12 h, with SiC balls and a ball-to-powder ratio of 2:1. The SLS printing experiment was conducted by using equipment (Hunan Farsoon High-Technology Co., Ltd, China) with a continuous and TEM00 Gaussian distributional CO_2 laser. The laser beam diameter was 510 μm, while the laser power was up to 60 W. The scanning speed, layer thickness, hatch distance and part bed temperature were 7620 mm · s^{-1}, 0.1 mm, 0.08 mm and 80 °C, respectively. The power was spread by using a roller at a speed of 180 mm · s^{-1}.

The SLS printed green bodies were carbonized into porous C_f/C preforms in a vacuum at 900 °C for 0.5 h, at a heating rate of 2 °C · min^{-1}. The pyrolyzed samples were subject to RFSC to form SiC ceramic composites, in which liquid Si was infiltrated into the porous C_f/C preforms, through capillary force and *in-situ* reaction with carbon to form SiC. The porous preforms and Si powder with particles of 1–3 mm were placed in a graphite crucible that was coated with BN to prevent the corrosion of liquid Si. The RFSC reaction lasted for 0.5 h in vacuum at 1550 °C.

Figure 4.82 shows fractured surface SEM images and representative photographs of the green bodies and C_f/C preforms processed at a laser power of 45 W. The green bodies were of layered porous structure with connected pores, thus allowing for the subsequent liquid Si infiltration. The fibers were wrapped by PR, while there was also PR in between the fibers. In addition, the fibers were randomly distributed in the

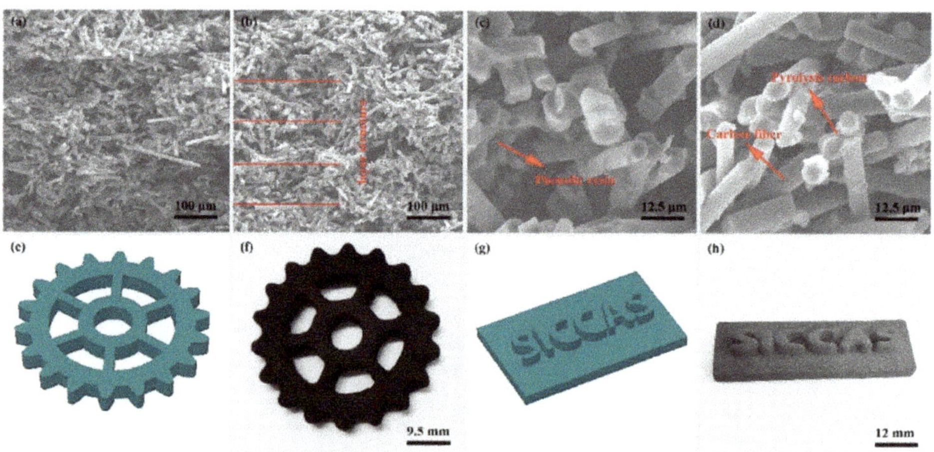

Figure 4.82. Fractured surface SEM images and photographs of representative printed structures: (a) C_f/PR green body, (b) C_f/C preform, (c) high-magnification image of the C_f/PR green body, (d) high-magnification image of the C_f/C preform, (e, f) printed model and the printed product with gear structures and (g, h) printed model and the printed product with SICCAS structures. Reprinted from [164], Copyright (2022), with permission from Elsevier.

samples. Meanwhile, the aspect ratio of the fibers acted as connectors to link the adjacent layers through the pinning effect [165]. Due to the thermoplastic nature, PR was melted by the laser due to the photothermal effect, thus combining the carbon fibers. As the second layer was printed, the residual heat induced partial melting of the PR in the first layer, so that the two successive layers were strongly connected.

As the scanning speed and hatch distance were fixed, the laser energy density was increased with increasing laser power. Consequently, the content, surface tension and viscosity of the liquid phase due to melting of the PR were influenced, which thus would directly affect the microstructure and performances of the final products. At high temperatures, PR was converted to carbon, accompanied with the formation of pores. The multilayer structure of the green bodies was retained in the C_f/C preforms. In addition, the fibers in the C_f/C preforms exhibited smoother surface than those in the green bodies, as seen in figure 4.82(d), due to the reconstruction of chemical bonds [166]. Photographs of representative C_f/PR green bodies with gear and SICCAS complex structures are depicted in figure 4.82(e, f), demonstrating feasibility of the SLS printing technique.

With increasing laser power, the open porosity of the green body first decreased and subsequently increased. The sample printed at a laser power of 45 W displayed the lowest porosity of 72.7%, along with the highest density. Owing to its low glass transition temperature, T_g =98.95 °C, PR absorbed laser energy, followed by softening and melting. The flowing melt triggered rearrangement of the particles. After the laser was switched off, the liquid phase was solidified, so that the fibers were combined together [167].

Laser power determines the spot temperature and hence the degree of liquification of the polymer. For instance, when the power was 35 W, PR was incompletely melted, thus leading to a sample with a high porosity of 75.9%. In comparison, the porosity was reduced to 72.7% in the sample printed at a laser power of 45 W. As the laser power was further raised, the heat would be accumulated, so that the temperature at the center of the laser spot was increased abruptly. In this case, decomposition of the PR occurred, thus producing gases, so that the porosity was increased. For example, porosity of the sample printed at 50 W was 76%. After carbonization, the porosity of the C_f/C preforms was increased by 2.1%, while the density was decreased by 3.4%.

The green bodies could be taken out from the powder bed, while their mechanical strength was not sufficiently high for cutting and grinding. The porous C_f/C preforms should withstand about 0.4 MPa loading when subject to LSI. As the laser power was increased from 35 to 45 W, flexural strength of the C_f/C preforms was increased by about 13% and maximized at 45 W with a level of 7.7 MPa. The strength was declined to 6.59 MPa at the laser power of 50 W. Too high laser energy density would increase porosity of the preforms, owing to the over-heat induced decomposition of PR. RFSC process occurred at temperatures of over the melting point of Si (T_m = 1414 °C), the Si melt infiltrated into the porous C_f/C preforms through capillary forces and then reacted with carbon to form SiC.

Figure 4.83 shows SEM images and element distribution profiles of the SiC ceramic composites. The three main components included newly reaction-formed

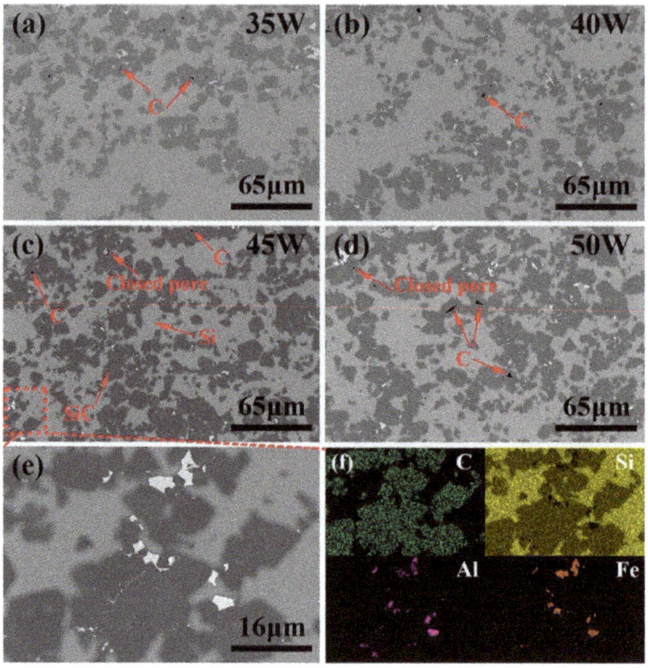

Figure 4.83. Polished surface SEM images of the SiC ceramic composites printed at different laser powders: (a) SiC-35 W, (b) SiC-40 W, (c) SiC-45 W, (d) SiC-50 W and (e) high magnification image of SiC-45 W. (f) EDS elemental distribution profiles of SiC-45 W. Reprinted from [164], Copyright (2022), with permission from Elsevier.

SiC, residual Si and residual carbon, present as dark gray, light gray and black areas, respectively. There were pores inside the SiC grains, which were formed owing to the reaction between Si and carbon. The reaction was accompanied by volume expansion, thus blocking the channels of Si infiltration. The SiC ceramic composites printed at laser powers of 35 and 40 W possessed relatively high content of residual Si, as observed in figures 4.83(a) and (b), because they had high open porosity, large pore size and random pore distribution.

In comparison, for the two samples processed at 45 and 50 W, the distribution of the main phases SiC and Si was uniform, as seen in figures 4.82(c) and (d). This was because their C_f/C preforms exhibited homogeneous pore distribution, reduced pore size and decreased porosity. Meanwhile, residual carbon was also present as a minor phase. Additionally, the bright white region was an Al–Fe alloy, as demonstrated in figure 4.84(e) and (f), which was derived from the impurities in the C_f and PR. As revealed by the high-magnification SEM image in figure 4.83(e), the SiC grains had contrast changes at the edges, which were related to the difference in hardness of SiC and Si. The SEM observations were confirmed by XRD characterization results.

Carbon fiber reinforced silicon carbide (C_f/SiC) composites were manufactured through liquid Si infiltration, with the preforms to be fabricated by using SLS printing, from phenolic resin (PR) coated carbon fibers with submicron Si powder

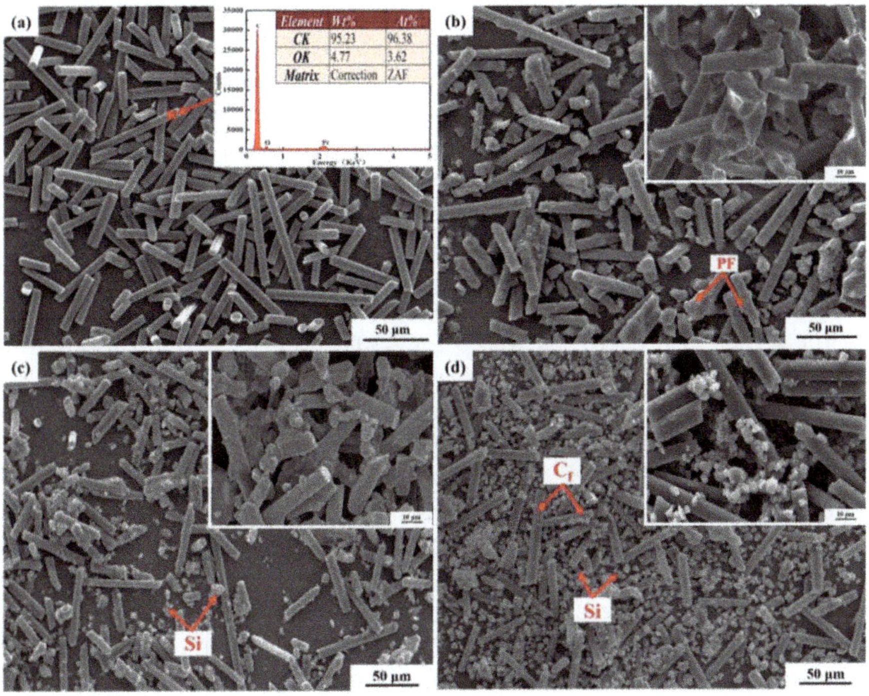

Figure 4.84. (a) SEM images and EDX profile of the raw carbon fibers. SEM images of the composite powders and the corresponding green bodies (insets): (b) PF/C_f-0% Si, (c) PF/C_f-15%Si and (d) PF/C_f-30%Si. Reprinted from [168], Copyright (2019), with permission from Elsevier.

[168]. The introduction of Si reduced porosity and pore size of the carbon preforms, thus resulting in C_f/SiC composites with microstructural homogeneity. The optimized C_f/SiC composite sample exhibited density of 2.89 g $\cdot$ cm^{-3}, flexural strength of 237 MPa and fracture toughness of 3.56 MPa $\cdot$ $m^{1/2}$. Coefficient of thermal expansion (CTE) of the C_f/SiC composite was about 5.5×10^{-6} K^{-1} in the temperature range of 25 °C–900 °C, while it had thermal conductivities of 74–84 W $\cdot$ m^{-1} $\cdot$ K^{-1} at room temperature and 35–40 W $\cdot$ m^{-1} $\cdot$ K^{-1} at 900 °C.

The composite powder for SLS printing was prepared by using a solvent evaporation method. PR powder (PF2123, Wuxi Mingyang Bonding Material Co., Ltd, China) was dissolved in absolute ethanol, into which carbon fiber and Si powders (Hunyuan Fuhong Mineral Products Co., Ltd) were dispersed with the aid of stirring. The suspensions were then mixed through ball milling for 1 h, followed by drying at 60 °C. PR/C_f-Si composite powder was obtained from the mixture after crushing and sieving. The volume ratio of PR to C_f was 3:7. Four groups of composite powders with an Si volume fraction of 0%–30% were studied, which were denoted as PF/C_f-0%Si, PF/C_f-5%Si, PF/C_f-15%Si and PF/C_f-30%Si.

The SLS printing experiment was conducted with a commercial 3D printer (HK P320™, Wuhan Huake 3D Technology Co., Ltd, China), with processing parameters including laser power of 8 W, scanning speed of 2000 mm $\cdot$ s^{-1}, single layer thickness of 0.1 mm, scanning spacing of 0.15 mm and bed temperature of 60 °C. The green bodies were post-cured and carbonized in flowing Ar. The carbonized samples were further densified to adjust porosity by impregnating, curing and carbonizing, with boron-modified phenolic resin (THC-400, Shaanxi Taihang Impede fire Polymer Co. Ltd, China). For impregnating, the samples were immersed in the solution of phenolic resin in vacuum. The impregnated samples were cured at 120 °C, 150 °C and 180 °C for 1, 1 and 3 h, respectively, to minimize the residual stress and shrinkage distortion.

After that, the samples were carbonized at 900 °C for 1 h to develop C_f/C preforms, which were named CP0, CP5, CP15 and CP30, corresponding to PF/C_f-0%Si, PF/C_f-5%Si, PF/C_f-15%Si and PF/C_f-30%Si. For Si infiltration, the C_f/C preforms were placed in a graphite crucible coated with boron nitride. Si powder with particle sizes of 1–3 mm was poured to cover the preforms. The infiltration was conducted at 1550 °C for 1 h, leading to C_f/SiC composites accordingly denoted as SC0, SC5, SC15 and SC30.

Figure 4.84 shows SEM images of the carbon fibers, PF/C_f-Si composite powders and SLS printed green bodies. The carbon fibers had a smooth surface and uniform length, as seen in figure 4.84(a), with an average length of 47.5 μm and diameter of 7.0 μm. The EDX result suggested that the compositions of carbon and oxygen were 95.2 and 4.77 wt%, respectively. The presence of oxygen-containing groups would have a positive effect on interfacial bonding between the carbon fibers and PR. With the coating of PR, surface roughness of the carbon fibers was increased, as revealed in figure 4.84(b). The inset in figure 4.84(b) depicts an SEM image of the SLS printed green body, where the PR was melted and the carbon fibers were bonded by the melt as a porous structure.

After Si powder was introduced, the carbon fibers were coated with less PR, owing to the reduction in relative content, as observed in figure 4.84(c). As the content of Si was increased to 30 vol%, individual Si particles became visible, while the carbon fibers were smooth, as presented in figure 4.84(d). The insets in figures 4.84(c) and (d) demonstrated that the Si particles were uniformly distributed and attached onto the carbon fibers through the viscous flow of PR. With increasing content of Si, porosity of the green bodies was decreased.

Figure 4.85(a) shows SEM image of the C_f/C preform without the presence of Si (CP0), where the carbon fibers (gray color) were bonded by the residual carbon after the carbonization of PR, along with visible pores (dark color). The microstructure was varied in the preforms with Si. As illustrated in figure 4.85(b), Si particles (bright color) were uniformly dispersed in the porous structure (CP5). The carbon phase continuously filled in the spaces in between the carbon fibers. There were cracks between carbon fibers and carbon, which were caused by the shrinkage during the curing and carbonization processes. The density of the sample CP5 was slightly increased, while the pore size was reduced, as compared with the sample CP0. With increasing level of Si, the Si particles tended to aggregate, while the pore size was further deceased, as illustrated in figures 4.85(c) and (d).

The as-printed green bodies had porosities of around 70%, with a slight increase with increasing composition of Si. After carbonization of PR, the porosity of the

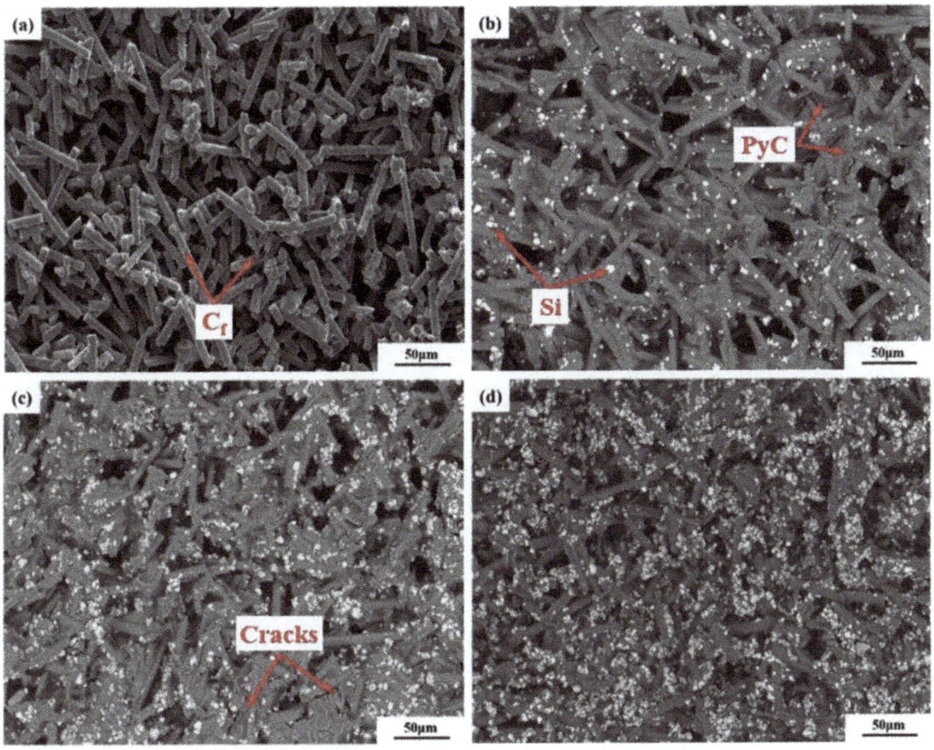

Figure 4.85. SEM images of the porous C_f/C preforms with different compositions: (a) CP0, (b) CP5, (c) CP15 and (d) CP30. Reprinted from [168], Copyright (2019), with permission from Elsevier.

C_f/C preforms was largely decreased to 50.4%–56.8%, with a similar trend regarding the content of Si, in agreement with the SEM observation in figure 4.84. When 1 mol amorphous carbon reacts with liquid Si to produce 1 mol SiC, the volume is expanded by about 60% [169]. As a result, the porosity of the carbon preforms would be about 40%, to support the effective infiltration of the liquid Si. Therefore, the porosity of C_f/C preforms was within the reasonable range.

The C_f/C preforms displayed a relatively wide range distribution of pore size, from microns to 30 μm. The C_f/C preform without Si possessed inhomogeneous internal pore channels, with an average pore size of 20 μm. The C_f/C preforms, with 5%, 15% and 30% Si, were 19.0, 16.7 and 12.1 μm, respectively. During the impregnation process, the dispersed Si particles in the green bodies intercepted PR, thus reducing the capillary diameters, which was transferred to carbon that filled the spaces.

Figure 4.86 depicts polished cross-sectional surface SEM images of the C_f/SiC composites. The C_f/SiC composites all consisted of three phases, including reaction-formed SiC, residual Si and remaining carbon, in colors of dark gray, light gray and black, respectively. The phase distribution in the sample SC0 was somehow nonuniform, with unreacted carbon to be wrapped in the matrix of SiC, as illustrated in figure 4.86(a). This observation was consistent with that of the C_f/C

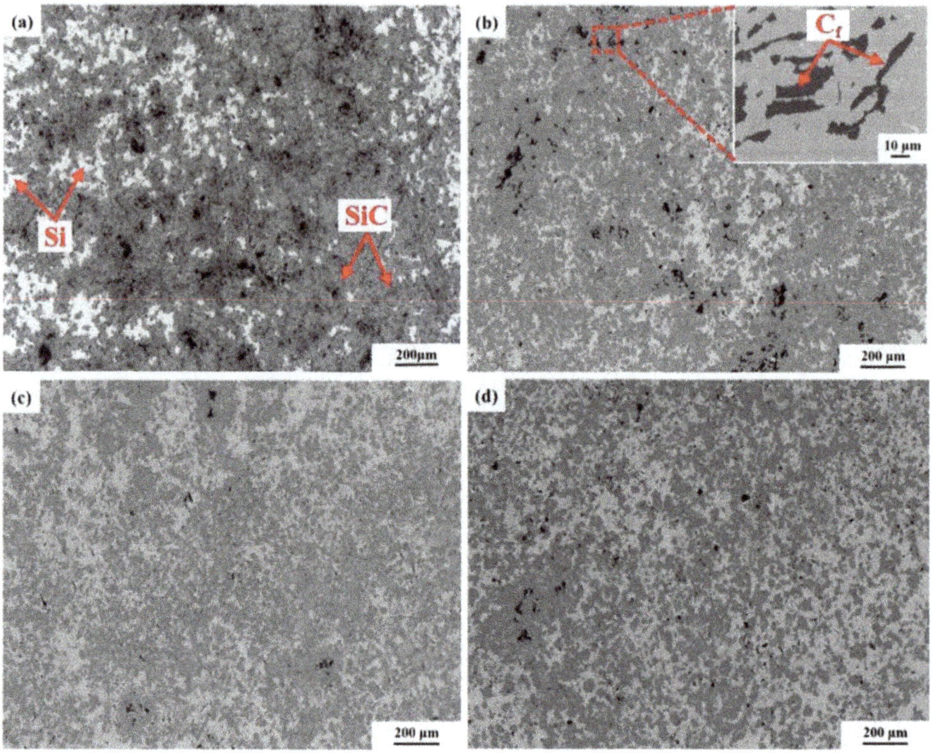

Figure 4.86. Polished cross-sectional SEM images of the C_f/SiC composites: (a) SC0, (b) SC5, (c) SC15 and (d) SC30. Reprinted from [168], Copyright (2019), with permission from Elsevier.

preforms. With addition of Si, the microstructure of the C_f/SiC composites was more homogeneous, as observed in figures 4.86(b)–(d). The sample SC0 had the highest content of residual carbon, while the level of SiC was also the lowest at 62 vol%. In contrast, with 5% SiC, the content of SiC was increased to 71.3 vol%, whereas the content of residual carbon was significantly reduced. Eventually, almost no residual carbon was present in the C_f/SiC composites, but the content of residual Si was pretty high, as revealed in figures 4.86(c) and (d).

SiC has various crystalline structures, with α-SiC and β-SiC to the major two phases. α-SiC is stable at high temperatures, whereas β-SiC is present below 1700 °C. The sample SC0 consisted of β-SiC and Si, with carbon being nearly invisible, due to its low content. The reaction between Si and C is exothermic, causing an increase in localized temperature. As the temperature is above 2100 °C, a phase transition from β-SiC to α-SiC takes place. Since the infiltration was conducted at the relatively low temperature of 1550 °C and the C_f/C preforms had micron-sized pore channels, no α-SiC was detected.

A direct-selective laser sintering (D-SLS) was employed to process SiC ceramics [170]. The processing parameters for the D-SLS of SiC were optimized through numerical simulation. SiC samples were effectively printed with the predicted parameters, demonstrating promising reliability. Evaluations were conducted with layer thicknesses of 22–40 μm and scanning speeds of 100–500 mm · s^{-1}. SiC samples with relative density of 82% could be directly printed on metal substrates, by using low scanning speeds and layer thicknesses of 22 and 30 μm. The relative density of the SiC ceramics was increased to 87%, with optimized processing parameters, including a laser power of 45 W, scanning speed of 100 mm · s^{-1} and hatching space of 40 μm.

The printing experiment was carried out with a Phoenix 3D printer (ProX® DMP 200, 3D Systems, USA), which was equipped with a continuous and single-mode fiber laser, having a maximum output power of 300 W at the collimator and laser spot size of up to 200 μm. The laser system had a flat field F-Theta lens with a focal length of 420 mm. Additionally, a compaction cylinder was adopted for the 3D printer to compact the layer powder after layering to increase the packing density of the powder bed, so that the density of the printed samples could be increased.

α-SiC powder (98.5%, D_{50} = 20 μm, Mersen Boostec®, France) was used as the raw materials. The SiC powder has an absorptivity to fiber laser of 70%, sufficiently high for D-SLS experiment. SiC circular plates (Mersen Boostec®, France) with a diameter of 65 mm were employed as substrates to guarantee strong adhesion of the first printed layer to the printer bed. The circular SiC plates were stuck to the metallic baseplates with Permabond 105 instant adhesive (Permabond, UK). Although the SiC powder consisted of irregular particles, it could be effectively deposited on the powder bed.

Laser power and scanning speed are two critical parameters for printing effect. The combination of low laser power and high scanning speed could not ensure sintering of the powders, while that of high laser power and low scanning speed would trigger decomposition of SiC. Accordingly, the two parameters should be optimized. Predictions were made through numerical modelling, to identify the appropriate laser power range for different scanning speeds, including 100, 250 and 500 mm · s^{-1}, along with layer thicknesses of 22, 30 and 40 μm. The maximum temperature of an appropriate laser power should be below the decomposition point of SiC. Generally, SiC starts to decompose to liquid Si and carbon at 2527 °C [171].

With layer thickness of 22 μm and scanning speed of 100 mm · s^{-1}, as the laser power was in the range of 20–30 W, the temperature would be below the decomposition temperature of SiC. However, these power levels were not sufficient to ensure full densification of the printed structures. Therefore, the laser power should be increased from 30 to 40 W. Accordingly, in the power range of 30–40 W, nearly all the requirements for effective printing were met. Nevertheless, the corresponding temperature was slightly higher than the decomposition temperature of SiC. Fortunately, owing to the transient behavior of laser irradiation, the temperature was insufficient to trigger the decomposition of SiC.

As the scanning speeds were 250 and 500 mm · s^{-1}, the effective laser powers should be in the ranges of 40–50 and 65–75 W, respectively. However, according to theoretical prediction, the laser power levels evaluated for the scanning speeds of 250 and 500 mm · s^{-1} were not able to realize strong adhesion between the printed layers, thus posing a negative effect on efficiency of the printing process. In this regard, the scanning speed should be controlled below a certain level.

Specifically, for a scanning speed of 250 mm · s^{-1}, the suitable laser power range was 45–55 W. In this case, the maximum temperature was above the decomposition point of SiC. Because the high temperature area was relatively small, almost no surface defect was formed. Meanwhile, the predicted laser powers ensured the penetration of the layer beam into the scanned layer, so as to guarantee strong adhesion between adjacent layers. Therefore, too high a scanning speed had a negative effect on efficiency of the printing process. For a scanning speed of 500 mm · s^{-1}, the laser power should be ranging over 65–85 W, whereas the maximum temperature was also higher than the decomposition limit of SiC.

For the layer thickness of 40 μm, the predicted results were similar to those obtained for the thickness of 30 μm. When the scanning speed was 100 mm · s^{-1}, the layers with entire thicknesses could be well sintered, with adhesion to the underneath layer, while the issue of too high a temperature still remained. The laser power range supported high scanning speeds of 250 and 500 mm · s^{-1}. Hatching space was crucial for the connection between adjacent paths, thus influencing mechanical strength of the final products. Too large a hatching distance led to unconnected scanning paths, while a small hatching distance meant longer printing.

The sintering contours for multiple scanning paths on the top surface of the layers printed at scanning speeds of 100, 250 and 500 mm · s^{-1} with different hatching space values were simulated. For hatching spaces of 100, 75 and 50 μm, the connection of the adjacent paths was not sufficiently strong. Only when the hatching space was reduced to 35 μm, the adjacent paths could be well connected for all the evaluated scanning speeds. Therefore, the promising hatching space was 35 μm, which was used for all experimental studies.

According to the simulation results, experimental parameters to be studied included scanning speeds of 100, 250 and 500 mm · s^{-1}, along with layer thicknesses of 22, 30 and 40 μm. Cubic SiC samples with dimensions of 10 × 10 × 5 mm^3 were printed. For low scanning speeds, e.g., 100 mm · s^{-1}, the laser power range was 40–45 W. Consequently, the samples could all be printed with a definite cube shape. However, they were not sufficiently strong for handling. As seen in figure 4.87(a), if the laser power was above the limit simulated with the numerical model, such as 48, 55 and 60 W, the samples were highly porous and mechanically weak. At the same time, there were SiC particles on surfaces of the samples, closely related to the slow scanning speed.

Density of the samples, printed at laser powers of 40–48 W, was about 2.5 g · cm^{-3}, corresponding to a relative density of 78%. As the laser power was raised to 55 and 60 W, the density was increased to 2.7 g · cm^{-3}, equivalent to a relative density of 84%. This increase in density was attributed to the decomposition of SiC into Si and carbon. Additionally, more defects were present in the samples. Photographs of the samples printed at laser powers in the range of 50–60 W and scanning speed of 250 mm · s^{-1} are depicted in figure 4.87(b). The top surface of the samples were quite rough. However, even when the laser power was lowered to be below the limit predicted by the numerical

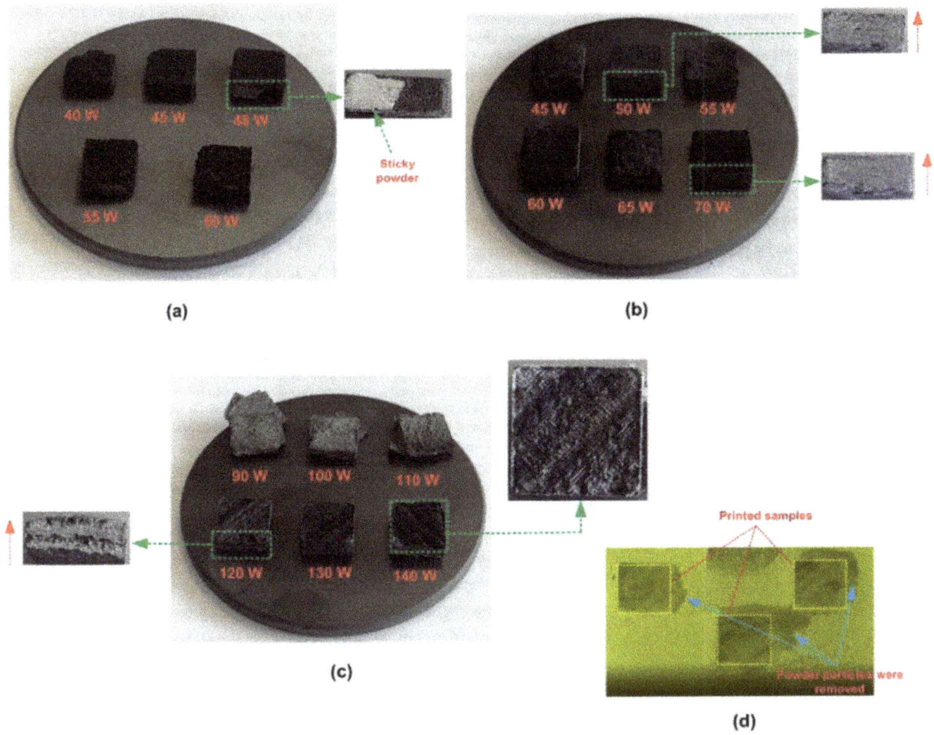

Figure 4.87. SiC samples with layer thickness of 40 μm at different scanning speeds: (a) 100 mm · s^{-1}, (b) 250 mm · s^{-1} and (c) 500 mm · s^{-1}. (d) Powder particles removed due to the high laser beam inertia. Reprinted from [170], Copyright (2023), with permission from Elsevier.

simulation, e.g., 45 W, the samples still had defects. For laser powers over the predicted values, e.g., 65 and 70 W, defects and flaws were also observed.

Samples printed at high scanning speed of 500 mm · s^{-1} with layer thickness of 40 μm are shown in figure 4.87(c). Unfortunately, the printing was not successful, where the first layer could not be properly adhered to the baseplate. Moreover, if the laser power was beyond the limit of the numerical result, all the printed layers were not connected and the dop surfaces were serious defective. Additionally, the high inertia of the laser beam induced ejection of the particles from the powder bed, as observed in figure 4.87(d). Therefore, the layering quality was negatively affected. When a new layer was deposited by the recoater, a portion of the recoated powder would fill the voids in the previous layer, so that the quantity of the powder to fill the current layer was insufficient. In other words, the layer thickness of 40 μm was not optimal for D-SLS of SiC.

Samples with a layer thickness of 30 μm were printed, while the scanning speeds were the same as those used for the ones with a layer thickness of 40 μm. As observed in figure 4.88(a), after printing at scanning speed of 100 mm · s^{-1}, the samples possessed a well-defined cubic shape, without the presence of defects, as the simulated laser powers of 35 and 40 W were used. The samples had a density of

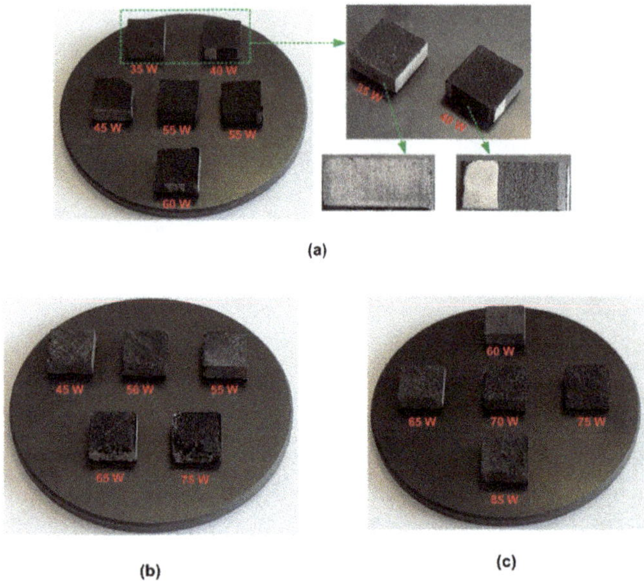

Figure 4.88. SiC samples with laser thickness of 30 μm at different scanning speed: (a) 100 mm · s^{-1}, (b) 250 mm · s^{-1} and (c) 500 mm · s^{-1}. Reprinted from [170], Copyright (2023), with permission from Elsevier.

2.65 g · cm^{-3}, corresponding to a relative density of 82.5%. For medium and high scanning speeds, i.e., 250 and 500 mm · s^{-1}, the samples would contain defects, such as layer degradation and top surface damage, as presented in figures 4.88(b) and (c).

D-SLS printed SiC samples with layer thickness of 22 μm at different scanning speeds are shown in figure 4.89. As demonstrated in figure 4.89(a), SiC samples could be readily printed at a scanning speed of 100 mm · s^{-1}, with laser powers of 30, 35 and 40 W, as suggested by the numerical simulation. The samples possessed a well-defined cubic shape and were free of the flaws or defects, although there were particles stuck on side surfaces of the samples. Density of the samples was 2.64 g · cm^{-3}, corresponding to a relative density of 83%. In this case, too high a laser power would result in very porous and weak samples.

SiC samples could also be printed when the medium scanning speed of 250 mm · s^{-1} was used, while the laser powers were in the range of 40–50 W. In comparison, if the laser intensity was 35 W, layer degradation occurred in the samples, as illustrated in figure 4.89(b). Density of the samples reached 2.62 g · cm^{-3}, or a relative density of 82%. For a scanning speed of 500 mm · s^{-1}, the SiC samples had defects, especially layer degradation, as observed in figure 4.89(c).

On one hand, samples with layer thicknesses of 22 and 30 μm could be printed with desired performances, at low and medium scanning speeds, i.e., 100 and 250 mm · s^{-1}, respectively. On the other hand, high scanning speeds, e.g., 500 mm · s^{-1}, was not recommended. Meanwhile, optimal layer thickness was 30 μm. Compaction level of the powder bed also influenced density of the printed SiC items. For instance, the SiC samples printed at a layer thickness of 22 μm, scanning speed of

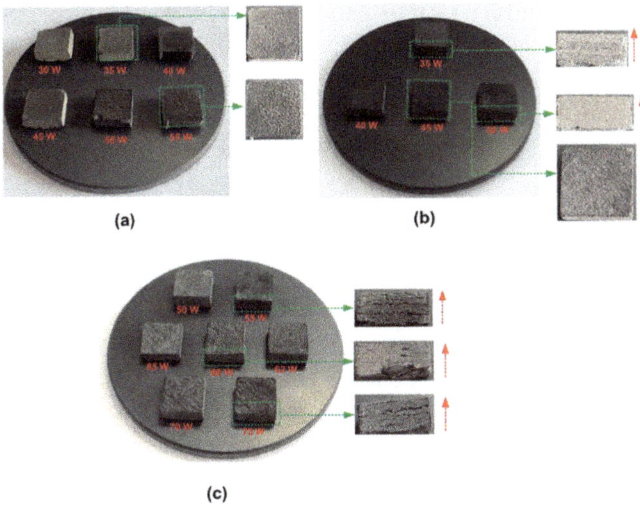

Figure 4.89. SiC samples with layer thickness of 22 μm at different scanning speeds: (a) 100 mm · s^{-1}, (b) 250 mm · s^{-1} and (c) 500 mm · s^{-1}. Reprinted from [170], Copyright (2023), with permission from Elsevier.

100 mm · s^{-1} and laser power of 35 W could have a relative density of 85%, when the compaction level was 300%.

Silicon nitride (Si_3N_4) has been printed by using the SLS process, with high strength for antenna windows [172]. The printed green bodies were debinded and then subject to cold isostatic pressing (CIP), while sintering aids were adopted to further enhance the densification process. Under optimal conditions, the final Si_3N_4 ceramics had a porosity of 18.7%, bulk density of 3.11 g · cm^{-3} and bending strength of 685 MPa. Elimination of the pores through the CIP process and the addition of sintering aids promoted the growth of rod-shaped β-Si_3N_4, thus bringing out interlocking structures.

Mixtures of Si_3N_4 (99.99%, D_{50} = 0.5 μm, Shanghai Chaowei Nanotechnology Co., Ltd, Shanghai, China) powder, with 5% Y_2O_3 (99.99%, (D_{50} = 0.5 μm, Shanghai Chaowei Nanotechnology Co., Ltd, Shanghai, China) and 3% Al_2O_3 (99.99%, D_{50} = 0.5 μm, Shanghai Chaowei Nanotechnology Co., Ltd, Shanghai, China), were used as starting materials. The mixtures were spray dried with a spray granulator (Kaiyide Drying Equipment Co., Ltd, Wuxi, China) in alcohol medium. For spray drying, the three powders were mixed with alcohol through ball milling, yielding slurries with a solid content of 70%. Polyvinyl alcohol (PVA) was used as a binder, with a mass content of 0.03% with respect to the solid content of the slurry. The granules were about 50 μm in diameter, with sphericity and aspect ratio of 0.959 and 0.949, respectively. E12 epoxy resin (EP12) with a particle size of 15 μm and content of 14% was incorporated with the mixed powder.

Samples were packed with polyethylene high-strength and high-toughness plastic bags and vacuumized to compact the samples. The plastic-sealed samples were placed into a cold isostatic press, at 125 MPa for 60 s (first level), 90 MPa for 30 s (second level) and then completely released. After that, the samples were debinded and then subject to secondary CIP at 280 MPa for 60 s (first level). SLS experiment was conducted with an eForm moulding machine (Hunan Farsoon High-Technology Co., Ltd), equipped with a CO_2 laser that has a spot diameter of 0.4 μm. The processing parameters included laser powers of 20–30 W, scanning speed of 1524 mm · s^{-1}, scanning pitches of 0.1–0.2 mm and single layer thicknesses of 0.07–0.15 mm.

The debinding process was carried out at 600 °C, while the sintering temperature was 1750 °C. The samples were sintered for 2 h in N_2 at a pressure of 3 MPa. Figure 4.90 shows a schematic diagram to fabricate high-strength Si_3N_4 antenna windows through the SLS printing. For a double-layer scanning experiment, to identify the optimum laser power, the scanning rate was 1524 mm · s^{-1}, the scanning distance was 0.15 mm and the single layer thickness was 0.09 mm. The laser powers were 20–30 W at an interval of 2 W.

Figure 4.91 shows results of the double-layer scanning experiment. When the laser power was raised from 20 to 30 W, the curing depth was gradually increased from 0.58 to 0.66 mm. At lower laser powers, there were more defects at the edge of the samples. Meanwhile, the filling was not complete, because the laser power was insufficiently high to ensure the curing process. As the laser power was 27 W, the density of defects was decreased. In this case, the thickness of the Si_3N_4 layer that could be densified was 0.65 mm, resulting in samples with fully dense microstructures. After further increase in laser powers to be over 27 W, the curing depth was not increased, but the edges became more defective. Additionally, the process at high powers was accompanied by severe smoking, which was not desirable.

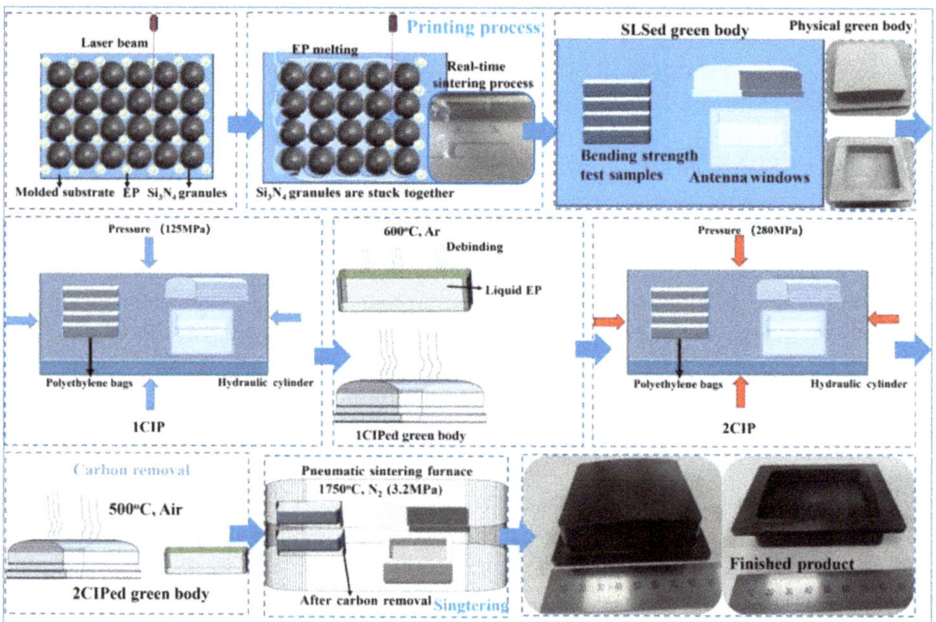

Figure 4.90. Schematic diagram of the SLS printing process to fabricate high-strength Si_3N_4 ceramic antenna window. Reprinted from [172], Copyright (2021), with permission from Elsevier.

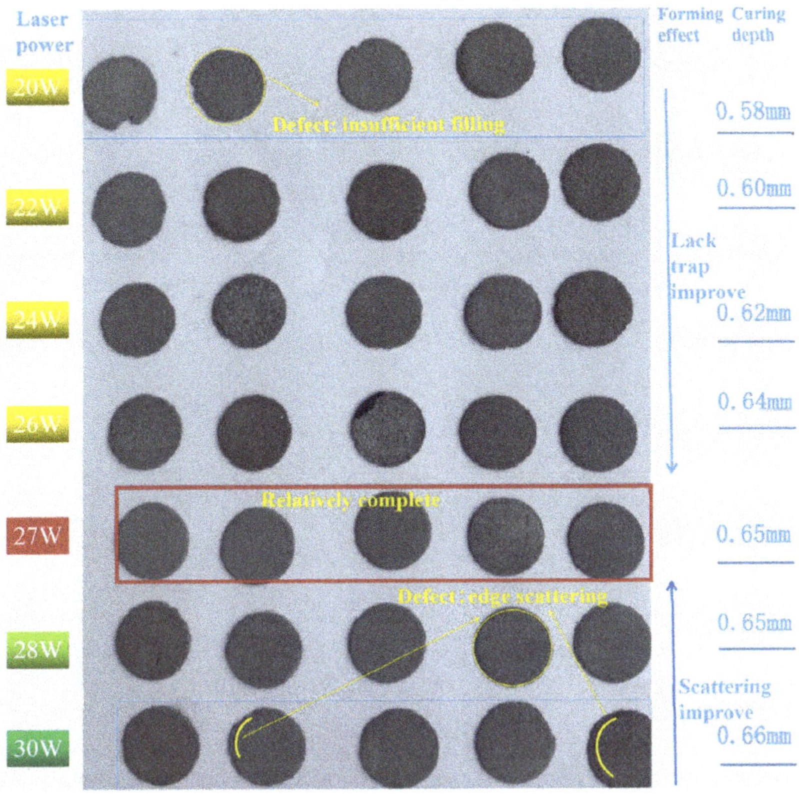

Figure 4.91. Double-layer scanning experiment results with the SLS printing process. Reprinted from [172], Copyright (2021), with permission from Elsevier.

At a laser power of 26 W, the surface of the Si_3N_4 sample exhibited a rough surface that was not flat; the specimen was thick in the middle and thin on the sides. At 27 W, the surface thickness was relatively uniform without obvious defects and scattering. When the laser power was increased to 28 and 30 W, the center of the samples was thicker than the periphery, due to the too-high power. Therefore, the optimal laser power was 27 W.

The effectiveness of the SLS to print Si_3N_4 was dependent on the scanning distance. At a laser power of 27 W, surface qualities of the Si_3N_4 samples were improved, as the scanning distance was increased from 0.05 to 0.15 mm. The sample printed at a scanning distance of 0.05 mm had surface scattering and the edge convex layer. The defects could be ascribed to the superposition effect of the curing times for two adjacent layers. The scattering height of the edge was obviously reduced, as the laser scanning distance was increased to 0.1 mm. However, serious delamination between adjacent layers occurred, as the scanning distance was increased to 0.2 mm. The sample processed at a scanning distance of 0.15 mm had a smooth surface and was nearly free of defects. Accordingly, the optimal scanning distance was 0.15 mm.

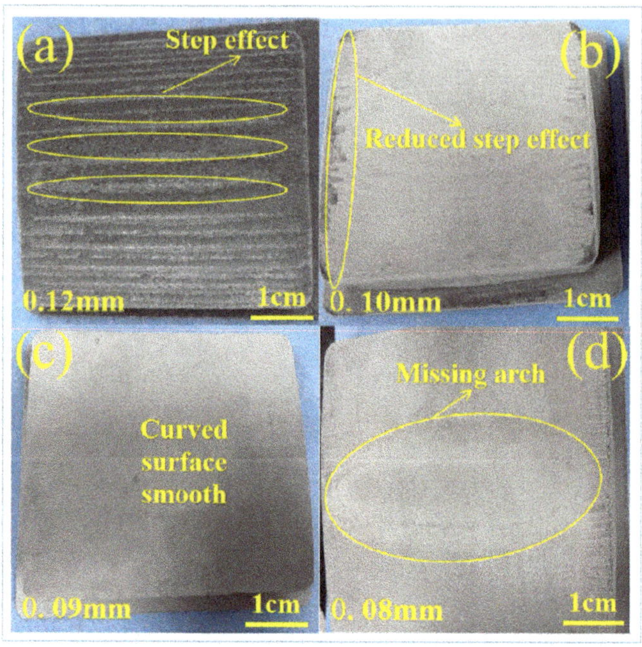

Figure 4.92. Effects of single layer thickness on printing effectiveness of the Si_3N_4 ceramics: (a) 0.12 mm, (b) 0.10 mm, (c) 0.09 mm and (d) 0.08 mm. Reprinted from [172], Copyright (2021), with permission from Elsevier.

Layer thickness had a strong effect on surface morphology of the SLS printed Si_3N_4 ceramics. If the layer was too thin, the roller could scrape off the solidified layer in the subsequent powder spreading, thus resulting in incomplete sheet-like areas in the samples. Too thick a layer would cause low bonding strength, incomplete structure, matching deviation and step phenomenon for adjacent layers. The arc structure should be well controlled in the processes of laser sintering. The effect of layer thickness on the printing effectiveness of arc-shaped structures is illustrated in figure 4.92.

The parallel strip grooves on the surface of the samples were distributed along the scanning direction of the laser, for a single layer thickness of 0.12 mm. The step effect was attributed to the too-thick layer. The sample with a layer thickness of 0.1 mm had much fewer defects, whereas edge defects were still visible on sides of the sample. The step effect could be minimized by reducing the single layer thickness. As the single layer thickness was decreased to 0.09 mm, the curved surface of the sample displayed a smooth transition, without the presence of serious defects. When the layer thickness was further reduced to 0.08 mm, the roller scraped off the uppermost curved surface, leading to arch defects on the surface of the sample, since the resin in between the spherical Si_3N_4 granules was excessively burned. Therefore, the bonding of the granules was weakened, especially for the curved surface. As a consequence, the optimal single layer thickness was 0.09 mm.

The SLS printed Si_3N_4 ceramics without CIP treatment were mechanically weak, which needed to be soaked with dimethicone to avoid collapse during debinding.

Such Si_3N_4 samples had a high porosity of 52.3%. The Si_3N_4 granules were weakly bonded by the resin, without densification. Also, function of the sintering aids was not realizable. Therefore, there was no effect on densification of the ceramics when increasing the content of sintering aids. After the treatment with CIP, the spherical granules were significantly overlapped, so that the porosity of the ceramics was reduced. Meanwhile, the CIP processed green bodies would not experience cracking and collapse during the debinding process.

In fact, pores inevitably were produced after the resin debinding, leading to Si_3N_4 ceramics with pretty high porosity. The CIP treatment helped to reduce the porosity, so that the final Si_3N_4 ceramic had desirable density. After the second CIP treatment, the sintering aids could function to promote the densification. The porosity of the ceramics was constantly decreased with increasing content of sintering aids, after the samples were subjected to the second CIP treatment. The porosity of the sample with 2% Y_2O_3 was 18.7%, which stayed almost unvaried with more Y_2O_3, simply because densification approached the limit of the materials.

The densities of the samples without sintering aids and CIP (S_1), without sintering aids but with first CIP (S_4) and with 2% Y_2O_3 and second CIP (S_8) were 1.70, 2.56 and 3.11 $g \cdot cm^{-3}$, respectively. The density was increased by 50.6% with one CIP process. With a second CIP process, the increase in density was 71.8%. The density of sample S_8 was increased by 82.9% as compared with that of sample S_1.

Bending strength of the Si_3N_4 ceramic without CIP treatment was 12 MPa. After the first and second CIP treatments, the bending strength was increased to 225 MPa and 592 MPa, respectively. The highest bending strength was 692 MPa, achieved in the sample with second CIP treatment and 4% Y_2O_3 (S_9). The enhancement in the bending strength was directly attributed to the reduction in porosity of the Si_3N_4 samples. Since the Si_3N_4 granules were damaged during the CIP process, direct contact between the adjacent granules was facilitated. As a result, densification of the ceramics was effectively promoted. Without CIP treatment or with first CIP treatment, the effect of sintering aids on densification and hence bending strength of the Si_3N_4 ceramics was neglectable. As the ceramic particles were too far apart, the sintering aids would not effectively function. Therefore, the increase in bending strength with increasing content of sintering aids was only observed in the samples treated with twice CIP process.

Fractured surface microscopic morphologies of the Si_3N_4 samples after sintering were characterized by using SEM technology. For the sample without CIP treatment and sintering aids, the spherical granules were essentially intact and isolated, while the distance between adjacent particles was quite large. The spherical granules experienced internal sintering, but nearly no binding took place among them. Meanwhile, the initial α-Si_3N_4 grains were transformed to rod-like β-Si_3N_4 grains, together with formation of layered silicon oxynitride (Si_2N_2O). However, the dispersed granules and the rod-shaped crystals do not overlap or combine with each other. Therefore, high porosity, low density, and low strength are observed. In addition, the length-to-diameter ratio of the rod-shaped crystal is not particularly large on the granule surfaces.

The granules were broken during the CIP process. The broken granules would fill the spaces in between the rest ones. After first CIP treatment, the number of the damaged spherical granules was relatively limited. Therefore, the distance between the granules was almost unchanged. After the second CIP treatment, the spherical granules were nearly entirely damaged, so that the particles were closely packed. Accordingly, the distance between the granules was largely reduced after the second CIP treatment and hence the newly formed pores due to the debinding were absent.

For the samples with second CIP treatment and sintering aids, densification of the Si_3N_4 ceramic was tremendously promoted. The rod-like β-Si_3N_4 grains on the surface of the granules grew in the longitudinal direction, which were connected to link the spherical granules. Consequently, the densification and bonding strength of the Si_3N_4 ceramics were increased. At the same time, sintering aids further boosted the longitudinal growth of β-Si_3N_4 grains, which in turn greatly resulted in significant improvement in the bending strength of the final Si_3N_4 ceramics.

4.4 Concluding remarks

Selective laser sintering/melting (SLS/SLM) has been proved to be a very productive technology for the processing of ceramics, with great success in various materials, including alumina (Al_2O_3), zirconia (ZrO_2, usually doped with Y_2O_3 or other oxides), calcium phosphates, silicon carbide (SiC), silicon nitride (Si_3N_4) and their composites. As compared with other 3D printing approaches, SLS/SLM has several superior characteristics, such as higher fabrication rates, less requirement for post-processing and so on. However, the applicability of this technology has been less successful in ceramic materials than in metals and polymers, simply because ceramics have much higher melting temperatures.

From a technique point of view, understanding the laser–matter interactions is a critical requirement in order to control the printing process, such as regulating the energy density transferred to the powder beds, controlling porosity and micro-structure, reducing the thermal stresses and maximizing mechanical performance of the printed items. The occurrence of the heating process and temperature variation during SLS/SLM processes is still a black box. Contact measurements with thermocouples are not suitable for the building chambers. In this regard, contactless methods, *e.g.*, pyrometers, could be more realizable, which can be used to monitor the real-time fast variation in temperature, but the accuracy is an issue of these methods. Additionally, standardization of the SLS/SLM technologies for ceramics materials is a research topic in the future, in order to ensure high quality products.

References

[1] Han W, Kong L B and Xu M 2022 Advances in selective laser sintering of polymers *Int. J. Extreme Manuf* **4** 042002

[2] Olakanmi E O, Cochrane R F and Dalgarno K W 2015 A review on selective laser sintering/melting (SLS/SLM) of aluminium alloy powders: processing, microstructure, and properties *Prog. Mater Sci.* **74** 401–77

[3] Sercombe T B and Li X 2016 Selective laser melting of aluminium and aluminium metal matrix composites: review *Mater. Technol.* **31** 77–85

[4] Singh S, Sharma V S and Sachdeva A 2016 Progress in selective laser sintering using metallic powders: a review *Mater. Sci. Technol.* **32** 760–72

[5] Song X F and Zhang Y 2023 Progress of high-entropy alloys prepared using selective laser melting *Sci. China Mater* **66** 4165–81

[6] Wang Z, Ummethala R, Singh N, Tang S Y, Suryanarayana C, Eckert J *et al* 2020 Selective laser maelting of aluminum and its alloys *Materials* **13** 4564

[7] Yap C Y, Chua C K, Dong Z L, Liu Z H, Zhang D Q, Loh L E *et al* 2015 Review of selective laser melting: materials and applications *Appl. Phys. Rev.* **2** 041101

[8] Zhang J L, Li F L and Zhang H J 2019 Research progress on preparation of metallic materials by selective laser melting *Laser Optoelectron.* P **56** 100003

[9] Zhang M H, Zhang B C, Wen Y J and Qu X H 2022 Research progress on selective laser melting processing for nickel-based superalloy *Int. J. Miner. Metall. Mater.* **29** 369–88

[10] Kruth J P, Mercelis P, Van Vaerenbergh J, Froyen L and Rombouts M 2005 Binding mechanisms in selective laser sintering and selective laser melting *Rapid Prototyp. J.* **11** 26–36

[11] Arantes V L, Coutinho R B, Martins S S, Huang S and Vleugels J 2019 Solid state sintering behavior of zirconia–nickel composites *Ceram. Int.* **45** 22120–30

[12] Braginsky M, Tikare V and Olevsky E 2005 Numerical simulation of solid state sintering *Int. J. Solids Struct.* **42** 621–36

[13] Kwon Y S and Savitskii A 2001 Solid-state sintering of metal powder mixtures *J. Mater. Synth. Process.* **9** 299–317

[14] Dong W M, Jain H S and Harmer M P 2005 Liquid phase sintering of alumina, II. Penetration of liquid phase into model microstructures *J. Am. Ceram. Soc.* **88** 1708–13

[15] German R, Suri P and Park S 2009 Review: liquid phase sintering *J. Mater. Sci.* **44** 1–39

[16] Santos A C and Ribeiro S 2018 Liquid phase sintering and characterization of SiC ceramics *Ceram. Int.* **44** 11048–59

[17] Xiao B and Zhang Y W 2006 Partial melting and resolidification of metal powder in selective laser sintering *J. Thermophys Heat Transfer* **20** 439–48

[18] Chen B, Zhang R, Liu F Y, Wu C L, Zhang H M, Sun M *et al* 2024 Effect of full melt temperature sintering and semi-melt heat preservation sintering on microstructure and mechanical properties of Ti_3SiC_2/Cu composites *Mater. Res. Express* **11** 015505

[19] Zhang D Q, Cai Q Z, Liu J H and Li R D 2011 Research on process and microstructure formation of W-Ni-Fe alloy fabricated by selective laser melting *J. Mater. Eng. Perform.* **20** 1049–54

[20] Chen A N, Wu J M, Liu K, Chen J Y, Xiao H, Chen P *et al* 2018 High-performance ceramic parts with complex shape prepared by selective laser sintering: a review *Adv. Appl. Ceram* **117** 100–17

[21] Grossin D, Montón A, Navarrete-Segado P, Özmen E, Urruth G, Maury F *et al* 2021 A review of additive manufacturing of ceramics by powder bed selective laser processing (sintering/melting): calcium phosphate, silicon carbide, zirconia, alumina, and their composites *Open Ceram* **5** 100073

[22] Han Z, Cao W B, Lin Z M, Li J T and Feng T 2004 Progress on rapid prototyping of ceramic parts by selective laser sintering *J. Inorg. Mater* **19** 705–13

[23] Sing S L, Yeong W Y, Wiria F E, Tay B Y, Zhao Z Q, Zhao L *et al* 2017 Direct selective laser sintering and melting of ceramics: a review *Rapid Prototyp. J.* **23** 611–23

[24] Kruth J P, Levy G, Klocke F and Childs T H C 2007 Consolidation phenomena in laser and powder-bed based layered manufacturing *CIRP Ann-Manuf. Technol* **56** 730–59

[25] Tolochko N K, Laoui T, Khlopkov Y V, Mozzharov S E, Titov V I and Ignatiev M B 2000 Absorptance of powder materials suitable for laser sintering *Rapid Prototyp. J.* **6** 155–60

[26] Juste E, Petit F, Lardot V and Cambier F 2014 Shaping of ceramic parts by selective laser melting of powder bed *J. Mater. Res.* **29** 2086–94

[27] Liu J, Zhang B, Yan C Z and Shi Y S 2010 The effect of processing parameters on characteristics of selective laser sintering dental glass-ceramic powder *Rapid Prototyp. J.* **16** 138–45

[28] Krantz M, Zhang H and Zhu J 2009 Characterization of powder flow: static and dynamic testing *Powder Technol.* **194** 239–45

[29] Xiao Z H, Sun X Y, Zhang H F, Wang C H, Liu L, Yang Z H *et al* 2018 Low temperature sintered magneto-dielectric ferrite ceramics with near net-shape derived from high-energy milled powders *J. Alloys Compd.* **751** 28–33

[30] Song J L, Li Y T, Deng Q L and Hu D J 2007 Rapid prototyping manufacturing of silica sand patterns based on selective laser sintering *J. Mater. Process. Technol.* **187** 614–8

[31] Zhang J, Zheng W, Wu J M, Yu K B, Ye C S and Shi Y S 2022 Effects of particle grading on properties of silica ceramics prepared by selective laser sintering *Ceram. Int.* **48** 1173–80

[32] Kazemi A, Faghihi-Sani M A and Alizadeh H R 2013 Investigation on cristobalite crystallization in silica-based ceramic cores for investment casting *J. Eur. Ceram. Soc.* **33** 3397–402

[33] Kazemi A, Faghihi-Sani M A, Nayyeri M J, Mohammadi M and Hajfathalian M 2014 Effect of zircon content on chemical and mechanical behavior of silica-based ceramic cores *Ceram. Int.* **40** 1093–8

[34] Breneman R C and Halloran J W 2015 Effect of cristobalite on the strength of sintered fused silica above and below the cristobalite transformation *J. Am. Ceram. Soc.* **98** 1611–7

[35] Zheng W, Wu J M, Chen S, Yu K B, Hua S B, Li C H *et al* 2021 Fabrication of high-performance silica-based ceramic cores through selective laser sintering combined with vacuum infiltration *Addit. Manuf.* **48** 102396

[36] Garrido L B and Aglietti E F 2004 Reaction-sintered mullite–zirconia composites by colloidal processing of alumina–zircon–CeO_2 mixtures *Mater. Sci. Eng. A-Struct* **369** 250–7

[37] Yu P C, Tsai Y W, Yen F S and Huang C L 2014 Thermal reaction of cristobalite in nano-SiO_2/α-Al_2O_3 powder systems for mullite synthesis *J. Am. Ceram. Soc.* **97** 2431–8

[38] Chen X, Zheng W L, Zhang J, Liu C Y, Han J Q, Zhang L *et al* 2020 Enhanced thermal properties of silica-based ceramic cores prepared by coating alumina/mullite on the surface of fused silica powders *Ceram. Int.* **46** 11819–27

[39] Li C H, Hu L, Zou Y, Liu J A, Xiao J H, Wu J M *et al* 2020 Fabrication of Al_2O_3–SiO_2 ceramics through combined selective laser sintering and SiO_2–sol infiltration *Int. J. Appl. Ceram. Technol.* **17** 255–63

[40] Shahzad K, Deckers J, Boury S, Neirinck B, Kruth J P and Vleugels J 2012 Preparation and indirect selective laser sintering of alumina/PA microspheres *Ceram. Int.* **38** 1241–7

[41] Liu K, Shi Y S, Li C H, Hao L, Liu J and Wei Q S 2014 Indirect selective laser sintering of epoxy resin-Al_2O_3 ceramic powders combined with cold isostatic pressing *Ceram. Int.* **40** 7099–106

[42] Nazemosadat S M, Foroozmehr E and Badrossamay M 2018 Preparation of alumina/polystyrene core–shell composite powder via phase inversion process for indirect selective laser sintering applications *Ceram. Int.* **44** 596–604

[43] Dong Y, Jiang H Y, Chen A N, Yang T, Gao S and Liu S N 2021 Near-zero-shrinkage Al_2O_3 ceramic foams with coral-like and hollow-sphere structures via selective laser sintering and reaction bonding *J. Eur. Ceram. Soc.* **41** 239–46

[44] Spath S, Drescher P and Seitz H 2015 Impact of particle size of ceramic granule blends on mechanical strength and porosity of 3D printed scaffolds *Materials* **8** 4720–32

[45] Bao S, Tang K, Kvithyld A, Tangstad M and Engh T A 2011 Wettability of aluminum on alumina *Metall. Mater. Trans.* B **42** 1358–66

[46] Shen Z L, Su H J, Liu H F, Zhao D, Liu Y, Guo Y N *et al* 2022 Directly fabricated Al_2O_3/$GdAlO_3$ eutectic ceramic with large smooth surface by selective laser melting: rapid solidification behavior and thermal field simulation *J. Eur. Ceram. Soc.* **42** 1088–101

[47] Boutaous M, Liu X, Siginer D A and Xin S H 2021 Balling phenomenon in metallic laser based 3D printing process *Int. J. Therm. Sci.* **167** 107011

[48] Zhou X, Liu X H, Zhang D D, Shen Z J and Liu W 2015 Balling phenomena in selective laser melted tungsten *J. Mater. Process. Technol.* **222** 33–42

[49] Yadroitsev I, Bertrand P and Smurov I 2007 Parametric analysis of the selective laser melting process *Appl. Surf. Sci.* **253** 8064–9

[50] Guo S, Wang M, Zhao Z, Zhang Y Y, Lin X and Huang W D 2017 Molecular dynamics simulation on the micro-structural evolution in heat-affected zone during the preparation of bulk metallic glasses with selective laser melting *J. Alloys Compd.* **697** 443–9

[51] Peña J I, Larsson M, Merino R I, de Francisco I, Orera V M, Llorca J *et al* 2006 Processing, microstructure and mechanical properties of directionally-solidified Al_2O_3–$Y_3Al_5O_{12}$–ZrO_2 ternary eutectics *J. Eur. Ceram. Soc.* **26** 3113–21

[52] Oliete P B and Peña J I 2007 Study of the gas inclusions in Al_2O_3/$Y_3Al_5O_{12}$ and Al_2O_3/$Y_3Al_5O_{12}$/ZrO_2 eutectic fibers grown by laser floating zone *J. Cryst. Growth* **304** 514–9

[53] Vonka P and Leitner J 2009 A method for the estimation of the enthalpy of formation of mixed oxides in Al_2O_3–Ln_2O_3 systems *J. Solid State Chem.* **182** 744–8

[54] Lee J H, Yoshikawa A, Durbin S D, Yoon D H, Fukuda T and Waku Y 2001 Microstructure of Al_2O_3/ZrO_2 eutectic fibers grown by the micro-pulling down method *J. Cryst. Growth* **222** 791–6

[55] Lee J H, Yoshikawa A, Kaiden H, Lebbou K, Fukuda T, Yoon D H *et al* 2001 Microstructure of Y_2O_3 doped Al_2O_3/ZrO_2 eutectic fibers grown by the micro-pulling-down method *J. Cryst. Growth* **231** 179–85

[56] Ren Q, Su H J, Zhang J, Yao B, Ma W D, Liu L *et al* 2016 Solid-liquid interface and growth rate range of Al_2O_3-based eutectic *in situ* composites grown by laser floating zone melting *J. Alloys Compd.* **662** 634–9

[57] Su H J, Zhang J, Yu J Z, Liu L and Fu H Z 2011 Directional solidification and microstructural development of Al_2O_3/$GdAlO_3$ eutectic ceramic *in situ* composite under rapid growth conditions *J. Alloys Compd.* **509** 4420–5

[58] Medeiros I S, Andreeta E R M and Hernandes A C 2007 Al_2O_3/$GdAlO_3$ eutectic fibers of high modulus of rupture produced by the laser heated pedestal growth technique *J. Mater. Sci.* **42** 3874–7

[59] Jian Z Y, Kuribayashi K and Jie W Q 2004 Critical undercoolings for the transition from the lateral to continuous growth in undercooled silicon and germanium *Acta Mater.* **52** 3323–33

[60] Su H J, Wang E Y, Ren Q, Liu H F, Zhao D, Fan G Y *et al* 2018 Microstructure tailoring and thermal stability of directionally solidified $Al_2O_3/GdAlO_3$ binary eutectic ceramics by laser floating zone melting *Ceram. Int.* **44** 7908–16

[61] Ma W D, Zhang J, Su H J, Fan G R, Guo M, Liu L *et al* 2021 Phase growth patterns for $Al_2O_3/GdAlO_3$ eutectics over wide ranges of compositions and solidification rates *J. Mater. Sci. Technol* **65** 89–98

[62] Su H J, Ren Q, Zhang J, Wei K C, Yao B, Ma W D *et al* 2017 Microstructures and mechanical properties of directionally solidified $Al_2O_3/GdAlO_3$ eutectic ceramic by laser floating zone melting with high temperature gradient *J. Eur. Ceram. Soc.* **37** 1617–26

[63] Liu H F, Su H J, Shen Z L, Wang E Y, Zhao D, Guo M *et al* 2018 Direct formation of $Al_2O_3/GdAlO_3/ZrO_2$ ternary eutectic ceramics by selective laser melting: microstructure evolutions *J. Eur. Ceram. Soc.* **38** 5144–52

[64] Perrière L, Valle R, Carrère N, Gouadec G, Colomban P, Lartigue-Korinek S *et al* 2011 Crack propagation and stress distribution in binary and ternary directionally solidified eutectic ceramics *J. Eur. Ceram. Soc.* **31** 1199–210

[65] Mesa M C, Oliete P B, Pastor J Y, Martín A and Llorca J 2014 Mechanical properties up to 1900 K of $Al_2O_3/Er_3Al_5O_{12}/ZrO_2$ eutectic ceramics grown by the laser floating zone method *J. Eur. Ceram. Soc.* **34** 2081–7

[66] Wen S F, Li S, Wei Q S, Yan C Z, Sheng Z and Shi Y S 2014 Effect of molten pool boundaries on the mechanical properties of selective laser melting parts *J. Mater. Process. Technol.* **214** 2660–7

[67] Niu F Y, Wu D J, Ma G Y, Wang J T, Guo M H and Zhang B 2015 Nanosized microstructure of Al_2O_3–ZrO_2 (Y_2O_3) eutectics fabricated by laser engineered net shaping *Scr. Mater.* **95** 39–41

[68] Llorca J and Orera V M 2006 Directionally solidified eutectic ceramic oxides *Prog. Mater Sci.* **51** 711–809

[69] Orera V M, Peña J I, Oliete P B, Merino R I and Larrea A 2012 Growth of eutectic ceramic structures by directional solidification methods *J. Cryst. Growth* **360** 99–104

[70] Aoyama T and Kuribayashi K 2000 Influence of undercooling on solid/liquid interface morphology in semiconductors *Acta Mater.* **48** 3739–44

[71] Larrea A, Orera V M, Merino R I and Peña J J 2005 Microstructure and mechanical properties of Al_2O_3–YSZ and Al_2O_3–YAG directionally solidified eutectic plates *J. Eur. Ceram. Soc.* **25** 1419–29

[72] Aragón-Duarte M C, Nevarez-Rascón A, Esparza-Ponce H E, Nevarez-Rascón M M, Talamantes R P, Ornelas C *et al* 2017 Nanomechanical properties of zirconia–yttria and alumina zirconia–yttria biomedical ceramics, subjected to low temperature aging *Ceram. Int.* **43** 3931–9

[73] Han J M, Zhao J and Shen Z J 2017 Zirconia ceramics in metal-free implant dentistry *Adv. Appl. Ceram* **116** 138–50

[74] Hannink R H J, Kelly P M and Muddle B C 2000 Transformation toughening in zirconia-containing ceramics *J. Am. Ceram. Soc.* **83** 461–87

[75] Vichi A, Louca C, Corciolani G and Ferrari M 2011 Color related to ceramic and zirconia restorations: a review *Dent. Mater.* **27** 97–108

[76] Zhang X P, Wu X and Shi J 2020 Additive manufacturing of zirconia ceramics: a state-of-the-art review *J. Mater. Res. Technol.* **9** 9029–48

[77] Zhu D B, Song Y J, Liang J S, Zhang X X, Chu R Q and Wu M Q 2018 Progress of toughness in dental zirconia ceramics *J. Inorg. Mater* **33** 363–72
[78] Fergus J W 2006 Electrolytes for solid oxide fuel cells *J. Power Sources* **162** 30–40
[79] Huijsmans J P P 2001 Ceramics in solid oxide fuel cells *Curr. Opin. Solid State Mater. Sci.* **5** 317–23
[80] Jacobson A J 2010 Materials for solid oxide fuel cells *Chem. Mater.* **22** 660–74
[81] Kendall K 2005 Progress in solid oxide fuel cell materials *Int. Mater. Rev.* **50** 257–64
[82] Ormerod R M 2003 Solid oxide fuel cells *Chem. Soc. Rev.* **32** 17–28
[83] Guan S H, Zhang X J and Liu Z P 2015 Energy landscape of zirconia phase transitions *J. Am. Chem. Soc.* **137** 8010–3
[84] Moriya Y and Navrotsky A 2006 High-temperature calorimetry of zirconia: heat capacity and thermodynamics of the monoclinic-tetragonal phase transition *J. Chem. Thermodyn.* **38** 211–23
[85] Shibata N, Katamura J, Kuwabara A, Ikuhara Y and Sakuma T 2001 The instability and resulting phase transition of cubic zirconia *Mater. Sci. Eng.* A **312** 90–8
[86] Antou G, Montavon G, Hlawka F, Cornet A, Coddet C and Machi F 2004 Processing of yttria partially stabilized zirconia thermal barrier coatings implementing a high-power laser diode *J. Therm. Spray Technol.* **13** 381–9
[87] Binner J, Vaidhyanathan B, Paul A, Annaporani K and Raghupathy B 2011 Compositional effects in nanostructured yttria partially stabilized zirconia *Int. J. Appl. Ceram. Technol.* **8** 766–82
[88] Tsukamoto H 2022 Enhancement of transformation toughening of partially stabilized zirconia by some additives *Ceram. Int.* **48** 20675–89
[89] Curi M O, Ferraz H C, Furtado J G M and Secchi A R 2015 Dispersant effects on YSZ electrolyte characteristics for solid oxide fuel cells *Ceram. Int.* **41** 6141–8
[90] Hotza D, García D E and Castro R H R 2015 Obtaining highly dense YSZ nanoceramics by pressureless, unassisted sintering *Int. Mater. Rev.* **60** 353–75
[91] Khan M S, Lee S B, Song R H, Lee J W, Lim T H and Park S J 2016 Fundamental mechanisms involved in the degradation of nickel yttria stabilized zirconia (Ni-YSZ) anode during solid oxide fuel cells operation: a review *Ceram. Int.* **42** 35–48
[92] Shahzad K, Deckers J, Zhang Z Y, Kruth J P and Vleugels J 2014 Additive manufacturing of zirconia parts by indirect selective laser sintering *J. Eur. Ceram. Soc.* **34** 87–95
[93] Gençaslan M and Keskin M 2009 Global phase diagrams for a compressible polymer–solvent system using the full Tompa model *J. Chem. Phys.* **131** 244112
[94] Guillen G R, Pan Y J, Li M H and Hoek E M V 2011 Preparation and characterization of membranes formed by nonsolvent induced phase separation: a review *Ind. Eng. Chem. Res.* **50** 3798–817
[95] Liu Q, Danlos Y, Song B, Zhang B C, Yin S and Liao H L 2015 Effect of high-temperature preheating on the selective laser melting of yttria-stabilized zirconia ceramic *J. Mater. Process. Technol.* **222** 61–74
[96] Wilkes J, Hagedorn Y C, Meiners W and Wissenbach K 2013 Additive manufacturing of ZrO_2–Al_2O_3 ceramic components by selective laser melting *Rapid Prototyp. J.* **19** 51–7
[97] Ferrage L, Bertrand G and Lenormand P 2018 Dense yttria-stabilized zirconia obtained by direct selective laser sintering *Addit. Manuf.* **21** 472–8

[98] Urruth G, Maury D, Voisin C, Baylac V and Grossin D 2022 Powder bed selective laser processing (sintering/melting) of yttrium stabilized zirconia using carbon-based material (TiC) as absorbance enhancer *J. Eur. Ceram. Soc* **42** 2381–90

[99] Özmen E, Grossin D, Lenormand P and Bertrand G 2024 Direct powder bed selective laser processing of dense alumina-toughened zirconia parts *Prog. Addit. Manuf.* **10** 2369–81

[100] Florio K, Pfeiffer S, Makowska M, Casati N, Verga F, Graule T *et al* 2019 An innovative selective laser melting process for hematite-doped aluminum oxide *Adv. Eng. Mater.* **21** 201801352

[101] Ghasemi A, Fereiduni E, Balbaa M, Elbestawi M and Habibi S 2022 Unraveling the low thermal conductivity of the LPBF fabricated pure Al, AlSi12, and AlSi10Mg alloys through substrate preheating *Addit. Manuf* **59** 103148

[102] Koopmann J, Voigt J and Niendorf T 2019 Additive manufacturing of a steel-ceramic multi-material by selective laser melting *Metall. Mater. Trans.* B **50** 1042–51

[103] Zhang Y M, Wei C R, Fukui T, Sugita N and Ito Y 2024 Ultrafast processing of zirconia ceramics by transient and selective laser absorption *Ceram. Int.* **50** 25273–81

[104] Zhang Y M, Ito Y, Sun H J and Sugita N 2022 Investigation of multi-timescale processing phenomena in femtosecond laser drilling of zirconia ceramics *Opt. Express* **30** 37394–406

[105] Yoshizaki R Y, Miyamoto R and Sugita N 2018 Ultrafast and precision drilling of glass by selective absorption of fiber-laser pulse into femtosecond-laser-induced filament *Appl. Phys. Lett.* **113** 061101

[106] Yoshizaki R, Ito Y, Ogasawara K, Shibata A, Nagasawa I, Sano T *et al* 2022 High-efficiency microdrilling of glass by parallel transient and selective laser processing with spatial light modulator *Opt. Laser Technol.* **154** 108306

[107] Zhang H, Zhang F T, Du X, Dong G P and Qiu J R 2015 Influence of laser-induced air breakdown on femtosecond laser ablation of aluminum *Opt. Express* **23** 1370–6

[108] Samant A N and Dahotre N B 2009 Laser machining of structural ceramics—a review *J. Eur. Ceram. Soc.* **29** 969–93

[109] Wang H J, Lin H T, Wang C Y, Zheng L J and Hu X Y 2017 Laser drilling of structural ceramics—a review *J. Eur. Ceram. Soc.* **37** 1157–73

[110] Sola D and Peña J I 2013 Study of the wavelength dependence in laser ablation of advanced ceramics and glass-ceramic materials in the nanosecond range *Materials* **6** 5302–13

[111] Chen T C and Darling R B 2005 Parametric studies on pulsed near ultraviolet frequency tripled Nd:YAG laser micromachining of sapphire and silicon *J. Mater. Process. Technol.* **169** 214–8

[112] Wei C R, Zhang Y M, Sugita N and Ito Y 2024 Generation mechanism and temporal-spatial evolution of electron excitation induced by an ultrashort pulse laser in zirconia ceramic *Appl. Phys. A-Mater.* **130** 105

[113] Ho C Y and Lu J K 2003 A closed form solution for laser drilling of silicon nitride and alumina ceramics *J. Mater. Process. Technol.* **140** 260–3

[114] Dorozhkin S V 2010 Bioceramics of calcium orthophosphates *Biomaterials* **31** 1465–85

[115] Dorozhkin S V 2022 Calcium orthophosphate ($CaPO_4$)-based bioceramics: preparation, properties, and applications *Coatings* **12** 1380

[116] Ioku K 2010 Tailored bioceramics of calcium phosphates for regenerative medicine *J. Ceram. Soc. Jpn.* **118** 775–83

[117] Li Q P, Feng C, Cao Q L, Wang W, Ma Z H, Wu Y H *et al* 2023 Strategies of strengthening mechanical properties in the osteoinductive calcium phosphate bioceramics *Regen. Biomater.* **10** rbad013
[118] Shuai C J, Gao C D, Nie Y, Hu H L, Qu H Y and Peng S P 2010 Structural design and experimental analysis of a selective laser sintering system with nano-hydroxyapatite powder *J. Biomed. Nanotechnol.* **6** 370–4
[119] Shuai C J, Gao C D, Nie Y, Hu H L, Zhou Y and Peng S P 2011 Structure and properties of nano-hydroxypatite scaffolds for bone tissue engineering with a selective laser sintering system *Nanotechnology* **22** 285703
[120] Vallet-Regí M and Ruiz-Hernández E 2011 Bioceramics: from bone regeneration to cancer nanomedicine *Adv. Mater.* **23** 5177–218
[121] Kao F C, Chiu P Y, Tsai T T and Lin Z H 2019 The application of nanogenerators and piezoelectricity in osteogenesis *Sci. Technol. Adv. Mater.* **20** 1103–17
[122] Lang S B 2016 Review of ferroelectric hydroxyapatite and its application to biomedicine *Phase Transit.* **89** 678–94
[123] Tofail S A M, Haverty D, Cox F, Erhart J, Hána P and Ryzhenko V 2009 Direct and ultrasonic measurements of macroscopic piezoelectricity in sintered hydroxyapatite *J. Appl. Phys.* **105** 064103
[124] Shuai C J, Li P J, Liu J L and Peng S P 2013 Optimization of TCP/HAP ratio for better properties of calcium phosphate scaffold via selective laser sintering *Mater. Charact.* **77** 23–31
[125] Miao X, Lim W K, Huang X and Chen Y 2005 Preparation and characterization of interpenetrating phased TCP/HA/PLGA composites *Mater. Lett.* **59** 4000–5
[126] Ryu H S, Youn H J, Hong K S, Chang B S, Lee C K and Chung S S 2002 An improvement in sintering property of β-tricalcium phosphate by addition of calcium pyrophosphate *Biomaterials* **23** 909–14
[127] Legeros R Z, Lin S, Rohanizadeh R, Mijares D and Legeros J P 2003 Biphasic calcium phosphate bioceramics: preparation, properties and applications *J. Mater. Sci.: Mater. Med* **14** 201–9
[128] Navarrete-Segado P, Frances C, Tourbin M, Tenailleau C, Duployer B and Grossin D 2022 Powder bed selective laser process (sintering/melting) applied to tailored calcium phosphate-based powders *Addit. Manuf* **50** 102542
[129] Navarrete-Segado P, Frances C, Grossin D and Tourbin M 2022 Tailoring hydroxyapatite microspheres by spray-drying for powder bed fusion feedstock *Powder Technol.* **398** 117116
[130] Tourbin M, Brouillet F, Galey B, Rouquet N, Gras P, Chebel N A *et al* 2020 Agglomeration of stoichiometric hydroxyapatite: impact on particle size distribution and purity in the precipitation and maturation steps *Powder Technol.* **360** 977–88
[131] Demnati I, Grossin D, Combes C, Parco M, Braceras I and Rey C 2012 A comparative physico-chemical study of chlorapatite and hydroxyapatite: from powders to plasma sprayed thin coatings *Biomed. Mater.* **7** 054101
[132] Prashanth K G, Scudino S, Maity T, Das J and Eckert J 2017 Is the energy density a reliable parameter for materials synthesis by selective laser melting? *Mater. Res. Lett.* **5** 386–90
[133] Tonsuaadu K, Gross K A, Pluduma L and Veiderma M 2012 A review on the thermal stability of calcium apatites *J. Therm. Anal. Calorim.* **110** 647–59

[134] Gu D D and Shen Y F 2009 Balling phenomena in direct laser sintering of stainless steel powder: metallurgical mechanisms and control methods *Mater. Des.* **30** 2903–10

[135] Wang Z Q, Wang X D, Zhou X, Ye G Z, Cheng X and Zhang P Y 2020 Investigation into spatter particles and their effect on the formation quality during selective laser melting processes *Comp. Model Eng* **124** 243–63

[136] García-Tuñón E, Couceiro R, Franco J, Saiz E and Guitián F 2012 Synthesis and characterisation of large chlorapatite single-crystals with controlled morphology and surface roughness *J. Mater. Sci.: Mater. Med* **23** 2471–82

[137] Schappo H, Giry K, Salmoria G, Damia C and Hotza D 2023 Polymer/calcium phosphate biocomposites manufactured by selective laser sintering: an overview *Prog. Addit. Manuf* **8** 285–301

[138] Song X H, Li W, Song P H, Su Q Y, Wei Q S, Shi Y S *et al* 2015 Selective laser sintering of aliphatic-polycarbonate/hydroxyapatite composite scaffolds for medical applications *Int. J. Adv. Manuf. Technol.* **81** 15–25

[139] Salmoria G V, Fancello E A, Roesler C R M and Dabbas F 2013 Functional graded scaffold of HDPE/HA prepared by selective laser sintering: microstructure and mechanical properties *Int. J. Adv. Manuf. Technol.* **65** 1529–34

[140] Ramu M, Ananthasubramanian M, Kumaresan T, Gandhinathan R and Jothi S 2018 Optimization of the configuration of porous bone scaffolds made of polyamide/hydroxyapatite composites using selective laser sintering for tissue engineering applications *Bio-Med. Mater. Eng.* **29** 739–55

[141] Du Y Y, Liu H M, Shuang J Q, Wang J L, Ma J and Zhang S M 2015 Microsphere-based selective laser sintering for building macroporous bone scaffolds with controlled microstructure and excellent biocompatibility *Colloids Surf.*, B **135** 81–9

[142] Eshraghi S and Das S 2012 Micromechanical finite-element modeling and experimental characterization of the compressive mechanical properties of polycaprolactone-hydroxyapatite composite scaffolds prepared by selective laser sintering for bone tissue engineering *Acta Biomater.* **8** 3138–43

[143] Feng P, Peng S P, Shuai C J, Gao C D, Yang W J, Bin S Z *et al* 2020 In situ generation of hydroxyapatite on biopolymer particles for fabrication of bone scaffolds owning bioactivity *ACS Appl. Mater. Interfaces* **12** 46743–55

[144] Abouliatim Y, Chartier T, Abelard P, Chaput C and Delage C 2009 Optical characterization of stereolithography alumina suspensions using the Kubelka–Munk model *J. Eur. Ceram. Soc.* **29** 919–24

[145] Pan M L, Kong X D, Cai Y R and Yao J M 2011 Hydroxyapatite coating on the titanium substrate modulated by a recombinant collagen-like protein *Mater. Chem. Phys.* **126** 811–7

[146] Ha Y, Yang J, Tao F, Wu Q, Song Y J, Wang H R *et al* 2018 Phase-transited lysozyme as a universal route to bioactive hydroxyapatite crystalline film *Adv. Funct. Mater.* **28** 201704476

[147] Wang D, Jang J, Kim K, Kim J and Park C B 2019 "Tree to bone": lignin/polycaprolactone nanofibers for hydroxyapatite biomineralization *Biomacromolecules* **20** 2684–93

[148] Xu D, Xu Z X, Cheng L D, Gao X H, Sun J and Chen L Q 2022 Improvement of the mechanical properties and osteogenic activity of 3D-printed polylactic acid porous scaffolds by nano-hydroxyapatite and nano-magnesium oxide *Heliyon* **8** e09748

[149] Jouyandeh M, Vahabi H, Rabiee N, Rabiee M, Bagherzadeh M and Saeb M R 2022 Green composites in bone tissue engineering *Emerg. Mater* **5** 603–20

[150] Gupte M J, Swanson W B, Hu J, Jin X B, Ma H Y, Zhang Z P *et al* 2018 Pore size directs bone marrow stromal cell fate and tissue regeneration in nanofibrous macroporous scaffolds by mediating vascularization *Acta Biomater.* **82** 1–11

[151] Velu R and Singamneni S 2014 Selective laser sintering of polymer biocomposites based on polymethyl methacrylate *J. Mater. Res.* **29** 1883–92

[152] Khumalo V M, Karger-Kocsis J and Thomann R 2010 Polyethylene/synthetic boehmite alumina nanocomposites: structure, thermal and rheological properties *Express Polym. Lett.* **4** 264–74

[153] Khumalo V M, Karger-Kocsis J and Thomann R 2011 Polyethylene/synthetic boehmite alumina nanocomposites: structure, mechanical, and perforation impact properties *J. Mater. Sci.* **46** 422–8

[154] Caulfield B, McHugh P E and Lohfeld S 2007 Dependence of mechanical properties of polyamide components on build parameters in the SLS process *J. Mater. Process. Technol.* **182** 477–88

[155] Masri P 2002 Silicon carbide and silicon carbide-based structures—the physics of epitaxy *Surf. Sci. Rep.* **48** 1–51

[156] Mehregany M, Zorman C A, Roy S, Fleischman A J, Wu C H and Rajan N 2000 Silicon carbide for microelectromechanical systems *Int. Mater. Rev.* **45** 85–108

[157] Soltys L M, Mironyuk I F, Mykytyn I M, Hnylytsia I D and Turovska L V 2023 Synthesis and properties of silicon carbide (review) *Phys. Chem. Solid State* **24** 5–16

[158] Kar A, Kundu K, Chattopadhyay H and Banerjee R 2022 White light emission of wide-bandgap silicon carbide: a review *J. Am. Ceram. Soc.* **105** 3100–15

[159] She X, Huang A Q, Lucía O and Ozpineci B 2017 Review of silicon carbide power devices and their applications *IEEE Trans. Ind. Electron.* **64** 8193–205

[160] Song S C, Gao Z Q, Lu B H, Bao C G, Zheng B C and Wang L 2020 Performance optimization of complicated structural SiC/Si composite ceramics prepared by selective laser sintering *Ceram. Int.* **46** 568–75

[161] Song S, Bao C G, Ma Y N and Wang K K 2017 Fabrication and characterization of a new-style structure capillary channel in reaction bonded silicon carbide composites *J. Eur. Ceram. Soc.* **37** 2569–74

[162] Song S C, Lu B H, Gao Z Q, Bao C G and Ma Y N 2019 Microstructural development and factors affecting the performance of a reaction-bonded silicon carbide composite *Ceram. Int.* **45** 17987–95

[163] Zhang Z, Zhang Y J, Gong H Y, Guo X, Zhang Y B, Wang X L *et al* 2016 Influence of carbon content on ceramic injection molding of reaction-bonded silicon carbide *Int. J. Appl. Ceram. Technol.* **13** 838–43

[164] Chen X, Yin J, Liu X J, Pei B B, Huang J, Peng X L *et al* 2022 Effect of laser power on mechanical properties of SiC composites rapidly fabricated by selective laser sintering and direct liquid silicon infiltration *Ceram. Int.* **48** 19123–31

[165] Zhang H, Yang Y, Hu K H, Liu B, Liu M and Huang Z R 2020 Stereolithography-based additive manufacturing of lightweight and high-strength C_f/SiC ceramics *Addit. Manuf* **34** 101199

[166] Zhu W, Fu H, Xu Z F, Liu R Z, Jiang P, Shao X Y *et al* 2018 Fabrication and characterization of carbon fiber reinforced SiC ceramic matrix composites based on 3D printing technology *J. Eur. Ceram. Soc.* **38** 4604–13

[167] Liu J H, Shi Y S, Lu Z L, Xu Y, Chen K H and Huang S H 2007 Manufacturing metal parts via indirect SLS of composite elemental powders *Mater. Sci. Eng. A-Struct.* **444** 146–52

[168] Fu H, Zhu W, Xu Z F, Chen P, Yan C Z, Zhou K *et al* 2019 Effect of silicon addition on the microstructure, mechanical and thermal properties of C_f/SiC composite prepared via selective laser sintering *J. Alloys Compd.* **792** 1045–53

[169] Wang Y X, Tan S H and Jiang D L 2004 The effect of porous carbon preform and the infiltration process on the properties of reaction-formed SiC *Carbon* **42** 1833–9

[170] Abdelmoula M, Küçüktürk G, Grossin D, Zarazaga A M, Maury F and Ferrato M 2023 Direct selective laser sintering of silicon carbide: realizing the full potential through process parameter optimization *Ceram. Int.* **49** 32426–39

[171] Calderon N R, Martínez-Escandell M, Narciso J and Rodríguez-Reinoso F 2009 The combined effect of porosity and reactivity of the carbon preforms on the properties of SiC produced by reactive infiltration with liquid Si *Carbon* **47** 2200–10

[172] Wang K J, Bao C G, Zhang C Y, Li Y H, Liu R Z, Xu H M *et al* 2021 Preparation of high-strength Si_3N_4 antenna window using selective laser sintering *Ceram. Int.* **47** 31277–85

Ling Bing Kong, Zhuohao Xiao, Bin He and Yin Liu

Chapter 5

Extrusion freeforming fabrication (EFF)

This chapter aims to summarize the process in extrusion freeform fabrication (EFF) to fabricate ceramics and ceramic materials. Specifically, they include robocasting, contour crafting, fused deposition of ceramics, freeze-form extrusion fabrication (FEF), thermoplastic 3D printing, ceramic on-demand extrusion and so on.

5.1 Introduction

Extrusion freeform fabrication (EFF) represents a group of fabrication processes, including robocasting [1–5], contour crafting [6, 7], fused deposition of ceramics [8], freeze-form extrusion fabrication (FEF) [9–11], thermoplastic 3D printing [12–14], and ceramic on-demand extrusion [15], with which ceramic extrudates are deposited in a layer-by-layer manner [16, 17].

The fabrication of fully dense ceramics with complicated geometries requires precise control over the flow rate during printing. Inaccurate start-stop cycles and inconsistent extrudate flow rates often lead to pore formation in the printed items, a common issue in filament-based and paste-based freeform extrusion processes [18]. Such defects accumulate progressively, which could finally result in weakening of the mechanical strength or even failure of the products, especially for ceramics printed with freeform extrusion processes [19]. In freeform extrusion processes, pastes are usually extruded with ram extruders, which are realized through a combination of syringes and plungers. There are several methods to regulate the extrusion force and plunger velocity for the FEF printing.

5.2 Types of extrusion methods and extrusion mechanisms

In terms of pastes, there are two extrusion processes, i.e., steady-state and transient. In steady-state extrusion, continuous filaments are extruded at a constant rate. Transient extrusion involves varied flow rates during the starts and stops of extrusion.

doi:10.1088/978-0-7503-4831-7ch5

There are currently three main extrusion methods for the fabrication of ceramics through 3D printing, including the ram extruder-based method, shutter valve-based method, and Auger extruder-based method. The two main components of ram extruders are a ram-driven plunger and syringes, where the flow rate of pastes is controlled by regulating the movement of the plunger. The extrusion is started or stopped as the force on the plunger is generated or released. Shutter valve extrusion follows a similar mechanism, with a shutter needle inserted in the flow path acting as a valve. The tip of the shutter needle is positioned close to the outlet of the extrudates. A force is applied to the needle so that it is lifted up or pressed down, corresponding to the opening or closing of the flowing path. The flow rate of the extrusion is dependent on two factors, i.e., the velocity of the plunger and the force on the plunger, whereas the motion of the shutter needle is utilized to control the start or stop of the extrusion. In an auger extruder, there is also a syringe, where the pressure is pre-applied with compressed air. In this case, the paste is delivered to the auger chamber. Then, extrusion is facilitated, as the auger is rotated with a servo motor. The auger's angular velocity is used to control the flow rate. Once the auger stops rotating, the paste flow is shut down.

5.3 Examples of materials

Extrusion freeforming, as one of the linear solid freeforming technologies, has been used to fabricate electromagnetic crystals based on Al_2O_3 [20]. One example was reported as solvent-based extrusion freeforming to construct complex 3D ceramic photonic crystals, known as bandgap metamaterials, for millimeter wave applications [21]. High purity Al_2O_3 (99.992%, $D_{50} = 0.48$ μm, dielectric constant $\varepsilon_r = 9.6$ at 100 GHz, ex Condea Vista, Tucson Arizona), $La(Mg_{0.5}Ti_{0.5})O_3$ (LMT, D_{50} = 1.9 μm), $(Zr_{0.8}, Sn_{0.2})TiO_4$ (ZST, D_{50} = 1.8 μm) and silica (D_{50} = 2 μm, PI-KEM Co., UK) powders were used as the dielectric components.

Poly (vinyl butyral) (PVB) and poly (ethylene glycol) (PEG, molecular weight (MW) = 600) were mixed to serve as a binder. PVB and PEG powders were dissolved in the solvent propan-2-ol. After that, the dielectric powders were dispersed in the PVB–PEG solution, with the aid of ultrasonication (U200S, IKA Labortechnik). The volume ratio of the alumina powder to the mixture of dry polymers was 3:2. The suspensions were heated to evaporate the solvent to solid content of about 88%, resulting in pastes for extrusion experiment. Lattice structures were printed by using an extruder at room temperature [22]. The samples after extrusion were dried at room temperature and then calcined at 400 °C for 1 h, at a heating rate of 2 °C · min^{-1}.

The extrusion freeforming platform had four axes, i.e., *X*, *Y*, *Z*, and extrusion. The *XY* table (MX80L Miniature Stage, Parker Hannifin Automation, Dorset, UK) could work at a high acceleration of 39.2 m · s^{-2} and moving speed of 0.1 m · s^{-1}. A stainless steel syringe with an internal diameter of 9 mm was fixed on the *Z*-axis, while the substrate was laid on the *XY* table. A load cell (Flintec, Redditch, UK) on the extrusion axis was used to monitor the extrusion pressure. The filament pattern in each layer was controlled by the trajectory of the die relating to the movement of the *XY* table, whereas the *Z*-axis moved at steps equivalent to the single layer thickness, thus forming 3D structures.

The distance between the nozzle tip and substrate is usually set to 2–3 times the diameter of filaments during extrusion, while the accuracy of the vertical position of the syringe, nozzle, and the *XY* table strongly influences the extrusion process. For a tilted substrate, it could be touched by the tip of the nozzle, or at least the tip of the nozzle could touch the first layer during extrusion. Therefore, the sizes of the samples that can be printed are closely dependent on the angles of the substrate.

The performance of the printing is also affected by the nozzle path that governs the moving of the *XY* table. A change in direction is accompanied by acceleration or deceleration in the direction of the *X*- or *Y*-axes [23]. Even at a steady state and constant paste flow, under-fill or over-fill could be present, due to the acceleration or deceleration associated with the change in direction during the extrusion process. In this regard, the velocity of the extrusion ram should be adjusted to match the nozzle path. During the printing process of woodpile structures, over-fill is a key issue for 180° turns, because the *XY* table path velocity is first decreased from the operational velocity to zero and then increased to the operational velocity in another direction. Figure 5.1(a) shows a photograph of the sample with typical overfill at corner paths, where the filaments were obviously thickened.

Over-fill or under-fill could be present, as the extrusion velocity is higher than or lower than the path velocity of the *XY* table. Figure 5.1(b) depicts an example of over-fill, where the extrusion velocity is higher than that of the *XY* table. In contrast, figure 5.1(c) presents a case of under-fill, where the extrusion velocity is lower than the path velocity of the *XY* table. The corresponding consequences were filament thickening and thinning, respectively.

On one hand, PVB is strongly adhesive to the particles of the ceramic powders, while it could also enhance the flowability and dispersibility of the powders by preferentially adsorbing on the surface of particles and hence forming a steric barrier for the particles to be repulsed by one another [24, 25]. On the other hand, PEG is commonly used as a plasticizer with a PVB binder [26]. In this case, the molecular weight (MW) of PEG has a strong effect on the performance of the additive systems. PEG with a low MW shows high capability of plasticization, thus resulting in suspensions with low viscosity and hence allowing for high solid loading levels. However, low MW PEG also means low mechanical strength.

To achieve high-quality printing, the paste should be free of particle agglomeration, entrained debris, and air bubbles. Figure 5.2(a) shows a schematic diagram of possible agglomerates and air bubbles in a ceramic paste, which could cause anisotropic mechanical properties, so that the extruded filaments would contain defects [27, 28]. For instance, to extrude filaments with diameters of tens of microns, the particle size should not be larger than 2 μm. However, the smaller the particle, the more pronounced the tendency to form agglomerates. Figure 5.2(b) presents examples of the filament having defects caused by the particle agglomeration. Air bubbles could be brought in a paste during the evaporation of solvent with stirring, thus leaving voids in the extruded filaments, as observed in figure 5.2(c). The bubbles in pastes are usually removed through degassing in vacuum [29–31].

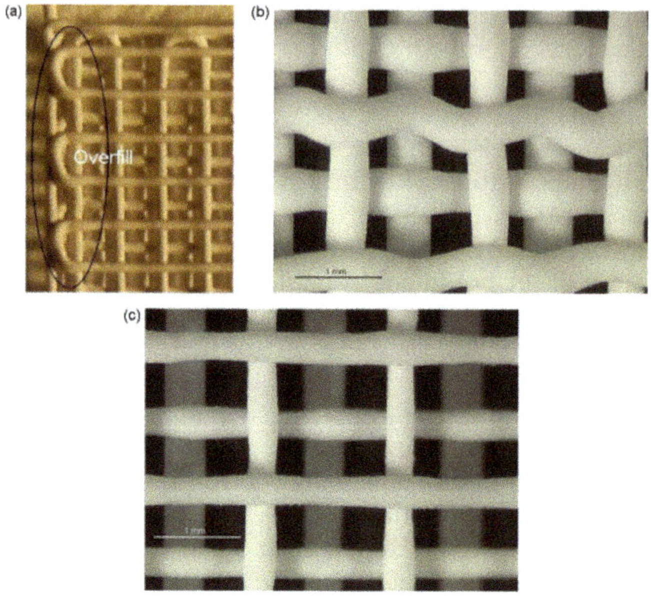

Figure 5.1. Photographs of representative samples with different issues: (a) over-fill caused by the change in path, (b) over-fill caused by a higher paste extrudate velocity than the *XY* table path velocity, and (c) under-fill caused by a lower paste extrudate velocity than the *XY* table path velocity. Reprinted from [21], Copyright (2009), with permission from Elsevier.

The solvent content has strong effects on the viscosity of the pastes and hence yield stress and shrinkage behaviors of the printed filaments. A high content of solvent is conductive in preventing high extrusion pressure. A low content of solvent results in rigid pastes, which helps retain the shape of the extruded filaments and prevents the sagging of the filaments between supports and the layers. In any case, the extruded filaments should be bent by 90° to allow the change in flow direction, because the extrusion is in the direction of the *Z*-axis while the filaments are laid on the *XY* plane, which could be realized by controlling the yield stress.

The circular cross-section of the Al_2O_3 filaments, with a diameter of 500 μm, was not retained, as illustrated in figure 5.2(c). The reason for the defects was the low rigidity of the paste. The extruded filaments also had wave distortion. Two possible reasons contributed for this defect. The filaments were not always extruded from the die exit in the exact vertical direction along the *Z*-axis. As the *XY* table moved faster than the extrusion rate of the paste, the angle between the filament and the substrate was not 90°, so the filament would have uneven stress distribution in all different directions. Meanwhile, the filaments were bent by 90° during the extrusion process. The bending would induce difference in stresses at the two sides of the filaments, thus producing wavy distortion on one side.

When the extrusion process was finished, the lattice structures were dried and subjected to debinding, followed by high-temperature sintering. Various defects, such as warps, voids, and cracks, could be retained or even worsened. Warping

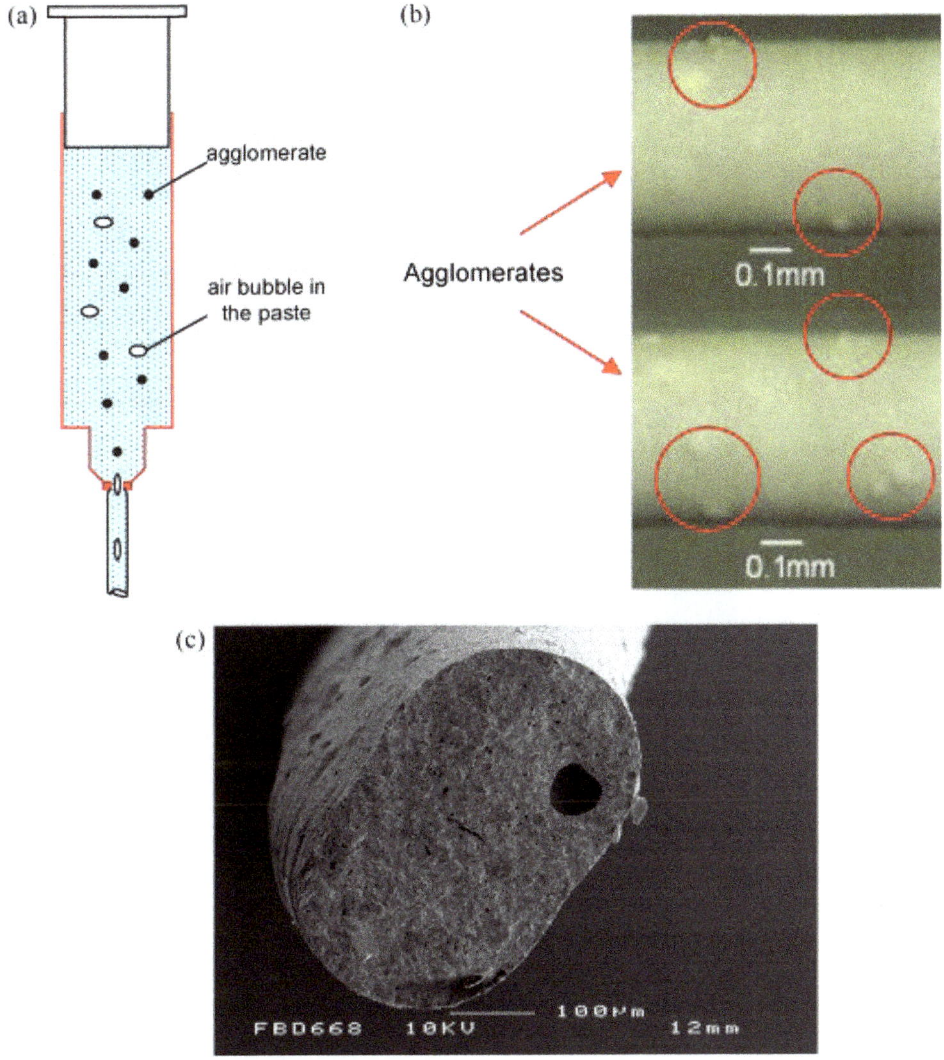

Figure 5.2. Defects generated during the preparation of the paste: (a) defects in the syringe barrel, (b) particle agglomerates, and (c) air bubble leaving void in the filament. Reprinted from [21], Copyright (2009), with permission from Elsevier.

occurs in conventional FDM, due to the requirement of cooling the thermoplastic filaments from the melting temperature to the glass-transition temperature. When thermoplastic filaments were deposited, adhered, and contracted, the resultant stresses within one layer would induce warping in a side-by-side or layer-by-layer manner [32, 33]. The change in volume of solvent-based pastes is attributed to the evaporation of solvents, although the process is relatively slow. When the first layer has strong adhesion to the substrate, no warping takes place. Otherwise, warping occurred due to the shrinkage of the filament, as demonstrated in figure 5.3(a).

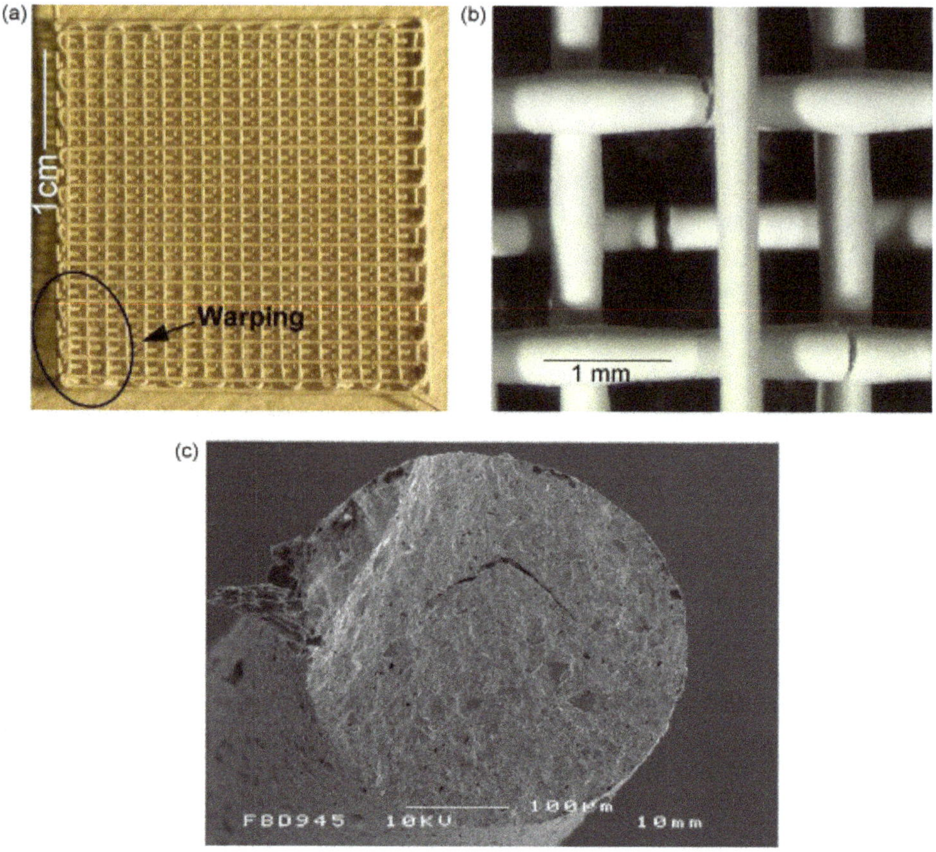

Figure 5.3. Defects generated during post-processing: (a) warping, (b) crack in cross-sections of the filaments, and (c) core–shell defects. Reprinted from [21], Copyright (2009), with permission from Elsevier.

Cracks tend to form in the filaments during the drying process. As the filaments are adhered to the substrate, they cannot shrink freely, thus resulting in tensile stress. Once the tensile stress is higher than the yield stress of the filament, cracks are generated. Figure 5.3(b) shows a sample with such cracks, which was derived from the paste with 63 vol% Al_2O_3. The formation of cracks is dependent on the composition of the polymer-ceramic binary system, where the elastic modulus increases sharply, as the solid loading level increases. Also, during the early stage of drying, sufficiently high solid loading levels would prevent particle rotation, so that higher packing and higher prefired strength could be achieved.

Cracks inside the filaments, as observed in figure 5.3(c), primarily formed at the debinding stage. Meanwhile, the formation of an internal core–shell structure could also be one of the reasons in such solvent-based systems. The process of cracking is actually the consequence of non-uniform shrinkage in volume during the burning out of the polymer components. When the green bodies were subject to debinding, the organic components were burned while producing gases, which escaped from the bodies through the surface of the filaments. Therefore, the effective volume fraction of particles at the surface was increased. The outer layer acted as a rigid exoskeleton at the early stage, while there were still polymer components and shrinkage of the system did not start. As a consequence, stresses were produced at this stage, thus causing a circumferential cut-like rupture [21].

To fabricate fully dense ceramics with complex structures by using extrusion freeforming technology, it is important to precisely control the start and stop of the extrusion process, so as to dispense materials on demand, a process known as extrusion-on-demand (EOD) [34]. Controlling the EOD process for slurries with high solid loading levels is a challenge, because ceramic slurries usually have characteristics, such as non-Newtonian behaviour, high compressibility, and compositional inhomogeneity. Li *et al* adopted three EOD methods, including ram extruder, shutter valve, and auger extruder, to process aqueous ceramic slurries with relatively high solid loading levels (>50 vol%) [35]. The extrusion effectiveness of the three methods was studied, such as accuracy of start and stop and stability of flow rate. In comparison, the auger extrusion method was superior in EOD performance, with the highest flow rate consistency.

A schematic diagram of the extrusion process and a photograph of the extrusion setup are shown in figure 5.4(a) [36]. Slurries with different properties would have different steady-state extrusion forces, at given plunger velocities. The plunger velocity could reach the steady state fast by simply modifying the plunger velocity with a normal tracking controller. Comparatively, the extrusion force had a slow response, taking several minutes to reach the steady-state level. As a result, reaching the steady-state extrudate velocity or steady-state slurry flow rate would take a quite long time. The main reason was the presence of air bubbles trapped in the slurries [29, 30]. The higher the solid loading levels of the slurries, the more difficult the degassing process in paste preparation process would be, owing to their high viscosity.

A hybrid force–velocity controller was reported to realize fast dynamic response of extrusion forces, which had settling times in the range of 0.8–1.6 s, for the start and stop of extrusion when extrusion force control was employed [37]. In the case of

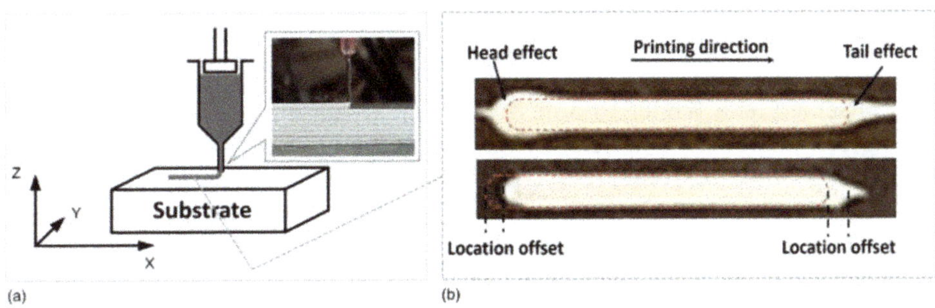

Figure 5.4. (a) Schematic diagram of the extrusion process, along with photograph of the experimental setup. (b) Schematic diagrams showing printing flaws owing to the inaccurate extrusion start and stop, with the dashed lines to indicate the designed outlines and locations. [34] (2017), reprinted by permission of the publisher (Taylor & Francis Ltd, http://www.tandfonline.com.)

steady-state extrusion, the plunger velocity control was used for the controller. The time delays for the start and stop of extrusion could be compensated by the dwell technique and look-forward technique, respectively [36]. With these techniques, the performances of EOD could be tremendously enhanced. Nevertheless, the processing parameters for ram extruders should be adjusted additionally to achieve the desired performances, because different batches of slurries could exhibit significantly different rheological behaviors.

If the extrusion parameters were not properly selected, the accuracy of the start and stop of extrusion could be sacrificed, inducing printing flaws or defects. As presented in figure 5.4(b), the white areas were the printed items, while the designed shapes and outlines were represented by the dashed rounded rectangles. Excessive slurries were extruded at the beginning of the extrusion, resulting in printed items with a wider head, known as the 'head effect'. The end of the printed structures exhibited a tail-like profile, which was attributed to the ineffective stopping of extrusion, denoted as the 'tail effect'. The head effect and tail effect are demonstrated in the upper panel of figure 5.4(b). The lower panel in figure 5.4(b) indicates the structure printed inaccurately at the designed locations, because of the time delays in extrusion, which were known as 'location offset'.

For high solid loading levels, achieving homogeneous dispersion of binders is a challenge in the preparation of slurries. If the binders are unevenly distributed in the suspension, ceramic particles will agglomerate, and the slurries will be inhomogeneous. Either under-fill or over-fill could happen, when the extrusion is conducted at constant velocities, reflecting inconsistent flow rate and inhomogeneity of the slurries. The ram extrusion process tended to suffer from the inhomogeneous properties of slurries and hence easily caused unstable printing effectiveness. In this regard, two other extrusion techniques could be adopted, which are shutter valve and auger extrusion.

The slurries were prepared with Al_2O_3 powder, DARVAN® C (ammonium polymethacrylate, Vanderbilt Minerals LLC, Gouverneur, NY), methocel (methylcellulose, Dow Chemical Company, Pevely, MO) and DI water. The Al_2O_3 powder and DARVAN® C were first dispersed in DI water by ball milling for 15 h to obtain homogeneous slurries. DARVAN® C was adopted as a dispersant for Van der Waals forces between the ceramic particles. The slurries were then heated at 70 °C, while methocel was dispersed in the system, aided by mechanical stirring for 10 min. Methocel served as a binder to adjust viscosity of the slurries, so as to ensure the printed green bodies with sufficiently strong mechanical strength after drying [34]. After cooling down to room temperature, slurries with the desired solid loading levels were developed. During the cooling down process, the slurries were subject to mechanical stirring. After that, the slurries were treated with a vacuum mixer (Model F, Whip Mix Corp., Louisville, KY) for 10 min to remove air bubbles.

Dash line extrusion testing was carried out with the three extrusion techniques. Location offset, tail effect, and head effect in printing dash lines are quantitative measures of the accuracy of extrusion start and stop. The experiments included three groups of samples, with nozzle sizes of 610, 406, and 610 μm, solid loading levels of 60%, 60% and 50%, denoted as Group 1, Group 2, and Group 3, respectively.

Meanwhile, the same paste flow rate, table speed, and layer thickness were used for the experiments.

Five dash line segments were printed to optimize the extrusion parameters, by varying start dwell (τ) of 10 ms and stop distance (d) of 0.1 mm. For the ram, shutter valve, and auger extrusion methods, the values of τ were 450, 70 and 0 ms, while the d values were 1.9, 0.3 and 0 mm, respectively. The same values of τ and d were adopted for Groups 2 and Group 3 in in the dash line printing experiments. The Group 1 and Group 2 were used to evaluate the effects of nozzle diameter, while the Group 1 and Group 3 were studied to identify the effects of paste solids loading.

The consistency of line width was evaluated by using a set of continuous line printing tests. A cap was put on the tip of the nozzle to control the height of the filament by fixing the height of the deposited slurry. The filament height (h) was fixed to 150 μm. The reference velocity of the plunger for the ram extruder was set to be 5 $\mu m \cdot s^{-1}$, corresponding to a paste flow rate of 0.198 $ml \cdot min^{-1}$, while the table speed was 660 $mm \cdot min^{-1}$. The filament height was controlled to be 150 μm, which was lower than the typical value of 450 μm, to print a wider nominal filament, thus being able to reduce the measuring error for the filament width. As the cross-section of the filaments was assumed to be a rectangle with a rounded geometry at the two ends, the width of the filament (w) was 2.02 mm.

The ram extrusion and the shutter valve methods were similar in terms of continuous line printing, which was a steady-state extrusion process. Therefore, the three extrusion methods were categorized into two groups. For clarity, shutter valve extrusion was chosen to compare with the auger extrusion in the actual experiment. Four groups of dash line printing tests were conducted, where Group 1, Group 2, Group 3 and Group 4 had nozzle diameters of 610, 406, 610, and 406 μm, respectively, with the first two being shutter valve extrusion and the last two being auger extrusion. The slurry used for the experiment had a solid loading level of 60%.

Additionally, ceramics with various structures were also printed to further evaluate the start and stop performances of extrusion and flow rate consistencies. For ceramic structures, under-fill would lead to high porosity, whereas over-fill would cause the slurry to accumulate, so that the nozzle might be interfered. Similarly, shutter valve and auger extrusions were conducted for comparison.

Ceramic green bodies, with dimensions of $30 \times 15 \times 4$ mm^3, were printed by using the shutter valve and auger extrusion processes with the same gantry and slurry, while the process parameters included paste flow rate of 280 $\mu l \cdot min^{-1}$, raster width of 600 μm, layer thickness of 400 μm, table speed of 21 $mm \cdot s^{-1}$ and nozzle diameter of 610 μm. The reference plunger velocity for the shutter valve extrusion group was derived from the reference flow rate of the slurry, while the average flow rate of the auger extrusion was close to the reference flow rate of 280 $\mu l \cdot min^{-1}$. After printing, the green bodies were all dried at 25 °C and 75% relative humidity for 20 h and then sintered at 1550 °C for 1.5 h.

The printing results of the dash line experiments by using the three extrusion processes are depicted in figures 5.5 and 5.6. The dash lines were all printed from right to left, whereas the dashed and solid vertical lines at the two ends stood for the designed start and stop points of the line segments, respectively. The values of the

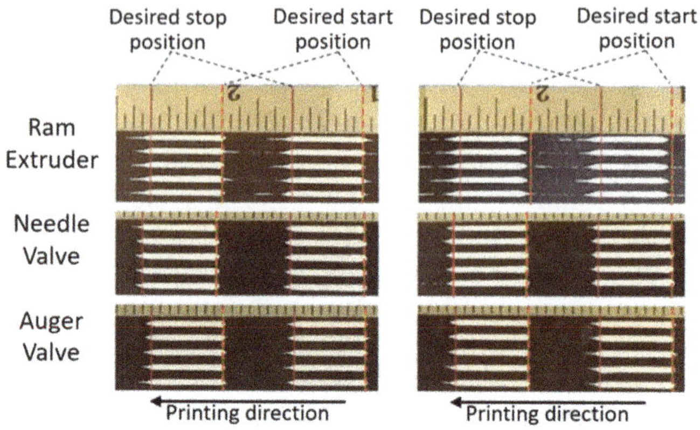

Figure 5.5. Dash line printing results with 610 μm diameter nozzle (left) and 406 μm diameter nozzle (right), using slurry with a solid loading level of 60%. [34] (2017), reprinted by permission of the publisher (Taylor & Francis Ltd, http://www.tandfonline.com.)

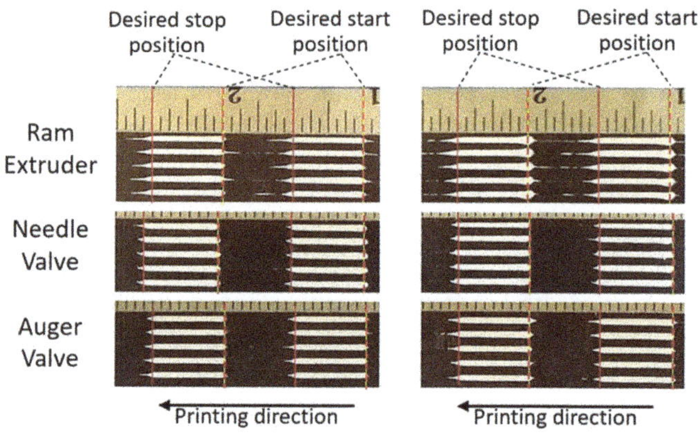

Figure 5.6. Dash line printing results derived from slurries with solid loading levels of 60% (left) and 50% (right), using nozzle with diameter of 610 μm. [34] (2017), reprinted by permission of the publisher (Taylor & Francis Ltd, http://www.tandfonline.com.)

start dwell time (τ) and early stop distance (d) for the shutter valve and auger extrusion processes methods were both shorter than those for the ram extrusion process. The calibrated τ and d for the shutter valve extrusion process were nearly zero, because the operation volume of the shutter valve extrusion could be neglected, as compared with that of the ram extrusion.

For the shutter valve extrusion process, the value of τ was 70 ms and that of d was 0.3 mm, which were detectable, owing to the pneumatic actuation effect of the needle of the shutter. The dash line segments printed by using the shutter valve and auger extrusion processes had sharper tails than those using the ram extrusion process, suggesting that the former two extrusion processes were more effective. According to the results in figure 5.5, the dash line segments printed by using the ram extrusion process suffered from a larger location offset, as the nozzle diameter varied from 610 μm (left) to 406 μm (right), implying that a larger diameter nozzle displayed higher performance.

As observed in figure 5.6, the dash line segments printed by using the ram extrusion process had an increase in size of the heads, as the solid loading level of the slurries was reduced from 60% to 50%. This observation suggested that a high solid loading was favorable in terms of extrusion performance. Meanwhile, the line segments printed by using the shutter valve and auger extrusion processes were almost unchanged when the diameter of the nozzle and solid loading level of slurry were varied. The optimal extrusion processing parameters for the ram extrusion process were calibrated for Group 2 and Group 3. The optimal extrusion parameters were strongly dependent on the diameter of nozzle and solid loading level of slurry.

For the four groups of samples printed using the shutter valve and auger extrusion processes, five serpentine lines with a length of 1778 mm were prepared in each group, with a nozzle diameter of 610 μm. A glass plate combined with a thick polymer tape was used in the substrate, whereas the tape was within the dash-lined rectangles. Because the areas outside the tape was lower than that within the tape, only the areas inside the dash-lined rectangles were subject to the measurement of diameters of the line segments.

Comparatively, in terms of paste flow rate, the shutter valve extrusion process had a larger fluctuation than the auger extrusion process. In addition, for the shutter valve extruded groups, the one serpentine line displayed large fluctuations in the line width, although the phenomenon was not very violent, indicating that the transient phases induced disturbance related to the compositional inhomogeneity of the slurry possessed a relatively slow response. In comparison, the samples in the auger extrusion groups exhibited a stable flow rate, because this process was relative insensitive to compositional inhomogeneity of the compressible slurry.

The sample printed by the shutter valve and auger extrusion processes had relative densities of 96.2% and 98.4%, respectively. Since the printed items had simple structures, no defects were produced in relation to the starts and stops of extrusion. The samples made by the shutter valve extrusion process had a slightly higher porosity, which could be ascribed to the formation of pores inside the bodies. It could be linked to the under-fill behavior owing to variation in the width of the printed lines, further indicating that the auger extrusion process was more feasible in terms of extrusion performances, such as the flow rate consistency of slurries.

Additionally, the auger extrusion process was capable of continuous printing at a large volume of slurry. If the requirements, such as compact machine size, high extrusion accuracies of the start and stop, high stability of flow rate consistency, large volume of feedstock, and low homogeneity of slurry are necessary, it is recommended to use the auger extrusion process. An example is the printing of pore-free large ceramic parts. For extrusion of abrasive materials, however, the wear of the auger will need to be considered. If the extruder is allowed to be large in size and weight, while the materials are inhomogeneous but the extrusion start and stop should be precisely controlled, the shutter valve extrusion process is more suitable. Otherwise, the ram extrusion process can be adopted, because it is cost-effective and simple in terms of processing. This is true especially for the extrusion with large nozzles.

The effect of rheology of ceramic slurries on processing efficiency and effectiveness of EFF was evaluated [38]. The rheological behaviors of a ceramic slurry are dependent on a series of parameters. Specifically, the relationships among formulation parameters, printing performances, and mechanical strength of the green bodies of Al_2O_3 ceramics were studied. The origins and reasons of the varied properties of the slurries to influence effectiveness of the EFF process were examined. Glycerine and boehmite with nanosized particulate needles were employed to increase printability of the slurries and mechanical strength of the green bodies. Mechanical properties were strongly dependent on the formulation parameters, as revealed in the green bodies, after the samples were partially sintered at 1000 °C. Both the shrinkage of the samples after drying and the printed lines could be decreased by controlling the formulations of the slurries. The effect of the formulation parameters on the printing efficiency was linked to the viscosity dependent on the flow rate of the slurries.

Al_2O_3 powder (Martoxid MZS-1, Martinswerk GmbH, Germany), with particle sizes of 0.65–5 μm was dispersed in DI water. To achieve slurries with high solid loading levels, the suspensions were stabilized through electrostatic stabilization by adding HNO_3 solution. Meanwhile, glycerin and needle shaped boehmite (AlO (OH)) nanoparticles (Nanoneedles, SASOL, Germany) were used as additives to modify the rheologic properties of the slurries.

The suspensions were ball milled (PM 400, Retsch, Germany) for 1 h at 400 rpm. 30 Al_2O_3 balls with a diameter of 10 mm were employed as the milling media. The slurries were filled into an array of cylindrical casting molds with a diameter of 6 mm and a depth of 4 mm, which were then dried at 50 °C for 2 h, thus forming green bodies. Figure 5.7(a) shows a photograph of an example of the green bodies. Half of the samples were sintered at 1000 °C for an additional 4 h.

The printing process was carried out with a custom-made extrusion 3D printer, as illustrated in figure 5.8. The machine consisted of a dosing screw unit (Preeflow ecoPEN300, Viscotec, Germany), which was mounted on a crossbeam between two parallel spindle drives, forming the z-axis of the printer. Two more spindle drivers were employed to control the movement of the platform in x- and y-directions. The slurries were carefully filled into the reservoir, in order not to produce entrapped air bubbles, noting that the slurries had a high viscosity. As the reservoir was subject to high pressures, the slurry continuously flew to the extruder.

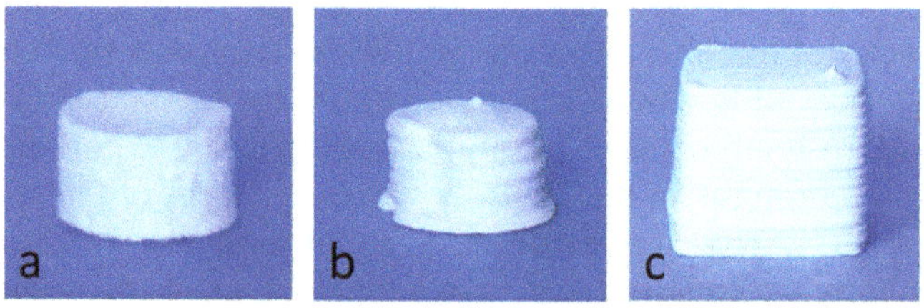

Figure 5.7. Photographs of the green bodies for the test: (a) mold-cast compression test sample, (b) printed compression test sample, and (c) cube-shaped test sample for estimating the layer thickness and shrinkage. Reprinted from [38], Copyright (2020), with permission from Elsevier.

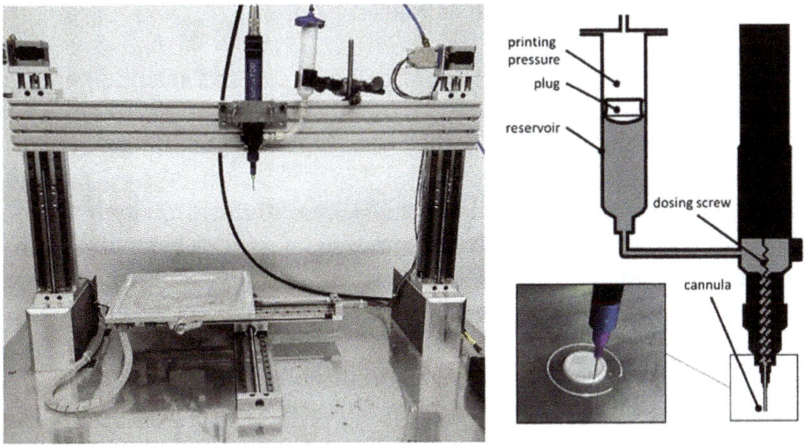

Figure 5.8. Photograph of the printer (left) and schematic diagram of the printing unit (right). Reprinted from [38], Copyright (2020), with permission from Elsevier.

After passing through a 0.84 mm cannula, the slurry was deposited on the printing platform. The axes and the extruder were controlled with a microcontroller board (Mega 2560, Arduino, USA), which was combined with a control module (RAMPS 1.4.2, German RepRap, Germany) and stepper motor driver carriers (DRV8825, Pololu, USA). A firmware (Repetier firmware 0.92, Hot-World, Germany) was used to derive the microcontroller, while the operation was facilitated with an open-source software (RepetierHost, Hot-World, Germany), which included a slicing tool (Slice3r 1.2.9, open source) to convert the digital models of the desired samples into g-code files.

The samples for compression test were all of cylindrical shape, with a diameter of 6 mm and a height of 4 mm, as presented in figure 5.7(b). Printing was performed with a cannula of 0.8 mm. The printing thickness of the single layer was 0.6 mm, while 100% infill was conducted at a feed rate of 720 mm $\cdot$ min^{-1}. Because of the small size of the test samples, their perimeter was set to be single contour. The shrinkage of drying was measured by using cube-shaped samples with aside length of 10 mm, as depicted in figure 5.8(c). The layer thickness of the perimeter was characterized with tactile profilometry (DEKTAK, Bruker). A needle was pulled over the surface of the test sample, which was perpendicular to the grooves. The surface profile was recorded, based on the displacement of the needle in *z*-direction. The lateral distance between the maxima was taken as the layer thickness.

Noting that the solid loading level of slurries for 3D printing should be as high as possible. To this end, micron-sized Al_2O_3 were used to prepare slurries, so as to achieve the behaviors like wet bulk solid instead of suspensions. With 1 wt% nitric acid added as stabilizer, the solid loading level could reach 81 wt%, corresponding to a volume concentration of 51.9 vol%. The rheological behaviors of the slurries are crucially important to achieve high efficiency and effectiveness of printing. On one hand, the slurries should have sufficiently low viscosity during the strong to allow smooth transportation. On the other hand, they should be extrudable and viscose after printing, in order to prevent materials running of and ensure the shape integrity.

In this case, shear-thinning rheological behaviors of the slurries were tailored to identify the formulation dependencies, by using two additives. As the content of glycerin was higher than 5 wt%, the viscosity was gradually increased with increasing level of glycerin at shear rate of $120\ 1 \cdot s^{-1}$. At $20\ 1 \cdot s^{-1}$, similar trend was observed, except for the sample with 20 wt% glycerin content. Although the addition of glycerin raised the viscosity, the shear-thinning behavior of the slurries was nearly not affected, because of the Newtonian fluid nature of both water and glycerin. The electrostatic interaction of the ceramic particles was not tremendously varied.

Compression tests were conducted for cylindrical samples to examine the effect of glycerin on the mechanical properties of the slurries. Meanwhile, breaking strengths were accordingly derived. For the green bodies dried at 50 °C, the breaking strength was increased with increasing content of glycerin. With a boiling temperature of 290 °C, glycerin was retained in the green bodies after evaporation of water, thus increasing the mechanical strength. After sintering at 1000 °C, the breaking strength of the sample from the micron-sized powder was largely enhanced by >16 times, owing to the partial sintering. At the same, the breaking strength was decreased with increasing content of glycerin. The decline in breaking strength after the addition of glycerin was caused by the damage of the fluid bridges at high temperatures.

In addition, needle-shaped boehmite nanoparticles were used to increase the performance of the slurries. Because the slurries already had sufficiently high solid loading of micron-sized Al_2O_3 particles, the content of boehmite was not higher than 2 wt%. In the section of high shear, the viscosity was slightly increased but was lower than the values of the samples with glycerin. At low shear regions, the increase in viscosity was more pronounced. More than double the viscosity was achieved at $20\ 1 \cdot s^{-1}$, with the addition of 2 wt% nanoparticles.

Such an increment in the shear-thinning behavior was favored for ensuring high quality printing. Shear-thinning behavior is commonly observed in suspensions with nanoparticulate components, because the strong particle interactions would be critical to the rheological behaviors at relatively low shear rates [39]. In addition, since the needle-shaped nanoparticles had a high aspect ratio, it offers further enhancement in the shear-thinning behavior [40, 41]. The strong particle interactions were stemmed from the high surface to volume ratio and the relatively close position between the nanoparticles, as compared with that for micron-sized particles with same solid loading levels.

Breaking strength of the samples after drying at 50 °C was increased with increasing content of the nanoneedle particles. The breaking strengths were higher than those of the samples from the slurries with glycerin. Figure 5.9 shows fractured surface SEM images of the green bodies, which could be used to explain the observation. The needles were concentrated in the region in between the adjacent micron-sized particles, instead of being on the surface, as seen in the bottom panel of figure 5.9. In this regard, the needle particles exhibited a bridging effect to connect

Figure 5.9. SEM images of the fractured surfaces of pure Al_2O_3 green body (top) and green body with 2 wt% nanoneedles (bottom). Reprinted from [38], Copyright (2020), with permission from Elsevier.

the micron-sized particles, thus leading to enhancements in the breaking strength of the green bodies.

The bridges were formed owing to the evaporation of water during the drying process. The surfaces of the micron-sized particles resembled convex surfaces, causing a reduction in boiling pressure. Therefore, these areas were dried in the last moment, whereas retreating boundaries were in between spaces of water and air, which forced the nanoparticles to migrate into areas in between the micron-sized particles. In the end of the drying process, the dried fluid bridges were left. It was responsible for the samples after sintering at 1000 °C to be much stronger than those dried at 50 °C.

As the content of the nanoneedles was less than 1 wt%, the green bodies were weakened. However, the strength was tremendously enhanced in the samples with 1 wt% nanoneedles, which was different from the observation in the samples with glycerin. Boehmite decomposed to Al_2O_3 at 400 °C through dehydration [42]. Nano-sized Al_2O_3 powder had high sinterability, with full densification at very low temperatures. However, an insufficient content of the nanoneedles was unable to entirely bridge the micron-sized particles. Comparatively, boehmite was superior to glycerin, in terms of mechanical strength of the green bodies and sintering products.

At the beginning, water drops were brought out of the cannula, while clogging of the slurry occurred. To address this issue, the composition of the suspension was modified to boost the water-binding performance. Dehydration took place as the pores in between the micron-sized particles could not withstand the compression force due to the external pressure, leading to the closure of voids in between the particles. The resistance against compression could be strengthened by increasing the capillary forces between the micron-sized particles. As the distance between the micron-sized particles was shortened, capillary forces would be increased. The nanoparticles filled into the spaces in between the micron-sized particles, reducing the interparticle distances.

Maximization of the solid loading level of slurries would effectively minimize the shrinkage the green bodies and sintered products. The introduction of nanoparticles made the slurries behave like wet bulk solids instead of viscous suspensions. The optimal composition of the slurries consisted of 72 wt% micron-sized particles and 6 wt% nanoneedles. In this case, the hydration and drying effects in the printing process could be prevented.

For achieving a high quality of printing, the volume flow during the extrusion is an important parameter, which directly influences over-filling or under-filling of the printed materials. To establish the relationships among formulation parameters, printing efficiency, and the final product properties, the effects of printing pressure and content of nanoneedles was systematically examined. The feeding rate and length of the path were kept unchanged. According to quantity of the materials to be extruded, the real flow rate was derived depending on the formulation properties. Significant deviations in flow rate were observed, suggesting that the relationship between the printing pressure and flow rate should be carefully monitored.

The volume flow increased with increasing pressure, but decreased with increasing content of nanoneedles, owing to the increase in viscosity. For instance, as the pressure was lower than 0.35 MPa, the slurry with 7 wt% nanoneedles was completely not extrudable, since no flow was achieved. For the sample with 6 wt% nanoneedles, over-filling occurred, and the pressure was not detectable. In fact, depending on the printing pressures, either over-filling or under-filling could take place. The set volume flow rate was not achievable at all for the 7 wt% slurry.

A slurry micro-extrusion process was used to prepare silica (SiO_2) ceramics for human dental restorations [43]. The dental ceramic slurry exhibited pseudoplastic characteristics and a moderate level of viscosity, allowing extrusion at relatively low pressures while maintaining promising shape keeping stability. The green tooth could be directly shaped through CAD digital model in 0.5 h. The green tooth had uniform shrinkage during sintering. The sintered tooth exhibited microstructure similar to that of the counterparts made by using the traditional dental restoration processes.

The dental ceramic powder was supplied by Dentsply Ceramco (Burlington, NJ), with 63.4% SiO_2, 16.7% Al_2O_3, 1.5% CaO, 0.8% MgO, 3.41% Na_2O, and 14.19% K_2O. The powder was ball milled at low speed with a ceramic vial and Al_2O_3 balls. The milled powder was treated through sedimentation to remove the particles with sizes of >1 μm, leaving a powder with an average particle size of 0.48 μm and a specific surface area of 16.33 $m^2 \cdot g^{-1}$.

To prepare ceramic slurries, the powder was dispersed in DI water, with solid loading levels of 40–45 vol%. The suspension was milled with high energy ball milling (SPEX Industries, Inc., Edison, NJ) for 5–10 min to break down particle

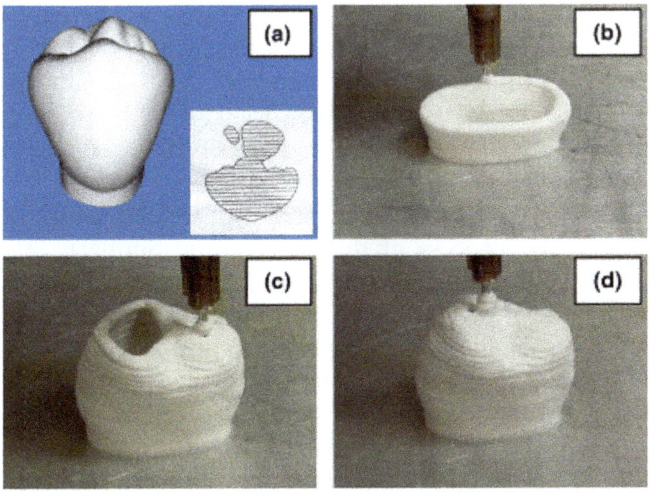

Figure 5.10. (a) STL digital model of an artificial tooth, with the inset showing a 2D slice from the 3D artificial tooth model with hatching lines and outlines to be followed by the micro-extruder. (b–d) Snapshots of a single-walled dental crown extruded at different stages, based on the digital model in panel (a). [43] John Wiley & Sons. © The American Ceramic Society.

agglomeration, leading to homogeneous and stable ceramic slurries. The ceramic slurry was extruded by using a scanning electron microscopy (SME) machine (University of Connecticut). The machine had three major components: (i) electric cylinders with precision force sensors to monitor the extrusion pressure, (ii) a positioning system with *X*–*Y* table and *Z* stage and (iii) three micro-extruders for the ceramic slurry with nozzles having diameters of 100–800 μm. The movement of *X*, *Y*, and *Z*-axes and the extrusion process were controlled to allow reading and slicing of the STL-format files, as illustrated in figure 5.10(a). The position and extrusion control signals were sent to the machine through a Galil DMC-1800 multi-axis motion control card (Galil Motion Control, Rocklin, CA). The ceramic slurry was extruded onto Cu substrates.

The as-extruded green bodies were dried at room temperature for 24 h in air and then sintered at 950 °C for 5 min. Photographs of a single-walled dental crown printed at different stages are shown in figure 5.10(b–d). The extrudate could stand without the requirement of supporting materials. The resistance to shear loading was sufficient at top of the crown, where the features had overhang angles of up to 45°. Meanwhile, the bottom part of the crown should be strong to prevent slumping, due to the weight of the accumulated extrudate. These could be achieved by adjusting the rheological properties of the ceramic slurries.

To ensure effective extrusion, the slurries should be of strong pseudoplasticity, i.e., shear-thinning behavior, which have high flowability and immediate gelation of extrudate as the slurry injected from the nozzle. Pseudoplasticity of slurries can be achieved by controlling their pH value. The viscosity of a slurry can be characterized with an iso-electric point (i.e.p., pH = 0.5) [44]. As the pH value of the slurry approached the i.e.p., its viscosity increased abruptly. In comparison, slurries with pH < 4.0 or pH > 8.0 possessed low viscosities. The slurry with pH = 5.7 exhibited strong shear-thinning behavior, with a decrease in viscosity by 9750 cps, as the shear rate increased from 1.5 to 60 s^{-1}. In contrast, the shear-thinning properties of the slurries with pH = 10.7 and 3.7 were relatively weak, with the viscosity decreasing by just 920 cps. Slurries with pH values in the range of 7.0–7.5 displayed intermediate shear thinning behaviors.

Figure 5.11 shows snapshots of green bodies of a solid tooth and multi-walled crowns during the SME process at different stages. Similarly, the capability of bridging and resistance to slumping and shear loading were essential requirements for the extrusion of the solid tooth and multi-walled crowns. Additionally, over-filling and under-filling of materials also significantly affected the qualities of the extruded samples, because they would cause swollen and porous products. To address these issues, extrusion parameters should be optimized [43].

The green bodies of the teeth experienced 3% uniform shrinkage after drying at room temperature in air for 24 h. After sintering for 5 min at 950 °C, the samples had shrinkages of 27% and 24% in the height and plane directions, respectively, as observed in figure 5.12(a). The shrinkages were almost uniform and repeatable, indicating dimensional stability of the dental structures. The sintering conditions were selected to minimize the 'stair' steps of the layer-by-layer manufacturing profile. The sintered ceramics consisted of two components, crystalline leucite embedded in a glass matrix, as demonstrated in figure 5.12(b), resulting in dental structures with sufficiently high mechanical strengths [45].

Polyetheretherketone (PEEK) is widely used in biomedical applications due to its biological activity, whose bone-implant interfaces and osseointegration could be improved through the incorporation of other bioactive materials, such as calcium phosphates bioglasses [46–49]. EFF 3D printing technique has been shown to be an

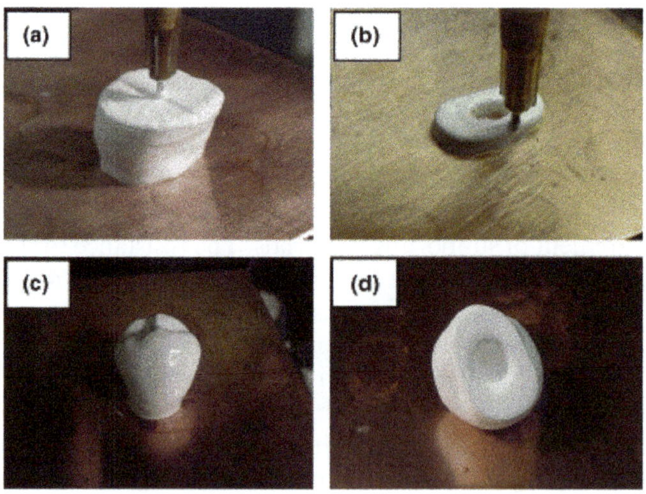

Figure 5.11. Snapshots of different samples in extrusion: (a) artificial tooth, (b) multi-walled dental crown, (c) side view of green body of the solid artificial tooth, and (d) bottom view of green body of the multi-walled crown. [43] John Wiley & Sons. © The American Ceramic Society.

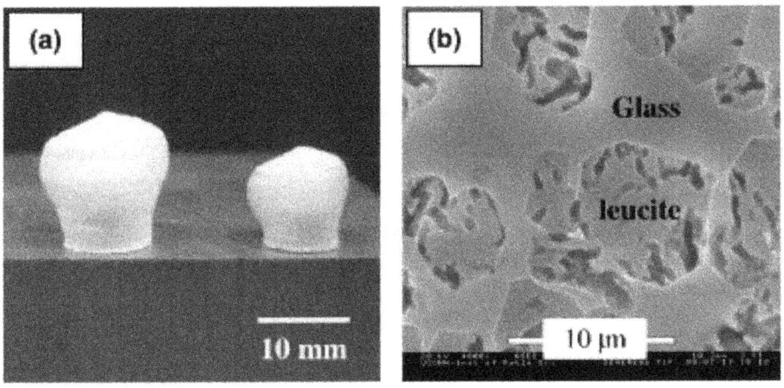

Figure 5.12. (a) Photographs of green and sintered bodies of a tooth. (b) Representative TEM image of the sintered teeth, with crystalline leucite particles embedded in a glassy matrix. [43] John Wiley & Sons. © The American Ceramic Society.

effective strategy to prepare bioactive PEEK/hydroxyapatite (PEEK/HA) composite with desired configurations, in which the distribution of the bioactive component could be well controlled inside the matrix of PEEK through computer programming [50]. The HA network in the biocomposite was entirely interconnective, thus being superior to its counterparts fabricated with other methods.

Meanwhile, porous PEEK structures could be produced by using this technology, with porosity, pore size, and size distribution being highly controllable, thus allowing for more effective and efficient cellular infiltrations and biological integration for implantation. The PEEK/HA biocomposites with 40% HA exhibited desirable static and cyclic modes, while the yield and compressive strengths were compatible with human cortical bones, demonstrating potential applications for load bearing. Moreover, the concern of biological safety of the composites was preliminarily addressed, while their biocompatibility was evidenced by experimental results, such as cell attachment, cell bridging, long-term viability, and matrix deposition.

The schematic diagram of the workflow to prepare bioactive PEEK/HA composites and porous PEEK structures is illustrated in figure 5.13(a). Porous bioactive HA scaffolds were first obtained by using the EFF 3D printing process, whereas PEEK melt was impregnated into the HA scaffolds through compression molding. The as-printed PEEK/HA composites could be used directly for bioactive applications, owing to the presence of HA. In addition, the HA network could be removed by soaking the composites in HCl solution to form entirely interconnected porous PEEK structures, as depicted in the dotted box in figure 5.13(a).

The photograph of the extrusion-based 3D printer is shown in figure 5.13(b), which was a lab-designed machine. The device comprises a three-axis table (Parker Hannifin Ltd, Warwick, UK), and a stepper motor (200 steps/rev, SL0-SYN step motor, Warner Electric, New Hartford, CT, USA) driving a 2 mm-pitch ballscrew to produce a continuous force on the extrusion syringe plunger. The extrusion syringe was preloaded with the HA paste. The speed of the stepper motor was controlled by using a 100–1 reduction gearbox, thus allowing finer tuning of the displacement. The HA filaments were precisely extruded, with small diameters of

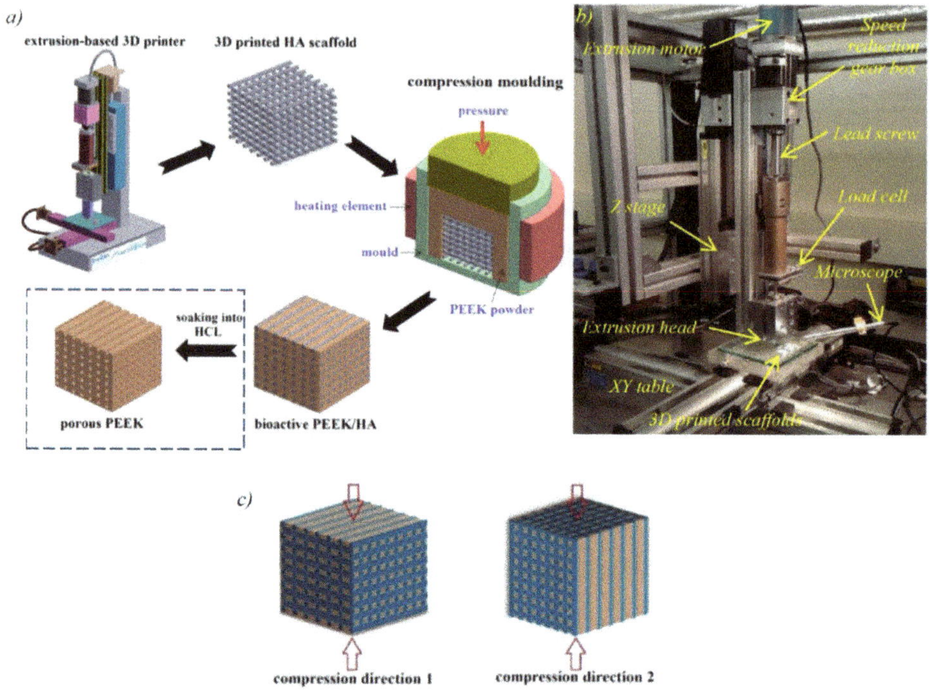

Figure 5.13. (a) Schematic diagram showing the process to print the bioactive PEEK/HA composites and pure porous PEEK with EFF technology and compression molding process. (b) Photograph of the experimental set-up for 3D printing of the bioactive HA scaffolds. (c) Directions for measurement of the samples using compression test. Reproduced from [50]. CC BY 4.0.

50 μm, by using customized nozzles with small die land length. The as-printed HA scaffolds were naturally dried at room temperature for a whole day and subject to debinding and sintering. The sintering was conducted at 1300 °C for 2 h.

The macroporosity was contributed by the spacing in between the HA filaments, which formed open pores. The microporosity was an internal characteristic of the filaments, owing to the incomplete sintering of HA, which formed closed pores. The microporosity was dependent on sintering parameters, including sintering temperature, time duration, and heating/cooling rates [51]. The details of the samples are listed in table 5.1.

The dried HA scaffolds with different filament dimensions and pore size were over-molded with PEEK-OPTIMA®LT3 UF powder (Invibio Ltd, Thornton–Cleveleys, UK), by using compression molding techniques, to obtain PEEK/HA composites, at various static loads. A mold with an internal diameter of 25 mm was prepared from tool steel, and it had a ventilation hole of size 0.5 mm on the bottom surface to avoid trapping air within the composite. The optimal molding temperature and pressure were 400 °C and 0.39 MPa, respectively, as the over-molding HA scaffolds had dimensions of 10 × 10 × 3 mm^3, with various filament diameter and pore size.

Table 5.1. Experimental macroporosity and microporosity of the HA scaffolds with different filament dimension and pore sizes. Reproduced from [50]. CC BY 4.0.

HA scaffold						
Filament (μm)	Pore (μm)	ms (g)	mw (g)	Vtotal mm^3	Macroporosity (%)	Microporosity (%)
240	250	0.5914	0.3892	382.1	56.32	7.17
240	400	0.5701	0.3746	494.3	60.45	7.42
400	250	0.9691	0.6335	505.3	33.50	8.84
240	550	0.4668	0.3035	509.5	67.95	9.25

Static loading was carried out with five aspects, (i) heating of the mold at 250 °C, (ii) applying load with pressure to be maintained until 400 °C, (iii) maintaining temperature and load for additional 20 min, (iv) stopping heating and cooling at the pressure to facilitate crystallization and solidification of the PEEK matrix, and (v) removing the composite samples from the mold at temperature just below the glass transition point (143 °C) of PEEK and cooling to room temperature.

Porous PEEK composites were separated from the over-molded samples, so the HA filaments were exposed. The composite samples were then soaked in HCl (37%, Fisher Scientific Ltd, Loughborough, UK) solution for 3 days. HA network in the PEEK/HA composites was etched out in the HCL solution, resulting in samples with interconnected channels. Surface roughness of the HA filaments and the hollow channels were characterized by using optical profilometry (Alicona Imaging GmbH, Raaba, Austria). PEEK/HA composites and porous PEEK structures were comparatively studied.

The PEEK/HA composite samples with 40 vol% HA, in which the HA filament size was 400 μm and the pore size was 700 μm, were represented for the unconfined uniaxial compression test, with an Instron 8032 test machine, at a strain rate of $3 \times 10^{-3}\ s^{-1}$. A 100 kN load cell was utilized, while the experimental data were recorded by using StrainSmart software (version 6200, Vishay Precision Group, Wendell, NC, USA). The samples were characterized in two directions, as demonstrated in figure 5.13(c). Pure PEEK samples were also examined with compression molding, for the comparison with the one with 40% HA.

All PEEK/HA samples were heated at 200 °C to increase the crystallinity of PEEK before static and cyclic compression measurement, by using the normal protocol for injection molded PEEK items. Firstly, the samples were subject to drying for 3 h at 150 °C and then heated up at 200 °C for 4 h, at a heating rate of $10\ °C \cdot h^{-1}$. Finally, the samples were cooled down to below 140 °C, at a cooling rate of $10\ °C \cdot h^{-1}$ and then naturally cooled down to room temperature.

Figure 5.14 shows the SEM images of representative HA scaffolds with different filament dimensions and pore sizes. The scaffolds exhibited a highly uniform structure, while the printing process was very productive and reproducible (figure 5.14(a)). As shown in figure 5.14(b), the high-magnitude SEM image revealed that the fractured surface of the filament displayed pores with sizes in the range of 1–5 μm, which were induced during the sintering process. The scaffolds were

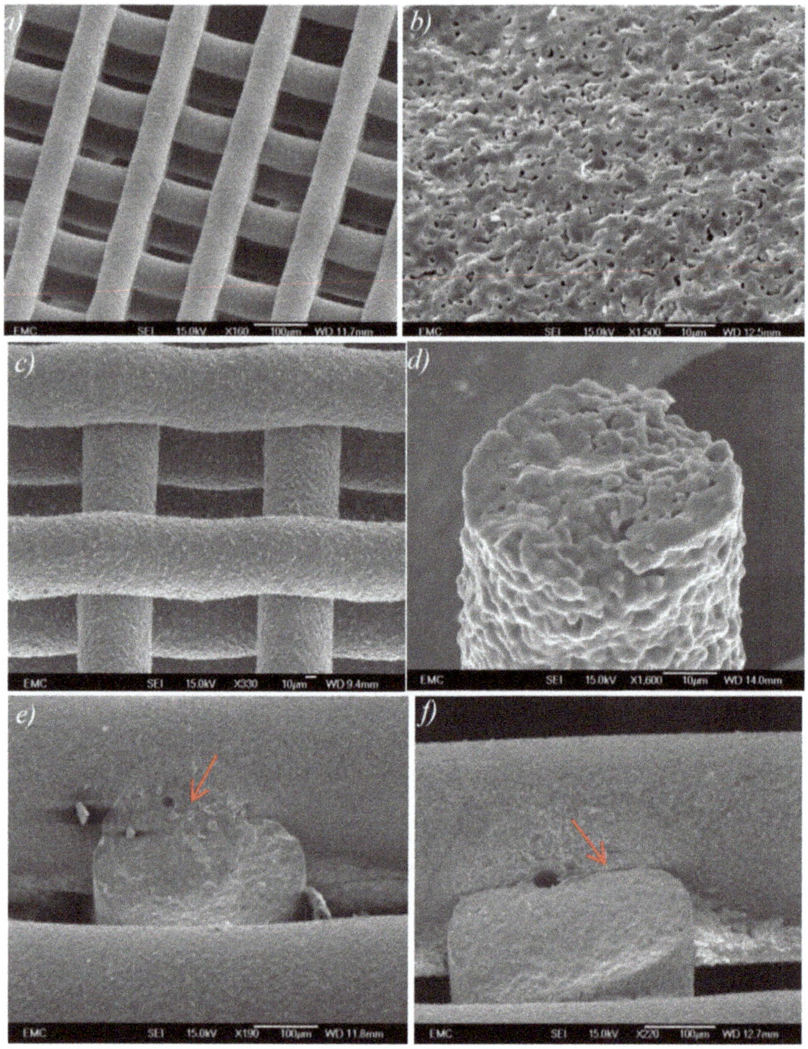

Figure 5.14. SEM images of the scaffolds: (a) sintered HA scaffold with uniform microstructure and microporosity, (b) high magnification fractured surface image of the HA filaments with micropores, (c) HA scaffold with filament deformation upon printing due to high solvent content in slurry, (d) high magnification image of a 50 μm HA filament, (e) strong inter layer bonding due to sufficient solvent content in the HA slurry and (f) weak layer bonding due to insufficient content of solvent. The red arrows in the panels (e) and (f) indicate the bonding between layers in the slurry. Reproduced from [50]. CC BY 4.0.

macroscopically porous, while the filaments were of microporosity, which could be well controlled during the EFF process, where the macropores were designed by the range of the filaments and the micropores of the filaments were associated with the sintering conditions [51].

Porosities of the HA scaffolds with different pore sizes and filament dimensions are summarized in table 5.1. The macroporosity was adjustable over a relatively wide span, ranging from 33% to 70%, while the macroporosity of the sample 400/250 was largely lower than that of the others, because it had thicker filaments. Theoretically, it is possible to increase the macroporosity up to 80%, by varying the diameters and configuration of the filaments.

Actually, the macroporosity should be appropriate for real applications. On one hand, too low a porosity meant high content of HA, so that less PEEK would be infiltrated and hence mechanical strength of the composites would be insufficient. On the other hand, the high porosity implied a low level of HA, leading to composites to be insufficient in the bioactive phase. The microporosity of the samples was in a relatively narrow range of 7%–10%, suggesting stable processing for the filaments.

Computed tomography (CT) analysis was used to characterize the PEEK/HA composites. Figure 5.15 shows a representative CT image, which was from an oblique cross section of the 3D-printed HA scaffolds, where air bubbles or voids were visible. Air bubbles inside the extruded filaments were generated during the preparation process of the slurries, such stirring rate, evaporation of solvent, and loading of slurry into syringe. In the PEEK/HA composites with HA ranging from 41–78 vol% HA, there was about 1.5 vol% air, as listed in table 5.2.

The content of HA was calculated by using the image processing software (VG Studio Max 2.1, Volume Graphics GmbH, Heidelberg, Germany). For samples with the same HA filament and pore sizes, the variation was less than 5%, as observed in table 5.2, implying reproducibility and accuracy of the technique. The pore size and formation of air bubbles were independent of each other. The sample with a pore size of 400 μm had the highest content of air (3.2 vol%), and a relatively large pore size of 700 μm corresponded to air bubbles of 5 vol%.

Figure 5.16 depicts 3D CT and SEM images of representative PEEK/HA composites. The air bubbles were present in the upper area of the images, corresponding to the bottom surface of the mold, where the air bubbles were

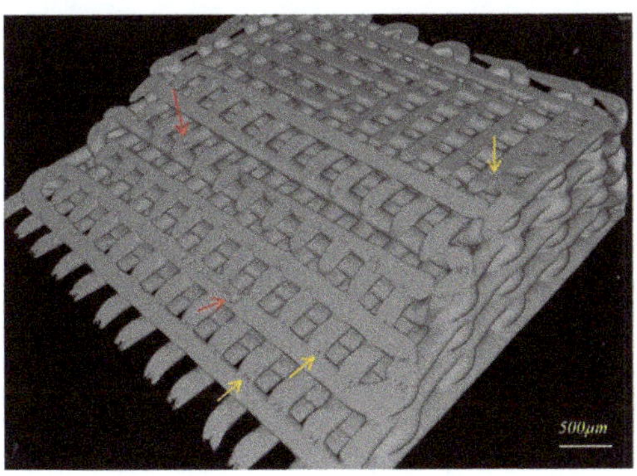

Figure 5.15. Computed tomography (CT) image of a representative 3D-printed HA scaffold sectioned by horizontal and oblique plans, microsized air bubbles inside the filaments (red arrows) and in the welding areas the filaments (yellow arrows). Reproduced from [50]. CC BY 4.0.

Table 5.2. CT analysis results of the PEEK/HA composites for compression test. Reproduced from [50]. CC BY 4.0.

HA scaffolds				
Filament (μm)	Pore (μm)	PEEK (Vol%)	HA (Vol%)	Air bubble (Vol%)
250	200	38.6	60.3	1.1
250	200	36.5	62.4	1.1
250	250	42.4	56.1	1.5
250	250	47.6	51.6	0.8
250	400	55.9	40.9	3.2
400	250	21.3	77.7	1.0
400	400	39.4	58.3	2.4
400	550	48.7	50.1	1.2

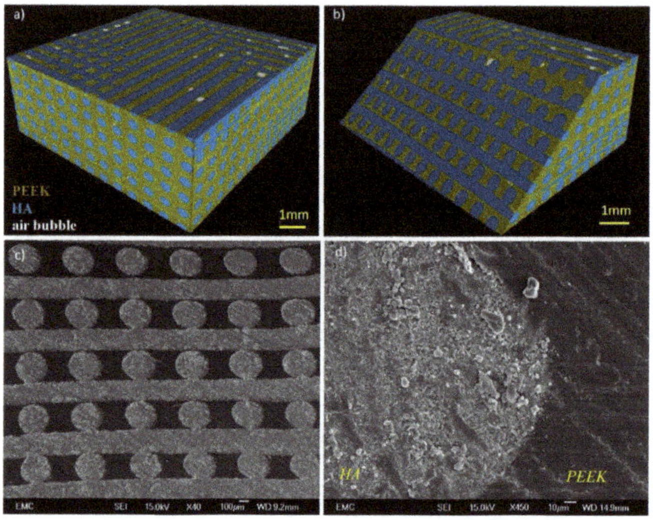

Figure 5.16. (a) 3D image constructed from CT scan of the PEEK/HA composites with HA filament sizes of 400 μm, pore size of 400 μm, HA content of 58%, air bubble of 2.4%, and PEEK content of 39% (total volume = 221 mm^3). (b) Oblique section image. (c) SEM image of a representative PEEK/HA composite, with scaffold size of 10 × 10 × 3 mm, HA scaffold filament size of 250 μm, pore size of 200 μm, molding temperature of 400 °C, dwelling time of 20 min, heating rate of 20 °C · min^{-1}, and static pressure of 0.39 MPa. (d) High magnification SEM image of the HA/PEEK interface. Reproduced from [50]. CC BY 4.0.

confined and blocked. The HA scaffold with a pore size of 200 μm was completely infiltrated with PEEK, in both the vertical and lateral directions, while the structural integrity and homogeneity of the HA network were well retained, as seen in figure 5.16(c). Due to the full interconnection between the bioactive HA and the PEEK matrix, such composites were advantageous for implant applications.

After complete dissolving of the HA filaments in HCl solution, the PEEK structures exhibited hollow channels, which were desirable for the attachment, infiltration, and proliferation of cells. For orthopedic tissue engineering, materials should have porous structures with strong interconnection infiltration and perfusion of nutrients, while the pore size should be appropriate for vascularization. Meanwhile, they should exhibit sufficiently high mechanical strengths. In this regard, the PEEK/HA composites made with the EFF printing process could be a candidate for such applications, owing to the unique porous structure, as observed in figure 5.17(a).

The surface characteristics of the HA filaments were copied onto the wall surface of the PEEK channels after the HA was etched, as revealed in figure 5.17(b). 3D surface profile maps of the surface of HA filaments and its replica in PEEK are depicted in figure 5.17(c, d). Although it is nearly impossible to fabricate filament-based structures with exactly the same channel interconnections, the filaments derived from the same slurry would have similar surface profiles. The surface of the HA filaments and the wall surface of the channel in the PEEK matrix exhibited almost the same average roughness of 0.4 μm, with standard deviations of 37 and 54 nm, respectively.

The incorporation of bioactive components into the PEEK matrix allowed the possibility to develop implants with desired biomechanical performances, while the mechanical strength of the implants could be sacrificed, due to the reduction in fracture energy and an increase in the brittleness of the materials [52–54]. Properties, including elastic modulus, compressive strengths, and micro-indentation hardness would be increased with the introduction of HA, whereas the tensile strength,

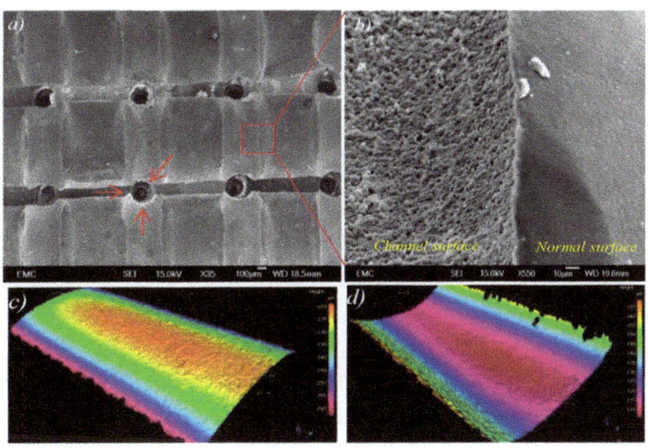

Figure 5.17. (a) Top view of the porous PEEK sample with interconnected channels, with red arrows indicating the crossed channels. (b) High-magnification SEM image from the surface of a channel in the porous PEEK. (c and d) 3D surface height maps of representative samples with 400 μm HA filaments and the channel produced with the same size HA filament: (c) profile of HA filament surface and (d) profile of PEEK channel surface. Reproduced from [50]. CC BY 4.0.

toughness, and strain to failure could be weakened. Therefore, the content of HA should be optimized to ensure mechanical strength for real applications.

The bioactive composites with 40 vol% HA and pure PEEK samples were subjected to compression tests in two directions, with the results depicted in figure 5.18(a). Sequential images of the PEEK/HA biocomposites in the compression test process in different directions are illustrated in figure 5.18(b). The compressive performance characteristics of the materials are listed in table 5.3, covering pure PEEK, PEEK/HA composites with 40 vol% HA, and human cortical bone [55].

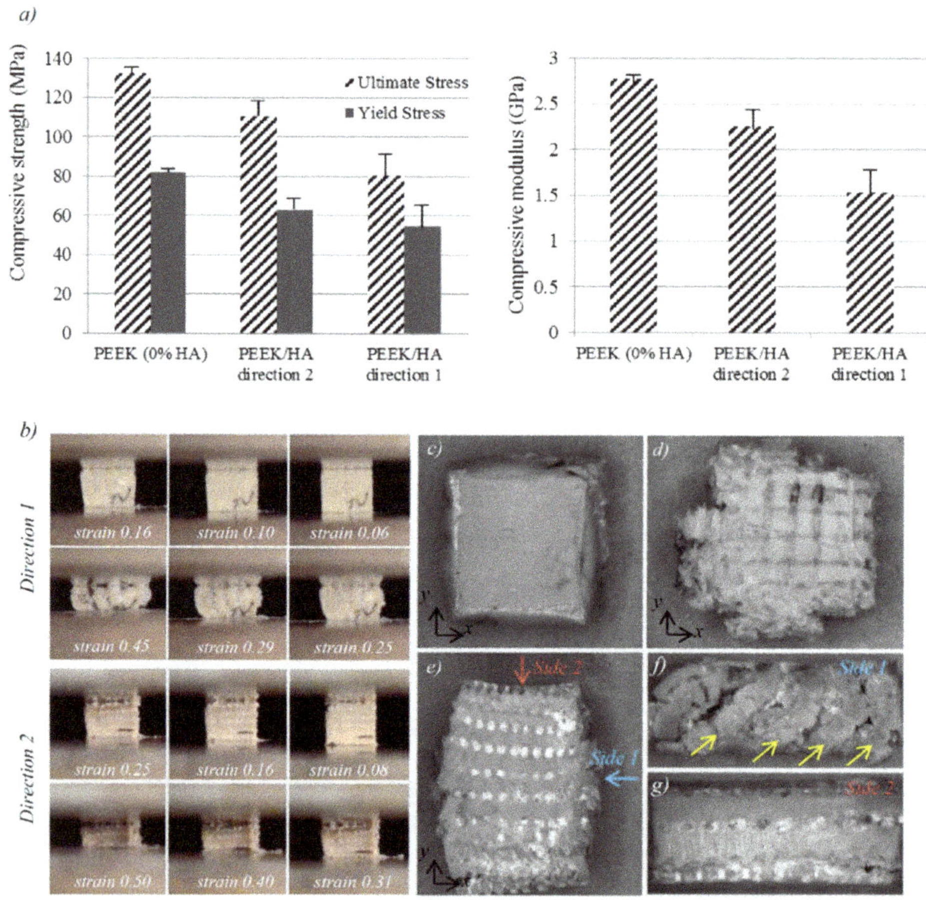

Figure 5.18. (a) Compressive elastic modulus, ultimate and yield strength of the pure PEEK and PEEK/HA composite samples compressed in different directions. (b) Sequential images of the PEEK/HA composites with dimension of 6 × 6 × 6 mm^3 with 40 vol% HA in the compression in different directions. (c–g) Optical images of the samples after compression test: (c) top view of the pure PEEK sample, (d) top view of the PEEK/HA compressed in direction 1, (e) top view of the PEEK/HA compressed in direction 2, and (f and g) front and side views of the PEEK/HA compressed in direction 2. The yellow arrows in the panel (f) show the plastic flow at pressures. Reproduced from [50]. CC BY 4.0.

Table 5.3. Compressive properties of pure PEEK, PEEK/HA composites with 40 vol% HA in direction 1, together with the human cortical bone [55].

	Humam cortical bone			
	Transverse	Longitudinal	Unfilled PEEK	PEEK/HA
Compressive moduli/GPa	N/A	4–22	2.8	1.6–2.5
Compressive yield strength/MPa	N/A	50–200	83	54–63
Compressive strength/MPa	50–70	70–280	134	80–110

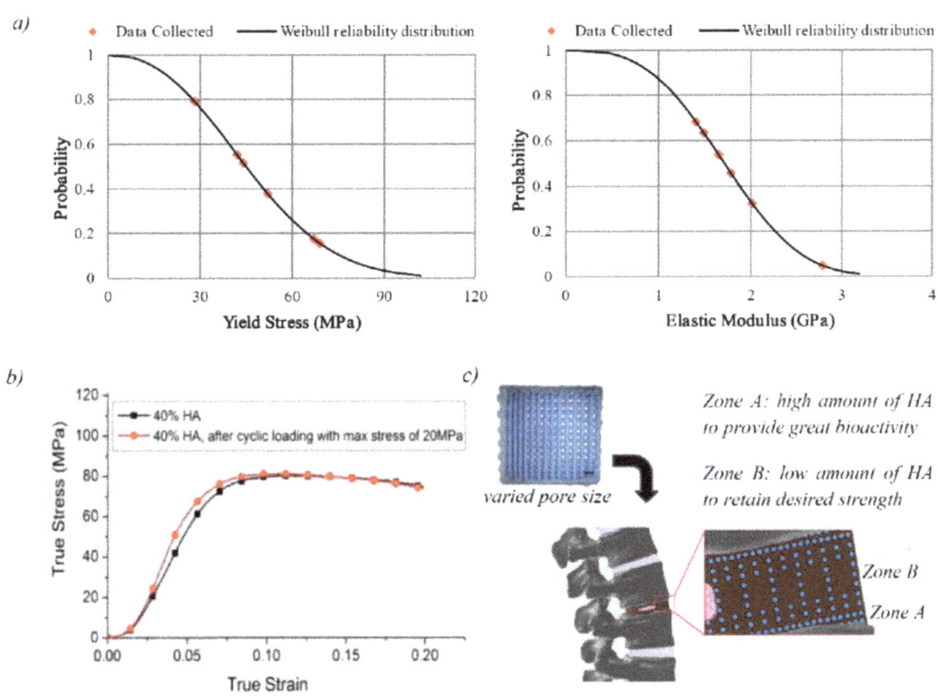

Figure 5.19. (a) Weibull reliability distribution profiles of yield stress and elastic modulus. (b) Compressive stress–strain plots of the PEEK/HA composites with 40 vol% HA before and after one million cyclic loading in direction 1. (c) Schematic diagram for the 3D-printed hierarchical HA scaffold with computer-controlled varied spacings suitable to make functionally graded PEEK/HA composites for spinal cage fusion. Reproduced from [50].

Weibull reliability distribution curves for yield stress and elastic modulus of the samples tested in direction 1 are illustrated in figure 5.19(a). According to the Weibull reliability curves, the compressive yield strength of the majority of the samples was higher than 20 MPa, while the elastic modulus was more than 0.8 GPa. Accordingly, samples with the same composition were tested for compressive–compressive cyclic loading, with maximum and minimum stresses of 20 and 2 MPa, for cycles up to one million.

Then, cyclic loading endurance of the biocomposites was evaluated by using the compression-to-failure test. Experimentally, compressive strengths of the cyclic loaded samples and the normal samples were almost the same, as observed in figure 5.19(b). The cyclic loaded samples had similar yields and ultimate strengths, while the yield strain was relatively smaller, but elastic modulus was higher. It was reported that HA particles with spherical morphology in PEEK/HA composites tended to detach from the PEEK matrix at cyclic loadings, owing to the weak interfacial adhesions [56]. Therefore, increasing interfacial interaction between PEEK and HA particles deserves to be further explored.

Fine ceramic lattices, with spatial resolution of <100 μm, precise dimensions and intricate hierarchical structures, were obtained by using the EFF technology [57]. The examples were calcium phosphate lattices with three structure levels, including submicron pores to ensure enhanced cell–surface interactions, pores on the scale of tens of microns to facilitate bone ingrowth, and corridors with sizes of hundreds of microns for vascularization. With controlled porous structures on the three scales, the ceramic lattices could be used for biological, mechanical, and geometrical applications.

The hydroxyapatite ($Ca_{10}(PO_4)_6(OH)_2$, HA, Grade P221 S, Plasma Biotal UK) and β-tricalcium phosphate ($Ca_3(PO_4)_2$, TCP, Grade P228 S, Plasma Biotal UK) powders were calcined at 900 °C and milled in water for 4 days. Powders with different ratios of HA/TCP were dispersed in a propan-2-ol solvent (GPR, VWR, UK), with the aid of ultrasonication (IKA U200S, IKA Labortechnik SATAUFEN, Germany) for 15 min. Polyvinyl butyral (PVB, grade BN18, Whacker Chemicals, UK) was used as a binder, together with polyethylene glycol (PEG, $M_{Wt} = 600$, VWR, UK).

Preliminary experiments were performed to establish the relationship between the composition of slurries and extrusion efficiency, with fine nozzles having diameters of 50–300 μm. Specifically, 4.3 g PVB and 1.4 g PEG600 were dissolved in 20 ml propan-2-ol to form a solution. Meanwhile, a 24.3 g mixture of 75 wt% HA and 25 wt% TCP was dispersed in 50 ml propan-2-ol through ultrasonication. The two systems were mixed to form a ceramic slurry, with a solid loading level of 60 vol%. These mixtures were thoroughly blended with a roller and zirconia media for 12 h. They were then heated to evaporate the solvent, in order to obtain slurries suitable for the extrusion experiment.

The extrusion machine had two micro-extruder screw lines, driven by independently controlled microstepper motors (ACP&D, Ashton-under-Lyne, UK), at 50 000 steps/rev, with a 64–1 reduction box to drive 1 mm pitch ball screws (Automotion, Oldham, UK). The drives were monitored by using load cells (Flintec, Redditch, UK), which also served as alarms for overpressures. The three-axis table (Parker Hannifin, Dorset UK, supplied through Micrometrics, Braintree, UK) and was driven with a 6k4 motion controller and Labview program.

The printing platform was enclosed to control the drying by regulating solvent evaporation, which was important for the filaments with diameters of <80 μm. An optional direct variable ratio drive was enabled between the printing table and extruder with the computer control system. The *XY* plane could be moved at a speed of 4 mm · s^{-1}. The ratio of the speed of extrusion to that of the *XY* motion was larger than the square of the ratios of the diameters of the extruder barrels to those of the extrudate, due to swelling of the extrusion dies.

The slurries were filled into a syringe (HGB81320 1 ml, Hamilton GB, Carnforth, UK), which was assembled with the extrusion dies with diameter of 50 μm (Model INZX0600060A, Lee Products, Gerrards Cross, UK) and sapphire water-jet cutting nozzles (Types 1708, 1710, 1715, Quick-OHM Kupper GmbH, Wuppertal, Germany), with diameters of 80, 100, or 150 μm. The lattice dimensions of the green bodies were retrodicted, by considering the shrinkage of the samples during sintering, which was 10.2% for sintering temperature of 1200 °C and designed dimension of the final ceramics. The dried green bodies were sintered at 1200 °C for 5 h, at a heating rate of 5 °C · min^{-1}.

HA has similar characteristics to carbonated apatite in bones, with insufficient in-vivo resorption [58–60]. TCP is highly degradable in animal bodies, thus offering an environment with high concentration Ca^{2+} and PO_4^{3-} for bone generation, whereas the degradation is too fast to induce collapse of the scaffolds before osteogenesis [61–64]. Therefore, an HA/TCP combination has been used to develop macro-porous structures for implantation, by adjusting the HA/TCP ratio to achieve ceramic-induced osteogenesis [65, 66]. The degradation rate is dependent on several factors, including the phase composition, specific surface area, porosity, and pore size [62]. Calcium phosphate ceramic architectures with a proper combination of macro- and micropores are able to induce the formation of bone in soft tissues. Therefore, the design and fabrication of multiscale HA/TCP structures are important for bone repairing.

The HA and TCP powders had compositional specifications meeting the requirements of surgical implantation standards (ASTM F1088 andASTMF1185-88), while the content of metals was well below the limited compositions. The powders both had particle sizes of <10 μm, with slight agglomeration. The values of D_{50} are 1.7 and 1.5 μm for HA and TCP, respectively. The as-received powders could be used to prepare slurries for extrusion of filaments through nozzles with a diameter of 80 μm. However, nozzles with a diameter of 50 μm would experience blockage, thus requiring milling and filtration for the powders.

The optimal solid loading level was 60 vol% in dry mass and 40 vol% in slurries. Slurries with too-high solid loading levels resulted in filaments without plastic deformation capability during the extrusion, so that they were brittle after drying. Also, the slurries should be extruded at higher pressures and dried more quickly, thus having inadequate weld formation. As the solid loading level was too low, the slurries exhibited low yield strengths, could not maintain shapes, and tended to slump between the printed filaments. For different powders and nozzles with different diameters, the solid loading of the slurries should be controlled, to optimize the extrusion process, welding effect, and drying rate.

Dilatency is an effect of slurries made from low molecular weight fluids and coarse powders, which should be prevented. In this regard, the solvent selected should allow for both drying of individual layers and welding of filaments in between layers, while the content is controlled for the slurries to ensure yield stresses that are sufficient for preventing the collapse of the filaments or lattices [67, 68].

Figure 5.20 shows images of the lattices printed through extrusion by controlling the movement of the machine and properties of the slurries, demonstrating their outstanding structural uniformity. The lattice in figure 5.20(a) consisted of 80 μm diameter filaments and 70 μm interstices (lattice 1), while that in figure 5.20(b) featured 150 μm diameter filaments and 50 μm interstices (lattice 2). Therefore, it could be concluded that structural parameters of the lattices were adjustable to allow designing structures with desired mechanical performances.

Figure 5.21 shows an example of lattice with nonuniform distribution of filaments. Bone formation is facilitated in microporous structure, while the structure has relatively low mechanical strength. By using the nonuniform distributed filaments, it is possible to produce structures that meet the requirements of both biological activity and mechanical strength. The larger channels with a size of 300 μm in the porous lattice offered space for blood vessels to grow deep into the scaffolds to carry nutrients and remove wastes, while the smaller pores with a size of

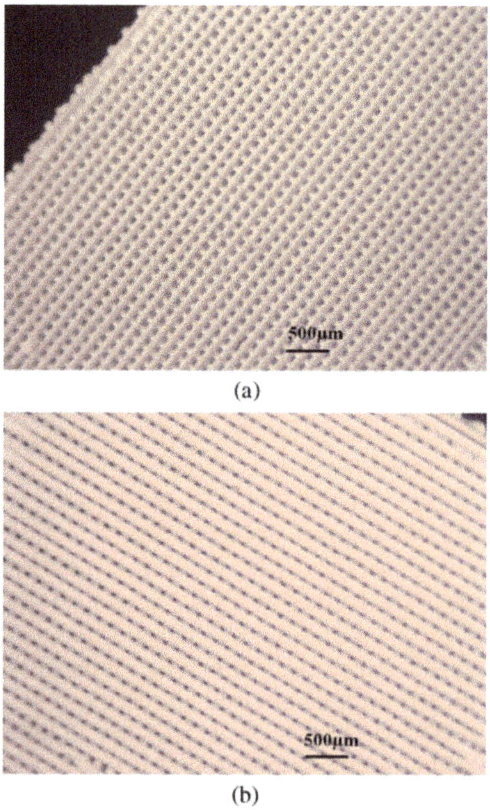

Figure 5.20. Optical images of fine ceramic lattices with 0–90 lay-up and 10 layers: (a) 80 μm filament and 70 μm interstices at $V_s = 0.419$ and (b) 150 μm filament and 50 μm interstices at $V_s = 0.589$. [57] John Wiley & Sons. Copyright © 2006 Wiley Periodicals, Inc.

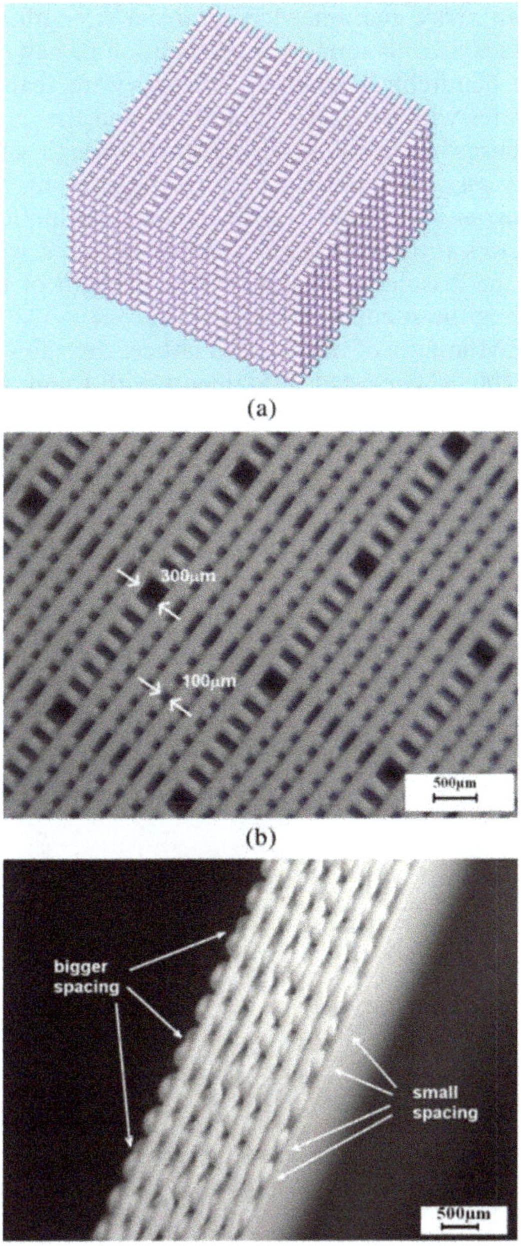

Figure 5.21. Optical images of the hierarchical lattices: (a) CAD model, (b) top view of green body lattice with 100 and 300 μm channels distributed among the lattice with 150 μm filaments, 16 layers, $V_s = 0.416$, and (c) side view of the lattice. [57] John Wiley & Sons. Copyright © 2006 Wiley Periodicals, Inc.

100 μm present as an array guaranteed mechanical strength of the scaffolds and sufficiently high surface area to support the formation of new bones.

Meanwhile, three hierarchical levels of porosity were realized in the extruded lattices. The large pores with diameters of 100 μm and the intermediate ones with sizes of tens of microns were accurately printed through computerization. The micron-sized porosity was generated by adjusting the sintering parameters and using pore agents in the slurries. After sintering, the filaments exhibited surfaces with high roughness and pore sizes at the micron and submicron scales, which are beneficial to bone growth. Both the dissolution behavior and strength of the lattices could be designed by controlling the submicron porosity.

Representative SEM images of the sintered lattices are shown in figure 5.22. The lattices sintered at 1200 °C consisted of filaments with rough and porous surfaces, whereas the interior porous structure was highly uniform. The welds formed in the slurry were retained after sintering, thus establishing strong bridges between the layers, as illustrated in figure 5.22(a). The porosity of the filaments was decreased with increasing sintering temperatures. After sintering at temperatures of ⩾1250 °C, the mechanical strength was increased, while the surface porosity was sacrificed.

(a)

(b)

Figure 5.22. Lattice produced with an 80 μm nozzle and 70 μm interstices after sintering at 1200 °C: (a) SEM image of fractured surface showing weld coherence and (b) SEM image showing surface porosity of the filaments. [57] John Wiley & Sons. Copyright © 2006 Wiley Periodicals, Inc.

Because the filaments had relatively small diameters, the gaseous products due to the burning of the polymer components in the green bodies could escape easily, while having no damage to the filament and lattice structures. Therefore, a high heating rate during the debinding process was not necessary for such printed structures before sintering, thus shortening the production time. Such achievements are believed to be applicable for other materials and applications.

Structures based on ferroelectric barium titanate ($BaTiO_3$ or BT) with different geometries were developed by using robocasting, a 3D printing technique through extrusion of shear-thinning pastes [69]. In this work, an environmentally friendly aqueous paste was developed, while the designed structures were layer-by-layer printed out. The sintered items had a relative density of 97%, with flexural strength of 40 MPa and hardness of 3 GPa, approaching the performances reported in the open literature [69]. Meanwhile, the piezoelectric coefficient (d_{33}) of the sintered products was 200 pC · N^{-1}.

Commercial BT powder (CAS-No: 12 047-27-7, D_{100} >3 μm, 99%, Sigma Aldrich, Germany) was dispersed in distilled water, at a solid loading level of 48.9 vol%, with the aid of 6.5 vol% Darvan 821 A (R.T. Vanderbilt Co. Inc., USA) as the dispersing agent. The suspension was thoroughly mixed at 2000 rpm for 2 min. Then, 3.4 vol% hydroxypropyl methylcellulose (Methocel F4M, Dow Wolff Cellulosics, Germany) as the binder and 2.9 vol% polyethyleneimine (PEI, Sigma Aldrich, Germany) as the coagulant were introduced, to further improve the printing efficiency of the slurries. After that, the suspension was further homogenized for 2 min at 2000 rpm, followed by degassing at 2200 rpm. Meanwhile, slurry with 5 vol% BT platelets (Lot: 020721, Entekno, Turkey) was prepared for comparison.

The slurries were used for the printing experiment with a modified multi-material Stone Flower printer (Stoneflower 3.0, Germany), with attachment of a volumetric print head (Vipro-Head 5, ViscoTec, Germany). Structures with various geometries, including bending bars, auxetic and hexagonal dimensions were printed, with nozzles with an inner diameter of 580 μm. The samples to be printed were immersed in oil bath (Hydraulic oil MR0 VG2, Fuchs Renolin, Germany) to avoid quick drying and prevent the nozzles from clogging. The as-printed samples were dried at room temperature, debinded at 700 °C for 2 h, and finally sintered at 1400 °C for 6 h at a heating rate of 5 °C · min^{-1}.

Photographs of representative printed samples with different geometries before and after sintering are illustrated in figure 5.23(a). The sintering shrinkage was about 17%. SEM images of the sintered samples are depicted in figures 5.23(b)–(d). Typical microstructures of BT ceramics with domain patterns are revealed in the SEM images [70–72]. The BT ceramics exhibited relatively homogeneous microstructure and possessed an average grain size of 30 μm. Mechanical and electrical properties of the printed BT ceramics were comparable with literature data [73, 74].

Specifically, the auxetic honeycomb structures exhibited a negative angle of −25°, while the hexagonal ones displayed a positive angle of +25°. The piezoelectric coefficients (d_{33}) of the auxetic and the hexagonal structures were 195.4 pC · N^{-1} and 202.4 pC · N^{-1}, respectively, which were close to those of the bulk BT sample made with 3D printing. However, the dielectric constant of the two structures was

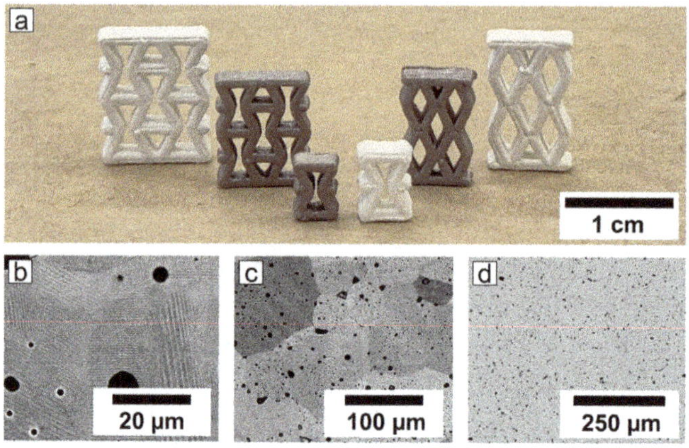

Figure 5.23. (a) Photographs of representative printed structures in green before (white samples) and after sintering (brown samples). (b–d) SEM images of the sintered BT ceramics. Reproduced from [69]. CC BY 4.0.

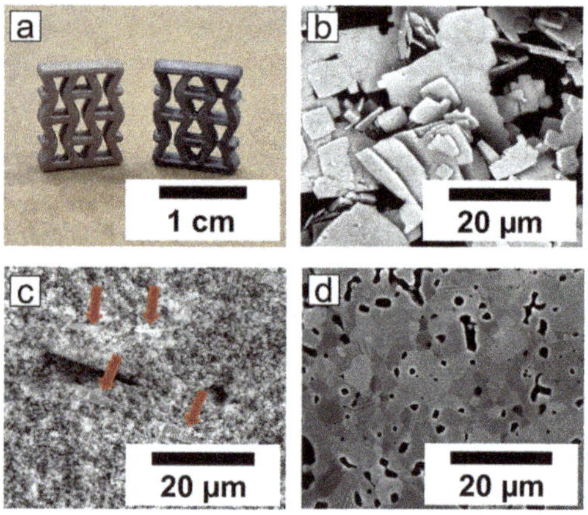

Figure 5.24. (a) Photographs of the auxetic structures with (right) and without (left) the addition of platelets. (b) SEM image of the pristine BT platelets. (c) Cross-sectional SEM image of the green body with the addition of the BT platelets (marked with arrows) embedded in the BT powder matrix. (d) SEM image of BT ceramics after sintering at 1400 °C and thermal etching. Reproduced from [69]. CC BY 4.0.

relatively low, with values of 858.8 and 973.6 for the auxetic and the hexagonal ones, respectively. For piezoelectric ceramics, the reduction in piezoelectric performance is usually caused by the increase in porosity, simply because of the entrapped air bubbles inside the piezoelectric matrix [75, 76].

Piezoelectric properties of piezoelectric materials can be enhanced by introducing anisotropic particles [77–80]. Aligned anisotropic particles could be formed under the action of shear forces, thus resulting in textured microstructures through heat treatment. During robocasting with nozzles, high shear force was induced, which in turn triggered the formation of anisotropic particles [81–83]. Figure 5.24(a) shows an

SEM image of the BT platelets, with lateral sizes of 2–40 μm and thicknesses of 0.5–1 μm. Figure 5.24(c) depicts fractured surface SEM image of the green body, clearly indicating the presence of the BT platelets in the small sized BT matrix.

A representative SEM image of the BT ceramics with the BT platelet particles after sintering is demonstrated in figure 5.24(d). By comparing the results in figures 5.23(b) and 5.24(d), it was revealed that the grain size was different in the samples with and without the BT platelets. The presence of the BT platelets hindered the grain growth of the BT ceramics, corresponding to a decrease in the average grain size from 30 to 3 μm, by one order of magnitude. The porosity of the two groups of samples was similar. While the Young's modulus was slightly reduced, the flexural strength was increased to about 60 MPa, by a magnitude of about 50%. Unfortunately, there was a pretty large reduction in piezoelectric coefficient with the addition of the BT platelet particles, which could be ascribed to the reduction in grain size, caused by the presence of bismuth in the BT platelet particles [84].

The EFF process combined with high temperature sintering was used to fabricate $YBa_2Cu_3O_{7-x}$ (YBCO) ceramic superconductors [85]. YBCO ceramics with a relative density of 93% were derived from the EFF printed green bodies that were sintered at 940 °C for 60 h. Phase compositions of the YBCO ceramics were closely related to the sintering temperature and time duration. A high critical transition temperature ($T_C = 92$ K) and promising magnetic levitation ability were achieved in the samples processed at optimal sintering conditions.

YBCO powder was synthesized by using the solid-state reaction process [86–88]. High-purity powders of $BaCO_3$, CuO, and Y_2O_3 (99.99%, Sigma–Aldrich) were mixed according to the stoichiometric composition, as described in figure 5.13. The powders were mixed with polyethylene glycol (PEG, Sigma–Aldrich), Solsperse 20 000 lubrizol), polyvinyl alcohol (PVA) and DI water as ceramic precursors to form slurries for the EFF printing experiment.

To prepare an extrudable paste with a high solid loading of YBCO precursor powder, PVA powder was dissolved in DI water at a weight ratio of 1:8, leading to a binder solution. Then, PEG-400 as plasticizer and Solsperse 20 000 as dispersant, with a binder/plasticizer/dispersant weight ratio of 15:5:4, were dispersed into the PVA solution, forming a dispersed solution. Finally, the YBCO precursor powder was dispersed into the polymeric solution, at a solid loading level of 71.5 wt%, resulting in homogeneous slurries. The slurries were filled into 10 ml plastic syringes, in order to prevent the evaporation of the solvent, as so not to induce any change in viscosity.

A 3Dison Multi printer equipped with a ROKIT. Inc. extruder was used for EFF printing, where the homogeneous slurry in the plastic syringe was extruded through a nozzle, as demonstrated in figure 5.25. The extruder could freely move within the *XY* plane, thus being able to print any patterns according to the predesigned CAD model, while extruding the slurries onto the surface of the substrate. As one layer was printed, the platform was lowered down in the *Z*-axis by a distance to be the same as thickness of the printed layer, with the process to be repeated until the desired green bodies were printed out.

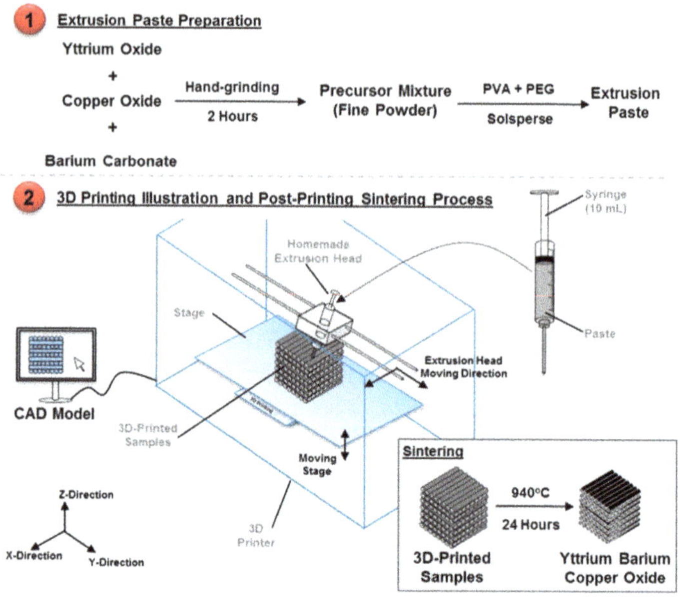

Figure 5.25. Schematic diagrams: (i) preparation of extrusion slurry and (ii) extrusion and post-sintering processes. Reprinted from [85], Copyright (2016), with permission from Elsevier.

The as-printed green bodies were dried in air at 70 °C. The dried samples were sintered at temperatures of 600 °C–1100 °C for 20 h, at a heating rate 1 $°C \cdot min^{-1}$. They were also sintered at 940 °C for different time durations, including 5, 20, 40, and 60 h. The sintered samples were then further annealed at 550 °C for 10 h for the purpose of oxygenation.

Figure 5.26 shows photographs and descriptions of the YBCO samples derived from the slurries with solid loading levels of 50–76 wt%. Generally, the behaviors of a polymer-ceramic slurry are of close relation to the contents of the polymer, ceramic powder, and solvent. At low ceramic solid loading levels, the slurries exhibited low yield strength, so that their shapes could not be maintained, with a tendency to slump sooner or later. However, as the ceramic loading was over 71.5 wt%, high pressures were required to extrude the slurries or even the slurries could not be extruded out the nozzles. Therefore, 71.5 wt% was the optimal ceramic loading level, in terms of continuous extrusion, and hence effective printing.

The green bodies had a relative density of 45%, which was increased to 50% after sintering at 600 °C for 20 h, owing mainly to the removal of the polymer additives and complete evaporation of solvent. As the sintering temperature was increased from 800 °C to 1100 °C, the relative density was raised from 63% to 94%. In other words, the sample sintered at 1100 °C was nearly fully densified, due to the entirely interparticle bonding promoted during the liquid phase sintering process.

As the sintering temperature was fixed at 940 °C, the relative density of the samples was increased almost linearly from 70% to 91%, when the time duration was prolonged from 5 to 60 h. Usually, it is aimed to achieve full densification for

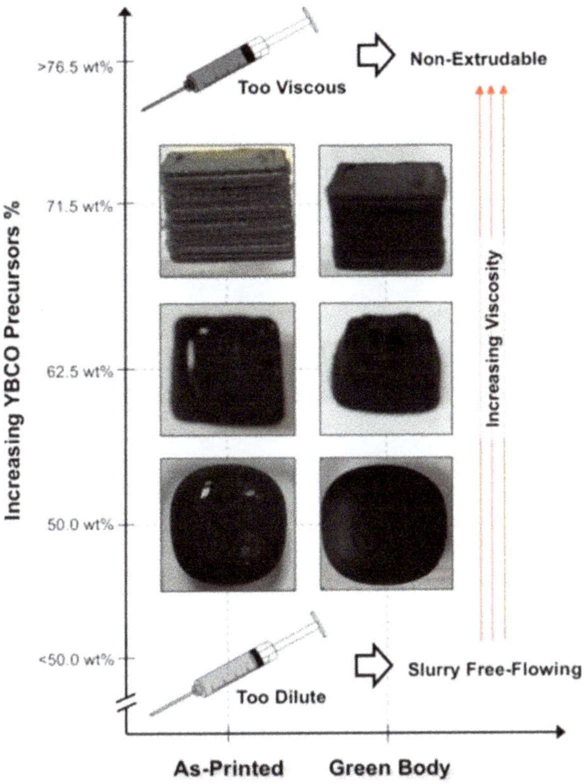

Figure 5.26. Photographs of the slurries demonstrating influence of the weight content of the YBCO precursor powders in the binder solution. Reprinted from [85], Copyright (2016), with permission from Elsevier.

processing ceramics, but it is not necessary for superconductor ceramics. For example, it was reported that porous YBCO ceramics with a relative density of about 88% offered the highest critical current density, because of the full oxygenation and clean grain boundaries [89].

Figure 5.27 shows SEM images of the YBCO ceramic samples sintered at different sintering temperatures for different time durations. The YBCO precursor powder was uniform in particles, with sizes in the range of 1–2 μm, as seen in figure 5.27(a). SEM images of the YBCO ceramic samples sintered at 800 °C, 940 °C and 1100 °C for 20 h are depicted in figures 5.27(b)–(d), indicating that the porosity was gradually reduced with increasing sintering temperature, while the grain size increased. At a sintering temperature of 940 °C, the densification was enhanced with increasing sintering time, as observed in figures 5.27(e)–(h). This observation suggested that sufficiently long sintering time could be used to achieve a high density. The elemental composition of the YBCO precursor powder was confirmed by the EDX analysis result, as demonstrated in figure 5.27(i).

Phase compositions of all the samples were characterized by using XRD. The sample sintered at 600 °C consisted of precursor powders of Y_2O_3, $BaCO_3$ and CuO,

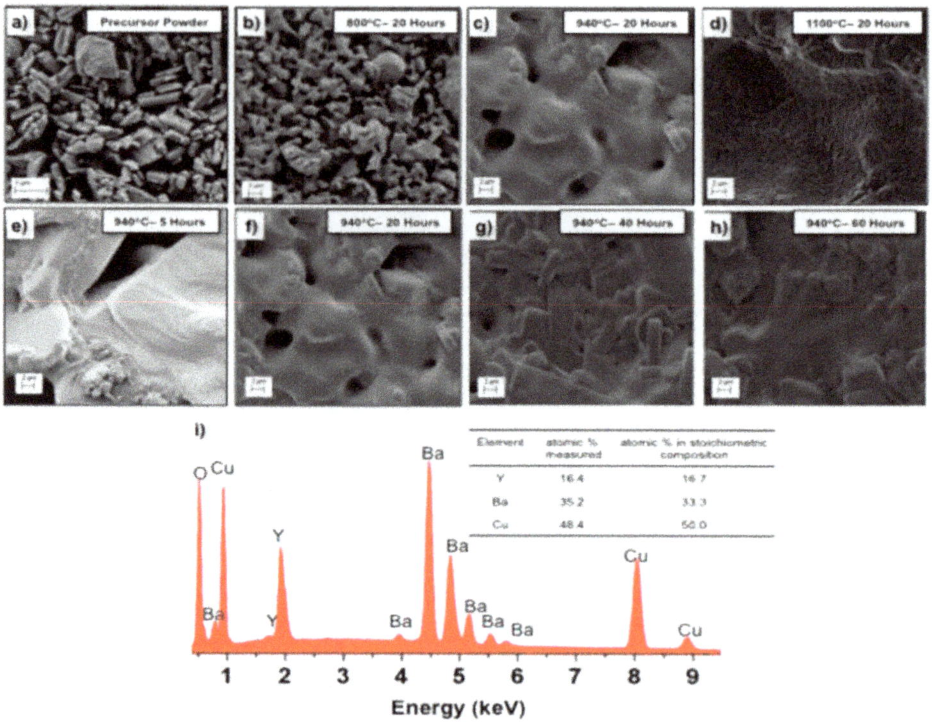

Figure 5.27. SEM images of the YBCO ceramic samples sintered for 20 h at different temperatures (scale bars = 2 μm for all samples): (a) YBCO precursor powder, (b) 800 °C, (c) 940 °C, and (d) 1100 °C. SEM images of the YBCO ceramics sintered at 940 °C for different time durations: (e) 5 h, (f) 20 h, (g) 40 h and (h) 60 h. (i) EDX spectrum for the precursor powder shown in panel (a), with the inset table listing the chemical composition. Reprinted from [85], Copyright (2016), with permission from Elsevier.

while the YBCO 123 orthorhombic phase was not formed. Even after sintering at 800 °C, the peaks from the precursor powders were still detectable, although the YBCO orthorhombic phase had been formed. In fact, phase pure orthorhombic YBCO was obtained after the sample was sintered at 940 °C. As the sample was sintered at 1100 °C, the YBCO 123 phase partially decomposed to Y_2BaCuO_5 (Y211) phase and liquid phases, such as $BaCuO_2$, because the sintering temperature was over the peritectic temperature of $T_P = 1015$ °C [90, 91]. The samples sintered at 940 °C were all of single orthorhombic phase, regardless of the sintering time duration.

$Li_2Si_2O_5$ ceramics, prepared by using subtractive manufacturing (SM) and additive/robocasting (AR), were compared, in terms of phase composition, microstructure, and mechanical properties [92]. For the AR group, $Li_2O_5Si_2$ powder was blended with ammonium polyacrylate, hydroxypropyl methylcellulose and polyelectrolyte to form slurries. The SM group samples were obtained by using the conventional ceramic fabrication process. The samples with pellet shapes were thermally treated at 840 °C for crystallization. Biaxial flexural strength (BFS) and

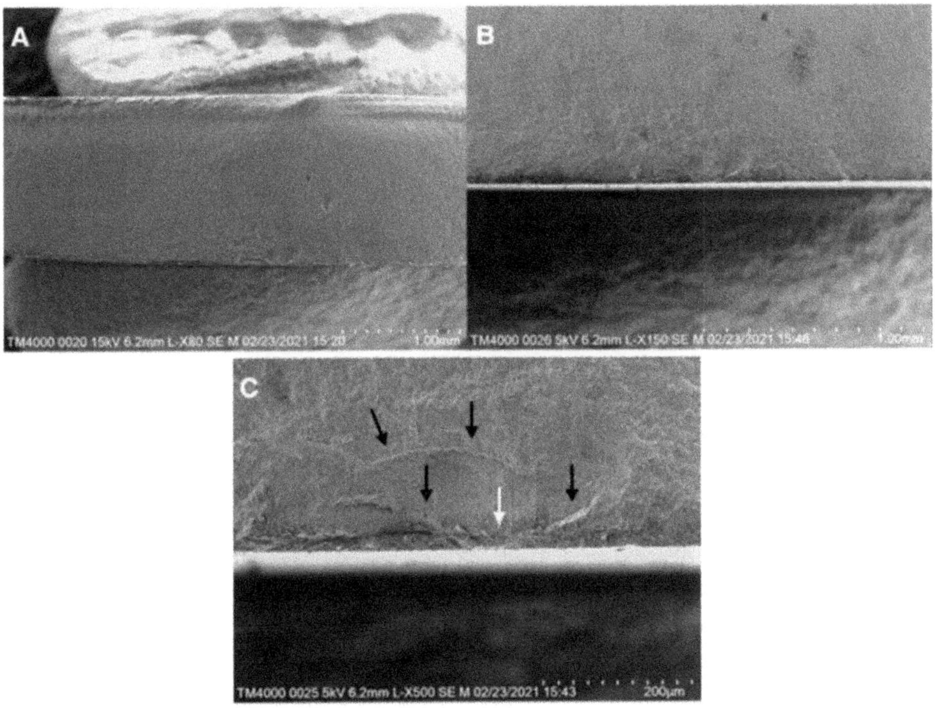

Figure 5.28. Cross-section SEM images of the SM group of samples after bending strength test at different magnifications: (A) 80×, (B) 150×, and (C) 500×. The white arrow points to the fracture origin, while the black arrows indicate arrest lines. Reprinted from [92], Copyright (2023), with permission from Elsevier.

hardness of the AR groups were 120.02 ± 34 MPa and 4.07 ± 0.3 GPa, while those of the SM group were 325.1 ± 64 MPa and 5.63 ± 0.14 GPa, respectively.

Cross-sectional SEM images of representative samples for the two groups are illustrated in figures 5.28 and 5.29. From the enlarged SEM images, it was observed that the origins of the fracture and directionality were clearly present. Meanwhile, the AR group samples exhibited a more porous microstructure, whereas the SM group samples exhibited more homogeneous and dense microstructures. In the AR group, the cracks were originated from pores in the central area of the samples. This observation suggested that more efforts should be made to improve the performances of the ceramics made by using 3D printing technologies.

Boron carbide (B_4C) with complex geometries was fabricated by using the robocasting technique, combined with spark plasma sintering (SPS) [93]. The as-printed B_4C green bodies had a near net-shape, as long as the inks possessed suitable rheological properties. After SPS processing, B_4C ceramics with promising mechanical strengths were obtained. To prepare the ceramic inks with 40 vol% B_4C, 1 wt% synthetic polyelectrolyte (Produkt KV5088, Zschimmer-Schwarz) as dispersant was dissolved in DI water at room temperature and at a neutral pH value, with constant stirring (ARE-250, Thinky) at 700 rpm for 10 min.

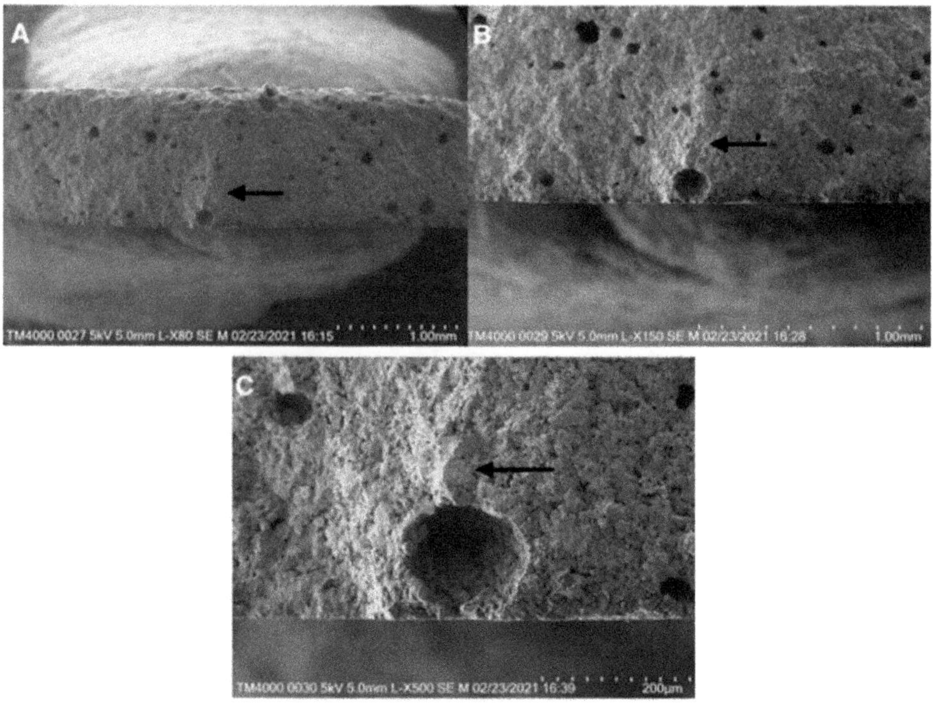

Figure 5.29. Cross-sectional SEM images of the AR group of samples after bending strength measurement at different magnifications: (A) 80×, (B) 150×, and (C) 500×. Black arrows indicate the crack beginning from inside the pores. Reprinted from [92], Copyright (2023), with permission from Elsevier.

Then, B_4C powder (Grade HD 20, H.C. Starck, $D_{10} = 0.1–0.36$ μm, $D_{50} = 0.3–0.6$ μm, $D_{90} = 0.9–1.5$ μm) was dispersed into the solution, with stirring at 800 rpm for 10 min. After that, 7 mg · ml^{-1} methylcellulose (Methocel F4 M, $M_w = 3500$ g · mol^{-1}, 5 wt%, Dow Chemical Company) was added to modify viscosity of the suspension, with stirring at 1000 rpm for 10 min. Finally, 4 vol% polyethylenimine (PEI) as flocculant (10% w/v in water, Sigma–Aldrich) was introduced to facilitate gelation of the suspension. The ink was ready for printing, after stirring at 12 000 rpm for 2 min and 700 rpm for 7 min.

Photographs of the B_4C green bodies and sintered ceramics with various complicated geometries are presented in figures 5.30(A) and (B). The square and hexagonal plates could be used to construct multi-segmented armor panels. The gears with different dimensions and shapes could find applications in the field of tribology. The conical nozzles with various diameters and tip sizes could be utilized for blasting and water-jet cutting, and the square and cylindrical porous elements could be applicable for application as filters or catalytic supports.

The printing parameters include a conical nozzle with a tip diameter of 410 μm and scanning speed of 20 mm · s^{-1}, corresponding to a flow rate of about 2.64 mm^3 · s^{-1}.

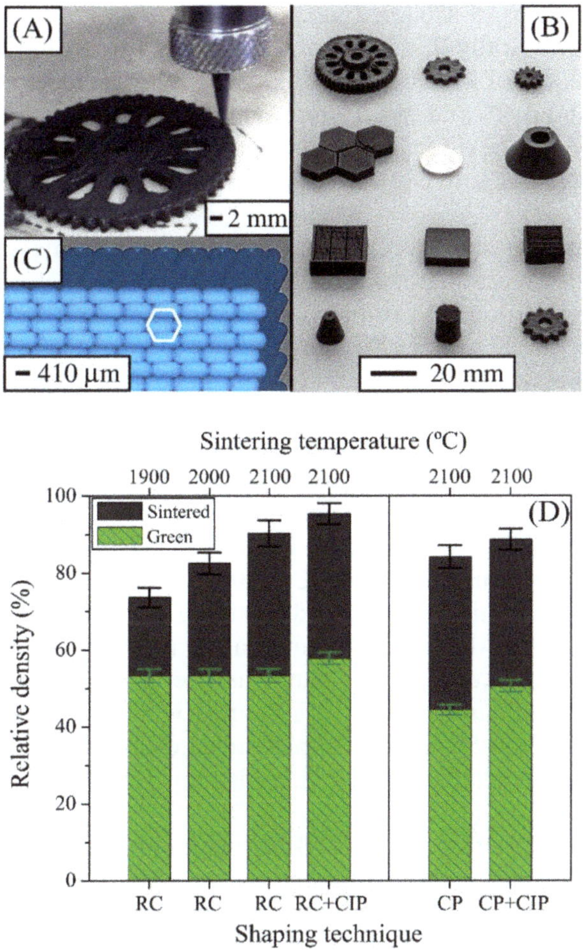

Figure 5.30. (A) A photograph captured *in situ* during the 3D printing of a circular gear. (B) Photographs of representative B_4C green parts shaped by RC and the dried ones. (C) Schematic diagram of a simple CAD model with parallel raster pattern. (D) Relative densities of the B_4C green bodies and sintered ceramics made with different shaping procedures and sintered at different temperatures. Reprinted from [93], Copyright (2018), with permission from Elsevier.

The items were printed according to parallel raster patterns, with an adjacent rod spacing of 342 μm, layer thickness of 356 μm, and in-plane shifting between adjacent layers of 205 μm, so that the filaments were hexagonally stacked, according to the designed structures, as observed in figure 5.30(C).

Conical structures were developed with the circular raster pattern, which consisted of three concentric rings that have the same rod spacing and layer thickness as those stated earlier, whereas in-plane layers were shifted following the designed angle of the cone. Porous samples were processed based on tetragonal meshes with rod spacing of 820 μm and layer thickness of 328 μm. After printing, all green bodies were naturally dried for two days, with lateral shrinkage of about 3% and vertical shrinkage of about 5%. If cold isostatic pressing was applied, the two shrinkages were about 4% and 7%. These slight differences in shrinkage between the in-plane and out-of-plane directions was ascribed to the shape distortion related to the printing process, due to the effect of gravity and overflow of excessive ink in the lateral directions.

The bulk density of the samples was measured according to their mass and dimensions in the form of square. Accordingly, relative densities of the dried samples and the cold isostatic pressed samples were about 53% and 58%, respectively, as observed in figure 5.30(D). In comparison, the disc samples made of the B_4C powder by using the dry pressing at 50 MPa and cold isostatic pressing at 200 MPa exhibited relative densities of 45% and 51%, respectively. Therefore, the B_4C green bodies had higher relatively density than their conventionally processed counterparts. This was because the printing inks were prepared by using a wet processing route, thus favoring the close packing of the B_4C particles.

The samples were SPS processed (HP-D-10, FCT Systeme GmbH) in vacuum of about 3 Pa, at temperatures of 1900 °C, 2000 °C, and 2100 °C, for 5 min, at a heating rate of 100 $°C \cdot min^{-1}$. Relative densities of the sintered samples were in the range of 74%–90%, over the sintering temperature range. With respect to the CAD models, the B_4C ceramics experienced lateral and vertical shrinkages in the ranges of 9%–12% and 17%–21%, respectively. Combined with cold isostatic pressing, the relative density reached 95% in the samples sintered at 2100 °C. Similarly, the 3D-printed B_4C green bodies had higher densification rates than the conventionally processed samples.

Figure 5.31 shows fractured surface SEM images of the B_4C ceramics, processed by using different methods and processing parameters. After sintering at 1900 °C, the 3D-printed samples were highly porous, at the early stage of the intermediate sintering regime, as seen in figure 5.31(A). The 2000 °C sintered sample reached the middle stage of the intermediate sintering regime, as illustrated in figure 5.31(B), while the early stage of the final sintering regime was achieved after sintering at 2100 °C, resulting in samples with highly dense microstructure, as demonstrated in figure 5.31(C).

If isostatic pressing was applied, the 3D-printed sample displayed nearly full densification, after sintering at 2100 °C, where the late stage of the final sintering regime was realized, as revealed in figure 5.31(D). However, after sintering at 2100 °C, the dry pressed samples without and with cold isostatic pressing were just approaching the middle stage of the intermediate sintering regime and the later stage of the intermediate sintering regime, as noticed in figures 5.31(E) and (F), respectively. At the same time, the 3D-printed B_4C ceramics also exhibited higher mechanical strengths, indicating advantages of the additive manufacturing over the conventional subtractive manufacturing technologies, at least for materials like B_4C.

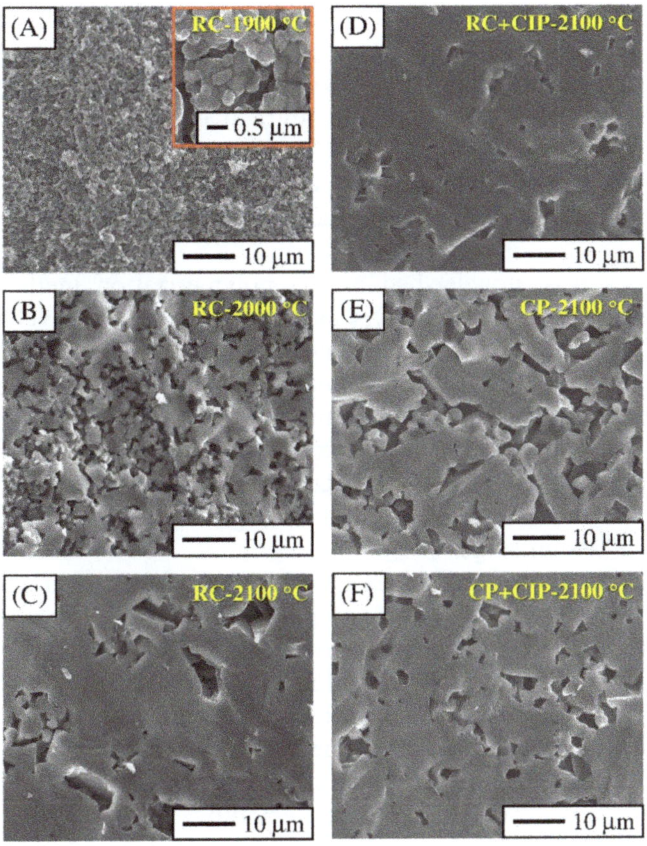

Figure 5.31. Fractured surface SEM images of the B_4C sintered ceramics prepared using the pairs of shaping procedure (PSPS) at the indicated temperatures. Reprinted from [93], Copyright (2018), with permission from Elsevier.

5.4 Concluding remarks

Extrusion freeforming, including direct ink writing (DIW), robocasting and so on, is an additive manufacturing technique that has been extensively applied in the fabrication of various ceramic materials, owing to its adaptability to high-viscosity materials and compatibility with diverse ceramic compositions.

One of the most important advantages of EFF is the ability to process ceramic slurries with relatively high solid loading levels, thus ensuring the printing efficiency and the properties of both the green bodies and sintered products. The layer-by-layer extrusion process enabled the manufacturing of structures with complex geometries, such as lattices, honeycombs, internal channels, and porous networks. This capability is particularly valuable for the fabrication of lightweight ceramic components, for aerospace and biomedical applications. Other advantages include low-temperature processing, high material efficiency, cost effectiveness, and multi-materials capabilities.

However, there are also various issues or problems that EFF technologies face. The minimum feature size in EFF is restricted by the diameter of the nozzles (typically 100–500 μm), resulting in relatively coarse surface finishes and textured structures, compared to conventional subtractive manufacturing or other 3D printing methods, e.g., stereolithography (SLA) or digital light processing (DLP). As a consequence, post-processing steps, such as grinding, polishing, or infiltration, are required for practical applications, leading to additional time consumption and cost.

EFF is slower than powder-bed or vat polymerization methods, due to the sequential extrusion process. High-viscosity inks lead to relatively low extrusion rates, to avoid nozzle clogging. Large-scale production remains a challenge. Developing stable ceramic inks with optimal shear-thinning behavior is critical for EFF, which is technically demanding. Similar to other various 3D printing methods, ceramic items printed with EFF are typically in the green body state, which should be subject to debinding and sintering processes to achieve final density and strength. These steps introduce shrinkages in the range of 15%–30%, which must be considered in the design stage.

Future research directions for EFF in ceramic fabrication should focus on increasing resolution, scalable production, material versatility and so on, while addressing current limitations and shortcomings. One critical topic is the refinement of ceramic ink formulations, in terms of high solid loadings and desired rheological properties, enabling finer feature resolutions (<100 μm) for applications in microfluidic devices or metamaterials. Real-time monitoring of the printing process could be a key point to increase the accuracy of such technology. Sustainable practices, such as recyclable inks, energy-efficient debinding and sintering, and exploring bioinspired hierarchical designs, represent promising avenues.

References

[1] Morissette S L, Lewis J A, Cesarano J, Dimos D B and Baer T Y 2000 Solid freeform fabrication of aqueous alumina-poly(vinyl alcohol) gelcasting suspensions *J. Am. Ceram. Soc.* **83** 2409–16

[2] Stuecker J N, Cesarano J and Hirschfeld D A 2003 Control of the viscous behavior of highly concentrated mullite suspensions for robocasting *J. Mater. Process. Technol.* **142** 318–25

[3] Miranda P, Saiz E, Gryn K and Tomsia A P 2006 Sintering and robocasting of β-tricalcium phosphate scaffolds for orthopaedic applications *Acta Biomater.* **2** 457–66

[4] Peng E, Zhang D W and Ding J 2018 Ceramic robocasting: recent achievements, potential, and future developments *Adv. Mater.* **30** 1802404

[5] Zhang D W, Jonhson W, Herng T S, Ang Y Q, Yang L, Tan S C *et al* 2020 A 3D-printing method of fabrication for metals, ceramics, and multi-materials using a universal self-curable technique for robocasting *Mater. Horiz* **7** 1083–90

[6] Khoshnevis B 2004 Automated construction by contour crafting-related robotics and information technologies *Autom. Constr.* **13** 5–19

[7] Khoshnevis B, Yuan X, Zahiri B, Zhang J and Xia B 2016 Construction by contour crafting using sulfur concrete with planetary applications *Rapid Prototyp. J.* **22** 848–56

[8] Lous G M, Cornejo I A, McNulty T F, Safari A and Danforth S C 2000 Fabrication of piezoelectric ceramic/polymer composite transducers using fused deposition of ceramics *J. Am. Ceram. Soc.* **83** 124–8
[9] Huang T, Mason M S, Zhao X Y, Hilmas G E and Leu M G 2009 Aqueous-based freeze-form extrusion fabrication of alumina components *Rapid Prototyp. J.* **15** 88–95
[10] Leu M C and Garcia D A 2014 Development of freeze-form extrusion fabrication with use of sacrificial material *J. Manuf. Sci. Eng.* **136** 061014
[11] Liu H J, Liu J, Leu M C, Landers R and Huang T S 2013 Factors influencing paste extrusion pressure and liquid content of extrudate in freeze-form extrusion fabrication *Int. J. Adv. Manuf. Technol.* **67** 899–906
[12] Moser I, Vuksic M, Dular M, Ivekovic A and Kocjan A 2024 Tuning the rheological properties of paraffin-wax ceramic feedstocks for deposition with thermoplastic 3D printing *J. Eur. Ceram. Soc.* **44** 7791–800
[13] Negi A, Goswami K, Diwan H, Agrawal G and Murab S 2025 Designing osteogenic interfaces on 3D-printed thermoplastic bone scaffolds *Mater. Today Chem.* **45** 102635
[14] Özden I, Scheithauer U, Ivekovic A and Kocjan A 2023 Effect of infill strategy on the flexural strength of 3Y-TZP bars fabricated by thermoplastic 3D printing (T3DP) *Open Ceram.* **14** 100367
[15] Li W B, Ghazanfari A, McMillen D, Leu M C, Hilmas G E and Watts J 2018 Characterization of zirconia specimens fabricated by ceramic on-demand extrusion *Ceram. Int.* **44** 12245–52
[16] Vaidyanathan R, Walish J, Lombardi J L, Kasichainula S, Calvert P and Cooper K C 2000 The extrusion freeforming of functional ceramic prototypes *JOM-J. Min. Met.* S **52** 34–7
[17] Lu X S, Lee Y, Yang S F, Hao Y, Evans J R G and Parini C G 2010 Solvent-based paste extrusion solid freeforming *J. Eur. Ceram. Soc.* **30** 1–10
[18] Scheithauer U, Schwarzer E, Richter H J and Moritz T 2015 Thermoplastic 3D printing-an additive manufacturing method for producing dense ceramics *Int. J. Appl. Ceram. Technol.* **12** 26–31
[19] Luo J J, Pan H and Kinzel E C 2014 Additive manufacturing of glass *J. Manuf. Sci. Eng.* **136** 061024
[20] Lu X S, Lee Y, Yang S F, Hao Y, Evans J R G and Parini C G 2009 Fabrication and evaluation of solid freeformed electromagnetic bandgap structures *J. Phys. D:Appl. Phys.* **42** 145107
[21] Lu X S, Lee Y, Yang S F, Hao Y, Evans J R G and Parini C G 2009 Fine lattice structures fabricated by extrusion freeforming: process variables *J. Mater. Process. Technol.* **209** 4654–61
[22] Lu X, Lee Y, Yang S, Hao Y, Ubic R, Evans J R G *et al* 2008 Fabrication of electromagnetic crystals by extrusion freeforming *Metamaterials* **2** 36–44
[23] Qiu D and Langrana N A 2002 Void eliminating toolpath for extrusion-based multi-material layered manufacturing *Rapid Prototyp. J.* **8** 38–45
[24] Tseng W J J and Lin C L 2002 Effect of polyvinyl butyral on the rheological properties of $BaTiO_3$ powder in ethanol-isopropanol mixtures *Mater. Lett.* **57** 223–8
[25] Park I, Ahn J, Im J, Choi J and Shin D 2012 Influence of rheological characteristics of YSZ suspension on the morphology of YSZ films deposited by electrostatic spray deposition *Ceram. Int.* **38** S481–S4

[26] Lim K Y, Kim D H, Paik U and Kim S H 2003 Effect of the molecular weight of poly (ethylene glycol) on the plasticization of green sheets composed of ultrafine $BaTiO_3$ particles and poly(vinyl butyral) *Mater. Res. Bull.* **38** 1021–32
[27] Russell B D, Blackburn S and Wilson D I 2006 A study of surface fracture in paste extrusion using signal processing *J. Mater. Sci.* **41** 2895–906
[28] Russell B D, Lasenby J, Blackburn S and Wilson D I 2003 Characterising paste extrusion behaviour by signal processing of pressure sensor data *Powder Technol.* **132** 233–48
[29] Li M Y, Tang L, Landers R G and Leu M C 2013 Extrusion process modeling for aqueous-based ceramic pastes—part 2: experimental verification *J. Manuf. Sci. Eng.* **135** 051009
[30] Li M Y, Tang L, Landers R G and Leu M C 2013 Extrusion process modeling for aqueous-based ceramic pastes—part 1: constitutive model *J. Manuf. Sci. Eng.* **135** 051008
[31] Mason M S, Huang T, Landers R G, Leu M C and Hilmas G E 2009 Aqueous-based extrusion of high solids loading ceramic pastes: process modeling and control *J. Mater. Process. Technol.* **209** 2946–57
[32] Panda B N, Shankhwar K, Garg A and Jian Z 2017 Performance evaluation of warping characteristic of fused deposition modelling process *Int. J. Adv. Manuf. Technol.* **88** 1799–811
[33] Wang T M, Xi J T and Jin Y 2007 A model research for prototype warp deformation in the FDM process *Int. J. Adv. Manuf. Technol.* **33** 1087–96
[34] Li W B, Ghazanfari A, Leu M C and Landers R G 2017 Extrusion-on-demand methods for high solids loading ceramic paste in freeform extrusion fabrication *Virtual Phys. Prototy* **12** 193–205
[35] Li W B 2019 Doctoral Dissertation, Freeform extrusion fabrication of advanced ceramics and ceramic-based composites (Rolla, MO: Missouri University of Science and Technology)
[36] Zhao X Y, Landers R G and Leu M C 2010 Adaptive extrusion force control of freeze-form extrusion fabrication processes *J. Manuf. Sci. Eng.* **132** 064504
[37] Deuser B K, Tang L, Landers R G, Leu M C and Hilmas G E 2013 Hybrid extrusion force–velocity control using freeze-form extrusion fabrication for functionally graded material parts *J. Manuf. Sci. Eng* **135** 041015
[38] Finke B, Hesselbach J, Schütt A, Tidau M, Hampel B, Schilling M *et al* 2020 Influence of formulation parameters on the freeform extrusion process of ceramic pastes and resulting product properties *Addit. Manuf* **32** 101005
[39] Knieke C, Steinborn C, Romeis S, Peukert W, Breitung-Faes S and Kwade A 2010 Nanoparticle production with stirred-media mills: opportunities and limits *Chem. Eng. Technol.* **33** 1401–11
[40] Ament K A, Kessler M R and Akinc M 2014 Shear thinning behavior of aqueous alumina nanoparticle suspensions with saccharides *Ceram. Int.* **40** 3533–42
[41] Amorós J L, Blasco E and Beltrán V 2015 Shear-thinning behaviour of dense, stabilised suspensions of plate-like particles: proposed structural model *Appl. Clay Sci.* **114** 297–302
[42] Yang Z H, Liu G H, Qi T G, Zhou Q S, Peng Z H, Shen L T *et al* 2024 Bayerite in aluminum hydroxide effecting the thermal decomposition pathways and reducing density of α-Al_2O_3 *J. Alloys Compd.* **1004** 175884
[43] Wang J W, Shaw L L and Cameron T B 2006 Solid freeform fabrication of permanent dental restorations via slurry micro-extrusion *J. Am. Ceram. Soc.* **89** 346–9
[44] Wang J W and Shaw L L 2005 Rheological and extrusion behavior of dental porcelain slurries for rapid prototyping applications *Mater. Sci. Eng.* A **397** 314–21

[45] Denry I L, Holloway J A and Colijn H O 2001 Phase transformations in a leucite-reinforced pressable dental ceramic *J. Biomed. Mater. Res.* **54** 351–9
[46] Basgul C, Desantis P, Derr T, Hickok N J, Bock R M and Kurtz S M 2024 Exploring the mechanical strength, antimicrobial performance, and bioactivity of 3D-printed silicon nitride-PEEK composites in cervical spinal cages *Int. J. Bioprinting* **10** 431–44
[47] Khallaf R M, Emam A N, Mostafa A A, Nassif M S and Hussein T S 2023 Strength and bioactivity of PEEK composites containing multiwalled carbon nanotubes and bioactive glass *J. Mech. Behav. Biomed. Mater.* **144** 105964
[48] Manzoor F, Golbang A, McIlhagger A, Harkin-Jones E, Crawford D and Mancuso E 2023 Effect of Zn-nanoHA concentration on the mechanical performance and bioactivity of 3D printed PEEK composites for craniofacial implants *Plast. Rubber Compos.* **52** 197–203
[49] Mostafa D, Kassem Y M, Omar S S and Shalaby Y 2023 Nano-topographical surface engineering for enhancing bioactivity of PEEK implants (*in vitro*-histomorphometric study) *Clin. Oral Investig.* **27** 6789–99
[50] Vaezi M, Black C, Gibbs D M R, Oreffo R O C, Brady M, Moshrefi-Torbati M *et al* 2016 Characterization of new PEEK/HA composites with 3D HA network fabricated by extrusion freeforming *Molecules* **21** 687
[51] Yang H Y, Yang S F, Chi X P, Evans J R G, Thompson I, Cook R J *et al* 2008 Sintering behaviour of calcium phosphate filaments for use as hard tissue scaffolds *J. Eur. Ceram. Soc.* **28** 159–67
[52] Abu Bakar M S, Cheang P and Khor K A 2003 Tensile properties and microstructural analysis of spheroidized hydroxyapatite-poly (etheretherketone) biocomposites *Mater. Sci. Eng.* A **345** 55–63
[53] Fan J P, Tsui C P, Tang C Y and Chow C L 2004 Influence of interphase layer on the overall elasto-plastic behaviors of HA/PEEK biocomposite *Biomaterials* **25** 5363–73
[54] Abu Bakar M S, Cheang P and Khor K A 2003 Mechanical properties of injection molded hydroxyapatite-polyetheretherketone biocomposites *Compos. Sci. Technol.* **63** 421–5
[55] Öhman C, Baleani M, Pani C, Taddei F, Alberghini M, Viceconti M *et al* 2011 Compressive behaviour of child and adult cortical bone *Bone* **49** 769–76
[56] Tang S M, Cheang P, AbuBakar M S, Khor K A and Liao K 2004 Tension-tension fatigue behavior of hydroxyapatite reinforced polyetheretherketone composites *Int. J. Fatigue* **26** 49–57
[57] Yang H Y, Yang S F, Chi X P and Evans J R G 2006 Fine ceramic lattices prepared by extrusion freeforming *J. Biomed. Mater. Res.* B **79B** 116–21
[58] Pearson J J, Gerken N, Bae C, Lee K B, Satsangi A, McBride S *et al* 2020 *In-vivo* hydroxyapatite scaffold performance in infected bone defects *J. Biomed. Mater. Res.* B **108** 1157–66
[59] Sepulveda P, Bressiani A H, Bressiani J C, Meseguer L and Konig B 2002 *In-vivo* evaluation of hydroxyapatite foams *J. Biomed. Mater. Res.* **62** 587–92
[60] Sun W T, Zhong J B, Gao B Y, Feng J L, Ye Z J, Lin Y L *et al* 2024 *In-vitro*/*in-vivo* evaluations of hydroxyapatite nanoparticles with different geometry *Int. J. Nanomed.* **19** 8661–79
[61] Bulina N V, Khvostov M V, Borodulina I A, Makarova S V, Zhukova N A and Tolstikova T G 2024 Substituted hydroxyapatite and β-tricalcium phosphate as osteogenesis enhancers *Ceram. Int.* **50** 33258–69

[62] Kurashina K, Kurita H, Wu Q, Ohtsuka A and Kobayashi H 2002 Ectopic osteogenesis with biphasic ceramics of hydroxyapatite and tricalcium phosphate in rabbits *Biomaterials* **23** 407–12

[63] Du M K, Kuang Z D, Ji H R and Mao K Y 2014 Enhanced degradation and osteogenesis of β-tricalcium phosphate/calcium sulfate composite bioceramics: the effects of phase ratio *J. Biomater. Tiss. Eng.* **4** 389–98

[64] Tian Y, Lu T L, He F P, Xu Y B, Shi H S, Shi X T *et al* 2018 β-Tricalcium phosphate composite ceramics with high compressive strength, enhanced osteogenesis and inhibited osteoclastic activities *Colloids Surf.* B **167** 318–27

[65] Oudadesse H, Derrien A C, Mami M, Martin S, Cathelineau G and Yahia L 2007 Aluminosilicates and biphasic HA-TCP composites: studies of properties for bony filling *Biomed. Mater.* **2** S59–64

[66] Zerankeshi M M, Mofakhami S and Salahinejad E 2022 3D porous HA/TCP composite scaffolds for bone tissue engineering *Ceram. Int.* **48** 22647–63

[67] de Sousa F C G and Evans J R G 2003 Sintered hydroxyapatite latticework for bone substitute *J. Am. Ceram. Soc.* **86** 517–9

[68] Michna S, Wu W and Lewis J A 2005 Concentrated hydroxyapatite inks for direct-write assembly of 3-D periodic scaffolds *Biomaterials* **26** 5632–9

[69] Wahl L, Köllner D, Weichelt M, Travitzky N and Fey T 2025 Environmentally friendly water-based robocasting of complex barium titanate structures *Open Ceramics* **22** 100773

[70] Lorenz M, Martin A, Webber K G and Travitzky N 2020 Electromechanical properties of robocasted barium titanate ceramics *Adv. Eng. Mater.* **22** 202000325

[71] Deng X Y, Wang X H, Chen L L, Wen H and Li L T 2006 Observation of ferroelectric domain patterns in nanocrystalline $BaTiO_3$ ceramics *Appl. Phys. Lett.* **89** 152901

[72] Ma N, Zhang B P, Yang W G and Guo D 2012 Phase structure and nano-domain in high performance of $BaTiO_3$ piezoelectric ceramics *J. Eur. Ceram. Soc.* **32** 1059–66

[73] Gadea C, Spelta T, Simonsen S B, Esposito V, Bowen J R and Haugen A B 2021 Hybrid inks for 3D printing of tall $BaTiO_3$-based ceramics *Open Ceram.* **6** 100110

[74] Kim H, Renteria-Marquez A, Islam M D, Chavez L A, Rosales C A G, Ahsan M A *et al* 2019 Fabrication of bulk piezoelectric and dielectric $BaTiO_3$ ceramics using paste extrusion 3D printing technique *J. Am. Ceram. Soc.* **102** 3685–94

[75] Zhang Y, Roscow J, Lewis R, Khanbareh H, Topolov V Y, Xie M Y *et al* 2018 Understanding the effect of porosity on the polarisation-field response of ferroelectric materials *Acta Mater.* **154** 100–12

[76] Li Z H, Roscow J, Khanbareh H, Taylor J, Haswell G and Bowen C 2023 A comprehensive energy flow model for piezoelectric energy harvesters: understanding the relationships between material properties and power output *Mater. Today Energy* **37** 101396

[77] Messing G L, Trolier-McKinstry S, Sabolsky E M, Duran C, Kwon S, Brahmaroutu B *et al* 2004 Templated grain growth of textured piezoelectric ceramics *Crit. Rev. Solid State Mater. Sci.* **29** 45–96

[78] Sabolsky E M, Maldonado L, Seabaugh M M and Swartz S L 2010 Textured-$Ba(Zr,Ti)O_3$ piezoelectric ceramics fabricated by templated grain growth (TGG) *J. Electroceram.* **25** 77–84

[79] Yan Y K, Zhou H P, Zhao W and Liu D 2007 Study on textured NBT-6BT ceramics fabricated by (reactive) templated grain growth *Rare Met. Mater. Eng.* **36** 479–83

[80] Zate T T, Abdurrahmanoglu C, Esposito V and Haugen A B 2025 Textured lead-free piezoelectric ceramics: a review of template effects *Materials* **18** 477

[81] Wahl L, Weichelt M, Goik P, Schmiedeke S and Travitzky N 2021 Robocasting of reaction bonded silicon carbide/silicon carbide platelet composites *Ceram. Int.* **47** 9736–44

[82] Walton R L, Brova M J, Watson B H, Kupp E R, Fanton M A, Meyer R J *et al* 2021 Direct writing of textured ceramics using anisotropic nozzles *J. Eur. Ceram. Soc.* **41** 1945–53

[83] Walton R L, Fanton M A, Meyer R J and Messing G L 2020 Dispersion and rheology for direct writing lead-based piezoelectric ceramic pastes with anisotropic template particles *J. Am. Ceram. Soc.* **103** 6157–68

[84] Mahajan S, Thakur O P, Bhattacharya D K and Sreenivas K 2009 Ferroelectric relaxor behaviour and impedance spectroscopy of Bi_2O_3-doped barium zirconium titanate ceramics *J. Phys. D: Appl. Phys.* **42** 065413

[85] Wei X X, Nagarajan R S, Peng E, Xue J, Wang J and Ding J 2016 Fabrication of $YBa_2Cu_3O_{7-x}$ (YBCO) superconductor bulk structures by extrusion freeforming *Ceram. Int.* **42** 15836–42

[86] Benavidez E R and Oliver C 2005 Sintering mechanisms in $YBa_2Cu_3O_{7-x}$ superconducting ceramics *J. Mater. Sci.* **40** 3749–58

[87] Imayev M F, Kabirova D B, Sagitov R I and Churbaeva K A 2012 Relation between change of porosity and parameters of grains during annealing of the superconducting ceramics $YBa_2Cu_3O_{7-x}$ *J. Eur. Ceram. Soc.* **32** 1261–8

[88] Prayoonphokkharat P, Jiansirisomboon S and Watcharapasorn A 2013 Fabrication and properties of $YBa_2Cu_3O_{7-x}$ ceramics at different sintering temperatures *Electron. Mater. Lett.* **9** 413–6

[89] Suasmoro S, Khalfi M F, Khalfi A, Trolliard G, Smith D S and Bonnet J P 2012 Microstructural and electrical characterization of bulk $YBa_2Cu_3O_{7-\delta}$ ceramics *Ceram. Int.* **38** 29–38

[90] Rao Q L, Fan X L, Shu D and Wu C C 2008 In-situ XRD study on the peritectic reaction of YBCO thin film on MgO substrate *J. Alloys Compd.* **461** L29–33

[91] Rao Q L, Fan X L and Yao X 2007 The preferred orientation study on YBCO thin film during its peritectic reaction *Scr. Mater.* **57** 663–6

[92] de Abreu J L B, Hirata R, Witek L, Jalkh E B B, Nayak V V, de Souza B M *et al* 2023 Manufacturing and characterization of a 3D printed lithium disilicate ceramic via robocasting: a pilot study *J. Mech. Behav. Biomed. Mater.* **143** 105867

[93] Eqtesadi S, Motealleh A, Perera F H, Miranda P, Pajares A, Wendelbo R *et al* 2018 Fabricating geometrically-complex B_4C ceramic components by robocasting and pressureless spark plasma sintering *Scr. Mater.* **145** 14–8

IOP Publishing

Additive Manufacturing of Ceramics

Ling Bing Kong, Zhuohao Xiao, Bin He and Yin Liu

Chapter 6

Laminated object manufacturing (LOM)

Laminated object manufacturing (LOM) is a powerful 3D printing technology to fabricated ceramics and ceramic materials. LOM is especially useful in developing structures with laminated configurations, consisting of sheets, papers, films, foils and others, while the types of materials could be more than one. The final 3D structured objects were manufactured through layer-by-layer lamination process. Representative ceramics made with LOM process will be overviewed and discussed in this chapter.

6.1 Introduction

The history of laminated object manufacturing (LOM) can be traced back to the 1970s, as presented by several patents [1]. Specifically, a computerized tool was developed to cut 2D items from sheet materials that had layered structures. Cutting started from the topmost sheet, while the cut sheets were subsequently stacked and strongly bonded together. This approach enabled the fabrication of structures with complicated desired shapes. Initially, the sheet materials used were normal papers or waxes. Early LOM technology was mainly utilized to build 3D topography relief maps.

The original photo sculpturing and relief mapping craft techniques in the LOM process were replaced by computer-controlled part recognition and design processes, leading to automated multilayer structure construction. The designed 3D structures were projected onto multiple layers of the sheet materials. There are two approaches to achieve the goal: cut-then-bond or cut-off-the-stack and bond-then-cut or cut-on-the-stack [2].

The cut-then-bond LOM process is also known for computer-aided manufacturing of laminated engineering materials (CAM-LEM) [3]. For this approach, sheets are first laminated and then cut, which helps minimize errors. The CAM-LEM process has been widely applied to polymer-ceramic composites and metallic materials, with a processing temperature of 80 °C and pressures between 0.34 and

doi:10.1088/978-0-7503-4831-7ch6

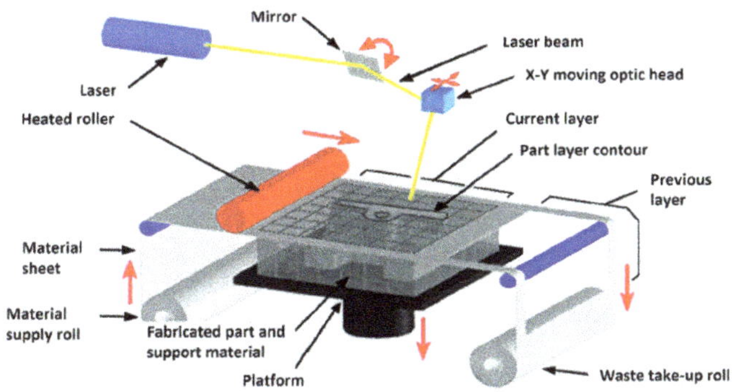

Figure 6.1. Schematic diagram of typical type of LOM process. Reprinted from [5], Copyright (2012), with permission from Elsevier.

0.68 MPa. Because the surfaces of stacked sheets are not perfectly flat, this variant is called curved LOM, reflecting its ability to fabricate complex structures. After each sheet is cut, the excessive materials should be taken away before the lamination process. Various ceramic devices have been fabricated using the CAM-LEM technique, with layer thicknesses ranging from 30 μm to 1.3 mm^{-1} [4].

Commercially available LOM machines need continuous feeding of materials. The cutting step is typically performed using lasers and multiple sensors to ensure precise position during printing. Figure 6.1 shows a schematic diagram of the LOM process [5]. In this process, the layer contour or cross-sectional outline is cut into a sheet, either before or after lamination onto the stacks. Additionally, a patterned cut should be made on the top layer to facilitate easy removal of waste after printing. When support materials are required, they are usually designed with cubic structures to simplify waste removal, a process known as decubing.

6.2 Characteristics of LOM process

6.2.1 Special aspects of LOM

LOM allows printing with nontoxic, relatively cost-effective and continuous materials, such as papers or polymer tapes, which are laminated layer by layer, resulting in 3D multilayer structures with complex shapes. Compared to other methods, the LOM process is most suitable for creating large-scale products [6]. It can be readily upgraded by incorporating with other sophisticated items, such as robotic sheet manipulation processes to achieve higher levels of automation [7]. Thus, prefabricated components can be integrated into multilayer stacks. Meanwhile, the printing resolution of LOM is comparable to other 3D printing techniques.

The LOM process has strong tolerance to raw materials, while no toxic chemicals or complicated chemical reactions are involved. Desktop LOM facilities have been commercialized, owing to the large variability in size of the printed structures and the wide availability of papers. Such LOM machines are advantageous in building models of implants based on papers and polymers. Moreover, the sheet materials can be coated, printed and embossed, before being integrated into the printed LOM structures. However, the LOM printing process also has disadvantages. For instance, producing structures with internal cavities is challenging [6, 8, 9]. Nevertheless, complex structures ranging from planar to full 3D features can still be manufactured.

6.2.2 Descriptions of the LOM technique

6.2.2.1 Processes of data input

The LOM process begins by converting CAD files into STL files. The CAD files present 3D objects with rounded geometrical shapes, while the surfaces may be either curved or flat. STL files represent the geometry using wireframes consistent with original CAD objects. The CAD to STL file conversion process is known as tessellation, where the continuous object surfaces are taken by triangular meshes. With the corresponding tessellation algorithms, product can be precisely and accurately fabricated.

Optimal results of the tessellation can be achieved through refining the mesh to obtain maximum accuracy. To make the triangular meshes smoother, they are modified through numerical processing [10]. This strategy is especially useful in constructing surfaces at bends and folds of the objects. There are three mesh partitioning methods that can be used to handle complex features: vertex-based, edge-based, and face-based methods [11].

In addition, there is an alternative file format known as additive manufacturing file format (AMF). While STL files only convey the basic design information in the form of triangular meshes, AMF files carry much more information. AMF files contain more accurate mesh data as well as specifications for colors, textures, substructures, and materials. In AMF files, curved triangles are enabled as facets in the tessellation meshes of the STL files, thus increasing the accuracy for the shapes of the printed structures.

During the LOM printing, the 2D layer images at a given build position are extracted from either STL or AMF files. The cross-sectional images of the 2D objects can be stored as common files of layer interface (CLI), or directly transferred to printers via G-code [12, 13]. In each printed layer, the cutting pattern must precisely follow the vertical cross-sectional outlines. Besides the cross-sectional outlines, an extra cutting pattern is needed to assist in waste removal after printing.

When using the bond-then-cut method, a square crosshatch pattern should be added to each layer outside the cross-sectional outlines. The crosshatches should align carefully with the outer areas of the sheet layers. The actual item outlines are called the inside area, which is enclosed by the outside area. The crosshatch patterns serve as support for the printed structures and are usually designed as square structures with suitable sizes.

Crosshatch-based waste removal is suitable for the bond-then-cut method. For the cut-then-bond method, bridge supports are used. In this approach, areas adjacent to the outside area are removed individually for each layer before bonding. Therefore, the bridge supports within the inside area remain well. To minimize material waste, the fraction of sheet material used for bridge supports should be kept as low as possible.

6.2.2.2 Processing parameters to be pre-considered

In the practice of LOM printing, it is important to maintain the printer in an efficient working state by adjusting key parameters to minimize printing errors. First of all, a layer cutting technique is involved in the beginning, by using different cutting techniques, including tungsten carbide–cobalt (WC–Co) blades, CO_2 laser and water jetting [14]. Water jet cutting is mainly suitable for metal foils as sheet materials, but not applicable to paper sheets. In comparison, laser cutting is feasible to a wide range of sheet materials, although heat generation is a major concern.

Specifically, the energy of the laser beam and the number of runs for each cutting have strong influences on cutting effectiveness and efficiency. Laser beam cutting typically results in parabolic cutting profiles, which maintain the shapes of the outer layer edges [15]. The depth of the cutting profile should match the single layer thickness during laser cutting. This ensures the layers are not damaged at the selected cutting speed and laser energy. The layer might be cut through just one time or multiple runs could be required, depending on two parameters. On one hand, when a low laser energy is applied to prevent overheating, the number of runs must be increased. On the other hand, higher laser energy allows for fewer runs, but increases the risks of local overheating.

As the outline of the printed items is cut out with just one run, there are still waste materials that should be taken away from the horizontal surface to be present as exposed portions at the sloped surfaces in between adjacent two layers. There are two methods to address waste bonding. One is to use a laser with sufficiently high energy, so that the adhesive bonds could all be burned off. Another way is to put the pattern of crosshatches on the overlaps. The first approach is applicable to burn off the adhesive, which is also denoted as the burnishing or burning rule, while the second one is usually known as adaptive crosshatching [16–18].

Surface roughness of the printed structures is another key parameter influencing LOM printing efficiency and quality. Surface profiles formed during stepwise printing processes should be examined separately based on orientations [19, 20]. Another effect on geometry of the printed structures is linked to the bonding process. A general solution is to use a heating roller. In practice, which is adopted is dependent on the spatial distribution of the induced stress and heat in the multilayer structures that have been printed. As a roller is utilized for the layer bonding, the stress behaves like line loadings, while pressing head would compress the multilayers in the cross-sectional direction. The second bonding process is especially effective for the cut-then-bond LOM printing, where bridge supports rather than cubes are utilized.

Both the parallel flat deposition of new layers and their uniform connection with the multilayers are critical to prevent warping of the LOM printed products. When rollers are employed for the layer bonding, the mechanical stresses induced in the multilayer stacks should be thoroughly analyzed. For example, the rolling process can be simulated using finite elemental method (FEM) [21]. Thus, the local compression and adjacent relaxation of the bonded layers can be theoretically modeled, allowing prediction of structural instabilities or layer deformations. For the rolling process, large rollers ensured more uniform laminations, while small ones offered rapid printing owing to the more concentrated mechanical stress under the rollers [22].

6.3 LOM printed ceramic materials and structures

6.3.1 Brief description

The freedom of fabrication with LOM printing techniques has been significantly enriched, due to the development of highly filled papers, preceramic polymeric tapes, frozen slurries and so on. New sheet materials and frozen slurry layers facilitated ceramic-based materials to have functionally graded microstructures (FGMs), in which the porosity distributions could be realized through design. In such FGMs, the materials are arranged with a desirable gradation between two different components, such as in an Al_2O_3 multilayer structure where ZrO_2 content gradually varies through the thickness.

Additionally, ceramic matrix composites (CMCs) could be developed by introducing reinforcement fibers or particles. CMCs consist of a matrix phase and a second phase, where the second phase is uniformly or directionally embedded. For instance, ZrO_2 fibers or particles are dispersed within an Al_2O_3 matrix. Generally, the reinforcement agents should be distributed as uniformly as possible. If the fibers are aligned in a single direction, textured structures with anisotropic properties are formed. If fibers are randomly oriented, the resulting CMCs exhibit isotropic behaviors.

LOM technique has been extensively applied to fabricate oxide ceramics and silicon-containing ceramics [23, 24]. To print silicon-based ceramics, the papers could be pyrolyzed and infiltrated with silicon melts [25]. If curved structures were made with LOM printing, the paper sheets were just employed to prepare the supports with necessary curvatures. Glass-ceramics could also be obtained by using the LOM process, where ceramic-based materials were derived from the preceramic tapes that were infiltrated with LZSA (Li_2O–ZrO_2–SiO_2–Al_2O_3) glass phase [26–28].

Similarly, preceramic papers can also be involved in the LOM process. The preceramic papers have relatively high contents of ceramic fillers, in the range of up to 90%. The fillers in preceramic papers could either oxides or nonoxides. In the drying process, the preceramic papers could be subject to a certain degree of compression to reduce the porosity or coated with ceramic slurries to fill up the pores inside the papers [29–33]. Moreover, layer structures made from frozen slurries have been reported for incorporation into the LOM technique [34]. A subsequent freeze-drying post-process enables the production of nearly fully ceramic green bodies.

6.3.2 Ceramics printed with LOM process

An early example of Al_2O_3 ceramics fabricated with the LOM technique was reported by Zhang *et al* [35]. The green tape with a thickness of 0.7 mm was prepared by using the roll-forming method [36–38]. The as-printed green body items after LOM printing were sintered by the simple pressureless sintering process, forming Al_2O_3 ceramic products. To prepare the green tapes, Al_2O_3 powder with a purity of 96% and an average particle size of 2 μm was used as the raw material, while polyvinyl butyral (PVB) was used as the organic binder at a content of 7 wt%. The resulting green tapes had a thickness of 0.76 mm and a density of 2.34 $g \cdot cm^{-3}$, corresponding to a relative density of about 60%.

An M RPMS-II LOM printer (Tshinghua University) was adopted to fabricate the ceramic green body structures. 3D CAD files (solid or wire-form models) were converted into STL format, and then slicing into thin cross-sections was conducted using a separate computer program. Noting the relatively low mechanical strength, the ceramic tapes were all manually handled. Polyvinyl acetate was utilized as the glue to connect the ceramic tapes. To determine the optimal conditions for binder removal, TG-DTA analysis was conducted on the Al_2O_3 green tapes, as shown in figure 6.2.

After binder removal, the green bodies retained the shapes and sizes without deformation and cracking. All the samples were sintered under optimized conditions. The shrinkage in the thickness direction was larger than in the plane directions. To avoid distortion and cracking for the sintered products, the cooling rate should be carefully selected. The mechanical strengths of the sintered ceramic samples were measured using an Instron-1186 machine, following the three-point bending method, while hardness was tested using an HV-120 sclerometer.

The green bodies of the Al_2O_3 ceramic layers were closely stacked due to the smooth surfaces of the thick tapes. After sintering, the Al_2O_3 samples experienced grain growth, while the ceramic grains became more and more faceted, appearing with polyhedron morphology. Meanwhile, no gaps were present between the adjacent layers, whereas the sintered products exhibited homogeneous microstructure, as revealed in figure 6.3. Owing to the anisotropic characteristics, the Al_2O_3 ceramics prepared by using LOM printing were mechanically weaker than those produced by using the conventional solid-state reaction method. Nevertheless, the LOM printing is of an advantage in fabricating ceramic items with small batches, because no modules are used in the process. Relative densities of the sintered samples at 1580 °C reached 97%, with a HV hardness of 391 under a load of 5 kg.

LOM was combined with frozen slurry (FS-LOM) to fabricate porous ceramics [34]. The slurry was prepared using DI water, alumina powder, organic binders, and

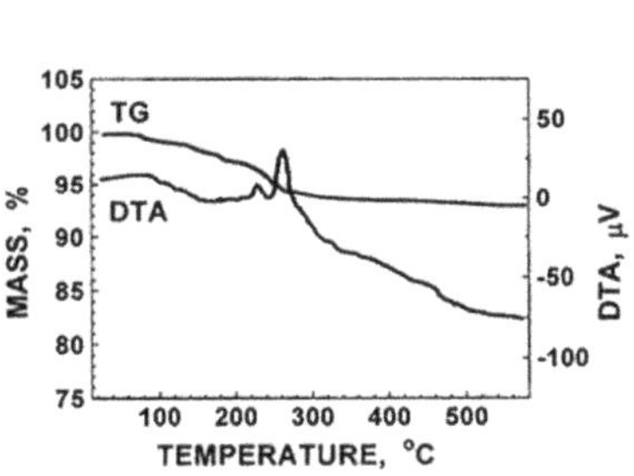

Figure 6.2. TG-DAT curves of the green tape. Reproduced from [35], with permission from Springer Nature.

Figure 6.3. Fractured surface SEM image of the Al_2O_3 ceramic structure printed with LOM technique. Reproduced from [35], with permission from Springer Nature.

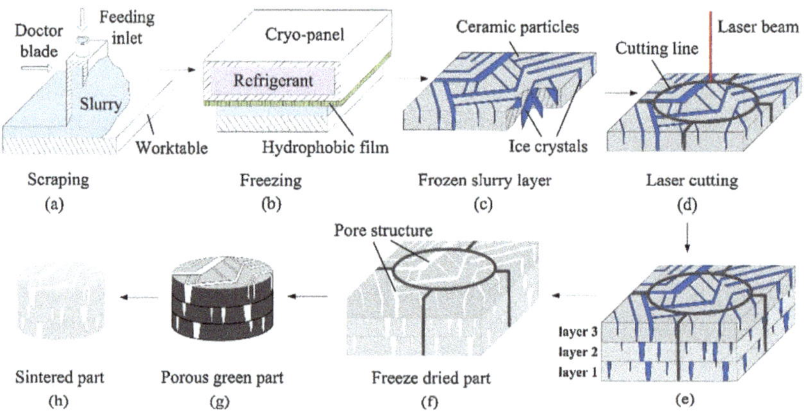

Figure 6.4. Schematic diagram of the FS-LOM printing process. Reprinted from [34], Copyright (2018), with permission from Elsevier.

additives. The water in each fresh slurry layer was crystallized to form a strong support. 2D patterns were created by cutting the outline with a laser beam, during which the ice crystals and organic binders were gasified. As the ice crystals were freeze dried, the stacks became highly porous structures. The ice crystals were grown vertically as lamellar during the layer-by-layer freezing process. As a result, the printed structures exhibited highly uniformity, oriented pores, and sufficiently high compressive strength. Meanwhile, the frozen slurry provided strong supports, effectively preventing deformation of the green bodies.

Figure 6.4 shows a schematic diagram of the FS-LOM printing process, including steps of paving, freezing, cutting, freeze-drying and sintering. The printing table with vertical motion was kept at a temperature of −20 °C during the printing process. As the printing table was moved down for each layer, the slurry was scraped over with a doctor blade, during which it flew from the feeding outlet, as seen in figure 6.4(a).

Meanwhile, the samples were subject to freezing through a cryo-panel, in which refrigerant was internally circulated. A thin hydrophobic film was attached to the bottom of the cryo-panel to prevent bonding between the frozen slurry and the panel, as demonstrated in figure 6.4(b).

Once the slurry was cooled down to the temperature lower than the eutectic point, it was frozen. As the ice crystals were entirely formed, the ceramic particles were held together as a strong entity, as presented in figure 6.4(c). Then, 2D patterns were incised through laser beam scanning, triggering the gasification of the ice and binders, as illustrated in figure 6.4(d). The frozen state of the slurry outside the scanned pattern was intact. After a new layer was laid out, the ice crystals on surface of the previous frozen slurry were molten. Because the new layer was frozen, the water released due to the melting was crystallized together with that present in the new layer, so that all the adjacent layers were strongly adhered.

Finally, a 3D green body was produced, which was enveloped by the frozen slurry, as depicted in figure 6.4(e). After printing, the sample was freeze-dried in a vacuum to remove all ice crystals and resulting in a porous green ceramic, as shown in figure 6.4(f–g). The sample was then sintered under certain conditions to obtain porous ceramic products, as shown in figure 6.4(h).

The laser irradiation triggered an instant rise in temperature of the materials. As the temperature was above the boiling point of water, the ice crystals and polymer binder were sublimed as gas molecules, allowing the ceramic particles to be released from bonding. Ejection of the ceramic particles was possible, owing to the strong dragging force induced by the escaping gases. In the areas of gasification, the polymer binders could be carbonized by the laser beam, due to the presence of high temperature. As the laser penetration depth gradually increased, energy attenuation became more and more pronounced. Since the laser beam could reach the area beneath the gasification area, the ice crystals there were not sublimed. However, the ice crystals were molten, resulting in a transient zone, in which the ceramic particles were released to induce redistribution. Eventually, the transient zone was refrozen again, owing to the low temperature of the surrounding frozen slurry.

The eutectic point of the ceramic slurry, with a solid loading level of 50 wt%, was measured by using a differential scanning calorimeter (DSC823e, Mettler Toledo, Switzerland), over the temperature range from −50 °C to 20 °C, at cooling rate of 10 °C · min^{-1}, resulting in a value of −14 °C. A slurry layer with a thickness of 2 mm was spread onto the printing table and then frozen with the cryo-panel down to −20 °C, within about 5 s. The laser parameters tested for printing included a spot size of 500 μm, laser powers of 30–41 W and scanning speeds of 100–250 mm · s^{-1}. Drying of the samples was conducted by using a vacuum freeze dryer (LGJ-10, Songyuan, China) for 18 h at the temperature of 10 °C and vacuum level of 3 Pa). Microstructures of the samples were characterized by using SEM (VEGA-II XMU, TESCAN, Czech Republic).

The slurries with 50, 55, 60 and 65 wt% alumina were used for the printing experiment, with a layer thickness of 200 μm. The freeze-dried samples were sintered in a muffle furnace (KSL-1700X, Kejing, China) at 1650 °C in air for 2 h, at a heating rate of 2 °C · min^{-1}. Selected samples were completely soaked in boiling water for 3 h to evaluate their porosity according to the weight differences before and after soaking. Samples with dimensions of 15 mm × 15 mm × 15 mm were cut using a diamond wire saw (STX-202A, Kejing, China) for the measurement of vertical compressive strength, with a microcomputer controlled electronic universal testing machine (CTM2500, Xieqiang, China), with the pressure loading rate controlled at 1 mm · m^{-1} in.

Figure 6.5 shows cutting line profiles of the frozen slurry after laser irradiation. It was experimentally demonstrated that both the cutting width and depth decreased, with increasing scanning speed at given laser powers. They both increased with increasing laser power, at fixed scanning speeds. This was simply because the increase in laser power and the decrease in scanning speed corresponded to increase in energy density of the laser. The higher the laser energy density, wider and deeper lines would be produced.

The accuracy of laser cutting is highly dependent on the cutting width. The narrower the cutting width, the higher the cutting accuracy would be. However, the prerequisite is that the cutting depth should be ensured for practical applications. For instance, for two set of laser parameters, $P = 33$ W, $V = 150$ mm · s^{-1} and $P = 41$ W, $V = 250$ mm · s^{-1}, both samples could achieve the depth of 200 μm, but the first group had a width larger than the second one by nearly 100 μm. In one word, the laser power and scanning speed should be optimized to realize high accuracy printing of the 2D patterns for LOM process. Porous structures with high integrity could be obtained with laser parameters of $P = 41$ W, $V = 250$ mm · s^{-1} and layer thickness of 200 μm, from thc slurry with a solid loading level of 50 wt%, after sintering at 1650 °C.

It has been reported that the growth of ice crystals occurs preferentially along the direction of freezing [39]. Meanwhile, it was also confirmed that a higher freezing rate results in smaller ice crystals produced [40–43]. In order to achieve structures with uniformly distributed ice crystals, the freezing rate should be as uniform as possible [44]. In the freezing process, the freezing occurred at the surface of the samples, which propagated from the outside to the interior, while the freezing was gradually slowed down. Finally, the crystallization stopped at the center of the sample, with the crystals

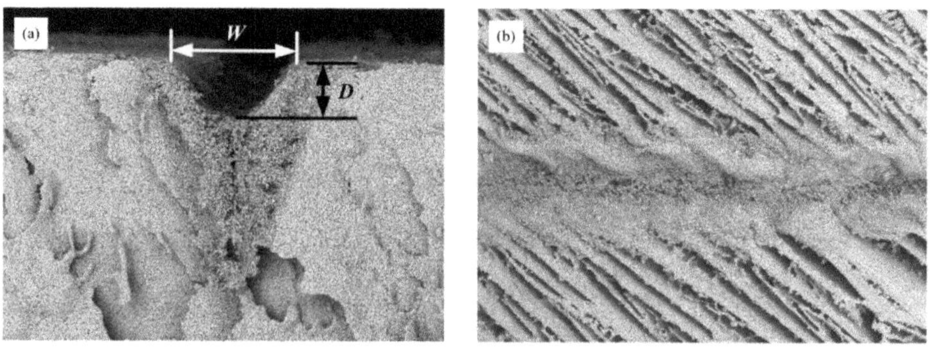

Figure 6.5. Morphologic characteristics of the cutting line of the frozen slurry with the laser beam: (a) cross-sectional view and (b) surface profile. Reprinted from [34], Copyright (2018), with permission from Elsevier.

from all directions met there. The ice crystals had integrated morphology, but with ununiform size distribution and random orientation.

The cryo-panel used to freeze the slurry layer could generate freezing in a top-to-bottom direction. Owing to the relatively small thickness, the materials inside the slurry layer could be very quickly frozen, taking a time duration of less than 5 s. This unidirectional freezing process facilitated the formation of ice crystals, with unique fine and uniform lamellar structures that were parallel to the direction of Z-axis. The principle of freeze drying indicated that the structure of the pore resembled the structure of the ice crystals. Accordingly, the sintered products exhibited similar inhomogeneous and disordered pore distribution to the as-printed samples. In addition, the sample derived from the slurry with 50 wt% solid had larger pores than that from the one with 60 wt%. In other words, the higher the water content, the larger the size of the pores would be. In fact, both the apparent porosity and closed porosity were decreased with increasing solid loading level of the slurries.

It was found that compressive strength of the samples in the Z-axis direction increased with increasing solid loading level. The structures made through the layer-by-layer freezing exhibited higher compressive strength than those processed through holistic freezing. In principle, the samples made with holistic freezing supported full growth of the ice crystals, thus enabling the formation of pores with large sizes after the freeze-drying. Accordingly, the latter could not withstand large compressive forces, because the large pores tended to be broken more easily. Moreover, the random orientation of the pores gave rise to isotropic behaviors of the porous structures. In contrast, the layer-by-layer freezing resulted in structures with ordered pore orientation and absence of large pores, thus leading to significant improvement in compressive strength, especially in the direction of Z-axis. Photographs of representative structures made with the FS-LOM process are shown in figure 6.6.

More recently, a new LOM process based on aqueous slurry to fabricate porous ceramics was reported [45]. Mesh sheets of polymer were first prepared, serving as templates for pore-forming, over which slurry layers were scraped. Then, the 2D patterns were printed out by laser cutting through the dried mesh-ceramic composite layer. Eventually, porous structures with predesigned pores were developed after

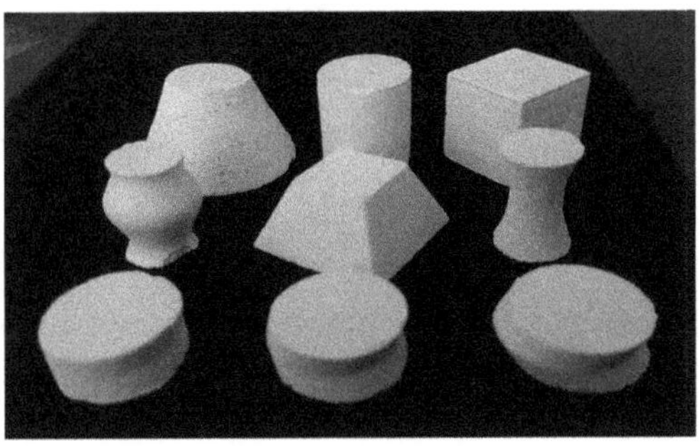

Figure 6.6. Photographs of the ceramic items processed by using the FS-LOM process. Reprinted from [34], Copyright (2018), with permission from Elsevier.

debinding and sintering. Alumina ceramic items, with porosities of up to 51.5% and round hole diameters of 80 μm were fabricated from slurry with 70 wt% alumina, combined with a 100 mesh net of nylon. Since the polymer meshes not only acted as a template but also as a supporting framework, they prevented damage to the green bodies and ensured integrity, uniformity, and connectivity of the micron-sized pore networks. Furthermore, by using a layer-by-layer drying process, the delamination phenomenon was effectively avoided, while the paving density was significantly improved.

Alumina powder with $D_{50} = 1$ μm (Sinopharm Chemical Reagent, China) as the ceramic component, carboxymethyl cellulose sodium (Yajuli Pure Phemical, Japan) as the organic binder, ammonium polyacrylate as the dispersant (Sinopharm Chemical Reagent, China and nylon net (PA 6, 100 mesh, Shangshai Bolting Cloth Manufacturing Co., Ltd, Shanghai, China) as the pore forming sacrificial template were used in the experiment. The bond-then-cut LOM process was adopted for the three-dimensional forming strategy, while the organic meshes were employed as the sacrificial template to form desired porous structures.

The fabrication process of the new method is schematically shown in figure 6.7. A layer of nylon mesh sheet was laid on the work table, on which a layer of slurry was spread. After the layer was dried with an IR heater, a composite layer was formed. Then, the peripheral contours of a layer-related 2D slice were cut from an object model using a CO_2 laser with beam diameter of 0.1 mm. After that, the work table was lowered a distance equivalent to the thickness of the printing layer, followed by repeating the previous steps, until the designed printing process was finished. Finally, green bodies were taken out by removing all the excessive materials, which were then subjected to debinding and sintering, resulting in the desired porous alumina ceramic products.

The pore formation mechanism is schematically illustrated in figure 6.8. The slurry layer was formed using a silicone blade, which scraped the slurry over the surface of the mesh, ensuring all holes were entirely filled. In this case, thickness of the composite layer, i.e., the slice thickness of the 3D model, was determined by the thickness of the mesh sheets. 70 wt% alumina powder, 1 wt% binder, 1.4 wt%

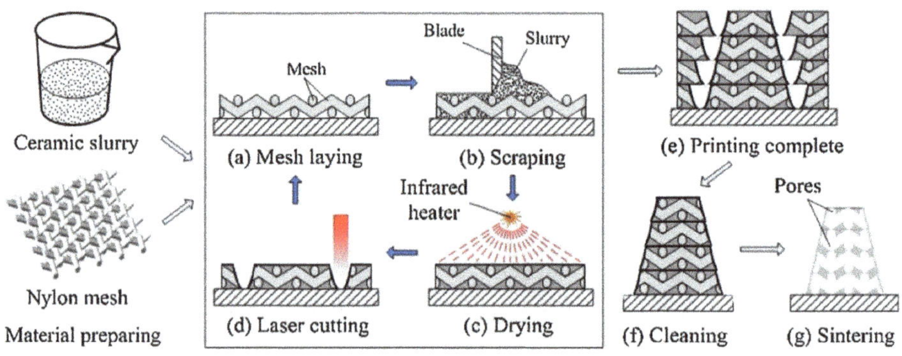

Figure 6.7. Schematic diagram of the LOM printing process. Reprinted from [45], Copyright (2021), with permission from Elsevier.

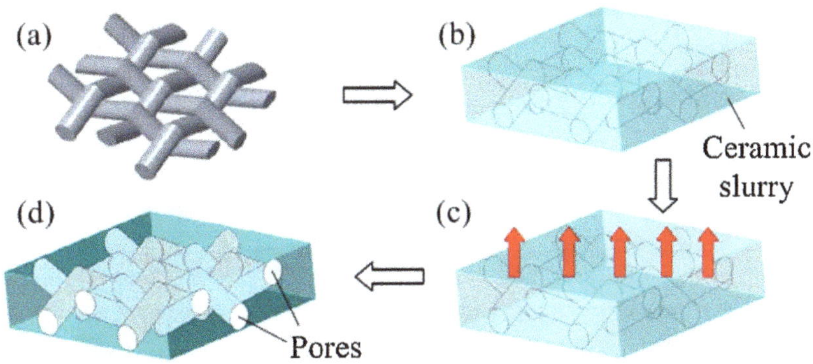

Figure 6.8. Ore forming mechanism in the process: (a) organic mesh sheet, (b) mesh-ceramic composite layer, (c) removal of organic binders through heat treatment and (d) network of pores. Reprinted from [45], Copyright (2021), with permission from Elsevier.

dispersant and 27.6 wt% DI water were thoroughly blended through ball milling for 12 h. After printing, the green bodies were sintered at 1550 °C in air for 2 h.

The nylon filaments had a diameter of about 100 μm, while the nylon mesh sheet had a thickness of 190 μm. The regular mesh structure was woven with wavy filaments that were of circular cross-section. After the slurry was spread, a composite layer was formed, with a thickness of 210 μm, which was reasonably thicker than the mesh sheet. The as-printed samples were optimally dried at 200 °C for 40 s. Too-high temperatures would cause deformation of the green bodies, because the melting point of nylon was 220 °C.

With increasing laser energy density, the cutting depth was gradually increased. In order to easily remove the green bodies after printing, the composite layer should be cut completely, i.e., the cutting depth was $D > 210$ μm. For example, at $P = 30$ W, the laser could not cut through the layer. For layer powers of 40–50, 60–80 and 90 W, the corresponding allowed maximum scanning speeds were 200, 300 and 400 $\text{mm} \cdot \text{s}^{-1}$, respectively. Finally, the optimized cutting parameters included $P = 40$ W and $V = 200$ $\text{mm} \cdot \text{s}^{-1}$, corresponding to D and W to be 219 and 282 μm, respectively.

Figure 6.9 shows photographs of the representative samples at different stages. The layer-by-layer drying process enabled strong bonding between adjacent layers. In addition to structural support from the organic binder, layers were also adhered through hydrogen bonding and capillary force generated during the drying process. The green bodies had sufficient mechanical strength, due to the nylon mesh, so that they could be readily taken out after printing and removal of the excessive materials. By using a sufficiently slow heating rate, the green bodies experienced no deformation or damage in the debinding process, porous ceramic products could be successfully obtained after sintering at appropriate conditions.

Circular pores were uniformly present in the bulk of the sintered products, while delamination in between the layers was not observed. Both obliquely upward and downward channels were formed, with wavy shapes, when viewing from the vertical section. The channels had an average diameter of 80 μm, which was smaller than the

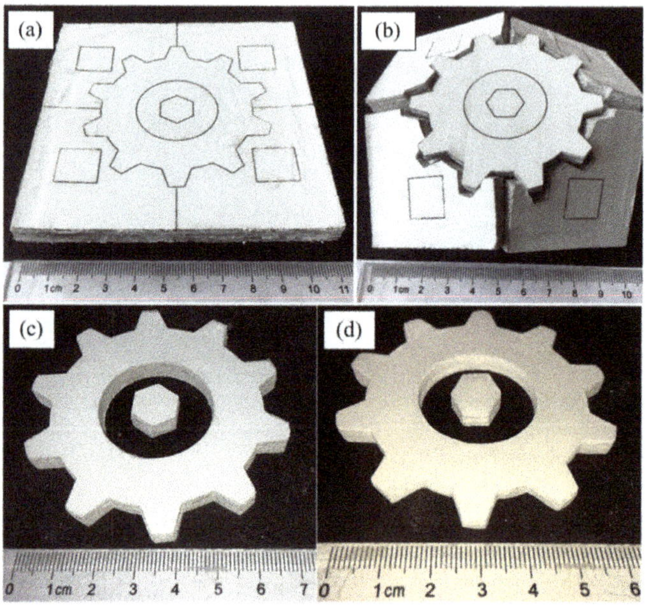

Figure 6.9. Photographs of the representative printed samples: (a) green body, (b and c) after removal of excess materials and (d) sintered sample. Reprinted from [45], Copyright (2021), with permission from Elsevier.

diameter of the nylon filament, because of dimensional shrinkage during the sintering process. The channels were entirely connected at the intersection of the horizontal and longitudinal ones. The porous alumina ceramics had a density of 1.93 $g \cdot cm^{-3}$, with open and closed porosities of 51.1% and 0.4%, respectively.

SiC ceramics have also been fabricated using LOM process with preceramic polymer tapes [23]. To prepare curved LOM structures for body armor applications, SiC-filled tapes were coated onto paper mandrels, allowing the samples to achieve desirable curvatures. As the curvatures matched the designed ceramic structures, the overlapped areas could be minimized, leading to minimal material removal after printing. The slurry was made of bimodal SiC powder, carbon black powder and graphite powder, which were mixed and blended with organic binders, at contents of 15–20 wt%. The tapes were 250 μm in thickness. The as-printed green bodies were calcined at 325 °C to ensure the structural integrity and then pyrolyzed 700 °C in Ar. The pyrolyzed samples were subjected to silicon infiltration at 1600 °C, thereby converting them into fully dense SiC products, which had a flexural strength of 142–165 MPa.

LOM was combined with pressureless sintering to prepare complex-shaped SiC ceramic products [46]. The SiC green tapes with a thickness of 0.15 mm were used for the LOM printing, which were developed by using SiC powder and an organic binder through tape casting. The final SiC products exhibited a relative density of 98.2%, along with bending strength, hardness, toughness and elastic modulus of 402 MPa, 19.86 GPa, 3.32 $MPa \cdot m^{1/2}$ and 393 GPa, respectively.

Pyrolyzed filter papers were fabricated for LOM printing processes to develop Si–SiC composite ceramic structures [25]. The multiplayered components were then subjected to post-infiltration with silicon melts, leading to dense Si–SiC laminar composite ceramics. Commercial filter papers of cellulose fibers (Type 2992, area density = 200 g · m^{-2}, sheet thickness = 430 μm, Hahnemühle Fine Arts, Germany) were employed for the LOM printing. The paper sheets were pyrolyzed in a muffle furnace (Gero GLO 40, Gero, Germany) in N_2, at 350 °C for 1 h, at a heating rate of 1 °C · min^{-1}, to prevent the papers from cracking and curling.

Then, the samples were heated at 800 °C for 1 h, at a heating rate of 2 °C · min^{-1}. After that, the samples were cooled to room temperature at a cooling rate of 5 °C · min^{-1}. The weight loss and linear shrinkage were 75 wt% and 23%, respectively. Adhesive tapes were obtained using tape casting, with the slurries added with phenolic resin, polyvinyl butyral (Solutia Inc., St. Louis, MO, USA), benzyl butyl phthalate (Brenntag, Germany) and ethanol (BfB, Germany). The adhesive tapes were bonded to the pyrolyzed paper sheets to form biocarbon papers, one side of which were coated with phenolic resin adhesive.

After coating, the biocarbon paper sheets were printed using an LOM printer (Kira PLT A4, Kira Corporation, Japan). Laminating was carried out at 180 °C for 20 s, which was used for the curing cycle of the adhesive layers. The samples with 18 layers had a thickness of 4.5 mm, with a density of 0.86 g · cm^{-3}. The as-printed green bodies were calcined at 800 °C in N_2, so that the phenolic resin was carbonized in the adhesive bonding tapes. After carbonization, the porous samples were infiltrated with Si melts at 1500 °C for 1 and 7 h in vacuum. The entire process is shown schematically in figure 6.10.

Figure 6.11 shows cross-section SEM images of the printed samples before Si melt infiltration. The layered structures of pyrolyzed carbon paper and the phenolic resin were clearly visible, which were present as highly porous structures. Slight delamination was detectable, owing to the localized excessive adhesive between

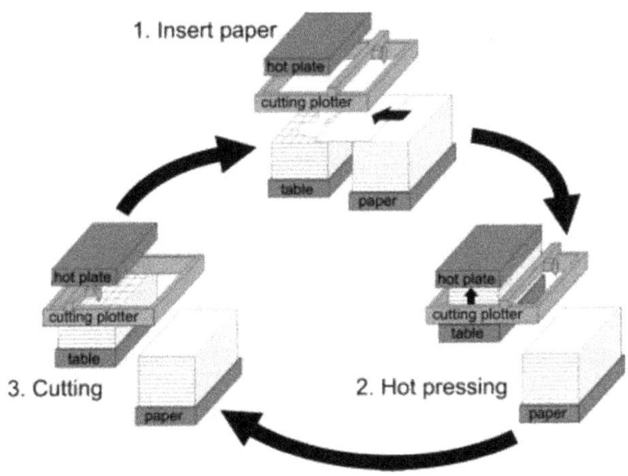

Figure 6.10. Schematic diagram of the LOM printing process. [25] John Wiley & Sons. Copyright © 2004 WILEY-VCH Verlag GmbH & Co. KGaA, Weinheim.

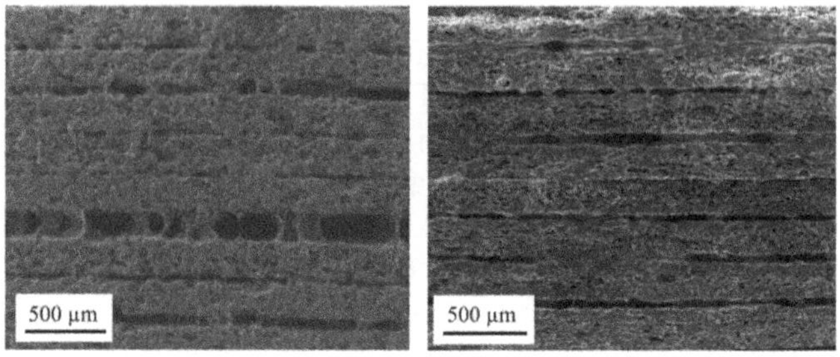

Figure 6.11. Multilayer structural morphologies of the samples before infiltration with Si: (a) as-printed green body and (b) after debinding. [25] John Wiley & Sons. Copyright © 2004 WILEY-VCH Verlag GmbH & Co. KGaA, Weinheim.

the pyrolyzed layers after the carbonization process. Specifically, the mass loss was about 34%, while the in-plane and out-plane shrinkages were 2% and 1.5%, respectively. The density of the samples was about 0.6 g · cm^{-3} and the porosity was as high as 69%.

Figure 6.12 shows SEM images of the Si–SiC composite ceramic samples after Si melt infiltration and sintering at 1500 °C for time durations of 1 and 7 h. Obviously, residual carbon was still observed in the microstructure. The SiC grains had sizes of 5–12 μm. The volumetric contents of carbon in the samples sintered for 1 and 7 h were 26% and 16%, respectively. The contents of Si were 33% and 23%, while those of SiC were 41% and 61%. The results suggested that prolonged sintering time was beneficial to the conversion of SiC. The final composite ceramics displayed a bending strength of 130 MPa after sintering for 1 h, while it decreases to 123 MPa in the sample sintered for 7 h, implying that prolonged sintering time had a negative effect on mechanical strength of the composite materials fabricated using LOM printing technique.

Similarly, another group of preceramic papers, SiC-filler-loaded cellulosic papers, were prepared for LOM printing to fabricate dense Si–SiC ceramics with complex structures [47]. The printing process involved preparation of preceramic paper, adhesive coating, LOM processing and pyrolysis and Si infiltration. The preceramic paper was made of SiC powder (Mikro F 1200 D, D_{50} = 4.5 μm, ESK-SiC, Germany), pulp (Celbi PP, Celulosa Beira Industrial, Portugal) and retention agent and binder (Catiofast VFH, Bayer, Germany), at weight concentrations of 76.8%, 20% and 3.2%, corresponding to volume concentrations of 30%, 18% and 7%, respectively, using Rapid Köthen aqueous hand-sheet-forming processes (DIN EN ISO 5269–2). The as-prepared paper sheets were subjected to calendaring treatment to minimize the porosity and smoothen surfaces.

The adhesive coating was derived from powder mixture consisting of 76.9 wt% (61.4 vol%) non-crosslinked polysiloxane (methylphenyl poly(silsesquioxane), Silres H44, Wacker, Germany), 15.4 wt% (6.4 vol%) novolacphenolic resin (0222 SP 04, Bakelite, Germany) and 7.7 wt% (32.2 vol%) fumed silica (OX 50, Degussa, Germany) as a dispersion agent, at weight contents of 76.9%, 15.4% and 7.7%, corresponding to volume contents of 61.4%, 6.4% and 32.2%, respectively, through

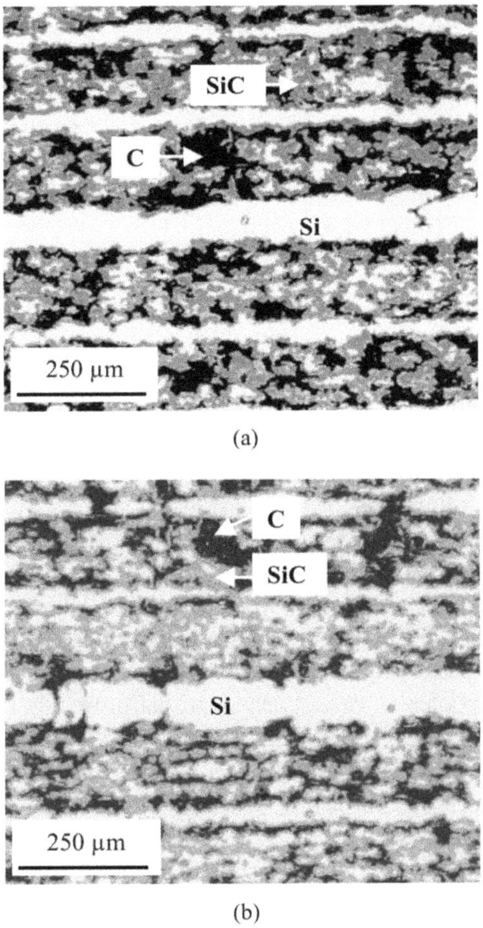

Figure 6.12. Microstructures of the Si–SiC composite ceramics after sintering at 1550 °C for different time durations: (a) 1 h and (b) 7 h. [25] John Wiley & Sons. Copyright © 2004 WILEY-VCH Verlag GmbH & Co. KGaA, Weinheim.

ball milling for 24 h with alumina balls as the milling media. The milled mixture was sprayed on the paper sheets through a 200 mesh sieve. After that, the preceramic paper sheets were dried at 90 °C for 10 min to homogenize the adhesive coating.

The laminated preceramic paper sheets were processed using the LOM printer to obtain preforms, with the machine having a heated lamination roller and a laser cutter (Helisys 1015plus, Helisys Inc., USA). The laser cutting system was equipped with a 25 W continuous wave (CW) CO_2 laser. The laminating process was conducted at 140 °C, at both the forward and backward heater speeds, to be 40 mm · s^{-1}. By retracting the printing table by 0.1 mm, layer bonding was guaranteed in the lamination process. Samples with rectangular shapes were printed with every two adjacent layers to be arranged with different orientations, as schematically demonstrated in figure 6.13. The samples had layer numbers of 15, 28 and 104 for the orientations (a), (b) and (c), respectively.

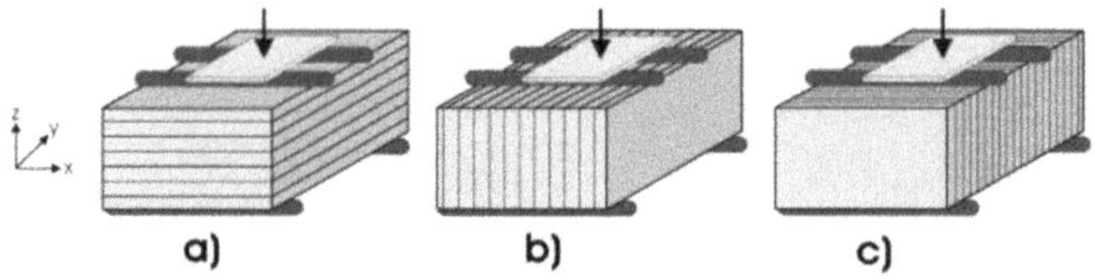

Figure 6.13. Schematic diagram presenting the layer orientations and loading directions. [47] John Wiley & Sons. Copyright © 2007 WILEY-VCH Verlag GmbH & Co. KGaA, Weinheim.

After removal of excessive materials, the as-printed green bodies were calcined at 350 °C and 800 °C for 1 h each in N_2, at both the heating and cooling rates of 1 °C · min^{-1}. The calcined samples were infiltrated with Si melts in a graphite crucible coated with BN (Sintec, Buching, Germany), to separate the graphite crucible and the Si melts. Si powder (Silgrain HQ coarse, 99.4% Si, Elkem ASA, Norway) was used to surround the samples. The infiltration experiment was conducted at 1500 °C in a vacuum of <10 Pa for 1 h, at both the heating and cooling rates of 10 °C · min^{-1}.

The preceramic paper sheets were prepared using the Rapid Köthen aqueous hand-sheet-forming process. The surfaces of the filler particles and cellulose fibers were negatively charged in solutions with pH ≈ 7 [48]. The retention of the filler in the fabrication process involved three steps, i.e., flocculation, adsorption and filtration. In this case, the retention agent should be positively charged, so as to flocculate the filler particles to be adsorbed on cellulose fibers. The retention agent was bonded onto the cellulose fibers through flocks of the filler particles. Flexibility and strength of the paper sheets were modified with a latex binder. The fibers in the sheets were of no preferential orientation. Meanwhile, the sheets were calendered to decrease the porosity and surface roughness.

Figure 6.14 depicts SEM images of representative SiC-loaded preceramic papers, which had been produced by the sheet forming process. The SiC particles were filled in the spaces among the cellulose fibers, followed by coating with polymer adhesive. The adhesive was softened at about 60 °C, thus allowing for the LOM printing experiment. The as-prepared preceramic paper sheets were heated at 90 °C for 10 min, while the adhesive mixture was applied to form binder dots, with diameter of 400 μm and area weight of 31 g · cm^{-2}.

The samples calcined at 800 °C had a ceramic phase content of about 75 wt%, while the pyrolyzed product of the adhesive facilitated interfacial bonding between the layers in the paper sheets. Because the weight losses of the adhesive and preceramic paper were only slightly different, the calcined samples experienced very little dimensional change. After calcination at 800 °C, shrinkage of the laminates was almost isotropic, with values of 3.5% and 3.1%, in the x- and y-directions, confirming that the fibers were randomly oriented in the plane of the paper sheets. In comparison, the shrinkage in the z-direction (out-of-plane) was only slightly higher, with a level of 3.9%.

Geometrical and skeletal densities of the calcined sheets were 1.13 and 2.93 g · cm^{-3}, respectively, corresponding to a volumetric porosity of 61.5%. The samples exhibited a peak porosity in the range of 3–5 μm, which reflected the capillary cell diameters of the cellulose fibers. After calcination, capillary diameter was reduced, owing to the shrinkage

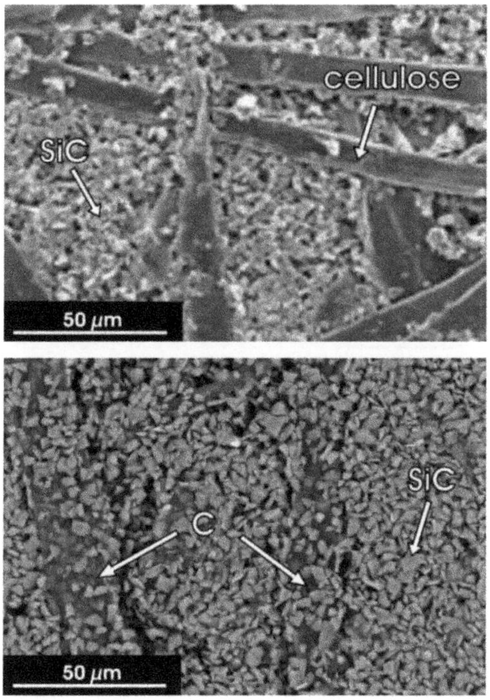

Figure 6.14. Surface SEM images of the SiC-filler-loaded preceramic papers, with top and bottom panels to be the green body and pyrolyzed sample, respectively. [47] John Wiley & Sons. Copyright © 2007 WILEY-VCH Verlag GmbH & Co. KGaA, Weinheim.

in the radial direction off the cellulose fibers, with typical values of 30%–40%. The pore size peaked at 1–2 μm, due to the spacings among the filler particles. The large pores of >10 μm were closely associated with the interfacial spacings between adjacent layers in the paper sheets.

Owing to the high wettability of Si melt on carbon, the porous SiC–C preforms derived from pyrolysis could be readily infiltrated with Si at 1500 °C [49]. Figure 6.15 shows SEM images of representative Si–SiC composites. After Si infiltration, the Si–SiC composite samples had a density of 2.6 g · cm^{-3}. The samples exhibit no residual carbon in the matrix, suggesting that the Si melt completely reacted with carbon. As revealed by the cross-sectional SEM image, uniform and coplanar Si-rich layers were observed with thicknesses ranging from 5–20 μm. According to image analysis results based on the SEM images in figure 6.15, the total volumetric fractions of Si and SiC in the composites were 54.9% and 45.1%, respectively. The values were slightly different from those measured in the paper sheets, due possibly to the existence of Si in the bond layer.

Mechanical properties of the Si–SiC composites were anisotropic, simply because of their layered microstructural characteristics. The samples exhibited bending strength of 150 MPa in *c*-direction, while the average fracture toughnesses were 3.8 and 3.5 MPa · $m^{1/2}$ the *a*- and *b*-directions, respectively, which were comparable with those reported in open literature [1]. The fracture toughness in the *c*-direction

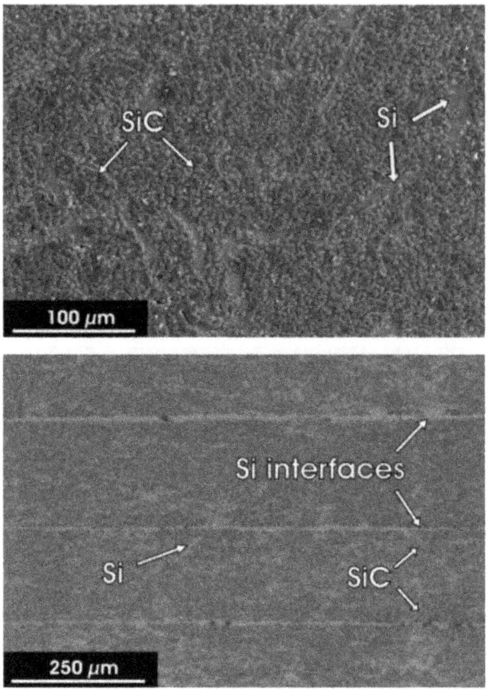

Figure 6.15. Representative SEM images of the printed Si–SiC composite ceramics. Top panel: relief-polished top surface indicating that the SiC filler particle distribution can be correlated to the statistical fiber packing during the paper fabrication. Bottom panel: relief-polished cross-sectional SEM image of the Si–SiC layers. [47] John Wiley & Sons. Copyright © 2007 WILEY-VCH Verlag GmbH & Co. KGaA, Weinheim.

was relatively low, mainly because of the presence of Si in the bonding layers. Hardness was nearly constant at the loadings of 10–100 N. In addition, Young's modulus was 252 GPa, which was also within the expectation.

Uniform Si_3N_4 green sheets with 48.7 vol% solid loading were fabricated using tape casting, combined with LOM to develop Si_3N_4 ceramics with complicated shapes [50]. The slurry exhibited typical shear-thinning behavior without precipitation during the tape casting process, as evidenced by the energy dispersion spectrum analysis results. Ceramic structures with complicated shapes were obtained through layer-by-layer stacking of the green sheets, followed by pressureless sintering at 1800 °C. Phase transition from α-Si_3N_4 to β-Si_3N_4 was confirmed by the XRD characterization results. The final Si_3N_4 ceramics exhibited full densification and promising mechanical strengths.

Cost-effective commercial α-Si_3N_4, Al_2O_3 and Y_2O_3 powders were used as the raw materials. Al_2O_3 and Y_2O_3 powders were mixed at mass ratio of 3:5, as the sintering aid, at a content of 10 wt% with respect to Si_3N_4. Polyacrylic acid (PAA), polyvinyl alcohol (PVA), glycerol and n-butyl alcohol were adopted as dispersant, binder, plasticizer and defoamer to prepare the slurry. The dispersant and the ceramic powders were dispersed in water, followed by mixing for 12 h. After that, the binder and plasticizer were introduced into the suspension, followed by milling for additional 2 h to form the slurry. Then, n-butyl alcohol was incorporated, while the slurry was degassed for 30 min at vacuum level of 0.1 Pa. The slurry was coated on glass substrate through tape casting (LYJ, Beijing Dongfang Co. Ltd China) at a speed of 0.2 m · min^{-1}. The as-casted tapes were naturally dried at room temperature in air. The dried tapes had a thickness of 150 μm.

The dry tapes were cut into pieces, which were stacked layer-by-layer, forming green bodies with different shapes and structures. The as-stacked samples were pressed at 50 MPa for 2 min with isostatic pressing at room temperature. The samples were then subjected to debinding, at 650 °C in air at a heating rate of 0.5 °C · min^{-1}, followed by pressureless sintering at 1800 °C for 1 h N_2 at 50 Pa, with a gas sintering furnace (SGM/VB/8–18, Shanxi Bohua Co. Ltd China).

For tape casting, the ceramic slurries should be of pseudoplastic behaviors with proper viscosities. Viscosity of the slurries decreased while passing through the slit below the blade because of the shear thinning effect, whereas the recovery of viscosity after the blade prevented unexpected flow or sedimentation [51, 52]. Viscosity of the slurry was 600 MPa·s, as the shear rate was at the level of 30 s^{-1}. Noting that the viscosity of the slurry decreased with increasing shear rate, the slurry was confirmed to have the shear-thinning behaviors for pseudoplastic fluids. At the same time, the viscosity of the slurry was time independent, being desirable for the tape casting experiment.

The dried tapes displayed apparent density of 2.4 g · cm^{-3} and bulk density of 1.47 g · cm^{-3}, corresponding to 61 vol% in total for the ceramic powder and the organic components. The green tapes after debinding had porosity of 51.3%, i.e., the content of the ceramic phase was 48.7 vol%. Such a high solid content was beneficial for densification of the silicon nitride ceramics. Small pores with an average size of 130 nm were present in the dried tapes. The top and the bottom surfaces of the dried tapes had same elemental compositions, as revealed by the EDS results indicated identical chemical composition for the compounds. The sintering aids of Y_2O_3 (5.01 g · cm^{-3}) and Al_2O_3 (3.95 g · cm^{-3}) were not precipitated in the slurry during the during tape casting process.

For the dried tapes, the top surface was smoother than the bottom one, since the concentration of PVA in the slurry varied throughout the drying process of the newly prepare tapes in air [53, 54]. Specifically, the maximum roughness values were 0.850 and 0.737 μm, for the top and the bottom surfaces, respectively. The relatively small surface roughness ensured effective stacking of the tapes to form green bodies with strong layer cohesions.

The green tapes could be cut into structures with different shapes, as illustrated in figure 6.16(a). The green tapes were stacked into green bodies, as seen in figure 6.16 (b). After sintering at 1800 °C for 1 h, the samples were densified, without variation in shape, as depicted in figure 6.16(c). Meanwhile, deformation and warping were not observed in the sintered structures. A total linear shrinkage was 23.2%, confirming high sintering behavior of the green bodies. The sample was of a density of 3.0 g · cm^{-3}, with a low porosity of 6.3%. Figure 6.17 shows XRD patterns of the sintered silicon nitride ceramic structures. The main phase was β-Si_3N_4, with a trace of $YAlO_3$, while α-Si_3N_4, Al_2O_3 and Y_2O_3 were absent.

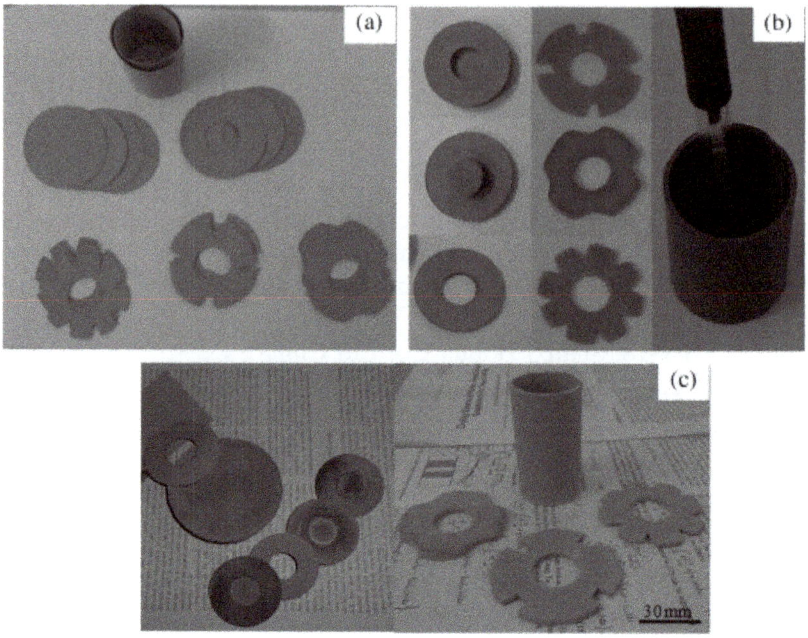

Figure 6.16. Photographs of the Si_3N_4 samples with complex shapes: (a) cut tapes, (b) stacked green bodies and (c) sintered ceramics. Reprinted from [50], Copyright (2015), with permission from Elsevier.

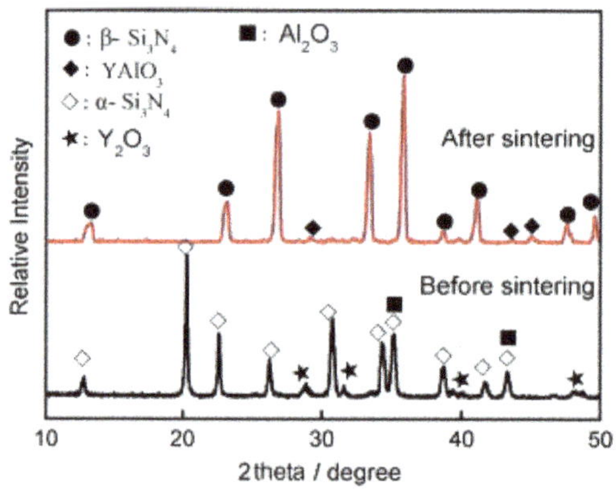

Figure 6.17. XRD patterns of the silicon nitride samples before and after sintering. Reprinted from [50], Copyright (2015), with permission from Elsevier.

The silicon nitride parts after sintering exhibited a bending strength of 475 ± 34 MPa. Such mechanical performance was comparable with that of silicon nitride ceramics processed using the conventional ceramic method [55–57], while they were higher than those derived using gel casting method. Figure 6.18 shows typical SEM images of the silicon nitride ceramics, in which anisotropic growth and pull-out profiles of the β-Si_3N_4 grains were clearly present. The rod-like β-Si_3N_4 grains were responsible for the high mechanical performance of final ceramic samples, owing to

Figure 6.18. Typical fractured surface SEM images of the silicon nitride ceramics after sintering at 1800 °C for 1 h. Reprinted from [50], Copyright (2015), with permission from Elsevier.

the reinforcement effect through the mechanisms of crack deflections, crack bridgings and grain pulling out.

MAX phase Ti_3SiC_2 was obtained through an *in situ* synthesis process, from laminated TiC and SiC tapes via LOM processes, combined with pyrolysis and infiltration of liquid silicon [58]. Three-dimensional ceramic gears were fabricated with defect-free structure, while the linear shrinkage after densification was <3%, which was close to the near-net-shaping process of ceramic components with complicated structures. Commercial ceramic powders, TiC (HC Stark, Goslar, Germany, D_{10} = 0.9 μm, D_{50} = 2.4 μm, D_{90} = 5.4 μm) and SiC (α-SiC, ESK-SIC GmbH, Frechen, Germany, D_{10} = 1.8 μm, D_{50} = 3.9 μm, D_{90} = 6.7 μm), were mixed with various mass to make samples with different compositions, including TiC/SiC volumetric ratio = 30:70, 50:50 and 70:30, denoted as samples #3, #5 and #7, respectively.

The solid loading of the ceramic powder was 30 vol% for all slurries. Firstly, the TiC and SiC powders were dispersed in a mixed solvent consisting of 68 wt% ethanol and 32 wt% toluene. These mixtures were then deagglomerated using atumbling mixer (Turbula, Willy A. Bachofen AG, Muttenz, Switzerland), milling for 1 day with Al_2O_3 ceramic balls as the milling media. After that, binder (polyvinyl butyral, B-98, Solutia Inc., St. Louis, USA) and plasticizer (Santicizer 9280, Ferro, USA) were introduced. Finally, the mixtures were milled for one more day to ensure homogeneity of the slurries, followed by sieving through screen with hole size of 200 μm, to separate the milling balls. All the slurries were degassed at 210 mbar for 0.5 h to remove the bubbles.

The tape-casting machine was equipped with a casting head consisting of two doctor blades. PET films coated with silicon (Mitsubishi Plastics, Inc., Japan) were used as carriers of the tapes, which were 100 μm in thickness. The front doctor blade was set with a gap of 1.0 mm, while the gap height of the rare blade was 1.2 mm. All tapes were cast at a speed of 70 cm min^{-1}. The dried tapes #3, #5 and #7 exhibited thicknesses of 550, 400 and 350 μm, respectively. Square samples with dimension of

40 × 40 mm^2 were cut from the dried tapes for the measurement of mechanical properties. At the same time, the tapes were stacked through thermal pressing (Polystat 200 t, Servitec GmbH, Wustermark, Germany) at 180 °C and 3.5 MPa for 10 min to form samples with thickness of 5 mm.

Meanwhile, the green tapes derived from slurry #3 were processed using a laser-equipped LOM machine (1015, Helysis Inc., MI, USA) to produce 3D samples with gear geometry. The speeds for the cutting and roller were set at 3 and 6 cm · s^{-1}, respectively. The roller was controlled to be at 60 °C. For lamination at relatively low temperature and low pressure, double-sided adhesive tapes (TESA, Norderstedt, Germany) were inserted between every two adjacent green tapes to ensure sufficiently strong adhesion.

The green laminates were calcined in Ar at 1000 °C for 1 h with a heating rate of 3 °C · min^{-1}, and then the samples were sintered at 1600 °C for 2 h at a heating rate of 4 °C · min^{-1}. The sintered samples experienced a linear shrinkage of <3%, resulting in porous microstructures. The porous samples were then infiltrated with liquid silicon, thus leading to a MAX phase without pores. The infiltration experiment was conducted in vacuum with a pressure of <100 Pa at 1450 °C for 2 h.

The dried tapes derived from slurries #3, #5 and #7 had porosities of 24%, 22% and 21%, respectively. After calcination and sintering in Ar, the porosities were increased to 43%, 39% and 40% for samples #3, #5 and #7, because the organic components in green tapes were burnt out during the thermal treatment process. After liquid silicon infiltration, sample #3 was free of defects and almost fully densified. In comparison, samples #5 and #7 encountered deformation in structure, such as cracks and delamination effects, which was attributed to the large change in volume related to the reaction between TiC and Si during the high temperature infiltration process [59].

Figure 6.19 shows XRD patterns of the samples before and after the infiltration process, confirming the aforementioned *in situ* reaction. The samples without

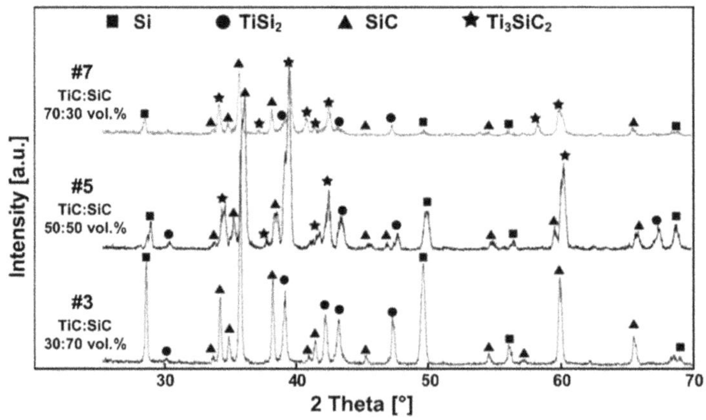

Figure 6.19. XRD patterns of the TiC/SiC laminates after infiltration with liquid silicon. Reprinted from [58], Copyright (2017), with permission from Elsevier.

infiltration had similar XRD patterns, consisting of SiC and TiC. After infiltration, diffraction peaks of SiC, Si and $TiSi_2$ were present. In samples #5 and #7, MAX Ti_3SiC_2 was formed, while it was not present in sample #3. Figure 6.20 depicts SEM images of the TiC/SiC laminated structures, with different compositions, before and after the liquid silicon infiltration experiment. After sintering at 1600 °C in Ar, the samples were composed of TiC, SiC and pores, appearing as white, dark and black areas, respectively, as shown in figures 6.20(a), (d), and (g).

Upon infiltration with liquid silicon, highly dense microstructures were formed, as observed in figures 6.20(b), (e), and (h), while TiC was absent. Obviously, sample #3 possessed more SiC, as illustrated in figure 6.20(b), as compared with samples #5 and #7, after infiltration. Due to the high temperature reaction, $TiSi_2$ was formed as the reaction product in sample #3, as demonstrated by the grey parts in figure 6.20 (b), whereas excessive Si was left as a residual phase at the interface between $TiSi_2$ and SiC. MAX phase was not detected in sample #3 infiltrated with liquid silicon.

In contrast, the MAX Ti_3SiC_2 was formed in samples #5 and #7, with relatively high content, as revealed by the brighter phase in figures 6.20(e) and (h). Nevertheless, both the samples #5 and #7 contained $TiSi_2$ (grey phase) and SiC (dark phase) as the major phases. The Si phase observed by using XRD in sample #5 and sample #7 was actually near the surface, as confirmed by the SEM and EDS

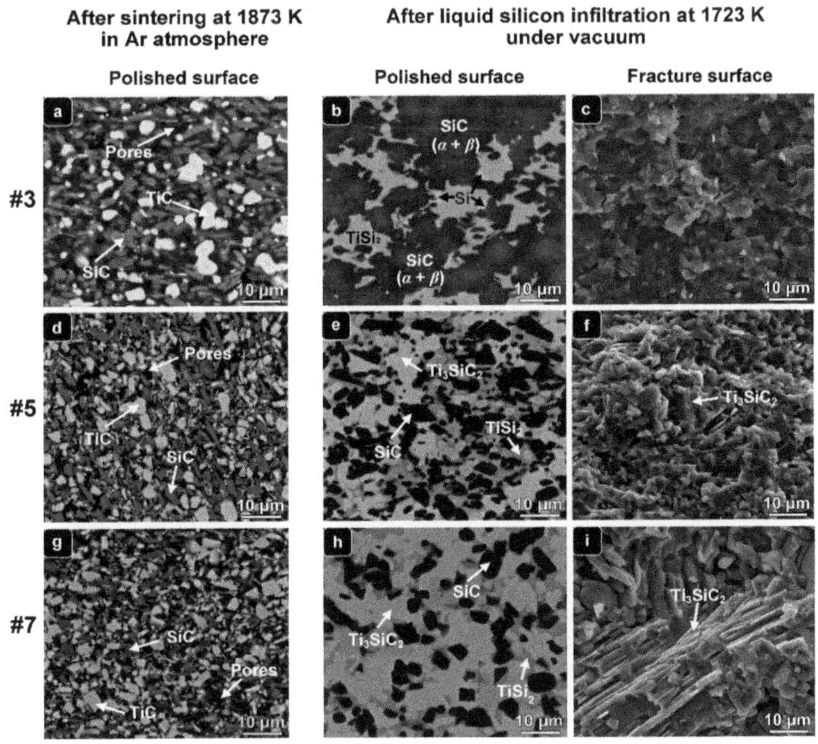

Figure 6.20. SEM images of the TiC/SiC laminates with different compositions before and after the infiltration process with liquid silicon. Reprinted from [58], Copyright (2017), with permission from Elsevier.

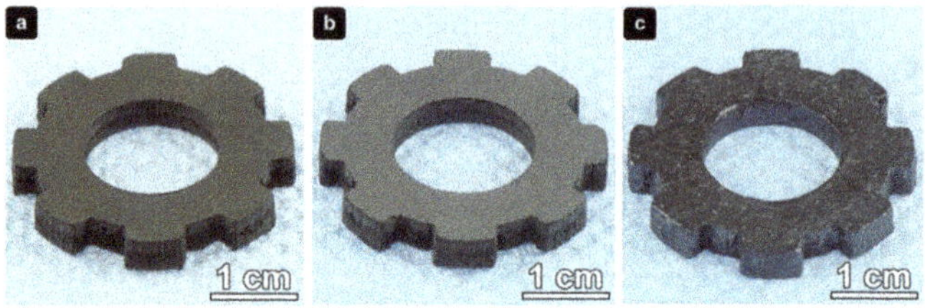

Figure 6.21. Photographs of the samples with gear geometry made of sample #3 at different stages: (a) green body, (b) after sintering in Ar and (c) after infiltration of liquid silicon. Reprinted from [58], Copyright (2017), with permission from Elsevier.

results. In addition, the particle size of SiC was tremendously increased after infiltration with liquid silicon. This was particularly evident in sample #3, as shown in figure 6.20(b), where SiC was present in two phases. One was an α-SiC phase from the raw materials, while the other one was a β-SiC phase, which was the product of the high temperature reaction during infiltration process. The MAX Ti_3SiC_2 was visible at the fractured surface of samples #5 and #7, as seen in figures 6.20(f) and (i). Therefore, the formation of MAX was dependent on the TiC content in the green tapes.

In practical experiments, a sufficiently large amount of silicon should be used to ensure that all the pores are filled during the infiltration process. For samples #5 and #7, the Si/TiC ratio was suitable for the simultaneous formation of $TiSi_2$ and Ti_3SiC_2. However, in sample #3, the content of Si was excessive. Therefore, to form Ti_3SiC_2 in sample #3, the amount of Si used for the infiltration should be reduced. In this case, the sample would exhibit higher porosity.

The pores in sintered samples #5 and #7 were filled with the liquid Si, while SiC was formed due to the reaction between Si and TiC. Three more TiC was consumed for the reaction to form one mole Ti_3SiC_2. Because the reaction was accompanied by large variation in volume, defects were induced in the sample #5 and sample #7. In other words, a balance must be maintained between the synthesis of MAX and the densification of the samples. Since the formation of one mole $TiSi_2$ corresponded with the consumption of one mole TiC, the infiltration of sample #3 resulted in microstructure with fewer defects. Figure 6.21 shows photographs of the samples derived from the green tape #3, with gear geometry, fabricated by using the LOM process. The near shrinkage was less than 3%, which was very close to near-net-shape process.

6.3.3 Glass-ceramics and ceramic composites

Glass-ceramics have also been used in the LOM printing for different applications. Glass-ceramics are a unique class of materials that combine the best properties of glass and ceramics [60–65]. They are initially formed as homogeneous glass through conventional glass-making techniques, such as melting raw materials like silica, alumina, and various metal oxides at high temperatures and then cooling them rapidly to avoid crystallization. When glasses are subjected to post-heat treatment, controlled crystallization is induced within the glass matrix, resulting in materials with a microstructure composed of fine grained crystals dispersed in a glassy phase, which are known as glass-ceramics. Glass-ceramics have unique mechanical properties, such as high strength and hardness. They also exhibit high thermal shock resistance, withstanding rapid changes in temperature without cracking or breaking [66]. Therefore, they have a wide range of applications in various fields.

Representatively, Li_2O–ZrO_2–SiO_2–Al_2O_3 (LZSA) glass green tapes were prepared by using aqueous tape casting, which were then applied to LOM printing [26, 27, 67]. Rheological behaviors of the slurries and mechanical properties of the green tapes were studied. The surface of the LZSA glass powder was acidic in nature. The solid loading was about 72 wt%, corresponding to 27 vol%. The glass particles were of anisometric characteristics, thus leading to increment in particle interaction. With the addition of dispersants, aqueous glass suspensions exhibited a shear-thickening effect. Slurries with three compositions were tested for LOM experiments. The higher the tensile strength of the green tapes, the higher the tensile strength of the laminated structure would be. The green bodies were calcined at 525 °C, sintered at 700 °C for 1 h and crystallized at 850 °C for 0.5 h. The total volume shrinkage was about 20%, while the products had a smooth surface with minimal flaws and a homogeneous microstructure.

LZSA glass frit was crashed with water and alumina grinding media. Two commercial ammonium polyacrylates (NH4PA) were employed as dispersants to prepare the glass suspensions (Darvan C, Vanderbilt, Norwalk, CT and Dolapix CA, Zschimmer & Schwarz, Lahnstein, Germany). PVA solution (31.5 wt%, Mowiol 4–88, Kuraray, Frankfurt am Main, Germany) was utilized as a binder, while polyethylene glycol (PEG, PEG 400, Synth, Karlsruhe, Germany) was used as a plasticizer, which was blended with modified fatty and alkoxylated compounds as an antifoamer (Agitan 354, Munzing, Heilbronn, Germany). Firstly, the parent glass powder was dispersed in distilled water together with the dispersant for 24 h. Then, the binder solution was added and stirred for 12 h. Finally, the plasticizer and antifoam were introduced and mixed for another 12 h.

The slurries were cast on polyethylene terephthalate (PET) films, by using a double doctor blade tape casting machine, at room temperature with a casting rate of 450 mm $\cdot$ min^{-1}. The tapes were naturally dried for two days before being removed from the PET carriers. The LOM printing was conducted with a continuous-wave (CW) CO_2 laser system (1015, Helisys, Rochester Hills, MI) at a power of 16.8 W. The speeds of cutting and roller were 50 and 25 mm $\cdot$ s^{-1}, respectively. The roller temperature was maintained at 80 °C. Lamination of the tapes was carried out with a 5 wt% aqueous binder solution to ensure adhesion, through brush painting on the side that was in contact with the PET films.

After ball milling, the LZSA glass powder was refined, with the particle size decreasing with increasing milling time. The milling duration was set to 13 days to ensure the particle size met the requirements of the tape casting experiment. The milled glass powder had particle sizes of 2–8 μm, with agglomeration and sharp-edged irregular morphology. Relative viscosity of the LZSA glass powder in aqueous suspension at 1200 s^{-1} increased with increasing solid loading level.

At relatively low solids loading level, e.g., 50 wt%, viscosity of the slurry was too low to process the tape casting, owing to the too large volume shrinkage [68, 69]. As the solid loading level was raised to 60 wt%, the viscosity was obviously increased. Once the solid content was increased to 70 wt%, the slurry displayed thixotropy and dilatant behaviors. Such time-dependent behaviors were present, as the concentration of the dispersion agent was as low as 1 wt%. The optimal concentration of the binder was 10.5 wt%, in terms of high quality tape casting. With the presence of the PVA binder, rheological behavior of the glass powder slurries transitioned from shear thickening to shear thinning, thus meeting the requirements of the tape casting process.

Three concentrations of organic additives were evaluated, i.e., 12.28, 13.28 and 16.18 wt%, which were denoted as slurry 1, 2, and 3, respectively. With increasing content of organic components, tensile strength of the tapes was reduced. Specifically, the tape samples derived from slurry 3 had the smallest tensile strength, due to the low binder-to-plasticizer ratio. In comparison, tapes made of slurries 1 and 2 had comparable tensile strengths, although the binder-to-plasticizer ratio was the same. Meanwhile, the tape from slurry 1 possessed higher porosity, probably due to poor dispersant efficiency. Figure 6.22 depicts a photograph of the green tape obtained with slurry 2, with high flexibility and relatively low strain.

The tapes were laminated in a layer-by-layer way, where every two adjacent layers were rotated by 90° to enhance the mechanical strength of the green bodies. At the same time, tensile strength of the laminated green bodies was proportional to the tensile strength of thc tapes. Therefore, it is important to optimize the tap performance. Figure 6.23 shows a representative SEM image of the LOP laminated green bodies. The small porosity was associated with air bubbles caused during the brushing of the binder. Figure 6.24 illustrates a photograph of two samples with gear wheel shapes, which were made of the tapes derived from slurry 2. After sintering, the gear profile was well retained.

Ceramic composites are advanced materials that are engineered to overcome the inherent brittleness and improve other properties of traditional ceramics. They are composed of a ceramic matrix, which can be based on oxides, such as alumina, or non-oxides, such as silicon carbide, within which various reinforcing elements are

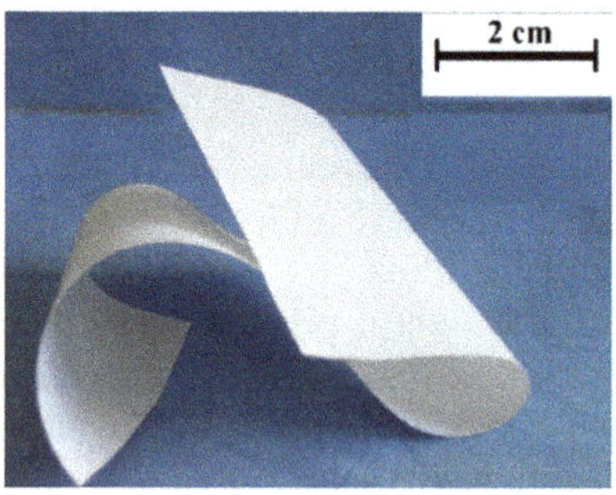

Figure 6.22. Photograph of the flexible tape derived from slurry 2. [26] John Wiley & Sons. © 2009 The American Ceramic Society.

Figure 6.23. Fractured surface SEM image of the laminated green bodies made of the tape derived from slurry 2. [26] John Wiley & Sons. © 2009 The American Ceramic Society.

Figure 6.24. Photographs of the Li_2O–ZrO_2–SiO_2–Al_2O_3 glass-ceramic samples with gear wheels before (left) and after sintering (right) sintered. [26] John Wiley & Sons. © 2009 The American Ceramic Society.

embedded. These reinforcements can take the form of fibers, particles or whiskers. The addition of these reinforcements significantly enhances the mechanical behavior of the ceramics.

Al_2O_3/Cu–O composites were developed with paper-derived alumina as the matrix, through infiltration with metallic Cu containing 3.2 wt% oxygen [70]. Alumina-loaded preceramic papers were sintered at 1600 °C for 4 h to form the paper-derived alumina preforms, which had open porosities in the range of 14%–25%. The infiltration was conducted at 1320 °C for 4 h at ambient pressure, leading to highly dense composites, which exhibited promising mechanical and electrical performances, including fracture toughness of 6 $MPa \cdot m^{1/2}$, four-point-bending strength of 342 MPa, Young's modulus of 281 GPa and electrical conductivity of 2 $MS \cdot m^{-1}$.

Al_2O_3/Cu structures have unique structural and electrical properties for various applications, especially in semiconductor packages [71]. When Cu metal was spread on an α-Al_2O_3 substrate, the wetting angle ranged between 160° and 170°, at high temperatures of 1100 °C–1250 °C [72]. In this case, it is difficult for Cu to infiltrate into Al_2O_3. To increase the wettability of Cu on Al_2O_3, and thereby facilitate pressureless infiltration, oxygen should be introduced [73]. The introduction of oxygen promoted the formation of $CuAlO_2$ through the interfacial reaction between Cu or Cu_2O and the Al_2O_3. The optimal content of oxygen was 3.2 wt%.

Preceramic papers with solid loadings of 75 and 80 vol% alumina (CT 3000 SG, Almatis GmbH, Frankfurt, Germany) were obtained from diluted aqueous suspensions, together with pulp fibers. The pulp mixture consisted of two non-refined pulps, i.e., 40 wt% thermomechanical softwood pulp (Orion ECF, Zellstoff Pöls AG, Pöls, Austria) with diameter and length of 22 and 1665 μm and 60 wt% hardwood pulp (Celbi S.A., Figueira da Foz, Portugal) with diameter and length of 15 and 657 μm. The contents of the pulp fibers were 14 and 18 vol%. The organic additives were totally at volume concentrations of 5.5% and 7%.

The preceramic paper sheets were prepared using a dynamic manual sheet making machine (Dynamic hand-sheet former D7, Sumet Systems GmbH, Denklingen, Germany). The pumped flow rate was 5450 ml $\cdot$ min^{-1}, while the wire speed was controlled at 1200 rpm. The resultant paper sheets were dried at 110 °C for 15 min and then calendered using a single-nip lab calendar (CA5/250–150–20, Sumet Systems GmbH, Denklingen, Germany), with hardened steel rolls that could be heated and had a diameter of 250 mm. The roll operated at a temperature of 80 °C and line load of 80 N $\cdot$ mm^{-1}. The papers were then shaped through LOM processes, followed by pressureless infiltration with Cu-alloy. Also, a sandwich structure was assembled, consisting of a face skin of Al_2O_3/Cu–O composite and a core of a dense Al_2O_3 substrate.

The preceramic papers coated with adhesive were laminated and pressed at a pressure of 10 MPa at 80 °C for 5 min (Polystat 200 T, Servitec Maschinenservice, Wustermark, Germany), leading to samples with dimensions of 30 × 30 × 0.2 mm^3. Meanwhile, corrugated boards with directed macroscopic channels of pores were obtained by combining two sheets of preceramic paper (liners) connected by a corrugated inner part (fluting), without the application of pressures.

The LOM machine (1015, Helysis Inc., MI, USA) was used for the experiments, equipped with a laser cutting kit and a lamination roller that could be heated. A 25 W continuous wave (CW) CO_2 laser was utilized for the laser cutting kit. The layers were stacked through the adhesive applied to the bottom of each paper sheet. The lamination experiment was carried out at 140 °C, while the roller heater was moved forwardly and reversely, at moving speeds of 40 mm $\cdot$ s^{-1}. The final samples had eight layers of paper sheets.

The green bodies of the preceramic papers were calcined and sintered, over the temperature ranges of 25 °C–200 °C, 200 °C–350 °C, 350 °C–700 °C and 700 °C–1600 °C, at ramping rates of 5, 0.5, 0.5, and 5 °C $\cdot$ min^{-1}, respectively. The samples were kept at 350 °C and 1600 °C for 1 h and 4 h, respectively, while the cooling rate was set to be 5 °C $\cdot$ min^{-1}. The pressureless infiltration was conducted at 1320 °C in Ar for 4 h, at the heating and cooling rates of 5 °C $\cdot$ min^{-1}. The sandwich structures were treated in a similar way.

The green bodies had open porosities in the range of 13.9%–25.2%, while the residual porosities were reduced to the levels of 1.9%–2.4% after being infiltrated with the Cu–O alloy. Phase compositions of the samples were characterized using XRD analysis, confirming the formation of $CuAlO_2$ after the infiltration experiment. The components, including the starting materials and the reaction product, were uniformly distributed in the samples, as presented in figure 6.25(a). Fractured surface SEM

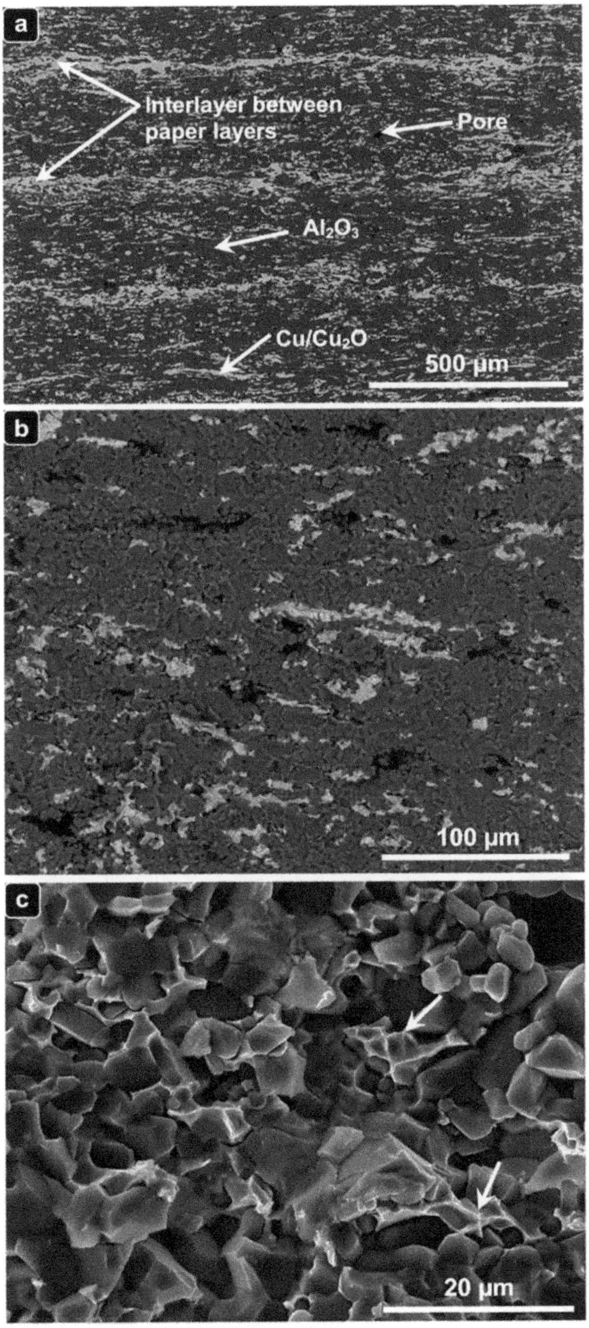

Figure 6.25. Cross-sectional SEM images of the infiltrated composites at different magnifications. Reprinted from [70], Copyright (2018), with permission from Elsevier.

images are depicted in figures 6.25(b) and (c), revealing plastic deformation for Cu particles. Meanwhile, the alumina grains were covered by cuprite networks.

In the sandwich structure, with dimensions of 20 × 20 × 4 mm^3, the middle Al_2O_3 layer had thickness of about 0.8 mm, while the two Al_2O_3/Cu–O layers were about 1.6 mm in thickness. Cross-sectional SEM images of the sandwich structure are illustrated in figure 6.26. The black area indicated the dense cast tape without infiltration, where the ceramics were bonded by the copper alloy, forming paper-derived Al_2O_3/Cu–O composites. After infiltration, the copper alloy flowed into the spaces between the ceramics, thus acting as a metallic solder. The copper alloy layer was formed during the first round of the infiltration process. The Cu_2O areas exhibited orientation towards the ceramic substrate. At the interface between the substrate and the copper alloy layer, a new layer formed due to the reaction at the interface, as demonstrated in figure 6.26(b). The total compositions included 51.3at% oxygen, 24.5at% Al and 24.2at% Cu, which is close to the formular of $CuAlO_2$. Figure 6.27 shows photographs of the structures before and after infiltration with the Cu–O alloy. The infiltration resulted in corrugated structures with nearly fully dense microstructures.

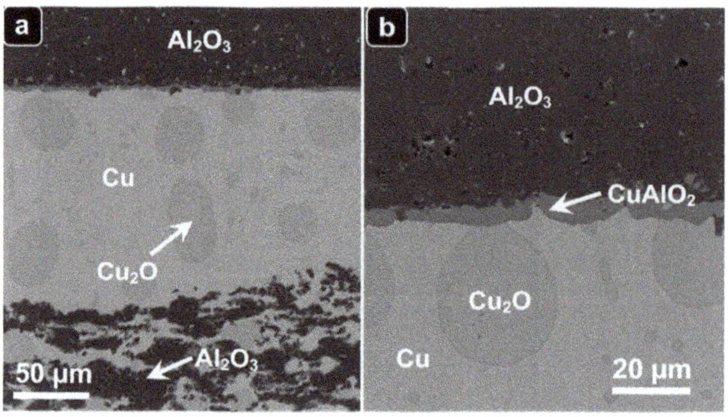

Figure 6.26. Cross-sectional interface SEM images of the copper alloy layer in the Al_2O_3/Cu–O composite: (a) low magnification and (b) high magnification. Reprinted from [70], Copyright (2018), with permission from Elsevier.

Figure 6.27. Photographs of the corrugated structures before (left) and after pressureless infiltration experiment (right). Reprinted from [70], Copyright (2018), with permission from Elsevier.

6.4 Conclusions and perspectives

Laminated object manufacturing (LOM) is a significant 3D printing technology known for its unique characteristics. In LOM, sheets of one or more materials, commonly in the form of papers, plastic films or metallic foils, are employed to construct 3D objects and structures in a layer-by-layer manner. The general LOM process involves bonding of consecutive layers, which could be facilitated by heating. When paper is used as the laminate, heat-activated adhesives are applied. The adhered layers are then cut using a mechanical cutter or laser, according to CAD models.

One of the major advantages of LOM is its relatively high production rate. Because only the outer contours of the layers need to be cut, large-scale models and production could be achieved in a short period. In addition, LOM is relatively cost-effective compared to other 3D printing technologies, especially when paper is used as the raw material. The process also produces models with precise dimensional stabilities and high structural integrities, which is especially evident in larger structures. Furthermore, no support structures are required in the LOM printing process, which greatly simplifies post-processing.

However, LOM faces several challenges. For instance, the range of usable materials is relatively limited, restricting its applications in areas that require materials with specific performances. LOM-processed structures typically have poor surface finishes, requiring additional post-processing treatments, which reduce productivity and increase both processing time and costs. Another problem is the possible delamination, i.e., the layers may separate over time, particularly in environments with high humidity or frequent variation in temperature. Moreover, the simple LOM process is not suitable for applications requiring complex geometries.

In the future, material innovation will be an important research topic in LOM technology. New materials are expected to exhibit higher mechanical performance, enhanced resistance to delamination, and suitability for a wider range of applications. Process optimization cannot be overemphasized. Refining the cutting and bonding processes is necessary to achieve greater precision, higher accuracy, and smoother surfaces. Integration of LOM with other technologies, such as injection molding, may open new possibilities for creating more complicated and functional ceramic materials. Last but not least, improvements in the software algorithms that control the LOM process should be considered, to better manage the layer-by-layer printing process and reduce printing errors.

References

[1] Dermeik B and Travitzky N 2020 Laminated object manufacturing of ceramic-based materials *Adv. Eng. Mater.* **22** 202000256

[2] Kim H, Lin Y R and Tseng T L B 2018 A review on quality control in additive manufacturing *Rapid Prototyp. J.* **24** 645–69

[3] Deckers J, Vleugels J and Kruthl J P 2014 Additive manufacturing of ceramics: a review *J. Ceram. Sci. Technol.* **5** 245–60

[4] Vaezi M, Seitz H and Yang S F 2013 A review on 3D micro-additive manufacturing technologies *Int. J. Adv. Manuf. Technol.* **67** 1721–54

[5] Ahn D, Kweon J H, Choi J and Lee S 2012 Quantification of surface roughness of parts processed by laminated object manufacturing *J. Mater. Process. Technol.* **212** 339–46

[6] Tofail S A M, Koumoulos E P, Bandyopadhyay A, Bose S, O'Donoghue L and Charitidis C 2018 Additive manufacturing: scientific and technological challenges, market uptake and opportunities *Mater. Today* **21** 22–37

[7] Bhatt P M, Kabir A M, Peralta M, Bruck H A and Gupta S K 2019 A robotic cell for performing sheet lamination-based additive manufacturing *Addit. Manuf.* **27** 278–89

[8] Colombo P, Schmidt J, Franchin G, Zocca A and Günster J 2017 Additive manufacturing techniques for fabricating complex ceramic components from preceramic polymers *Am. Ceram. Soc. Bull.* **96** 16–23

[9] Goh G D, Yap Y L, Agarwala S and Yeong W Y 2019 Recent progress in additive manufacturing of fiber reinforced polymer composite *Adv. Mater. Technol.* **4** 201800271

[10] Rypl D and Bittnar Z 2006 Generation of computational surface meshes of STL models *J. Comput. Appl. Math.* **192** 148–51

[11] Hao J B, Fang L A and Williams R E 2011 An efficient curvature-based partitioning of large-scale STL models *Rapid Prototyp. J.* **17** 116–27

[12] Akhoundi B, Jahanshahi A S and Abbassloo A 2024 G-code generation for deposition of continuous glass fibers on curved surfaces using material extrusion-based 3D printing *Eng. Res. Express* **6** 015401

[13] Rivet I, Dialami N, Cervera M, Chiumenti M and Valverde Q 2023 Mechanical analysis and optimized performance of G-code driven material extrusion components *Addit. Manuf.* **61** 103348

[14] Butt J, Mebrahtu H and Shirvani H Title: numerical and experimental analysis of product development by composite metal foil manufacturing *Int. J. Rapid Manuf.* **7** 59–82

[15] Paul B K and Voorakarnam V 2001 Effect of layer thickness and orientation angle on surface roughness in laminated object manufacturing *J. Manuf. Process.* **3** 94–101

[16] Travitzky N, Bonet A, Dermeik B, Fey T, Filbert-Demut I, Schlier L, Schlordt T and Greil P 2014 Additive manufacturing of ceramic-based materials *Adv. Eng. Mater.* **16** 729–54

[17] Chiu Y Y and Liao Y S 2003 Laser path planning of burn-out rule for LOM process *Rapid Prototyp. J.* **9** 201–11

[18] Cho I, Lee K, Choi W and Song Y A 2000 Development of a new sheet deposition type rapid prototyping system *Int. J. Mach. Tools Manuf.* **40** 1813–29

[19] Chiu Y Y, Liao Y S and Hou C C 2003 Automatic fabrication for bridged laminated object manufacturing (LOM) process *J. Mater. Process. Technol.* **140** 179–84

[20] Kechagias J 2007 Investigation of LOM process quality using design of experiments approach *Rapid Prototyp. J.* **13** 316–23

[21] Liao Y S, Chiu L C and Chiu Y Y 2003 A new approach of online waste removal process for laminated object manufacturing (LOM) *J. Mater. Process. Technol.* **140** 136–40

[22] Sonmez F O and H. Thomas Hahn H T 1998 Thermomechanical analysis of the laminated object manufacturing (LOM) process *Rapid Prototyp. J.* **4** 26–36

[23] Klosterman D, Chartoff R, Graves G, Osborne N and Priore B 1998 Interfacial characteristics of composites fabricated by laminated object manufacturing *Composites A: Appl. Sci. Manuf.* **29** 1165–74

[24] Cui X M, Ouyang S, Yu Z Y, Wang C G and Huang Y 2003 A study on green tapes for LOM with water-based tape casting processing *Mater. Lett.* **57** 1300–4

[25] Weisensel L, Travitzky N, Sieber H and Greil P 2004 Laminated object manufacturing (LOM) of SiSiC composites *Adv. Eng. Mater.* **6** 899–903

[26] Gomes C M, Rambo C R, de Oliveira A P N, Hotza D, Gouvêa D, Travitzky N and Greil P 2009 Colloidal processing of glass-ceramics for laminated object manufacturing *J. Am. Ceram. Soc.* **92** 1186–91

[27] Gomes C, Travitzky N, Greil P, Acchar W, Birol H, de Oliveira A P N and Dachamir D 2011 Laminated object manufacturing of LZSA glass-ceramics *Rapid Prototyp. J.* **17** 424–8

[28] Schindler K and Roosen A 2009 Manufacture of 3D structures by cold low pressure lamination of ceramic green tapes *J. Eur. Ceram. Soc.* **29** 899–904

[29] Kluthe C, Dermeik B, Kollenberg W, Greil P and Travitzky N 2012 Processing, microstructure and properties of paper-derived porous Al_2O_3 substrates *J. Ceram. Sci. Technol.* **3** 111–7

[30] Schlordt T, Dermeik B, Beil V, Freihart M, Hofenauer A, Travitzky N and Greil P 2014 Influence of calendering on the properties of paper-derived alumina ceramics *Ceram. Int.* **40** 4917–26

[31] Schultheiss J, Dermeik B, Filbert-Demut I, Hock N, Yin X W, Greil P and Travitzky N 2015 Processing and characterization of paper-derived Ti_3SiC_2 based ceramic *Ceram. Int.* **41** 12595–603

[32] Stares S L, Kirilenko A, Fredel M C, Greil P, Wondraczek L and Travitzky N 2013 Paper-derived bioactive glass tape *Adv. Eng. Mater.* **15** 230–7

[33] Menge G, Lorenz H, Fu Z W, Eichhorn F, Schader F, Webber K G, Fey T, Greil P and Travitzky N 2018 Paper-derived ferroelectric ceramics: a feasibility study *Adv. Eng. Mater.* **20** 201800052

[34] Zhang G, Chen H, Yang S B, Guo Y Z, Li N, Zhou H W and Cao Y 2018 Frozen slurry-based laminated object manufacturing to fabricate porous ceramic with oriented lamellar structure *J. Eur. Ceram. Soc.* **38** 4014–9

[35] Zhang Y, He X, Han J, Du S and Zhang J 2001 Al_2O_3 ceramics preparation by LOM (laminated object manufacturing) *Int. J. Adv. Manuf. Technol.* **17** 531–4

[36] Gong S P, Sang P G and Zhou D X 2003 Research on the manufacture of laminated $BaTiO_3$-based thermistor by roll forming *Mater. Sci. Eng. B: Solid State Mater. Adv. Technol.* **99** 425–7

[37] Kim Y S, Lee S Y, Hong S K and Jeon H J 2001 Formation of barrier ribs for plasma display panel via roll forming of green tapes *J. Am. Ceram. Soc.* **84** 1470–4

[38] Liu W, Zhang H J and Zhang W H 2009 Study on preparation technique and properties of zirconia wear-resistant ceramic balls manufactured by a roll-forming method *Rare Met. Mater. Eng.* **38** 198–201

[39] Deville S 2008 Freeze-casting of porous ceramics: a review of current achievements and issues *Adv. Eng. Mater.* **10** 155–69

[40] Deville S, Saiz E and Tomsia A P 2007 Ice-templated porous alumina structures *Acta Mater.* **55** 1965–74

[41] Krishnan P P R, Kumar P A and Prabhakaran K 2023 Preparation of macroporous alumina ceramics by ice templating without freeze drying using natural rubber latex binder *J. Porous Mater.* **30** 1499–507

[42] Munch E, Launey M E, Alsem D H, Saiz E, Tomsia A P and Ritchie R O 2008 Tough, bio-inspired hybrid materials *Science* **322** 1516–20

[43] Sabat S, Sikder S, Behera S K and Paul A 2022 Large scale alignment of alumina platelets en route to porous nacre-like alumina by ice-templating *Ceram. Int.* **48** 2893–7

[44] Preiss A, Su B, Collins S and Simpson D 2012 Tailored graded pore structure in zirconia toughened alumina ceramics using double-side cooling freeze casting *J. Eur. Ceram. Soc.* **32** 1575–83

[45] Zhang G, Guo J D, Chen H and Cao Y 2021 Organic mesh template-based laminated object manufacturing to fabricate ceramics with regular micron scaled pore structures *J. Eur. Ceram. Soc.* **41** 2790–5

[46] Zhong H, Yao X, Zhu Y, Zhang J, Jiang D, Chen J, Chen Z, Liu X-J and Huang Z 2015 Preparation of SiC ceramics by laminated object manufacturing and pressureless sintering *J. Ceram. Sci. Technol.* **6** 133–40

[47] Windsheimer H, Travitzky N, Hofenauer A and Greil P 2007 Laminated object manufacturing of preceramic-paper-derived Si–SiC composites *Adv. Mater.* **19** 4515–9

[48] Mandlez D, Koller S, Eckhart R, Kulachenko A, Bauer W and Hirn U 2022 Quantifying the contribution of fines production during refining to the resulting paper strength *Cellulose* **29** 8811–26

[49] Scheithauer U, Schwarzer E, Moritz T and Michaelis A 2018 Additive manufacturing of ceramic heat exchanger: opportunities and limits of the lithography-based ceramic manufacturing (LCM) *J. Mater. Eng. Perform.* **27** 14–20

[50] Liu S C, Ye F, Liu L M and Liu Q 2015 Feasibility of preparing of silicon nitride ceramics components by aqueous tape casting in combination with laminated object manufacturing *Mater. Des.* **66** 331 5

[51] Bitterlich B and Heinrich J G 2002 Aqueous tape casting of silicon nitride *J. Eur. Ceram. Soc.* **22** 2427–34

[52] Gutiérrez C A and Moreno R 2000 Tape casting of non-aqueous silicon nitride slips *J. Eur. Ceram. Soc.* **20** 1527–37

[53] Liu S C, Chen P, Li Y M, Li W J, Gao S X and Ye F 2016 Effect of stacking pressure on the properties of Si_3N_4 ceramics fabricated by aqueous tape casting *Ceram. Int.* **42** 16281–6

[54] Liu S C, Ye F, Hu S Q, Yang H X, Liu Q and Zhang B 2015 A new way of fabricating Si_3N_4 ceramics by aqueous tape casting and gas pressure sintering *J. Alloys Compd.* **647** 686–92

[55] Gal C W, Song G W, Baek W H, Kim H K, Lee D K, Lim K W and Park S J 2019 Fabrication of pressureless sintered Si_3N_4 ceramic balls by powder injection molding *Ceram. Int.* **45** 6418–24

[56] Wang H J, Yu J L, Zhang J and Zhang D H 2010 Preparation and properties of pressureless-sintered porous Si_3N_4 *J. Mater. Sci.* **45** 3671–6

[57] Penas O, Zenati R, Dubois J and Fantozzi G 2001 Processing, microstructure, mechanical properties of Si_3N_4 obtained by slip casting and pressureless sintering *Ceram. Int.* **27** 591–6

[58] Krinitcyn M, Fu Z W, Harris J, Kostikov K, Pribytkov G A, Greil P and Travitzky N 2017 Laminated object manufacturing of *in situ* synthesized MAX-phase composites *Ceram. Int.* **43** 9241–5

[59] Kero I, Tegman R and Antti M L 2010 Effect of the amounts of silicon on the *in situ* synthesis of Ti_3SiC_2 based composites made from TiC/Si powder mixtures *Ceram. Int.* **36** 375–9

[60] Chen X D, Tan Y, Yan H, Shi J, Wu J J and Ding B 2024 A review of cleaner production of glass-ceramics prepared from MSWI fly ash *J. Environ. Manage.* **370** 122855

[61] Höland W, Beall G H and Smith C M 2025 Glass-ceramics: from ideas to products *J. Am. Ceram. Soc.* **108** 200086

[62] Rüssel C and Wisniewski W 2025 Glass-ceramic engineering: tailoring the microstructure and properties *Prog. Mater Sci.* **152** 101437

[63] Montazerian M, Baino F, Fiume E, Migneco C, Alaghmandfard A, Sedighi O, DeCeanne A V, Wilkinson C J and Mauro J C 2023 Glass-ceramics in dentistry: fundamentals, technologies, experimental techniques, applications, and open issues *Prog. Mater. Sci.* **132** 101023

[64] Venkateswaran C, Sreemoolanadhan H and Vaish R 2022 Lithium aluminosilicate (LAS) glass-ceramics: a review of recent progress *Int. Mater. Rev.* **67** 620–57

[65] Xiao Z H, Luo M H, Han R L and Wang Y Z 2015 Crystallization behaviour of Y_2O_3 doped germanate oxyfluoride glass-ceramics *Glass Technol.-Eur. J. Glass Sci. Technol.* A **56** 126–31

[66] Xiao Z H, Dong X F, Luo M H, Liang H Y, Luo W Y, Yu X N, Yi W and Li X 2018 Preparation of cordierite glass-ceramics with low expansion coefficient *J. Ceram* **39** 239–43

[67] Gomes C M, Oliveira A P N, Hotza D, Travitzky N and Greil P 2008 LZSA glass-ceramic laminates: fabrication and mechanical properties *J. Mater. Process. Technol.* **206** 194–201

[68] Alazzawi M K, Beyoglu B and Haber R A 2021 A study in a tape casting based stereolithography apparatus: role of layer thickness and casting shear rate *J. Manuf. Process.* **64** 1196–203

[69] Heunisch A, Dellert A and Roosen A 2010 Effect of powder, binder and process parameters on anisotropic shrinkage in tape cast ceramic products *J. Eur. Ceram. Soc.* **30** 3397–406

[70] Pfeiffer S, Lorenz H, Fu Z W, Fey T, Greil P and Travitzky N 2018 Al_2O_3/Cu-O composites fabricated by pressureless infiltration of paper-derived Al_2O_3 porous preforms *Ceram. Int.* **44** 20835–40

[71] Shi Y G, Chen W G, Dong L L, Li H Y and Fu Y Q 2018 Enhancing copper infiltration into alumina using spark plasma sintering to achieve high performance Al_2O_3/Cu composites *Ceram. Int.* **44** 57–64

[72] Li F P, Wang W X, Dang W, Zhao K and Tang Y F 2020 Wetting mechanism and bending property of Cu/Al_2O_3 laminated composites with pretreated CuO interlayer *Ceram. Int.* **46** 17392–9

[73] Zheng J W, Gao D M, Qiao L, Ying Y, Li W C, Jiang L Q and Shenglei C 2016 Influence of the Cu_2O morphology on the metallization of Al_2O_3 ceramics *Surf. Coat. Technol.* **285** 249–54

Chapter 7

Applications of 3D printed ceramics

With the rapid development of 3D printing technology in the fabrication of ceramic materials, their corresponding applications have also attracted much attention. Currently, ceramics with various applications have been reported in the open literature in almost all fields, where examples, such as biomedical, piezoelectric, microwave, transparent ceramics and ceramic matrix composites, will be discussed in this chapter.

7.1 Brief introduction

3D printing technology, or additive manufacturing, has emerged as a powerful tool for fabricating and developing ceramic materials for a wide range of applications. For instance, 3D printing can be used in medical field to produce customized dental crowns and bridges that offer a better fit and enhanced aesthetics compared to traditional fabrication methods. 3D printed ceramic bone implants with tailored porosity can promote cell growth and tissue integration, facilitating faster healing and reducing the risk of complications. This chapter aims to provide a brief summary of recent progress in the development of ceramics using 3D printing, for applications across various fields, including biomedical, piezoelectric, microwave, transparent ceramics, and ceramic matrix composites.

7.2 Applications of 3D printed ceramics

7.2.1 Biomedical applications

Obviously, 3D printing technology has the potential to fabricate implants with patient-specific structures that match the geometry and size of individual human bones. In addition, it enables the creation of site-specific structures with functional gradient in composition, density, and mechanical properties [1]. Ceramics fabricated via 3D printing for biomedical applications, which currently under intensive study, include zirconia (ZrO_2), alumina (Al_2O_3), silicates, and phosphates [2–8].

doi:10.1088/978-0-7503-4831-7ch7

Representative examples are briefly presented below, and readers are encouraged to consult the original literature for more detailed information.

A zirconia crown was fabricated using direct inkjet writing (DIW) processes with zirconia suspensions, resulting in a 3D structure with characteristic occlusal surface topography [9]. The suspensions had a solid loading level of 27 vol%. Due to nozzle clogging during the DIW process, both the as-printed and sintered ceramics possessed some flaws and defects. Nevertheless, the final samples achieved a relative density of about 97%, with a bending strength of 763 MPa and a fracture toughness of 6.7 MPa · $m^{1/2}$. These values comparable to those of 3Y-TZP ceramics processed via conventional methods combined with cold isostatic pressing (CIP) [10, 11].

A root analogue implant (RAI) with a certain level of precision, was fabricated using stereolithography (SLA) 3D printing technology [12]. The printed RAI exhibited a 6.67% increase in surface aera. Compared to the CAD model, the printed RAI showed a greater change in surface area. Several factors influenced the precision and accuracy of the printing process, including the resolution of the digital mirrors and the composition of the ceramic slurries. Therefore, it is necessary to precisely control the composition of the slurries and use high-precision printing equipment to ensure the accuracy of the printed implants, which is especially important for biomedical applications.

Customized zirconia-based dental implants with sufficiently high dimensional accuracies were developed using the SLA technique [13]. The sintered samples exhibited a flexural strength of 943 MPa. Figure 7.1 depicts the digital file of the customized design of the implant and the 3D printed sample. Commercial powder TZ-3YS-E was blended with photocurable resin to prepare ceramic slurries. The digital files of the implant design were in Standard Tessellation Language (STL) format. The printing experiment was conducted with a commercial digital light processing (DLP) printer (ADMAFLEX 2.0, ADMATEC Europe BV, The Netherlands). The as-printed green bodies were calcined to remove the organic additives. The calcined samples were sintered at 1500 °C, reaching a relative density of 99.8%. Figure 7.2 shows representative SEM images of the implant samples.

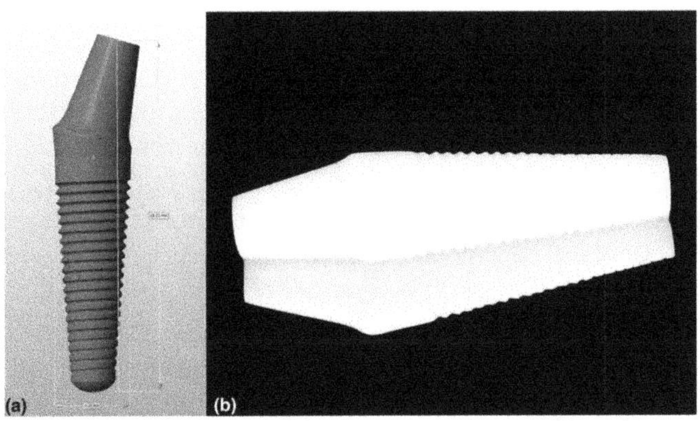

Figure 7.1. (a) 3D CAD model of the implant. (b) 3D printed zirconia implant. Reprinted from [13], Copyright (2017), with permission from Elsevier.

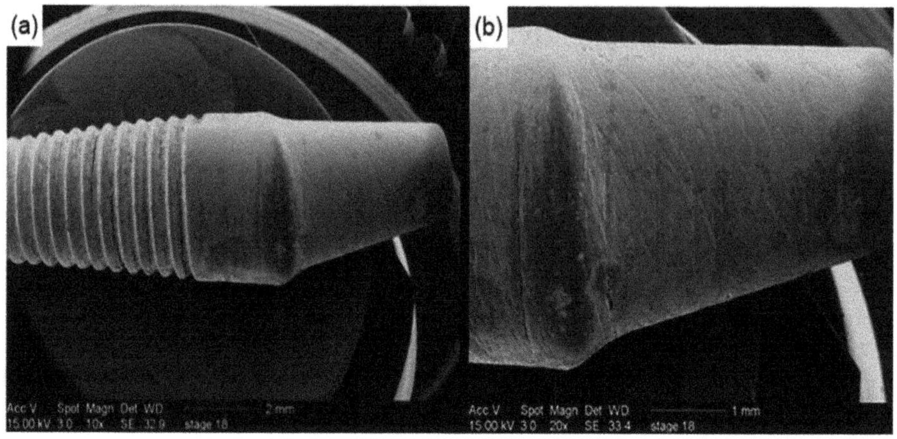

Figure 7.2. SEM images of the printed implants at different magnifications: (a) 10× and (b) 20×. Reprinted from [13], Copyright (2017), with permission from Elsevier.

Low-viscosity aqueous zirconia ceramic suspensions with a solid loading of 40 vol% were developed for vat photopolymerization to print dental crowns [14]. The slurries were prepared with DI water, incorporating of 3 wt% dispersant and 0.5 wt% photoinitiator. The 3D printed green bodies were easily water-washable to eliminate additives. After sintering at 1600 °C, the final zirconia ceramics exhibited relative density of 98.3%, with flexural strength of 708 MPa and Vickers hardness of 14.7 GPa. Dental crowns were obtained with dimensional accuracy of 92.8%.

Commercial dental zirconia doped with 3 mol% yttria (3Y-TZP, Jiaxing Ceramplus Tech. Ltd, China) with $D_{50} = 0.1$ μm was used to prepare ceramic slurries. Polymer resins included acryloyl morpholine (ACMO), hydroxybutyl vinyl ether (HBVE), and 15 ethoxylated trimethylolpropane triacrylate (15EO-TMPTA). Additional components included hydroxyethyl methacrylate (HEMA), polyethylene glycol diacrylate (PEGDA), urethane dimethacrylate (UDMA), while photoinitiators CPI01 (Jiaxing CeramPlus Tech. Ltd) and 819DW (Yuming Chemical Co., Ltd) were employed.

Figure 7.3 illustrates the fabrication process of the ZrO_2 ceramic crowns using vat photopolymerization from aqueous suspensions. The precision of vat photopolymerization is a significant factor in the fabrication of dental structures. The sintered ceramic dentures were canned and compared with the models using the Geomagic Control software, with the results to be presented in figure 7.3(g). For the green bodies, the error was less than 0.1 μm, corresponding to an accuracy of 92.8%, suggesting that the vat photopolymerization successfully replicated the originally designed model, while the finished products were close to the ideal models.

Yttria-stabilized zirconia (3Y-TZP) ceramics were 3D printed using DLP techniques for dental applications [15]. The sintered samples had a density of 6.031 g · cm^{-3}, with a flexural strength of 451.9 MPa, tensile strength of 143 MPa and compressive strength of 298.4 MPa. The slurries contained 85 wt% 3Y-TZP powder. A DLP 3D printer (Zipro, AON, South Korea) was used in the fabrication process. The printed

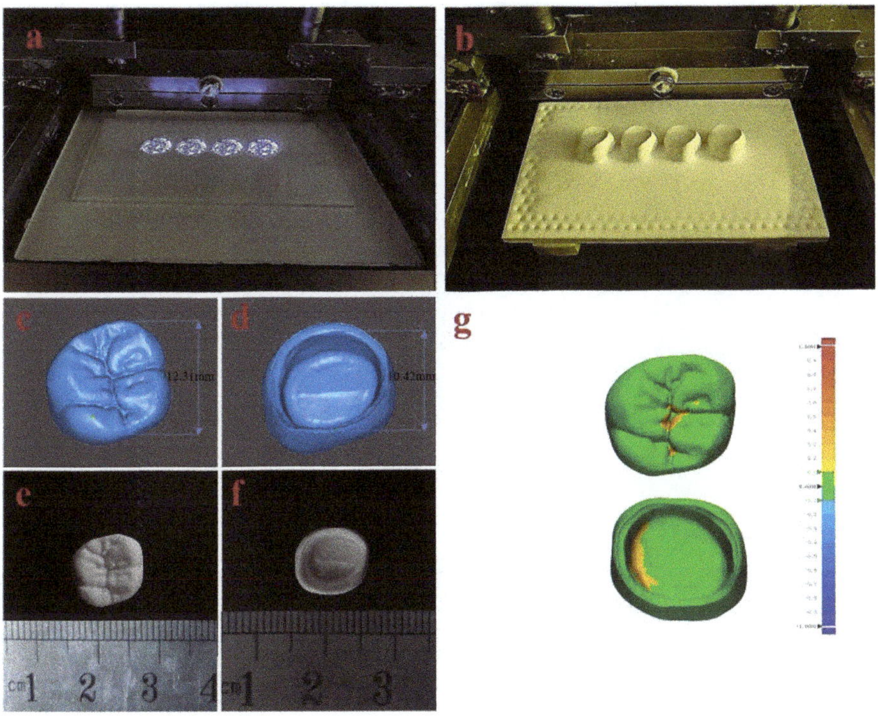

Figure 7.3. Fabrication process of the zirconia ceramic dentures: (a) printing, (b) green bodies, (c and d) 3D models, (e and f) photographs of the sintered samples and (g) precision analysis. Reprinted from [14], Copyright (2024), with permission from Elsevier.

green bodies were subjected to a debinding process. The samples were quickly heated to 200 °C, followed by slow heating to 320 °C at a rate of 0.025 °C · min^{-1} and then calcined at 320 °C for 1 h. From 320 °C to 490 °C, the heating rate was fixed at 0.25 °C · min^{-1} and the samples were further calcined at 490 °C for 1 h. Subsequently, the samples were sintered at 1500 °C for 2 h, at a heating rate of 10 °C · min^{-1}.

For cuboid samples, the average shrinkage in *x*-, *y*- and *z*-directions was 22.1%, 22.5%, and 25.3%, respectively, with corresponding scaling ratios of 1.2839, 1.2906, and 1.3390. The layer thickness of the dried green bodies was 50 μm, which was reduced to 37 μm after sintering. The green bodies exhibited a predominantly monoclinic crystal structure with a minor tetragonal phase, while the tetragonal phase became dominant and the monoclinic phase was reduced after sintering. Figure 7.4 presents photographs of the 3D printed dental prosthesis. Almost no defects or flaws were observed on the surfaces of the veneers and crowns.

Dental crowns were developed from 5-mol% yttria partially stabilized zirconia (5Y-PSZ) using DLP, which were mechanically strong, optically translucent and dimensionally accurate [16]. The 5Y-PSZ slurries had a solid loading level of 50 vol%. The final ceramics sintered at 1500 °C for 2 h exhibited a relative density of 99%, with 59.1% cubic phase. The samples displayed a flexural strength of 625.4 MPa, a Weibull modulus of 7.9 and an optical transparency of 31.4%. The crowns also demonstrated

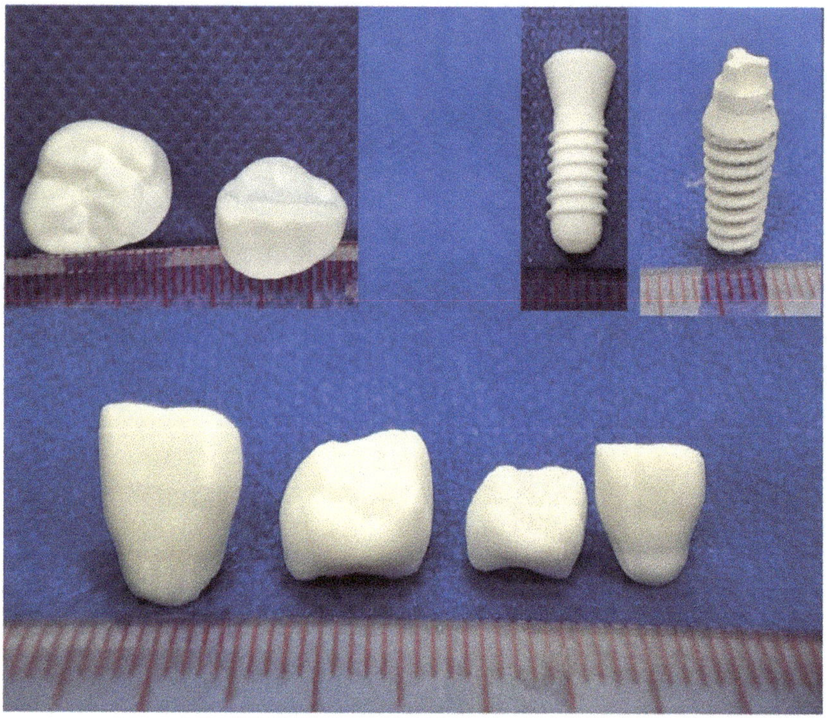

Figure 7.4. Photographs of the 3D printed dental prostheses. Reprinted from [15], Copyright (2024), with permission from Elsevier.

high dimensional accuracy, as indicated by a root mean square (RMS) marginal discrepancy of 44.4 μm and an RMS for internal gap of 22.8 μm.

The dimensions of the green bodies were oversized, which larger than the designed dental crowns by about 25.2%, 25.2% and 24.5% in the *x*-, *y*- and *z*-directions, respectively. Photographs of the 3D printed green body and sintered 5Y-PSZ dental crowns are shown in figure 7.5(A). The sample shrank after sintering at 1500 °C for 2 h. The results of μ-CT analysis indicated that no visible defects, such as voids, cracks or distortions, were detected in the dental crowns, as illustrated in figure 7.5(B). Additionally, the dental crowns were visibly translucent when exposed to light, as shown in figure 7.5(C).

Solvent-based slurry SLA was applied to fabricate zirconia ceramics for dental applications, using both pristine and recycled zirconia powders [17]. The sintered zirconia ceramics reached a relative density of greater than 99% for the pristine powder when printed with a layer thickness of 20 μm. In contrast, the recycled powder slurry was printed within a 40 μm layer thickness, resulting in a sintered density of 90%. Although the Vickers microhardness (~1300 HV) was similar for both types of ceramics, their flexure strengths were differed significantly, at 1057 and 389 MPa, respectively. The 3D printed zirconia from the pristine powder is suitable for use in dental prostheses, while the ones from the recycled powder may be more appropriate for non-critical applications.

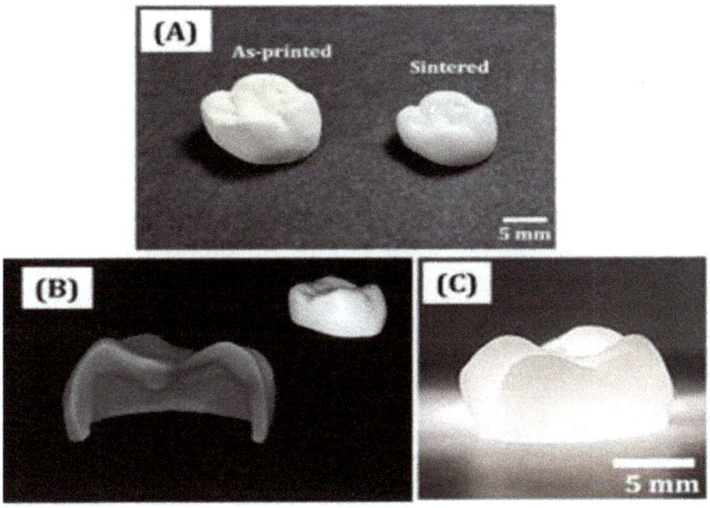

Figure 7.5. (A) Photographs of the 3D printed green body and sintered dental crowns. (B) Representative μ-CT image of the sintered dental crown. (C) Photograph demonstrating translucency of the sintered dental crown with light illumination. Reproduced from [16]. CC BY 4.0.

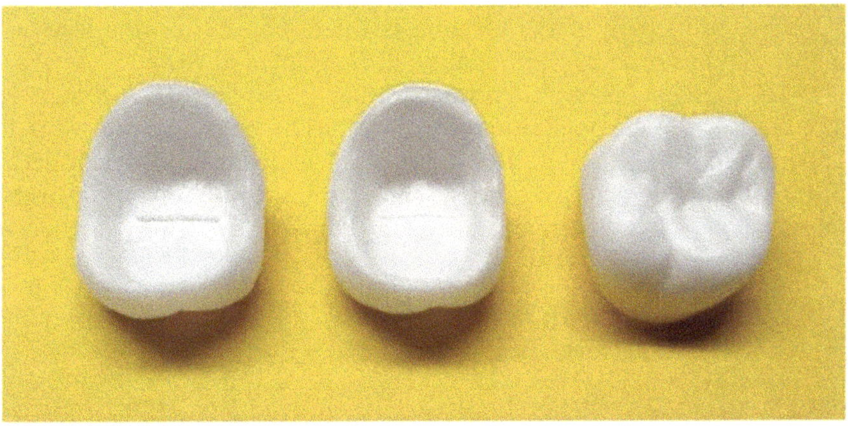

Figure 7.6. Photographs of the sintered zirconia prostheses printed with layer thickness of 40 μm from the slurry made of the recycled zirconia powder. Reprinted from [17], Copyright (2020), with permission from Elsevier.

The accuracy of prosthesis is typically high, with acceptable tolerance below 100 μm. In practice, the error should be controlled within 50 μm. Figure 7.6 shows photographs of the 3D printed dental prosthesis fabricated from the recycled zirconia slurry with a 40 μm layer thickness. Compared with the models, the errors of the 3D printed dental prosthesis ranged from 50 to 60 μm, meeting the requirements for practical dental applications.

The 3D printed zirconia prosthesis from the slurry made of pristine powder with printing thickness of 20 μm displayed an average error of 42.95 ± 8.14 μm, ranging from 28.9 to 58.9 μm. As the printing thickness was increased to 40 μm, the average error was raised to 53.73 ± 12.99 μm, with the values to be in the range of 36.8–74.5 μm. The error of the prosthesis sample from the slurry made of the recycled powder with printing thickness of 40 μm was larger, which was 67.29 ± 19.13 μm, over 42.1–92.6 μm. This error was smaller than 100 μm and can meet the clinical requirement.

Alumina ceramics were produced using SLA printing processes for applications as dental crown frameworks [2]. Two groups of alumina powders were used to prepare the printing slurries, namely small (S, 0.46 ± 0.03 μm) and large (L, 1.56 ± 0.04 μm) particles, while three solid loading levels (70%, 75%, 80%) were examined. The viscosity of the S80 slurry was too high to be measured using rheological methods. The samples printed with S75 and S80 slurries encountered structural deformations. The SLA samples derived from slurries (L75 and L80) with large particles and sufficiently high solid loading levels exhibited density and mechanical strength comparable to those fabricated by subtractive manufacturing.

The S75 and L70 slurries differed significantly in viscosity ($p < 0.05$), with values of 218.9 and 65.7 mPa · s, respectively. The samples derived from the S75 slurry exhibited macroscopic and microscopic deformations in the printed layers due to the relatively high viscosity, as shown in figures 7.7(C) and (D). In comparison, the S70, L70, L75, and L80 slurries showed high printing, as illustrated in figures 7.7(A) and (B). Therefore, properties of the ceramic powders had a strong effect on slurry performances and printability and quality of the printed parts.

Figure 7.8 presents examples demonstrating possibility of the SLA printed alumna ceramics for dental applications. Photographs of the crown framework inserted on an all-ceramic crown preparation (L80 slurry) is depicted in figure 7.8(C). The crown framework could be fitted onto the all-ceramic crown preparation, although a large marginal gap was generated, owing to the excessive oversizing during the

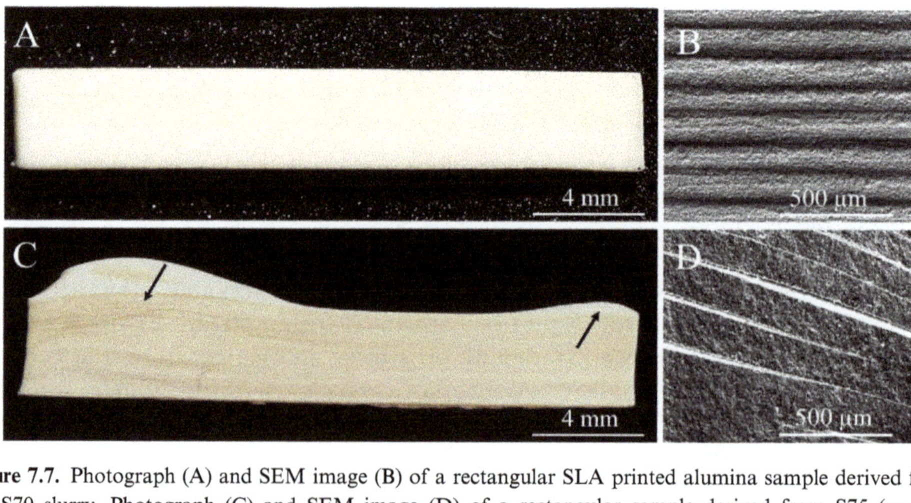

Figure 7.7. Photograph (A) and SEM image (B) of a rectangular SLA printed alumina sample derived from the S70 slurry. Photograph (C) and SEM image (D) of a rectangular sample derived from S75 (arrows: macroscopic deformations). Reprinted from [2], Copyright (2017), with permission from Elsevier.

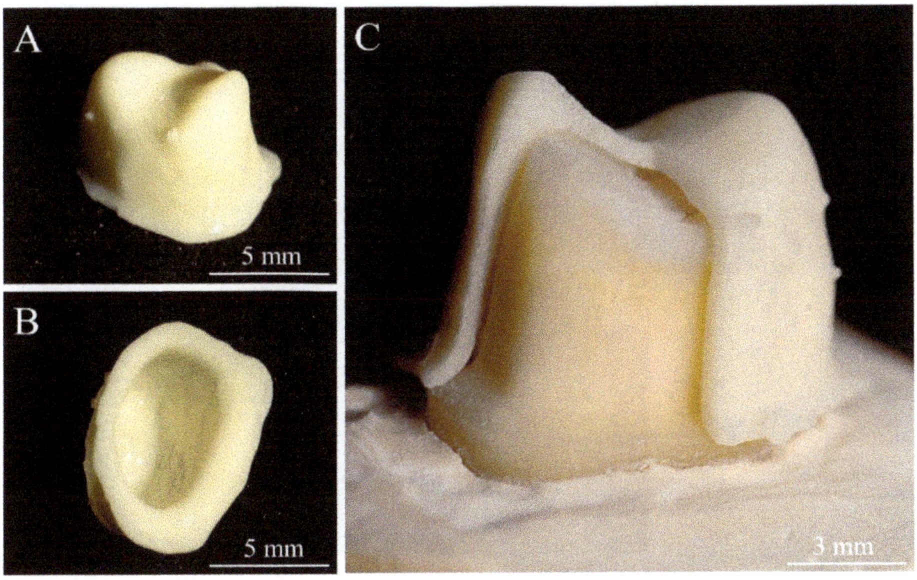

Figure 7.8. Photographs of representative SLA printed alumina dental crown framework: (A) external view and (B) internal view. (C) Crown framework in place on an all-ceramic crown preparation. Reprinted from [2], Copyright (2017), with permission from Elsevier.

design. As a consequence, residual pillars and marginal gaps were visible, as observed in figure 7.8(A). In this case, it is safe to conclude that the SLA printing technology could be used to fabricate ceramic dental crown frameworks.

Feldspar glass-ceramics were prepared using binder jetting technology for dental applications [18]. This 3D printing technology was demonstrated to be suitable for manufacturing of dental restorations meeting the class 1a requirements according to DIN EN ISO 6871:2019-01. The shrinkage behavior of the 3D printed parts was highly dependent on the printing orientation ($p < 0.001$). The x-direction displayed the smallest shrinkage (x–z: $p < 0.001$, x–y: $p = 0.025$), while the z-direction demonstrated the largest shrinkage (z–x: $p < 0.001$, z–y: $p < 0.001$). Figure 7.9 shows photographs of the printed dental items, including the as-printed green bodies, sintered ceramics and final products after machining and polishing.

Glass-ceramic (GC) based composites, incorporating 5 mol% yttria-stabilized zirconia (5Y-PSZ), were designed and fabricated using 3D printing for potential applications as functionally gradient ceramic dentures [19]. The composites were composed of 5Y-PSZ and FAp GC powders. The GC consisted of FAp grains embedded in a glass matrix, providing excellent properties and high F^--ion release rate. The GC was prepared by mixing the raw materials, including SiO_2, Al_2O_3, Na_2CO_3, K_2CO_3, $CaCO_3$, $CaHPO_4$, CaF_2 and others, at the weight percentages of 54.6% SiO_2, 14.4% Al_2O_3, 8.6% Na_2O, 4.2% K_2O, 4.0% P_2O_5, 1.5% ZrO_2, 0.7% F, 0.2% Li_2O, 6.0% CaO, 1.0% TiO_2, 0.8% CeO_2, 3.0% ZnO and 1.0% B_2O_3. The mixtures were melted and then water quenched to form glass [20]. The GC powder

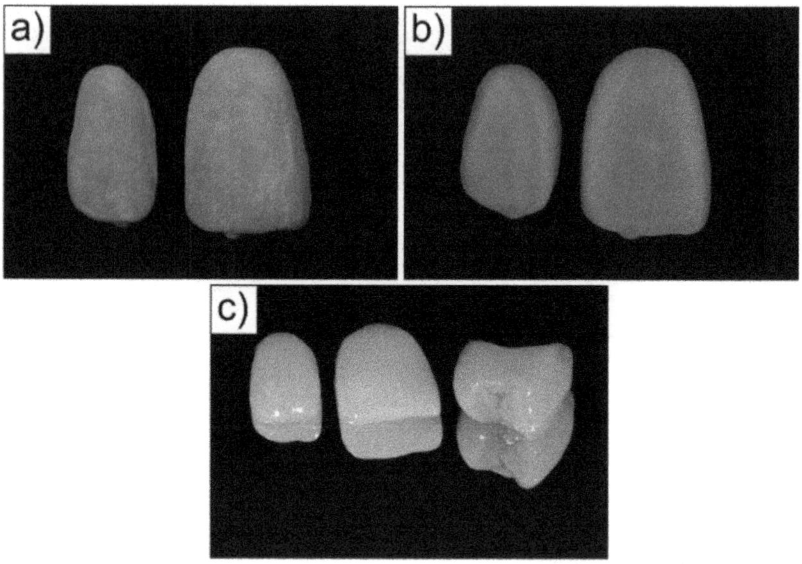

Figure 7.9. Photographs of the LSD-printed dental items (two veneers and a crown geometry): (a) green bodies, (b) as-sintered and (c) final products after dying and polishing. Reprinted from [18], Copyright (2025), with permission from Elsevier.

was then mixed with 5Y-PSZ powder (Shandong Guoci Functional Materials Co., Ltd) using ball milling for 2 h.

The mixed powders were dispersed in ethanol with the assistance of ultrasonication, and then dried at 120 °C for 2 h. The content of the 5Y-PSZ powder was up to 15 wt% in the composites. The samples were named according to their compositions, such as Z0 for the pristine GC, Z5-Z15 for GC/PSZ composite samples with 5–15 wt% 5Y-PSZ. Meanwhile, ZG represented functionally graded ceramic samples. To prepare slurries, the composite powders were blended with the prepared solution containing an acrylic acid monomer, a dispersant, an organic accelerator, and a photoinitiator in specific ratios. The solid loading level was experimentally optimized to be 55 vol%.

An SL 3D printer (CeraBuilder100, Wuhan Intelligent Laser Technology Co., Ltd, China) was used to print samples, with dimensions of 50 mm × 6 mm × 5 mm. The printing parameters included a scanning speed of 3000 mm $\cdot$ s^{-1} and a layer thickness of 40 μm. The curing depth was maintained to be 130 μm, while the laser scanning direction was set with alternating orientations at 90°, thus ensuring effective solidification of the printed layers. The green bodies were sintered at 900 °C for 2–4 h.

Figure 7.10 depicts photographs of the gradient ceramic dentures, including green bodies and sintered items. A gradual transition in light transmission was visible from top to bottom parts, resembling the natural appearance of human teeth, although distinct boundaries appeared between adjacent layers, indicating the limited precision and accuracy of the printing technology. It is necessary to further optimize the processing parameters and improve equipment performance to enable the

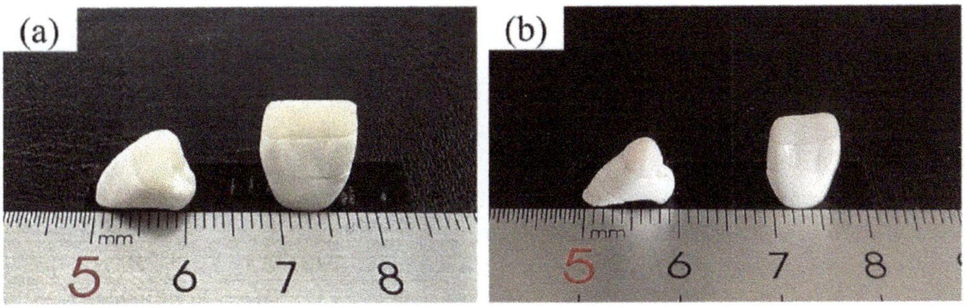

Figure 7.10. Photographs of the gradient ceramic dentures: (a) green bodies and (b) sintered items. Reprinted from [19], Copyright (2025), with permission from Elsevier.

development of materials and structures suitable for real-world applications as natural tooth replacements.

Bioactive scaffolds based on tricalcium phosphate (TCP) were fabricated by using extrusion printing technology, i.e., robocasting [21]. Critical mandibular segmental defects could be restored in the scaffolds, reaching the levels that were close to the native bones, after implanting in an adult rabbit mandibular defect model for eight weeks. Directional bony ingrowth into the scaffold interstices was observed, with the tracking healing pathway that was originated from the defect walls and marrow spaces. Moreover, new bone was grown and the scaffolds were resorbed at the bone/scaffold interfaces.

Concentrated β-tricalcium phosphate (TCP) colloidal slurry, with solid loading level of 46 wt%, was prepared by mixing the TCP ceramic powder, ammonium polyacrylate (14.5 mg · g^{-1} ceramic powder, Darvan 821 A, R.T. Vanderbilt, Norwalk, CT) solution, DI water, hydroxypropyl methylcellulose (7 mg · ml^{-1} of ceramic powder, Methocel F4M, Dow Chemical Company, Midland, MI) as the thickening agent, and polyethylenimine (150–200 mg per 30 ml colloidal slurry, Sigma–Aldrich, St. Louis, MO) to gel the colloidal ink suspension [22].

β-TCP scaffolds were printed using the robocasting process, with a 3D direct-write micro printer gantry robot system (Aerotech Inc., Pittsburgh, PA), in which the colloidal inks were extruded to form designed structures. The models of the 3D scaffolds were designed using a computer-aided design system (RoboCAD 4.3; 3D Inks LLC, Tulsa, OK), while the structures had dimensions of 11 mm in length, 9 mm in major axis diameter, 4.5 mm in minor axis diameter, 250 μm in struts and 330 μm in pore spacing. The colloidal inks were filled into a syringe (Nordson Corp., Westlake, OH), while the extrusion nozzle (Nordson Corp.) had a diameter of 250 μm. The printing rate was 8 mm · s^{-1}. The samples were soaked in a tray with low-viscosity paraffin oil to prevent drying during the printing process. After printing, the dried scaffolds were calcined at 400 °C and 900 °C for burning out the organic components and then sintered at 1100 °C for 4 h.

3D scaffolds with desired porous structures, with pore sizes spanning from nanometer to millimeter scales, supporting the reconstruction of centimeter-sized osseous defects, were fabricated by using voxel-wise 3D printing processes [23].

Nano-porous hydroxyapatite granulates were used for the printing. The cylindrical design was aimed to resemble the hollow bone with high densities at the periphery, while millimeter-wide central channels aligned with the symmetry axis and connected the perpendicularly arranged micro-sized pores.

The mean distances to the materials were overestimated according to the 2D distance maps, by 33%–50% compared to the 3D analysis results. The scaffolds contained 70% μm-sized pores that were entirely interconnected. Virtual spheres were adopted to characterize cell migration within the pores demonstrating that the central channel was accessible to spheres as large as 350 μm. Nearly isotropic shrinkage of about 27% was observed after sintering of the 3D printed matrix. The designs and tomographic data were quantitatively compared, thus allowing assessment of scaffold quality. Clinical potentials of the 3D printed scaffolds were confirmed by the histological analysis seeded with osteogenic-stimulated progenitor cells.

Three groups of HA scaffolds were fabricated, which were denoted as A, B and C, with pixel-wise and layer-wise designs. Cubic voxels had an edge length of 240 μm, corresponding to the printer resolution of 106 dpi and the layer thickness of 240 μm. The scaffolds were totally 30 voxels in height, consisting of 15 alternative double-layers. A millimeter-wide central channel was built axially through the scaffold, with a width equal to the length of 4 voxels. It was connected to a micro-channel network, with the periphery to be opened by 8 (Design A) or 12 (Designs B and C) pores per double-layer. To have azimuthal orientation, an additional block with 3 × 3 voxels was put on the fourth double-layer on surface of the scaffolds.

The HA powder was sprayed into granules with an average size of 22 μm. A water-based polymer binder was used in the form of solution, which was eliminated through calcination at 450 °C. HA scaffolds with sufficiently high mechanical strength were obtained after sintering (Chamber Furnace RHF 17/10E, Carbolite GmbH, Ubstadt-Weiher, Germany) at 1250 °C for 2 h. A central opening consisting of HA grains was observed using SEM, as illustrated in figure 7.11. Cells were adhered to the granules forming a network-like structure between the open hollow spheres.

A selected *H* and *E* stained histological slice of the cell-scaffold constructs after 28 days of cultivation is depicted in figure 7.12(a), showing soft tissues composed of cells with fibroblastic morphologies. Near the dissolved HA, the cells exhibited cuboidal osteoblast-like morphology. The heterogeneous cell morphologies indicated the development of vital cells from the dNC-PCs through the distinctive differentiation paths. Figure 7.12(b) shows an optical image of a partially selected slice of the non-decalcified brittle HA scaffold (Design C), with the cells stained with toluidine blue. The micro-sized pores were filled with cells displaying various morphologies. After 28 days of cultivation, the differentiated dNC-PCs contributed to the formation of connective tissues.

7.2.2 Piezoelectric applications

Biomedical devices continuously demand materials with enhanced performances. Piezoelectric ceramics are important materials for biomedical device applications due to their capability to mutually convert mechanical and electrical energy.

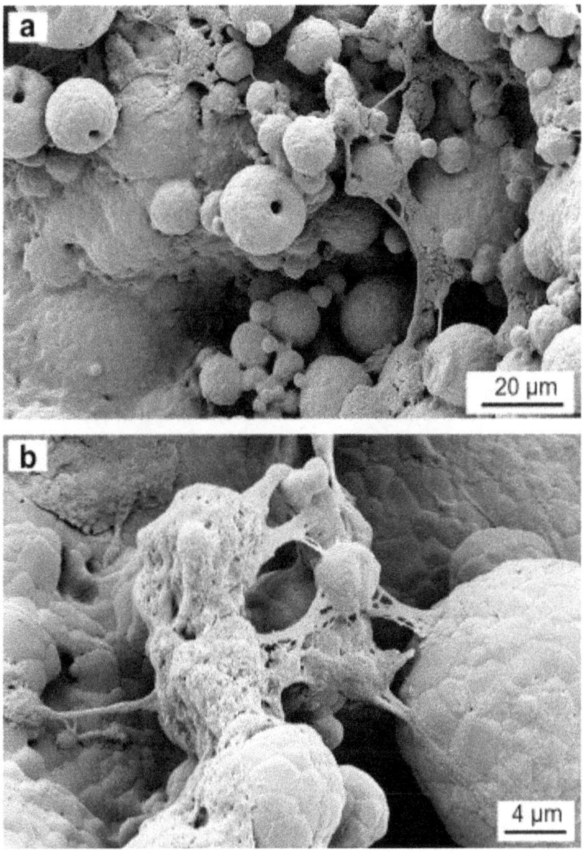

Figure 7.11. Surface SEM images of the scaffolds with cells: (a) several granules with central opening and (b) nanometer-sized orifices at the grain boundaries. Reprinted from [23], Copyright (2008), with permission from Elsevier.

3D printing technology offers more flexibility in the fabrication of piezoelectric ceramics in terms of structural complexity and geometric design, thus attracting interests of researchers and industries worldwide. Significant progress has been achieved in the development and exploration of 3D printing for the fabrication of piezoelectric ceramics [24–26].

Piezoelectric-composite slurries containing $BaTiO_3$ (BT) nanoparticles (100 nm) were 3D printed using mask-image-projection-based SLA (MIP-SLA) technique [27]. The sintered samples had a density of 5.64 g $\cdot$ cm^{-3}, corresponding to relative density of 93.7%. The samples possessed a piezoelectric constant of 160 p/CN and a dielectric constant of 1350. The 3D printed ceramics were used to assemble an ultrasonic transducer for energy focusing and ultrasonic sensing applications. A 6.28 MHz ultrasonic scan conducted using the transducer, which was capable of visualizing the structure of a porcine eyeball.

Methylethylketone (66 v/v%, 99%, MEK, Sigma–Aldrich, Saint Louis, MO) and ethanol (34 v/v%, 99.5%, Sigma–Aldrich, Saint Louis, MO) were azeotropically

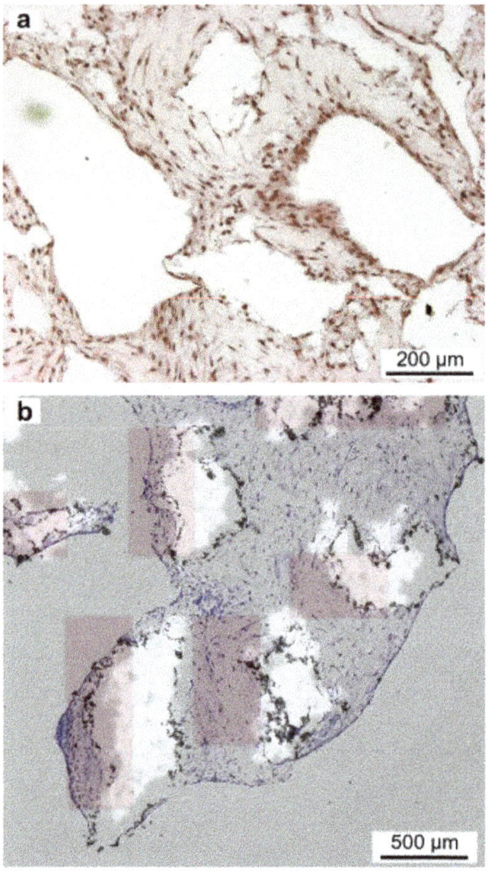

Figure 7.12. (a) Image of the decalcified section demonstrating the cells with different morphologies. (b) Histological slice of the non-decalcified section (HA scaffold Design C with toluidine blue stained cells) manually registered with the tomogram (white) and printing matrix (transparent red). Reprinted from [23], Copyright (2008), with permission from Elsevier.

mixed. BT powder (25 v/v% solid loading, 100 nm, Sigma–Aldrich St. Louis, MO) was dispersed in the azeotropic mixture, together with dispersant (Triton *x*-100, 0.5–0.8 wt%) using planetary ball milling (pulverisette 5, FRITSCH Idar-Oberstein, Germany) for 12 h at 200 rpm, with stainless steel balls as the milling media. The milled mixtures were dried at 50 °C for 12 h. Dry BT powder with dispersant molecules adsorbed on the surface of the ceramic particles was finally formed.

The processed BT powder was dispersed in photocurable resin (SI500, EnvisionTec Inc., Ferndale, MI) through ball milling for 1 h, forming slurries with a solid loading level of 70 wt%. A volume of 0.045 ml of slurry was filled to film collector and spread to form a thin layer of 50 μm in thickness using a doctor blade. A computer-aided design (CAD) model was sliced into two-dimensional (2D) images, which were then projected onto the bottom surface of the film using a digital micromirror device (DMD). Meanwhile, the photocurable resin in the slurry was polymerized, thus producing a cross-linked matrix under visible light irradiation. In this case, the piezoelectric particles and polymer network were strongly bonded to form a network structure. When each single layer was cured, the platform was moved up, so that the film was separated from the surface of the collector with PDMS coating. The printed green bodies were calcined in Ar at 600 °C for 3 h to eliminate the organic additives. The calcined samples were sintered in air at 1330 °C for 4–6 h.

The 3D printed segment annular array (3D-SAA) and focused concave-shaped piezoelectric element (PF-CPE) are depicted in figures 7.13(b) and (e). Optical microscopy images of the 3D-SAA with 64 pillars (1 mm in height) in the shape of a fan on the base with a size of 1 mm are displayed in figures 7.13(a) and (c). The 3D printing technique offers higher flexibility for producing structures with more complex geometries. Detailed structures of the PF-CPE are illustrated in figures 7.13 (d) and (f). The aperture of the working area was 5 mm, while the arc length and thickness were 5.2 mm and 390 μm, respectively. Such structures can generate and focus ultrasonic waves for ultrasonic sensing applications.

A similar SLA printing process was reported for the fabrication of $BaTiO_3$ (BT) ceramics for transducer applications [28]. The BT ceramics exhibited a piezoelectric constant (d_{33}) of 166 pC/N. A 1.4 MHz focused ultrasonic array was prepared, showing –6 dB bandwidth of 40% and insertion loss of 50 dB at center frequency.

BT powder with an average particle size of 500 nm was blended with a photosensitive resin to form slurries with solid loading levels of 70–86 wt%, using ball milling. The photosensitive resins consisted of oligomers, a monomer and UV

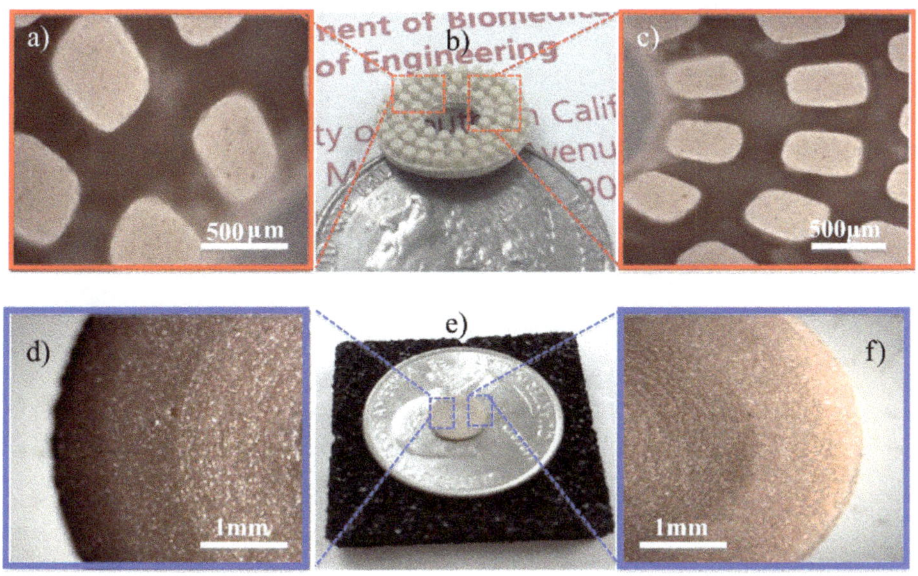

Figure 7.13. (a–c) Optical microscopy images of the 3D-SAA with the 64 pillars annular segment array. (d–f) Optical microscopy images of the concaved-shaped piezoelectric element (PF-CPE). Reprinted from [27], Copyright (2016), with permission from Elsevier.

initiator. Meanwhile, a dispersant was used to prevent aggregation of the ceramic particles. Green bodies, with a disk size of φ12 × 3 mm, were prepared for materials characterization. The samples were calcined and then sintered at 1290 °C for 2 h.

Figure 7.14 depicts photographs of the 3D printed BT ceramic array and transducers. The BT array had a curvature radius of 20.3 mm and a thickness of 1.8 mm, while the pitch and kerf were 4.5 and 0.5 mm, respectively. The prototype of the focused ultrasonic array was made of BT ceramics as active elements, together with a backing layer and cables. The backing layer was 10 mm in thickness and was coated on the ceramics with epoxy (Epo–Tek 301, Epoxy Technology, Billerica, MA, USA) that was mixed with W powder and microbubbles. The backing layer had an acoustic impedance of 5 MRalys. The kerfs between the elements were filled with epoxy to minimize the cross-talking effect. The array was tested in the water tank towards a quartz target. An ultrasound pulser/receiver (JSR Ultrasonics DPR500, Pittsford, NY, USA) and an oscilloscope (Keysight DSOS054A) were connected to the element. Figure 7.15 displays a pulse-echo waveform and frequency spectrum of the focused array.

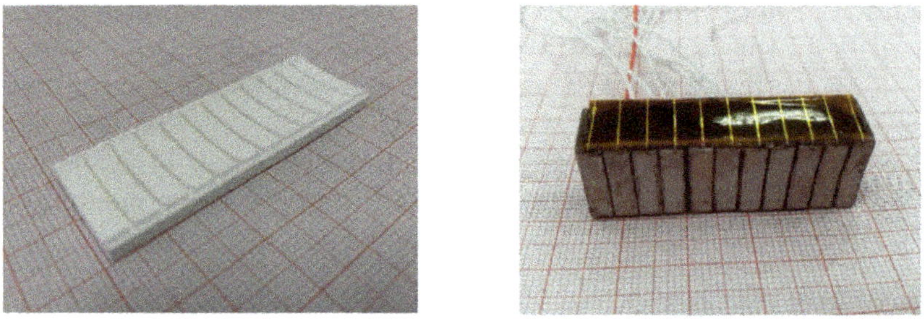

(a) (b)

Figure 7.14. Photographs of the 3D-printed BT ceramics: (a) ceramic sample and (b) ultrasonic arrays. Reproduced from [28]. CC BY 4.0.

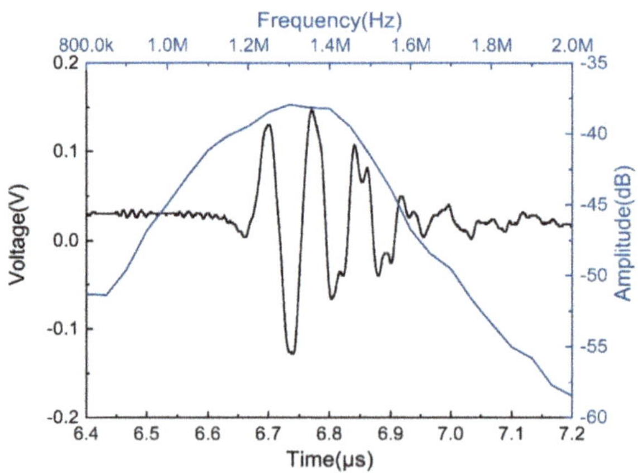

Figure 7.15. Pulse-echo waveform and frequency spectrum of the 3D printed BT focused array. Reproduced from [28]. CC BY 4.0.

Lanthanum oxide doped PZT ceramics were prepared by using 3D printing, with enhanced electrical performances. In this process, lead carbonate rather than lead oxide, was used to achieve ceramic slurries with high curability and efficient printability [29]. The final piezoceramics exhibited outstanding dielectric and piezoelectric properties, with high transduction coefficient, electromechanical coupling factor and output voltage/current, given that the compositions of the samples were optimized.

Commercial powders of $PbCO_3$ (99.7%), ZrO_2 (99%), TiO_2 (98%) and La_2O_3 (99.99%) were supplied by the Sinopharm Chemical Reagent Co., Ltd TPO (BASF, Germany), polyethylene glycol-200 (Sinopharm Chemical Reagent Co., Ltd) and solsperse 41000 (Lubrizol, Spain) were selected as photoinitiator, plasticizer and dispersant to prepare the printing slurries. The photosensitive resins consisted of O-phenylphenoxyethyl acrylate (OPPEOA), 1,6-hexanediol diacrylate (HDDA), tripropylene glycol diacrylate (TPGDA) and ethoxylated trimethylolpropane triacrylate (ETPTA), were prepared at specific mass ratios.

The powders were mixed according to the formula $Pb_{1-x}La_x (Zr_{0.52}Ti_{0.48})O_3$, with $x = 0$, 0.01, 0.03, 0.05. The mixtures were blended with the photosensitive resins, by stirring at 2800 rpm for 6 min. Then, TPO, PEG-200 and Solsperse 41 000 were introduced, followed by homogenization at 2800 rpm for 3 min, leading to $Pb_{1-x}La_x (Zr_{0.52}Ti_{0.48})O_3$ slurries for printing experiments.

A DLP printer (Autocera-M, Beijing Ten Dimensions Technology Co., Ltd) was used to print samples under 405 nm UV light irradiation. Green bodies of $Pb_{1-x}La_x(Zr_{0.52}Ti_{0.48})O_3$ were obtained through repeated layer-by layer printing. Figure 7.16 depicts photographs of the $Pb_{1-x}La_x(Zr_{0.52}Ti_{0.48})O_3$ samples with different geometries and structures. According to the TG-DTG analysis results, all the green bodies were calcined at 238 °C, 296 °C and 377 °C for 2 h to burn out the organic additives, at a heating rate of 0.2 °C · min^{-1}. The calcined samples were

Figure 7.16. Photographs of the 3D printed $Pb_{1-x}La_x(Zr_{0.52}Ti_{0.48})O_3$ green bodies with different geometries. Reprinted from [29], Copyright (2024), with permission from Elsevier.

sintered at 1250 °C for 2 h to form densified $Pb_{1-x}La_x(Zr_{0.52}Ti_{0.48})O_3$ ceramics through solid-state reaction.

Piezoelectric energy harvesting performances of the $Pb_{1-x}La_x$ $(Zr_{0.52}Ti_{0.48})O_3$ ceramics were evaluated, by periodically applying pressing-releasing measurement. The periodic load followed a sine wave pattern with a frequency of 0.5 Hz, while the output electric signals were measured using an electrometer. At a given periodic load along the polarization axis, the output voltage (V_{out}) and current (I_{out}) increased first and then decreased, with increasing content of La^{3+}, whereas the peak values were present in the samples with $x = 0.03$. Specifically, the values of V_{out} and I_{out} of the $Pb_{0.97}La_{0.03}(Zr_{0.52}Ti_{0.48})O_3$ ceramics were 0.54 V and 1.81 nA, respectively, as presented in figure 7.17.

Piezoelectric ceramics, $0.71Pb(Mg_{1/3}Nb_{2/3})O_3$–$0.29PbTiO_3$ (PMN-29PT), were developed using SLA 3D printing techniques [30]. The sintered PMN-29PT ceramics possessed a dielectric constant and a piezoelectric coefficient (d_{33}) of 2698 and 487 pC · N^{-1}, respectively. Meanwhile, focused piezoelectric elements with bowl-like structures were fabricated with the SLA printing process. The ultrasonic transducer based on the 3D printed focused elements exhibited strong device performances, with a center frequency of 2.85 MHz and 45% –6 dB bandwidth.

$(MgCO_3)_4 \cdot Mg(OH)_2 \cdot 5H_2O$ and Nb_2O_5 were thoroughly mixed and calcined at 1000 °C to obtain $MgNb_2O_6$. PbO and TiO_2 were mixed with $MgNb_2O_6$, according to the composition of PMN-29PT. The mixtures were calcined at 850 °C to prepare the perovskite PMN-29PT. The PMN-29PT powder was blended with dispersant, photocurable resin, and photoinitiators, through ball milling at 160 rpm for 4 h, resulting in ceramic slurries for printing with a solid loading level of 78 wt%. The 3D printed green bodies were calcined at 600 °C to eliminate the organic components and then sintered at 1235 °C for 2 h.

The 3D printed bowl-like ceramic samples were coated with 300 nm thick Au electrodes. Copper wires were connected to the backing of the bowl-like ceramics using E-solder 3022. The wire connected ceramics were attached onto a copper

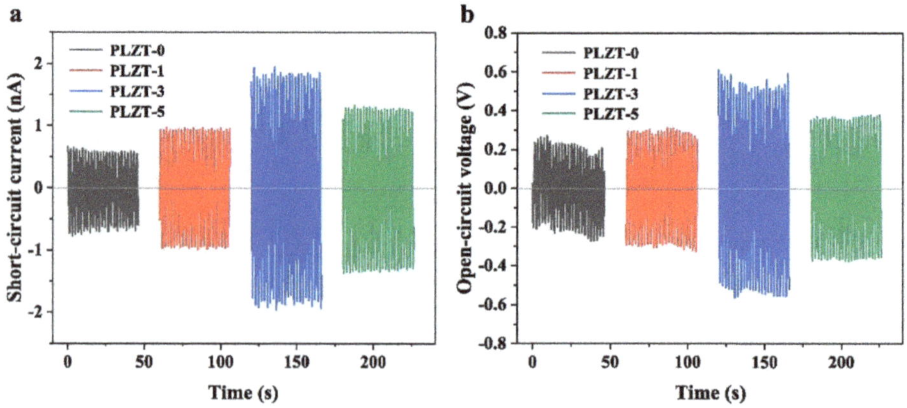

Figure 7.17. Output voltages and currents of the $Pb_{1-x}La_x(Zr_{0.52}Ti_{0.48})O_3$ ceramics. Reprinted from [29], Copyright (2024), with permission from Elsevier.

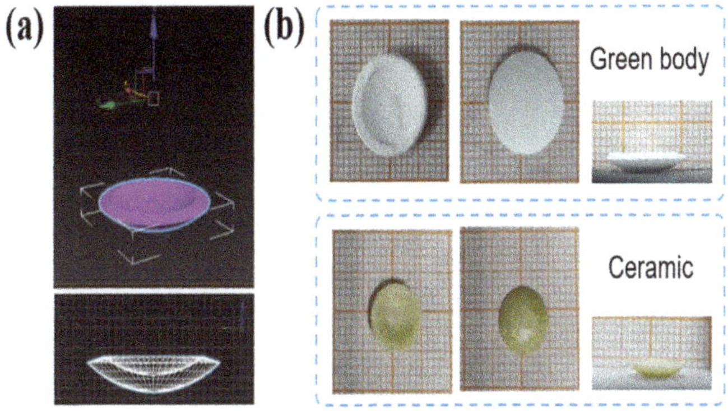

Figure 7.18. (a) Schematic diagram of the bowl-like structures. (b) Photograph of the 3D printed bowl-like green bodies and the sintered ceramics. Reprinted from [30], Copyright (2024), with permission from Elsevier.

housing with epoxy, while the wires were linked to SMA connectors. After that, the surface of the transducer was coated with 300 nm thick Au electrodes. The transducers were protected by a parylene C layer.

Figure 7.18 shows schematic diagrams and photographs of the bowl-like piezoelectric ceramics. The bowl-like/focused structures were well retained after sintering. The green bodies experienced a geometric shrinkage of about 33.8% during the sintering process. The sintered ceramics were free of macrocracks, while single-phase perovskite was formed, without the presence of pyrochlore phase, thus ensuring piezoelectric performances. The samples exhibited a dense microstructure with sufficiently low porosity.

Figure 7.19 presents electrical properties of the ultrasonic transducers made of the 3D printed bowl-like ceramics. The mechanism that the bowl-like piezoelectric ceramics could serve to narrow the ultrasound beam is schematically illustrated in figure 7.19(a). Through this mechanism, the lateral resolution of the transducer during the object detection could be significantly increased. A photograph of the ultrasonic transducers assembled with the bowl-like ceramics is demonstrated in figure 7.19(b), in which the 3D printed ceramics with Au electrodes were attached to the front surface of the transducers.

Pulse-echo results are plotted in figure 7.19(c), indicating transmitting and receiving performances of the ultrasonic transducers. Both the center frequency and the −6 dB bandwidth could be estimated with the pulse-echo data. The transducers could be used to transmit and receive ultrasound signals, thus be able to detect objects through imaging. Frequency-dependent impedance characteristics are depicted in figure 7.19(d). The impedance of the transducers was about 50 Ω at the center frequency of 2.85 MHz, which maximized the transmission of electric energy between the pulse/receiver and the transducers.

7.2.3 Microwave applications

Microwaves refer to electromagnetic waves with frequencies in the gigahertz range. Devices and systems-based microwave materials have a wide range of applications

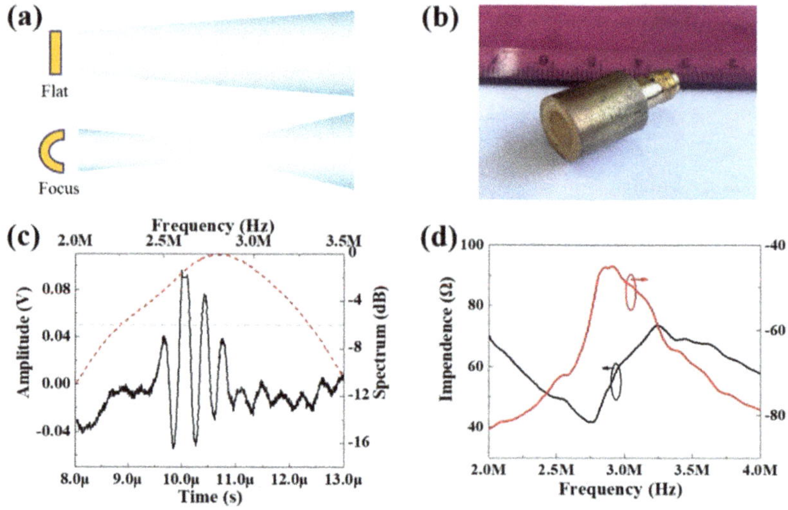

Figure 7.19. (a) Schematic diagram of the beam focusing experiment. The ultrasonic transducer: (b) photograph, (c) pulse-echo wave and frequency spectrum and (d) electrical magnitude and phase angle. Reprinted from [30], Copyright (2024), with permission from Elsevier.

in communications, medical treatments, engineering and architecture, textile industry, remote sensing, wearable electronics, and automotive industries [31–33]. Microwave dielectric ceramics, with an appropriate dielectric constant and sufficiently low dielectric loss, have been extensively utilized in radar systems, satellite communications, and 5 G wireless mobile communications. The applications of 3D printing technology to develop microwave materials and devices have been well reviewed in the open literature, while selected samples will be presented in this subsection as references and potential guidance for the readers [34, 35].

Microwave dielectrics, $(1-x)MgTiO_3-xCaTiO_3$ ceramics, with $x = 0, 0.1, 0.2$, and 0.4, were developed using SLA 3D printing technology [36]. $MgTiO_3$ and $CaTiO_3$ ceramic powders were prepared via the traditional ceramic process, and subsequently mixed with ultraviolet curable resins, along with the addition of a dispersant and photoinitiator. The printable slurries achieved a loading of 80 wt%. A flat Luneburg lens antenna was printed with the $MgTiO_3$–$CaTiO_3$ ceramics.

Commercial powders, including MgO (99.5%), $CaCO_3$ (99.8%) and TiO_2 (99.5%), were used as the raw materials. To synthesize $MgTiO_3$ powder, MgO and TiO_2 were mixed through ball milling with ZrO_2 balls in distilled water for 6 h, followed by drying and calcination in air at 1100 °C for 3 h, with a heat rate of 5 °C · min^{-1}. $CaTiO_3$ was prepared with $CaCO_3$ and TiO_2 in a similar way. After that, the two powders were mixed to form the $(1-x)MgTiO_3-xCaTiO_3$ precursor powders.

Two acrylates, bisphenol A epoxy acrylate (EA) and 2-(2-ethoxyethoxy)ethyl acrylate (EOEOEA), were used as UV curable monomers. Irgacure 2022 was prepared by mixing Irgacure 819 and Darocur 1173 as the photoinitiator, while Byk110 was utilized as a dispersant. The organic additives were blended with hydroquinone, forming thermal polymerization inhibitor. The 3D printed green bodies were calcined at 220 °C and 400 °C for 6 h and 600 °C for 12 h in N_2, followed by heating in air at 600 °C for 2 h at a heating rate of 2 °C · min^{-1}. Finally, the samples were sintered at 1350°C for 3 h.

The Luneburg lens is a passive microwave device characterized by a gradient dielectric constant that decreases from the center outward. Similar to the effect of a convex lens, Luneburg lens can be used to construct antennas with gains that are tremendously improved. A Luneburg lens was fabricated by using the $MgTiO_3$–$CaTiO_3$ microwave dielectric materials. Photographs of a flat Luneburg lens based on the $MgTiO_3$–$CaTiO_3$ dielectrics before and after sintering are illustrated in figures 7.20(a) and (b). The device had a diameter of 70 mm and thickness of 8 mm. The measured antenna gain with the Luneburg lens exhibited promising antenna radiation efficiency, in good agreement with the simulated results, as observed in figure 7.20(c), implying that 3D printing processes had sufficiently high precision and the ceramic materials were of high Q value.

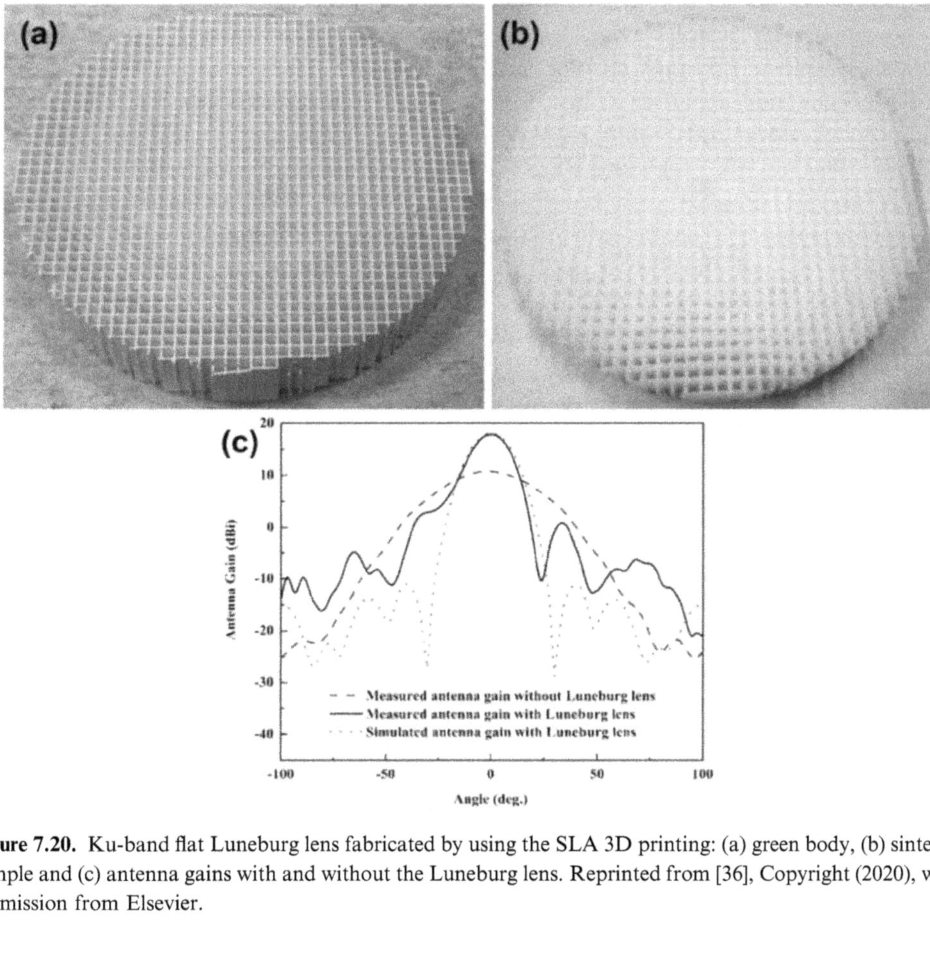

Figure 7.20. Ku-band flat Luneburg lens fabricated by using the SLA 3D printing: (a) green body, (b) sintered sample and (c) antenna gains with and without the Luneburg lens. Reprinted from [36], Copyright (2020), with permission from Elsevier.

The 3D printing process and polymer-derived ceramics (PDC) approach were incorporated to develop SiCN honeycomb structures with strong microwave absorption performances [37]. The slurries were based on UV curable polymeric precursors, containing polysilazane (PSZ) and multifunctional acrylates. The SiCN ceramic honeycomb structures had a hardness of 14.3 GPa and specific compressive strength of 333.3 MPa $\cdot$ g^{-1} $\cdot$ cm^3. Multifunctional (meth)acrylate monomer, trimethylolpropane triacrylate (TMPTA), 1,6-hexanediol diacrylate (HDDA), di(trimethylolpropane) tetraacrylate (Di-TMPTA), UV free-radical photoinitiator bis(2,4,6-trimethybenzoyl) phenylphosphine oxide (PI 819), UV absorber hxamethylphosphoramide (HMPA) and free-radical inhibitor-IRGANOX 1330 were used for the experiment.

The samples, with PSZ:TMPTA ratios of 1:1, 2:1 and 3:1, were denoted as PT11, PT21 and PT31, respectively. Samples with a PSZ:HDDA ratio of 2:1 and PSZ:Di-TMPTA of 2:1 were also fabricated, denoted as PH21, PD21, respectively. 0.3 wt% HMPA was used as the UV absorber, while 1 wt% radical photoinitiator Irgacure 819 and 0.7 wt% of the radical scavenger IRGANOX 1330 were added as an inhibitor. The UV cured samples, PT11, PT21, PT31, PH21, PD21 and TMPTA, were called PT11-U, PT21-U, PT31-U, PH21-U, PD21-U and TMPTA-U, respectively. Thermal cured PSZ was denoted as PSZ-TC.

The as-printed samples were washed with isopropanol to remove the uncured precursors, followed by post-curing using a system (CL-1000 l) for 0.5 h. The samples were then calcined at 1100 °C for 2 h in Ar, to convert the polymers into SiCN ceramics, at heating and cooling rates of 1 °C $\cdot$ min^{-1}. The samples, PT11-U, PT21-U, PT31-U, PH21-U, PD21-U, and PSZ-TC were subsequently pyrolyzed at 1100 °C and denoted as PT11-P, PT21-P, PT31-P, PH21-P, PD21-P, and PSZ-TC-P, respectively.

Electromagnetic parameters, including real permittivity (ε') and imaginary permittivity (ε'), of the 3D printed SiCN bulk ceramics and ceramic honeycomb structures, are plotted in figures 7.21(a) and 7.20(c), respectively. The average values of ε', ε'' and loss tangent (tanδ) of the SiCN honeycomb ceramics were all lower than those of the SiCN bulk ceramics. The presence of free carbon contributes to electrical conduction, resulting in an increment in the values of ε' and ε'' of the PDCs [38]. This explained the higher values of ε' and ε'' for the SiCN bulk ceramics. In this regard, the honeycomb structures were advantageous in terms of impedance matching for microwave transmission.

Meanwhile, the dielectric loss tangent represents the microwave attenuation capability of microwave absorption materials. The SiCN bulk ceramics had a tanδ value of 0.37, while the tanδ of SiCN honeycomb ceramic structure was 0.21, as seen in figures 7.21(a) and (c). As a result, the bulk samples had a higher microwave absorption performance than the honeycomb sample, although the latter had a higher degree of impedance matching.

Microwave absorption materials are characterized using reflection loss (RL) in terms of absorbing performances [39]. SiC and Si_3N_4 based ceramics are typical non-magnetic microwave absorption materials, so that their RL values are dependent on the complex permittivity. Generally, 90% microwave absorption is used as a criterion to evaluate the absorbing performances, corresponding to RL value of –10 dB. Meanwhile, the effective absorption bandwidth (EAB) is another important

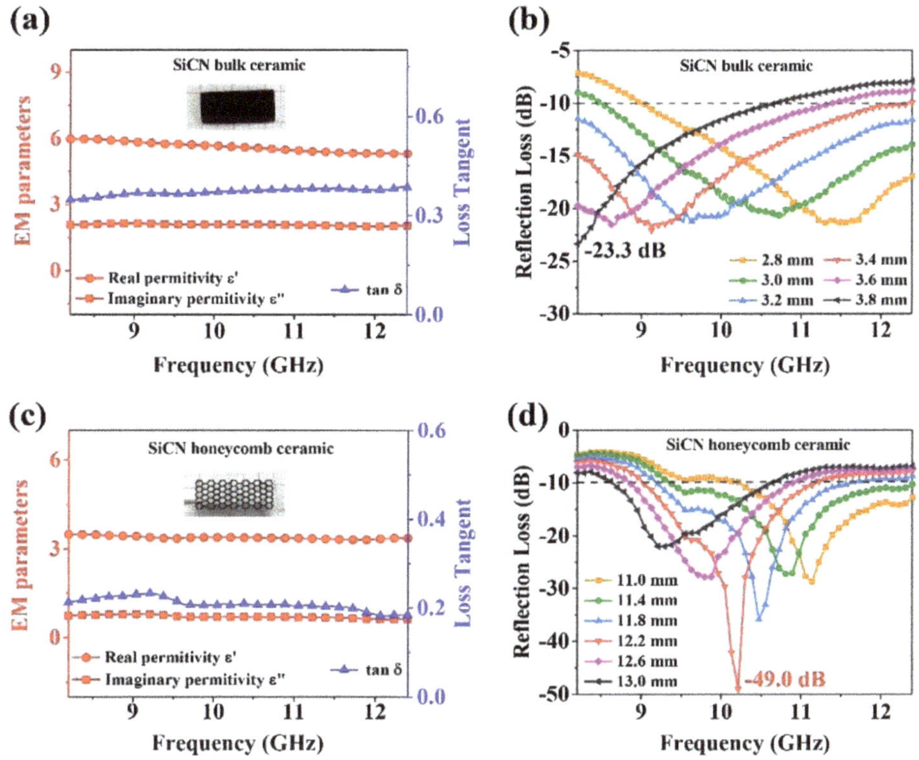

Figure 7.21. Electromagnetic parameters (a and c) and calculated RL curves (b and d) of the 3D printed SiCN bulk ceramics (a and b) and honeycomb structure (c and d). In panels (a and c), the insets are photographs of the SiCN bulk ceramic and SiCN honeycomb ceramic for electromagnetic measurement. Reprinted from [37], Copyright (2022), with permission from Elsevier.

parameter for microwave absorption materials, which is defined as the frequency range with RL below –10 dB [40].

Calculated RL curves of the SiCN bulk ceramics and honeycomb ceramic structures with different thicknesses over the frequency range of 8.2–12.4 GHz, are depicted in figures 7.21(b) and 7.20(d). For the SiCN bulk ceramics, the sample with a thickness of 3.8 mm displayed the minimum reflection loss (RL_{min}) of –23.3 dB at the frequency of 8.2 GHz, corresponding to an absorption of >99%. With the thickness of 3.2 mm, the SiCN bulk ceramics exhibited the widest EAB of 4.2 GHz, covering the X band. For the SiCN honeycomb ceramic structure, the RL_{min} was –49.0 dB at 10.2 GHz, suggesting that 99.99% of waves could be absorbed. Furthermore, a relatively wide EAB of 3.02 GHz was achieved when the sample was 11.4 mm in thickness.

7.2.4 Transparent ceramics

Transparent ceramics are advanced polycrystalline materials engineered with optical transparencies at certain wavelengths, combined with superior mechanical, thermal and chemical properties compared to conventional glass or single crystals [41–46]. Materials for transparent ceramics include yttrium aluminum garnet (YAG), magnesium aluminate spinel ($MgAl_2O_4$), aluminum oxynitride (AlON), aluminum oxide (Al_2O_3), and sesquioxides (Y_2O_3, Sc_2O_3, etc). Transparent ceramics have applications in defense (transparent armors, missile domes), optics (laser gain media in solid-state lasers, infrared windows) and aerospace (sensor protecters). They can also be used in medical devices and high-pressure lamps, owing to their high durability compared to traditional materials. In this subsection, several transparent ceramics processed using 3D printing technology for potential applications will be briefly described.

YAG transparent ceramics were prepared by using 3D printing technology, with 2.33 wt% ISOBAM per mixed powder to assist the aqueous slurry with a solid loading level of 72 wt% [47]. The 3D printing technique enabled the fabrication of complex YAG ceramic structures without the requirement of cold isostatic pressing (CIP) or hot isostatic pressing (HIP), thus effectively increasing the fabrication efficiency and reducing the cost of the products. The YAG transparent ceramics exhibited a relative density of 99.7% and in-line transmittance of 70%, close to the performances of the ones processed by using CIP, as presented in figure 7.22.

Commercial Al_2O_3 (99.99%, Taimei Chemicals Co. Ltd, Japan) and Y_2O_3 (99.999%, Stanford Advanced Materials, USA) were used as the raw materials, while TEOS (Sigma–Aldrich, USA) and MgO (99.99%, Alfa Aesar, USA) were selected as sintering aids. Glycerol (99+%, Alfa Aesar, USA) was employed as plasticizer, while ISOBAM 104 (Kuraray Co., Ltd, Osaka, Japan) was used as dispersant and binder.

Al_2O_3 and Y_2O_3 powders were mixed according to the composition of $Y_3Al_5O_{12}$ (YAG), with 0.4 wt% tetraethyl orthosilicate (TEOS) and 0.08 wt% MgO, by using ball milling for 24 h in anhydrous ethanol, followed by drying at 70 °C for one day. After grinding and sieving through 200-mesh grids, the powder mixture was dispersed in DI water, with the addition of glycerol and ISOBAM-104. The slurry was homogenized for 1 h and then degassed for 25 min. The extrusion-printed green bodies were kept in an auto-desiccator for two days to facilitate a slow drying process and prevent cracking. The dried samples were subjected to calcination at 500 °C for 4 h and 1000 °C for 3 h, at heating rates of 0.5 and 1 $°C \cdot min^{-1}$, respectively. The calcined samples were sintered in a graphite vacuum furnace at 1750 °C for 20 h, followed by annealing at 1450 °C for 10 h.

Scalable massive fabrication of 3D structures with complex geometries, based on highly transparent Nd:YAG ceramics, processed by using 3D printing technology, was demonstrated [48]. Nd:YAG ceramic green bodies were printed, with a printing layer thickness of 25 μm and lateral resolution of 40 μm, from photocurable slurries with solid loading level of 39 vol%. The sintered Nd:YAG transparent ceramics exhibited relative density of 99.9% and maximum in-line transmittance of 80% in the visible light.

The samples were subjected to pre-conditioning at 120 °C for 38 h, burning out at 500 °C for 2 h, pre-sintering at 1000 °C for 10 h, and vacuum sintering at 1750 °C for 20 h (for smaller-sized cup) or 30 h (for larger-sized slab). The sintered samples were

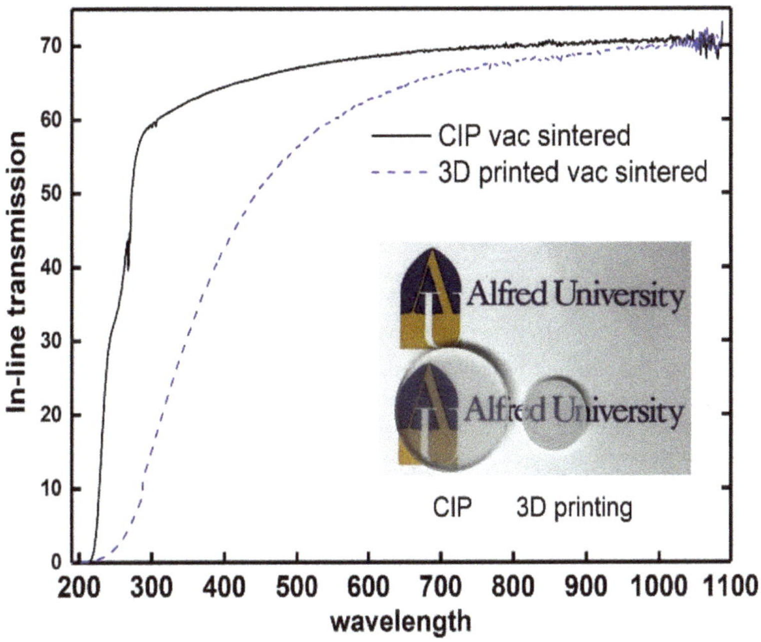

Figure 7.22. In-line transmittance curves and photographs (inset) the traditionally processed and the 3D printed YAG ceramics (with 2.33 wt% ISOBAM) at the thickness of 1.45 mm. Reprinted from [47], Copyright (2020), with permission from Elsevier.

annealed at 1450 °C in air for 10 h to eliminate coloration and oxygen vacancies that were induced during the vacuum sintering process. Finally, the transparent Nd: YAG ceramics were polished for optical measurement. To measure in-line transmission curves, the samples were polished to a thickness of 2.64 mm. Models and photographs of the printed items based on the Nd:YAG transparent ceramics are depicted in figure 7.23.

YAG transparent ceramic rods, serving as gain media for solid state lasers with an active Nd-doped core and an optically-clear cladding shell, were fabricated using a direct-ink-writing (DIW) process [49]. Hot isostatic pressing (HIP) was used to promote densification and enhance optical transparency of ceramics. Lu^{3+} and Gd^{3+} were selected as optically inert ions to co-dope the cladding layer to compensate for the increase in refractive index caused by the presence of Nd^{3+} in the core. Specifically, the variation in refractive index caused by 2% Nd^{3+} could be readily balanced with 11.6% Lu^{3+} or 3.8% Gd^{3+}. In practice, differences in the diffusion distances at the core-cladding interface induced fluctuation in index. DIW printing offers the potential to create a continuous gradient, rather than an abrupt compositional interface.

High purity oxide powders were used to prepare the ceramic inks for the printing experiment [50]. The slurries were extruded through a 500 μm nozzle and printed with controlled variation in composition. The printed items were dried at room temperature and then demolded. The samples were subjected to cold isostatic pressing (CIP) at 200 MPa and then calcined in air at 1000 °C to burn out the

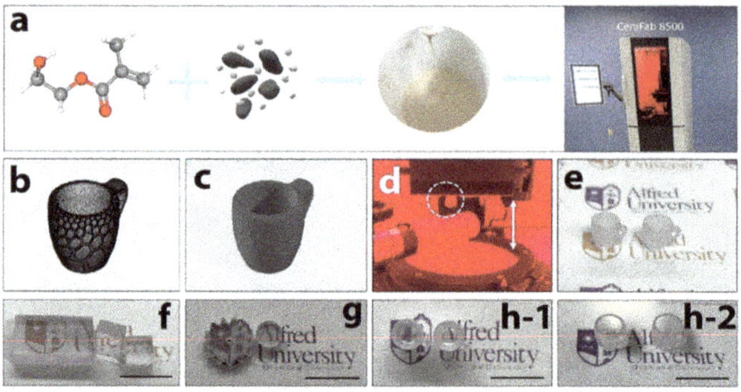

Figure 7.23. Schematic diagram and examples of the transparent Nd:YAG ceramics with 3D structures fabricated by using 3D printing technology. (a) Preparation of the slurry by mixing oxide powders with acrylate monomer-based resin ana photoinitiator. CeraFab 8500 ceramic printer was used for the printing experiment. (b–e) Overall fabrication processing steps: (b) generation of CAD model, (c) loading the 3D model to the printer for printing, (d) photograph showing the 3D printing process and (e) photographs of the final transparent Nd:YAG ceramic cups. (f–h) Photographs of the printed transparent Nd:YAG ceramics with different sizes and geometries after sintering, with the scale bar = 10 mm. Reprinted from [48], Copyright (2021), with permission from Elsevier.

organic additives. The rod samples were first sintered at 1600 °C for 8 h at vacuum level of about 10^{-6} Torr and then HIPed at 1850 °C for 4 h at 200 MPa in Ar.

To address the issue of sharp compositional interfaces and ensure a uniform refractive index along the diameter axis of the laser rods, a mixing nozzle was devised with two nozzles, to realize controlled mixing of two inks before the extrusion step, as demonstrated in figure 7.24(A). As a result, the YAG rods exhibited a smooth radial compositional gradient, eliminating the sharp core-cladding interface. The compositional gradient had roughly four stages, with gradual variation in the concentrations of different elements. The fluctuations in index were entirely absent, indicating advantages of 3D printing technology in fabrication of transparent ceramics for solid-state laser applications.

Y_2O_3 transparent ceramics were prepared by using direct ink writing (DIW) 3D printing process, from ceramic slurries with solid loading level of 45 vol% [51]. The sintered Y_2O_3 transparent ceramics displayed in-line transmittance of 71% at 850 nm, which was comparable to the samples processed by using CIP, as illustrated in figure 7.25. Noting that 3D printing allows for the fabrication of transparent ceramics with complicated structures and geometries, offering advantages over conventional ceramic approaches.

Commercial Y_2O_3 powder (JiaHua, Jiangyin, China, 5.2 μm, 2.4 $m^2 \cdot g^{-1}$, 99.99%) was mixed with ZrO_2 (Adamas-beta, Shanghai, China, 2.6 μm, 3.9 $m^2 \cdot g^{-1}$, 99.9%) and La_2O_3 (Macklin, Shanghai, China, 50 nm, 99.99%), according to the formula of $Y_{1.74}Zr_{0.2}La_{0.06}O_3$. The mixing was conducted via ball milling for 24 h in ethanol, with high-purity zirconia mill balls (diameter = 2 mm). The mixtures were blended with Isobam-104 and triammonium citrate TAC as dispersant. Glycerin was used as both water-retaining agent and lubricant, while DI water was selected to serve as the solvent to prepare the slurries. The green bodies were calcined at 700 °C

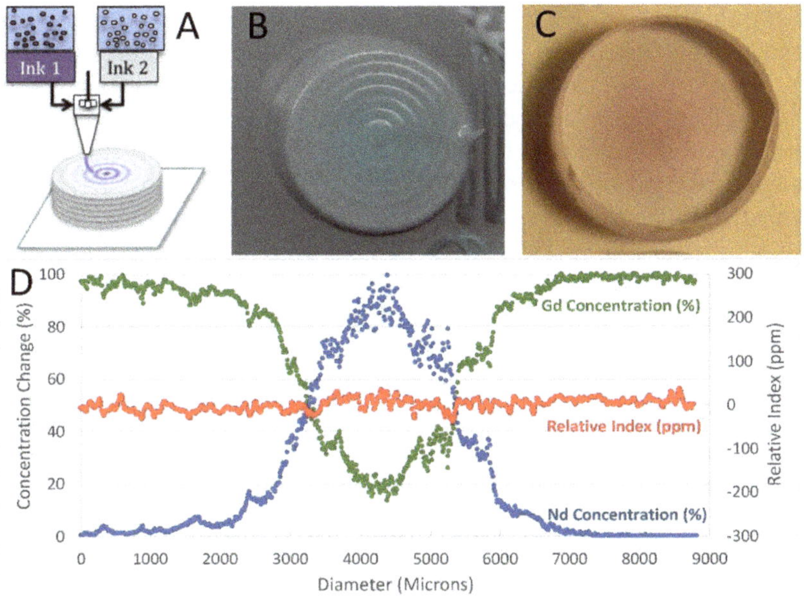

Figure 7.24. (A) Diagram of two nozzles for mixing inks to print radial gradient doping structures, (B) photograph of the printed green body, (C) sintered samples with doping gradient and (D) compositional profiles across the rod diameter showing the calculated index profiles. Reprinted from [49], Copyright (2020), with permission from Elsevier.

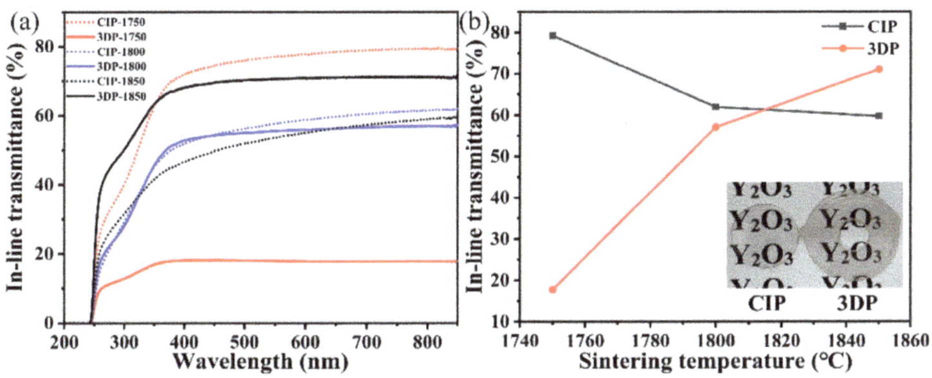

Figure 7.25. (a) In-line transmittance curves of the 3D-printed and CIP processed yttria ceramics. (b) In-line transmittance values of the 3D-printed and CIP processed ceramics at 850 nm versus sintering temperature. Reproduced from [51]. CC BY 4.0.

for 3 h to burn out the organic additives. The calcined samples were sintered at 1750 °C, 1800 °C and 1850 °C for 8 h, followed by annealing at 1250 °C for 6 h.

Optical transparency of transparent ceramics is critically dependent on their microstructures and surface roughness. Given the layer-by-layer construction manner of 3D printing, it is expected that the thickness of the printing layers would have strong impacts on microstructure and surface smoothness of the printed items.

Meanwhile, minimizing surface roughness of the 3D printed transparent ceramics is always challenging due to the presence of the stair-stepping phenomenon. In this regard, Y_2O_3 transparent ceramics were fabricated by using 3D printing, with an aim to reveal the effects of the layer thickness on performances of the final ceramics [52].

It was found that the larger the thickness of the printing layers, the higher the density of the ceramics and hence the higher the optical transparency would be. When the layer thickness was comparable to or smaller than the size of largest particle agglomerates in the powders, the as-printed green bodies were highly porous. For instance, the Y_2O_3 ceramic sample printed with the layer thickness of 45 μm possessed in-line transmittance reaching 97.7% of the theoretical transparency of Y_2O_3. At the same time, surface roughness of the Y_2O_3 transparent ceramics was found to decrease with increasing layer thickness.

To further increase optical transparency of the 3D printed Y_2O_3 ceramics, a chemical-mechanical polishing (CMP) process assisted with vibration was proposed, in which a SiO_2 colloidal suspension was used to replace water. By using this process, surface roughness of the Y_2O_3 transparent ceramics was reduced by a factor of 95.4%, owing to the elimination of the stair-stepping problem originating from 3D printing process. As a result, the in-line optical transmittance increased by 66.1%.

Commercial Y_2O_3 powder (99.99%, YeeYoung Cerachem, Korea), La_2O_3 and $ZrO(CH_3COO)_2$ were mixed via ball milling at 290 rpm for 24 h, with 2 mm ZrO_2 balls in anhydrous alcohol (99.9%, Samchun Pure Chemical, Korea). In this case, 10 mol% La_2O_3 and 3 mol% ZrO_2 served as sintering aids. The suspensions were dried and ten calcinated at 1200 °C for 4 h. The calcined powder was blended with 2 wt% Disperbyk 111 (Altana AG, Germany), trimethylolpropane triacrylate (TMPTA, Sigma–Aldrich, USA), 1,6-hexanediol diacrylate (HDDA, Sigma–Aldrich, USA), poly(propylene glycol) (PPG, $M_n = 425$, Sigma–Aldrich, USA) and a photoinitiator (Irgacure 819, BASF, Germany), to form printing slurries, with solid loading level of 44 vol% or 79.65 wt%, through ball milling at 2000 rpm for 10 min, using a centrifugal mixer (THINKY, Laguna Hills, CA, USA).

The 3D models were generated using SolidWorks software (SolidWorks Corp., USA). The UV light source had a wavelength of 405 nm and a power intensity of 10 mW $\cdot$ cm^{-2}. The printing thicknesses were 20–55 μm at an interval of 5 μm. The excessive slurry was washed out with HDDA. The green bodies were dried and calcined at 600 °C in N_2 to remove the organic components. The residual carbon in calcined bodies was burnt out by heating in air for 4 h. After that, the samples were sintered at 1800 °C for 10 h at a vacuum level of 10^{-3} Pa. The sintered samples were annealed at 1400 °C for 4 h in air and then polished for characterization.

Linear shrinkage rates of the 3D printed Y_2O_3 ceramics in different directions are depicted in figure 7.26(a). The shrinkage rate in the direction of Z-axis was much larger than those in the lateral directions. This was because in the Z-axis direction interlayer voids or spacings were produced during the printing process. Moreover, the shrinkage rates in the directions of the X- and Y-axes were independent of the printing thickness. In comparison, the effect of printing thickness on shrinkage rate in the direction of Z-axis was more significant, especially for the thicknesses of

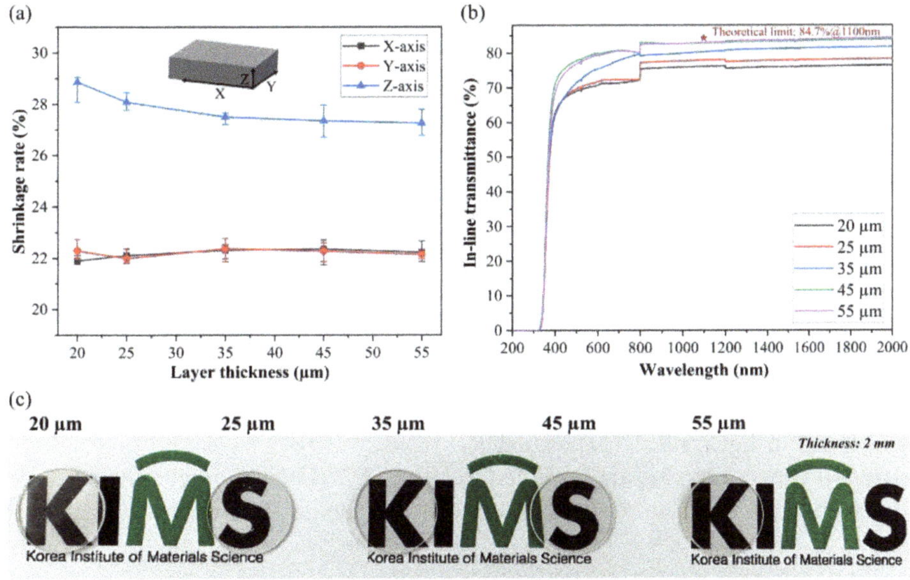

Figure 7.26. (a) Shrinkage rate, (b) in-line transmittance and (c) photographs of the Y_2O_3 ceramics 3D printed with different layer thicknesses. Reprinted from [52], Copyright (2025), with permission from Elsevier.

20–35 μm. In-line transmittance curves and photographs of the Y_2O_3 transparent ceramics with different printing layer thicknesses are displayed in figures 7.26(b) and (c). The line transmittance increased with increasing printing thickness. The samples with layer thicknesses of 45 and 55 μm exhibited similar transmittance.

Magnesium aluminate ($MgAl_2O_4$) spinel transparent ceramics, with a transparency of 97% of the theoretical level, were prepared using SLA 3D printing techniques [53]. Transparent lenses and microlattices could be printed with a resolution of 100–200 μm. The 3D printed lens was able to produce optical imaging, while the diamond microlattices served as transparent supports for photocatalyst TiO_2, leading to enhanced photocatalytic activity compared to that on the opaque supports. Commercial spinel nanosized powder (S30CR, Baikowski), with specific surface area of 28 $m^2 \cdot g^{-1}$, was used to prepare the printing slurry, with solid loading level of 55 wt%. Photosensitive trimethylolpropane triacrylate (TMPTA) resin, photoinitiator 2-hydroxy-2-methylpropiophenone (PI) and methyl orange (Sigma–Aldrich, Solsperse 85000, Lubrizol) were utilized as organic additives.

The as-printed green bodies were calcined to remove the organic components, using a tube furnace (GHA 12/450, Carb Lite Gero). The experiment was conducted in N_2 and then in air, at heating/cooling rates of 0.1 °C–1.0 °C $\cdot$ min^{-1}. The calcination temperatures were 250 °C, 350 °C, 450 °C and 600 °C, each for about 5h. The calcined samples were pre-sintered in a box furnace (LHT 04/18, Nabertherm) at 1600 °C–1700 °C for 25 h, followed by treatment with HIP (AIP10–30H, AIP) process at 1700 °C–1800 °C for 15 h at 180 MPa in Ar.

As compared with the conventional supports for photocatalysts, the 3D printed transparent ceramics offered larger surface area for illumination and low light absorption. Meanwhile, the mass flow could be well controlled, by designing the hollow channels inside the ceramics, thus promoting photocatalytic efficiency. Schematic diagrams of photocatalytic reactors made of the printed transparent ceramic lattices as the support of nanosized TiO_2 photocatalyst for water treatment and the working mechanism are illustrated in figure 7.27(A).

Different types of microlattices were printed for comparison and optimization. With dimensions of 6.3 mm × 6.3 mm × 18.9 mm, the diamond and gyroid microlattices had a solid volume fraction $V_f = 0.35$, while their surface areas were 1767 and 1569 mm^2, respectively. Photographs of the photocatalyst supports with groups of microlattices coated with TiO_2 nanosized particles are displayed in figures 7.27(B) and (C). Opaque spinel ceramic lattices with the same dimension were also examined, which had a volume fraction of 0.35, but featured larger surface areas of 1779 and 1699 mm^2. SEM images and EDX mapping profiles of the samples are presented in figures 7.27(D) and (E).

$\ln(A_0/A_t)$ versus t curves are depicted in figure 7.27(F), which were fitted to derive the reaction rate constant of the TiO_2 photocatalysts supported by the transparent

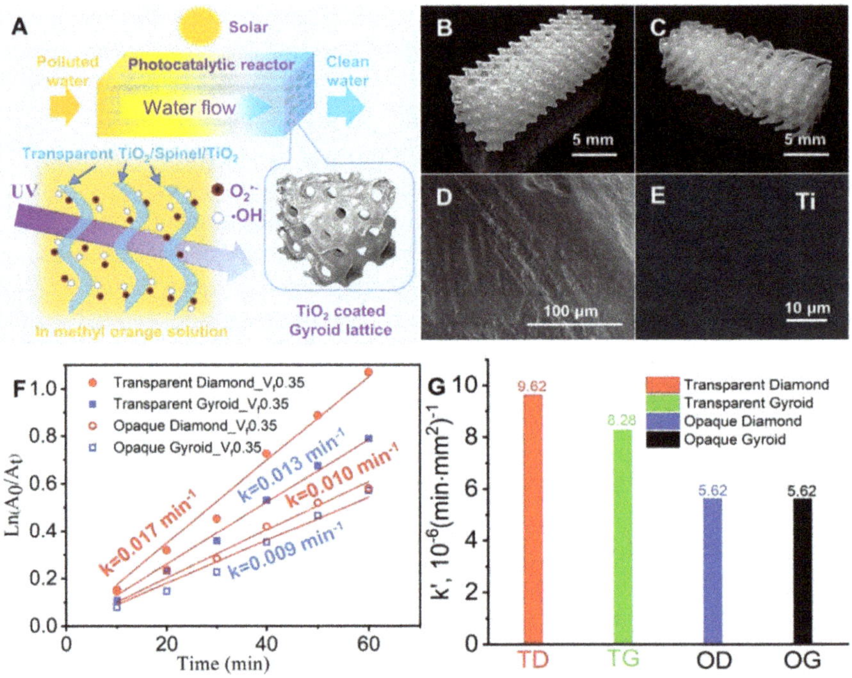

Figure 7.27. (A) Schematic diagram of photocatalytic reactors with the 3D printed transparent ceramics as support of the photocatalyst for water treatment. (B and C) Photographs of two types of ceramic photocatalyst supports coated with nanosized TiO_2 particles in the forms of diamond (B) and gyroid (C) lattice structures. (D) SEM image and (E) EDX mapping (Ti) of the TiO_2 films coated on the sheet surface of the ceramic lattices. (F) $\ln(A_0/A_t)$ versus irradiation time (t) and (G) Bar chart of the normalized rate constant (k') for different groups of ceramic photocatalyst supports (TD = transparent diamond, TG = transparent gyroid, OD = opaque diamond, OG = opaque gyroid). [53] John Wiley & Sons. © 2021 Wiley-VCH GmbH.

microlattices. The TiO_2 samples on the opaque gyroid and diamond lattice supports exhibited relatively low photocatalytic activity. The sample on the diamond microlattice presented the highest reaction rate constant of 0.017 min^{-1}. The gyroid microlattice resulted in relatively lower reaction rate constant of 0.013 min^{-1}, because of its smaller surface area. Photocatalytic efficiencies of the TiO_2 nanosized coatings on different supports, in the form of normalized reaction rate constants, are summarized in figure 7.27(G).

Alumina transparent ceramics were fabricated by using extrusion-based 3D printing technology [54]. The final alumina ceramics had relative density of 99% and total transmittance of 70% at 800 nm. Al_2O_3 powder (Baikalox High Purity CR 10D) and 625ppm MgO (Baikowski, Inc.) were mixed to prepare the printing slurries. An extrusion-based 3D printer (Hyrel 3D System 30 M) was used to print samples. The oxide powders were dispersed in DI water, forming slurries with solid loading levels were 68–74 wt%, together with 0.4–1.0 wt% Kuraray ISOBAM™-104 as dispersant.

The 3D printed green bodies were naturally dried at room temperature in air and at 70 °C in an oven each for 24 h. The dried samples were calcined at 500 °C for 4 h and 1000 °C for 3 h, at heating rates of 0.5 and 1 °C · min^{-1}, respectively. The calcined samples were sintered at 1650 °C–1875 °C for 6 h, at a heating rate of 10 °C · min^{-1}, during which the temperature was held at 1300 °C for 1 h. This was denoted as a conventional single-step sintering (SSS) process. Meanwhile, a two-step sintering (TSS) process was examined, where the final temperatures of 1850 °C or 1875 °C was immediately lowered to 1800 °C. The final sintering time was 8 h. Conventional CIP processed samples were prepared for comparison.

Figure 7.28 shows total transmission curves and photographs of the 3D printed (both SSS and TSS) and CIP processed alumina transparent ceramics at a thickness

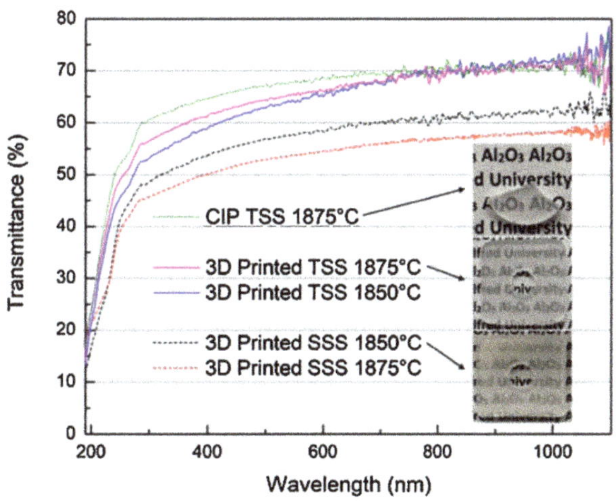

Figure 7.28. Total transmittance spectra of the 3D printed and CIP processed alumina ceramics with single-step (SSS) and two-step (TSS) sintering processes with the inset showing photographs of the corresponding samples. Reprinted from [54], Copyright (2021), with permission from Elsevier.

of 1.5 mm after polishing. The total transmittance of the samples was 70%–71% at wavelengths beyond 800 nm. In the visible wavelength range of 400–700 nm, the CIP samples were superior to the 3D printed ones, suggesting that further optimization is necessary to increase the performances of the alumina ceramics fabricated using 3D printing technology.

Transparent AlON ceramics were prepared by using the DIW printing process, with aqueous ceramic slurries [55]. Phase-pure AlON ceramic tiles, with dimensions of $10 \times 10 \times 0.9\ mm^3$, were achieved, which exhibited promising optical transparency and mechanical strength, after sintering at 1960 °C for 10 h in N_2. Al_2O_3 and AlN powders, with a molar ratio of 69:31, were mixed along with sintering aids of 0.1 wt% Y_2O_3 + 0.4 wt% $MgAl_2O_4$ + 0.12 wt% H_3BO_3. The mixtures were ball milled in anhydrous ethanol for 20 h, followed by drying and sieving. The dried powders were dispersed in DI water, with CE-64 (Dolapix CE-64, Zschimmer & Schwarz, Germany) as dispersant and HEC (HEC, 3400–5000 mPa·s, Shanghai Macklin Biochemical Co., Ltd, China) as a thickener.

The as-printed green bodies were dried at constant temperature and humidity. The dried green bodies were calcined at 650 °C for 6 h, at a heating rate of $1\ °C \cdot min^{-1}$ to burn out the organic additives. The samples were finally sintered at 1900 °C–1980 °C in N_2.

Figure 7.29 presents in-line transmittance curves of the samples sintered at different temperatures for different time durations. After sintering for 5 h, the in-line transmittance at 780 nm was increased from 44.8% to 71.0%, as the sintering temperature was raised from 1900 °C to 1960 °C, but then decreased sharply as the sintering temperature was further increased to 1980 °C, as seen in figure 7.29(a). For the samples sintered at 1960 °C, the optimal time duration was 10 h, with the in-line transmittance reaching 81.9%, as revealed in figure 7.29(b). Optimized mechanical properties were observed in the samples sintered at 1940 °C for 5 h, with a fracture toughness of $2.86\ MPa \cdot m^{1/2}$ and a Vickers hardness of 18.68 GPa.

7.2.5 Ceramic matrix composites

Ceramic matrix composites (CMCs) refer to materials composed of one or more reinforcements, in the forms of fibers, whiskers, rods ranging from carbon nanotubes (CNTs), carbon fibers, and graphene to oxide particulates and secondary polymeric or metallic phases, embedded in a ceramic matrix [56–61]. Besides their superior physical, chemical, thermal and mechanical performances, they could also offer functionalities that are not available in the components. Therefore, CMCs find applications in aerospace defense, automotive, energy, power, and electronics. It is naturally expected that such materials could be processed using 3D printing [62].

Vinyl-functionalized siloxane oligomers dispersed with SiC whiskers were used to prepare SiOC based composites via SLA 3D printing technology [63]. The 3D printed structures were pyrolyzed to form the composites. The composite ceramics exhibited a shrinkage of 37%, whereas samples without reinforcement experienced a shrinkage of 42%. Figure 7.30 shows photographs of a representative sample before and after pyrolysis. The ceramic matrix composites were microstructurally almost

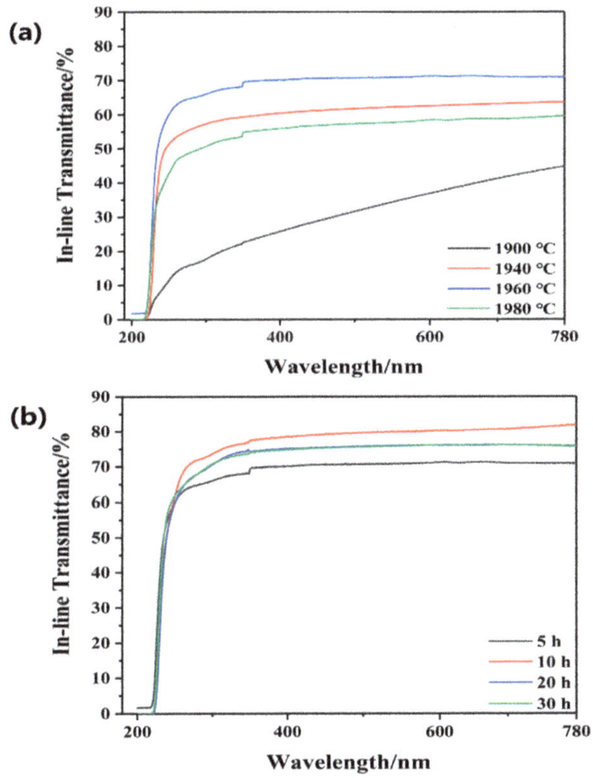

Figure 7.29. In-line transmittance curves of the AlON ceramics in the visible light range: (a) different sintering temperatures and (b) different time durations. Reprinted from [55], Copyright (2022), with permission from Elsevier.

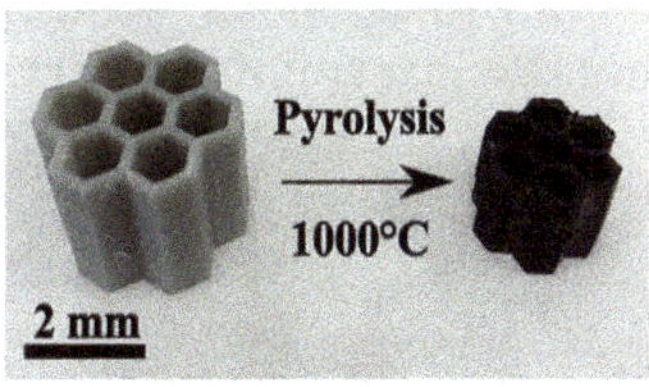

Figure 7.30. Photographs of the 3D printed preceramic polymer structure before and after pyrolysis at 1000 °C. [63] John Wiley & Sons. © 2018 WILEY-VCH Verlag GmbH & Co. KGaA, Weinheim.

free of porosity and cracks. Due to the incorporation of SiC whiskers, the hardness of the composites increased from 10.8 to 12.1 GPa, while the density decreased from 2.99 to 2.86 $g \cdot cm^{-3}$.

Vinylmethoxysiloxane, with an average molar weight of 500 $g \cdot mol^{-1}$ and a density of 1.10 $g \cdot cm^{-3}$ (Gelest, VMM-010), poly(ethyleneglycol)-diacrylate (PEGDA) with a molar weight of 258 $g \cdot mol^{-1}$ and density of 1.11 $g \cdot cm^{-3}$, photoinitiator Irgacure® 819 (Bis(2,4,6-trimethylbenzoyl)-phenylphosphineoxide) and free-radical scavenger

Irganox 1330™ (1,3,5-trimethyl-2,4,6-tris(3,5-di-tert-butyl-4-hydroxybenzyl)-benzene) were used to prepare the printing slurries. SiC whiskers (>99%, Sky Spring Nano Inc.) with a maximum diameter of 2.5 μm, a density of 3.213 g · cm^{-3}, and an aspect ratio (*L*/*D*) of 20, were used as the reinforcing agent.

To prepare the resin, the VMS and PEGDA were mixed at a mass ratio of 1:1 using a shaker table at 1000 rpm for 0.5 h. After that, the photoinitiator and free-radical scavenger were added at concentrations of 0.3 and 0.7 wt%, respectively, followed by milling at 1000 rpm for 0.5 h. For the samples containing SiC whiskers, the content was 0.5 wt%, corresponding to a volumetric concentration of 1.45 vol%. The mixture was further milled at 1000 rpm for 0.5 h, leading to a homogeneous mixture suitable for the printing experiment.

The VMS/PEGDA/Irgacure/Irganox resin mixture was polymerized under 405 nm UV irradiation. A FormLabs Form1+ stereolithographic (SLA) printer was utilized to print samples. The 3D printed preceramic polymer structures were pyrolyzed in a tube furnace under Ar at 500 °C and subsequently at 1200 °C for 1 h, at a heating rate of 1 °C · min^{-1}. After that, the temperature was reduced from 1200 °C to 200 °C, at a cooling rate of 2 °C · min^{-1} to minimize the risk of thermal-shock induced cracking. Figure 7.31 depicts photographs of the pyrolyzed SiC whisker reinforced SiOC ceramic composites with different structures and geometries.

New types of UV curable preceramic polymers were developed to fabricate SiOC ceramic composites reinforced with SiC whiskers through 3D printing [64]. To demonstrate this technique, 3D structures with complex geometries and high resolutions were printed. The pyrolyzed samples were of an amorphous state, with a dense microstructure and free of cracks, while the SiC whiskers were homogeneously distributed in the SiOC matrix. The incorporation of the SiC whiskers resulted in reduced shrinkage and mass loss after pyrolysis of the green bodies. More significantly, mechanical strength of the final ceramic matrix composites was largely enhanced, with compressive strength of 98.4 ± 12.3 MPa.

Raw materials for the experiment included γ-(methacryloxypropyl) trimethoxy silane (MATMS), ethyl alcohol (Beijing InnoChem Science & Technology Co., Ltd) and SiC whiskers (SiC_w, CAS Key Laboratory of Carbon Materials, Institute of

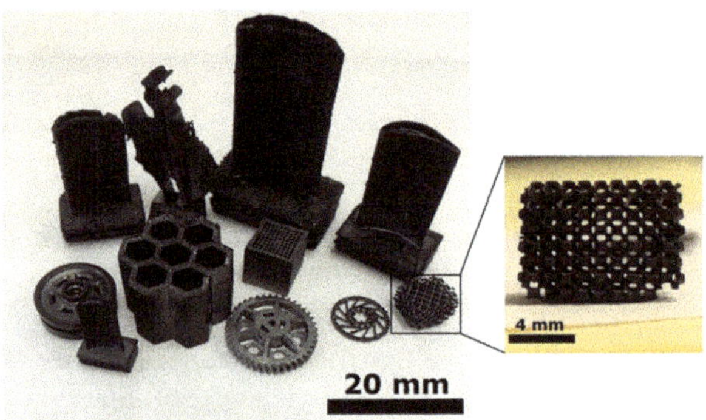

Figure 7.31. Photographs of polymer-derived ceramics with different structures and geometries by using the SLA 3D printing process. [63] John Wiley & Sons. © 2018 WILEY-VCH Verlag GmbH & Co. KGaA, Weinheim.

Coal Chemistry, CAS, China), as well as bis (2,4,6-trimethylbenzoyl) phenyl phosphine oxide (PI 819) and dicumyl peroxide (DCP, Energy Chemicals Co., Ltd). MATMS (1240 g, 5 mol) and methanol (310 g) were dissolved in distilled water (130.5 g) containing HCl (36.5%, 0.63 g), through slow addition in a drop-by-drop way.

The mixture was allowed to react for 5 h at room temperature. After that, the temperature gradually raised to 70 °C to complete the reaction for 2 h. Then, the mixture was further heated at 90 °C for 2 h to vaporize methanol at negative pressure. After removal of residual methanol, hydrolyzed MATMS (MATMS-H), with viscosity of 550 cp at 25 °C, was formed. Samples both with and without the SiC whiskers were prepared for comparison. The 3D printed green bodies were pyrolyzed in N_2, which was conducted at 170 °C for 2 h, 500 °C for 1 h and 1000 °C for 1 h, at heating rates of 1 °C $\cdot$ min^{-1}, followed by cooling to 200 °C at a rate of 2 °C $\cdot$ min^{-1}.

For the samples based on MATMS-H-n resin, the LED power was 80 μW, while the exposure time was 4 s. Photographs of the 3D printed green bodies and pyrolyzed samples, with geometries of hollow spheres, honeycomb structure and the chess pieces Rook and Pawn, from the MATMS-H-n resin, are illustrated in figure 7.32. All the samples had smooth surfaces and clearly defined features, suggesting that the photopolymer resins were highly printable, with sufficient resolution and accuracy. The pyrolysis process induced linear shrinkage of 32.8%, for the structures printed with the MATMS-H-n resin. The ceramic yield estimated from the weight ratio was about 48.5%.

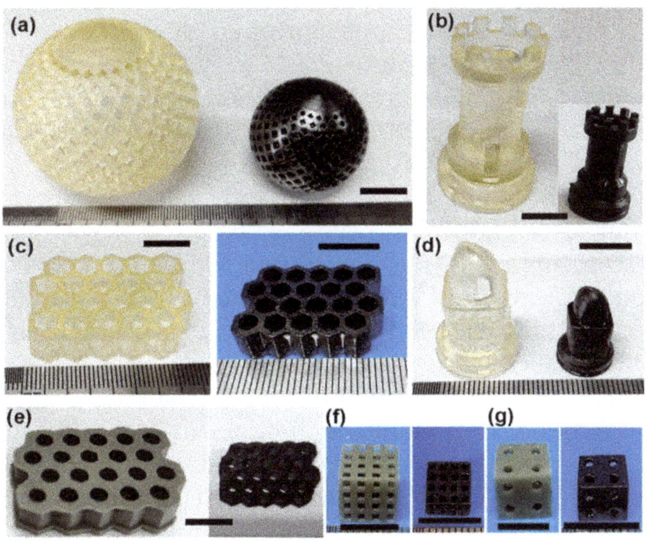

Figure 7.32. Photographs of the 3D printed green bodies and the pyrolyzed samples with the resins (scale bars = 1 cm): (a–d) MATMS-H-n and (e–g) MATMS-H-SiC$_w$ and corresponding pyrolyzed ceramic structures. Reprinted from [64], Copyright (2021), with permission from Elsevier.

SiC based ceramic matrix composites reinforced with SiC whiskers were derived from structures fabricated via direct ink writing (DIW), using SiC_w/polycarbosilane (PCS) slurries as the precursors [65]. The slurries with different solid loading levels were used to print 3D SiC_w/SiC lattice structures. Filaments with various morphologies, including accumulating, coiling, meandering, and straightening, were printed at different heights and speeds. Coiled filaments were obtained when printing height was sufficiently large. 3D lattices with straight filaments and strong shape retention were only achievable when the solid loading level of the SiC_w/PCS slurry reached 62.3 vol%. The final 3D lattices with porosity of about 62% displayed impressive bending strength of 33.2 MPa and compression strengths of 30.6 4.3 MPa.

SiC_w/SiC_p reinforced SiC based ceramic matrix composites were developed using polycarbosilane slurries and DIW 3D printing technology [66]. The content of SiC_p and sintering temperature had significant effects on the microstructure, pyrolysis behaviors, compositions and mechanical properties of the final ceramic matrix composites. SiC samples with complicated geometries could be printed out with the PCS/n-hexane/SiC_w–SiC_p slurries. The printed lines possessed core–shell structure, due to the reduction in shear stress in the radial direction of the nozzles. Specifically, the linear shrinkage was reduced to 8.3%, while the weight loss was minimized to 10.6%, by increasing the SiC_p/PCS ratios in the slurries. At SiC_p/PCS ratio of 0.7, the 3D-SiC lattices exhibited tensile strength of 21.3 MPa.

Chopped carbon fiber-reinforced ceramic matrix composites (C_f/SiC) were fabricated using SLA 3D printing technology, with C_f combined with liquid silicon infiltration [67]. The 3D structures exhibited high printing stability and accuracy, with structural deviations of <5%. The adjacent layers were strongly bonded due to the synergistic effect of the curing adhesion of photosensitive resins and the crossed pinning of the chopped carbon fibers. The as-obtained samples after liquid silicon infiltration had a high density with a flexural strength of 262.6 MPa.

Chopped carbon fibers with a density of $1.76\ g \cdot cm^{-3}$ and an average diameter of 7 μm (C_f, Shanghai Liso Composite Material Technology Co., Ltd, China) and photosensitive resin (PR, Shenzhen eSUN industrial Co., Ltd, China, 1.05–1.13 $g \cdot cm^{-3}$) were used as raw materials. C_f and PR were blended with high purity SiC balls through ball milling at speeds of $1500\ rad \cdot s^{-1}$ for 15 min. The C_f photosensitive slurries with different solid loading levels were subject to degassing before printing. The as-printed items were ultravioletly cured and then ultrasonically cleaned, resulting in C_f preforms. The dried samples were densified through liquid Si infiltration at 1650 °C for 1 h in Ar. Figure 7.33 presents photographs of the samples processed at different stages.

Eutectic ceramics are a special group of ceramic materials, consisting of two or more phases that coexist in a stable equilibrium at a specific composition and temperature, known as the eutectic point. These ceramics offer unique properties due to their unique microstructural characteristics. The eutectic microstructures are often of fine-scale intertwined phases, thus resulting in enhanced mechanical properties, such as high strength and toughness. Additionally, they may exhibit high thermal stability and resistance to wear and corrosion. Eutectic ceramics have applications in various fields, such as aerospace, cutting tools, etc. Meanwhile, the

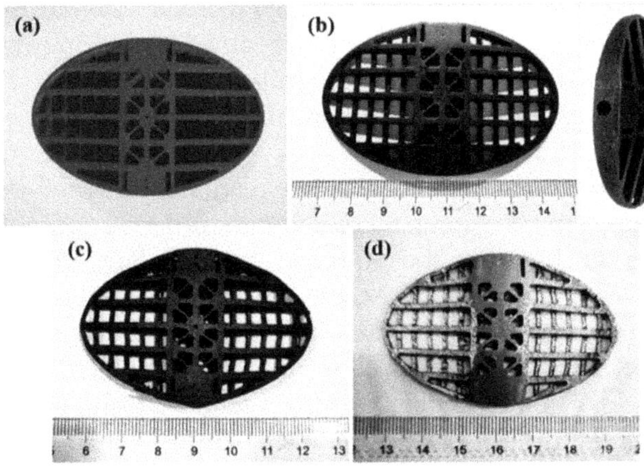

Figure 7.33. Photographs of the SLA 3D printed mirrors: (a) printing model, (b) printed green body and side view, (c) C_f preform after degreasing, (d) sintered C_f/SiC ceramic matrix composites. Reprinted from [67], Copyright (2020), with permission from Elsevier.

microstructures can be well tailored by controlling the fabrication processing. In recent years, 3D printing technology has been employed to fabricate various eutectic ceramics [68, 69].

Al_2O_3/ZrO_2 eutectic ceramics were prepared using 3D printing technology, while the phenomenon of crack formation was suppressed [70]. The average rod spacing (λ_{av}) depended on the scanning rate (V) with the relationship of $\lambda_{av}V^{0.5} = 1\ \mu m^{1.5}/s^{0.5}$. Typical eutectic microstructures, known as complex regular patterns, were modulated through the growth conditions. The concentration of defects could be effectively minimized by adjusting the processing parameters of the solidification process. The optimized samples exhibited promising mechanical properties, with a hardness of 16.7 GPa and fracture toughness of 4.5 $MPa \cdot m^{1/2}$.

The laser 3D printer consisted of a PRC2000 CW CO_2 laser, a four-axis working table with numerical controlling, a powder feeder with a nozzle installed in the lateral plane and a closed-loop controlled preheatable system with a thermal couple and temperature controller (Shimaden, SRS13A). The printing experiment was conducted in a glove box with a controlled environment. The laser was attached on a carriage to be overhead, while a window was opened on top of the chamber to direct the laser beam into the glove box. Ar was used as both the atmosphere gas in the glove box and the carrier gas to deliver the oxide powders.

Commercial Al_2O_3 and ZrO_2 powders were used as the precursors. The ZrO_2 powder was doped with 8 wt% Y_2O_3 to stabilize the tetragonal phase of ZrO_2 at room temperature. 58.5 wt% Al_2O_3 and 41.5 wt% ZrO_2 were mixed through ball milling in PVA solution. The mixture was dried at 200 °C for 1 h and then dry-pressed into bars with dimension of 70 mm × 10 mm × 5 mm. The samples were sintered at 1400 °C for 2 h to achieve sufficiently high mechanical strength for later processing. The Al_2O_3/ZrO_2 eutectic powder was granulated to form spherical particles with an average diameter of 150 μm.

The substrate was preheated to 1000 °C, onto which the laser beam was irradiated to generate a molten pool. Then, the eutectic powder was injected through the powder feed nozzle into the molten pool, followed by melting and then solidification, thus forming cladded layers. The resultant samples were crack-free, with dimensions of 20 mm × 8 mm × 8 mm.

Figure 7.34 shows longitudinal and transverse sectional SEM images of the Al_2O_3/ZrO_2 eutectic, revealing typical columnar colony-patterned microstructures. The colony microstructures were attributed to constitutional undercooling, which was associated with growth restrictions and faceted phases. The presence of the colonies in turn enhanced structural stability of the materials, owing to the release of the constitutional undercooling. The colonies were embedded in the coarse granular microstructures.

Al_2O_3–$Y_3Al_5O_{12}$ (AY) binary eutectics were prepared using laser engineered net shaping (LENS) processing [71]. The as-obtained samples were dominated by colony structures composed mainly of coupled irregular eutectics in the interior, while the growth behavior followed the Magnin–Kurz model. In each deposited layer, The $Y_3Al_5O_{12}$ phase in irregular eutectic competitive grew, with a transition from random characteristics to those with orientation, while the orientation of the

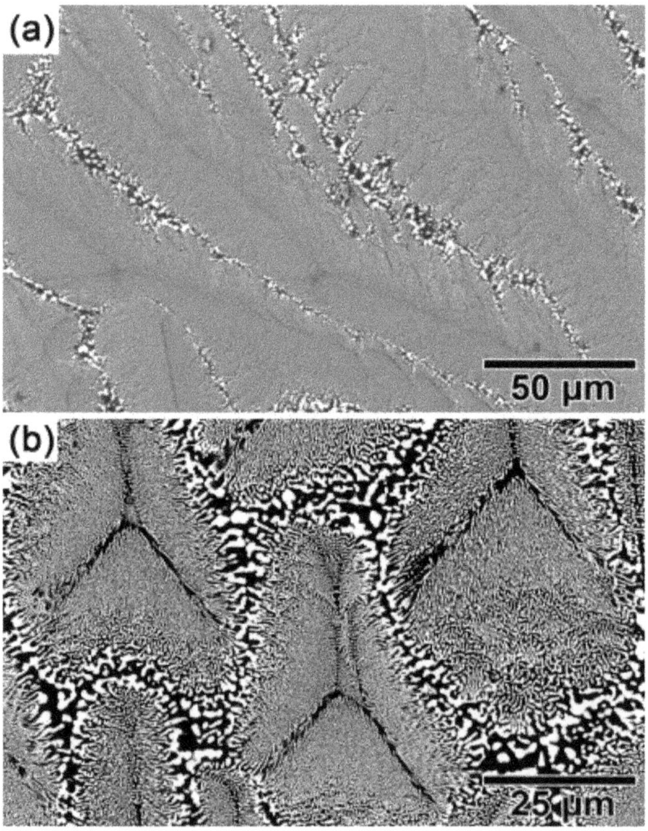

Figure 7.34. Representative SEM images of the Al_2O_3/ZrO_2 DSEC: (a) longitudinal cross-section and (b) transverse cross-section. Reprinted from [70], Copyright (2016), with permission from Elsevier.

Al_2O_3 phase remained unchanged. Crystallization of the colony-coupled eutectics occurred at the bottom of each layer, whereas the irregular eutectics transformed into regular ones in outer regions of the as-prepared samples.

The precursors were made of α-Al_2O_3 (>99.8%, averaged particle size = 45 μm) and Y_2O_3 (>99.9%, averaged grain size = 40 μm), through ball milling, according to the eutectic composition of 81.5 mol% Al_2O_3–18.5 mol% Y_2O_3 [72]. In the experiments, thin wall AY samples with dimensions of 10 × 10 mm^2 were printed in Ar using an OPTOMEC LENS® 450 printer, with Nd:YAG laser and power of 400 W. The substrate was Ti6Al4V. The printing parameters included a beam diameter of 1 mm, laser power of 200 W, scanning speed of 305 mm · min^{-1}, powder feeder rotation speed of 8 rpm and a layer thickness of 500 μm.

Figure 7.35 shows the eutectic morphological evolution of as-built AY composites. The regions of the AY thin-wall samples for SEM observation are

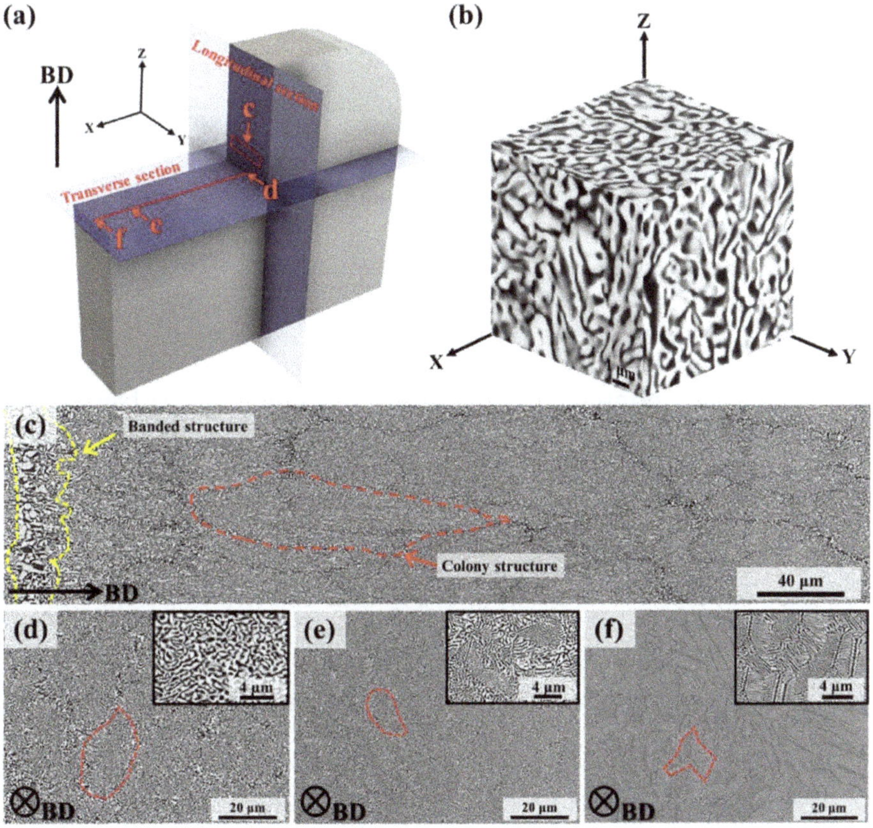

Figure 7.35. Microstructure evolution of LENS processed AY eutectic ceramics: (a) schematic diagram of the thin wall structure model, showing the areas for SEM observation, (b) isometric view showing the dominant microstructure, with the dark phase to be Al_2O_3 and the bright phase to be YAG, (c) stitched SEM image showing the eutectic morphology of a single layer along the longitudinal section (marked as Zone c in panel (a)), (d–f) SEM images along the transverse section marked as Zones d, e and f in panel (a), corresponding to the irregular interpenetrated eutectic structure(d), quasi-regular eutectic (e) and complex regular eutectic structure (f). Reprinted from [71], Copyright (2020), with permission from Elsevier.

schematically shown in figure 7.35(a). The building direction (BD) was in the direction of the Z-axis, while the laser was scanned in the direction (SD) X-axis. The microstructure was characterized by refined irregular eutectics, where the two phases were mutually interpenetrated, forming a network-like structure, as presented in figure 7.35(b). Because both constituent phases in the AY eutectics have high melting entropies, significantly anisotropic growth tended to occur, thus generating irregular eutectics [73, 74].

In fact, the solidification conditions of 3D printed samples varied spatially during the printing process. For instance, the thermal gradient at the front of the solidification would be decreased along the printing direction with each individual layer, leading to a corresponding increase in solidification rate. At the longitudinal section, periodic eutectic patterns were present in the printing direction. Typical microstructure of a single printed layer in Zone c in figure 7.35(a) is demonstrated in figure 7.35(c). The eutectics had banded coarsening microstructure at the bottom in the printing direction BD, as enclosed in the yellow dashed line region, which was transferred to colony structure with coupled irregular eutectics to be inside, as indicated by the red dashed lines.

The colonies tended to be elongated roughly along the printing direction, with slightly coarse microstructures at the boundaries. The formation of the colonies was ascribed to the decrease in thermal gradient, i.e., solidification rate ratio at the solid/liquid interface of the molten pools, which triggered the growth of cellular structure. Eutectic structures at Zones d–f are illustrated in figures 7.35(d)–(f), corresponding to the transverse section in figure 7.35(a). The eutectic features exhibited strong variation from the center to the edges, with irregular eutectic, quasi-regular eutectic and complex regular eutectic, in Zones d, e and f, respectively.

Besides binary systems, ternary systems of eutectic ceramics have also been explored by using 3D printing technology [75–79]. Hollow guide blades, based on Al_2O_3/YAG/ZrO_2 ternary eutectic ceramics, with almost full densification and homogeneous fine eutectic microstructures, were fabricated using a combined strategy involving laser floating zone melting, vat photopolymerization 3D printing and hot isostatic pressing (HIP) [75]. The final eutectic ceramics had a relative density of about 92%, after sintering at 1670 °C for 2 h. Further densification was achieved after hot isostatic pressing at 1550 °C at 200 MPa for 1 h, with relative density of about 99.3%. The highly dense eutectic ceramics displayed bending strengths of 352.99 and 299.38 MPa before and after the HIP processing, respectively, along with a hardness of 19.10 GPa and fracture toughness of 2.22 $MPa \cdot m^{1/2}$.

Commercial powders of Al_2O_3 (99.99%), Y_2O_3 (99.99%), and ZrO_2 (99.9%, stabilized with 3 mol% Y_2O_3) were utilized to prepare the precursors. The ternary composition was Al_2O_3:Y_2O_3:ZrO_2 = 65:16:19 (mol%). The oxide powders were mixed through ball milling in anhydrous ethanol (Tianjin Fu Yu Fine Chemical Co., Ltd, China) at 300 rpm for 24 h. The mixture was dried at 70 °C for 2 h, followed by pressing into precursors with dimension of $110 \times 5 \times 5$ mm^3 through dry pressing. The samples were sintered at 1300 °C for 2 h for precursors for laser processing. A photocurable resin (Intelligent Laser Technology Co., Ltd, China) was selected as the binder for slurries. The resin consisted of acrylate oligomers and monomers, a photoinitiator, a dispersant, and a plasticizer.

Laser floating zone melting (LFZM) was employed to conduct directional solidification of the precursors. The sintered precursors were melted using two fixed lasers at powers of 300–350 W to create a molten zone. The precursors were driven to move downwards at speeds of 1080 mm $\cdot$ h^{-1}, thus generating temperature gradient. As a result, directional solidification was triggered in the direction from bottom to top. Al_2O_3/YAG/ZrO_2 ternary eutectic ceramic rods exhibited smooth surfaces and uniform appearance, with lengths of 85–90 mm and diameters of 3.5–4 mm. Selected rods were crushed and ball milled into eutectic ceramic powders.

Ceramic slurries with a solid loading level of 53 vol% for the vat photopolymerization 3D printing were prepared by blending the ceramic powder with the photocurable resin. The ball milling was conducted at 300 rpm for 24 h. The resulting slurries were used to print using a UV 3D printer (CeraBuilder 160, China). The printing parameters included laser wavelength of 355 nm, layer thickness of 50 μm and the scanning speed of 40 mm $\cdot$ s^{-1}. The laser power was 2.26 W, while the UV light scanning rate was 2000 mm $\cdot$ s^{-1}, with the corresponding curing depth of 210 μm.

The printed green bodies were calcined at 100 °C, 200 °C and 320 °C for 3 h at a heating rate of 0.1°C $\cdot$ min^{-1}. They were then calcined at 420 °C and 550 °C for 2 h, at heating rates of 0.2 and 0.5 °C $\cdot$ min^{-1}, respectively. Finally, the samples were calcined at 1100 °C for 1 h, at a heating rate of 2°C $\cdot$ min^{-1}. The calcined samples were sintered at 1000 °C and 1550–1700 °C, for 10 min and 2 h, at heating rates of 8 and 1 °C $\cdot$ min^{-1}, respectively. For HIP processing, the temperature was kept at 1450 °C for 10 min at a heating rate of 15 °C $\cdot$ min^{-1} and then 1550 °C for 60 min, at a heating rate of 5 °C $\cdot$ min^{-1} and pressure of 200 MPa in Ar.

Eutectic microstructures of the transverse and longitudinal sections of the directionally solidified Al_2O_3/YAG/ZrO_2 eutectic ceramic rods are depicted in figure 7.36. The eutectic lamellae had an average spacing of 140 nm. The three components, Al_2O_3, YAG, and ZrO_2, were black, gray and white, respectively. In the transversal section, the eutectic structure was characterized by densely distributed colonies, as observed in figure 7.36(a). Within these colonies, two types of structures, lamellar and rod-like structures were observed, as shown in figure 7.36 (a1). The rod-like structure was enclosed by surrounding it in the regular lamellar structure. Because it was difficult to deviate from specific crystal orientations at the growth interface, ZrO_2 with a weak faceted crystalline structure was present in between Al_2O_3 and YAG, producing ternary lamellar structures, as seen in figures 7.36(b) and (b1).

Typical rod-like and lamellar structures within the eutectic colonies are demonstrated in figures 7.37(a) and (b). After annealing at 1500 °C for 100 h, the eutectic structures became significantly coarsened and deformed, as revealed in figures 7.37 (c) and (d). Diffusion in the lamellar structures occurred only in 2D spaces, while the rod-like structures remained more stable. Successive collapse was noticed for the layers of ZrO_2, Al_2O_3 and YAG, as the sintering temperature was gradually increased. Also, two types of microstructures in the Al_2O_3/YAG/ZrO_2 ternary

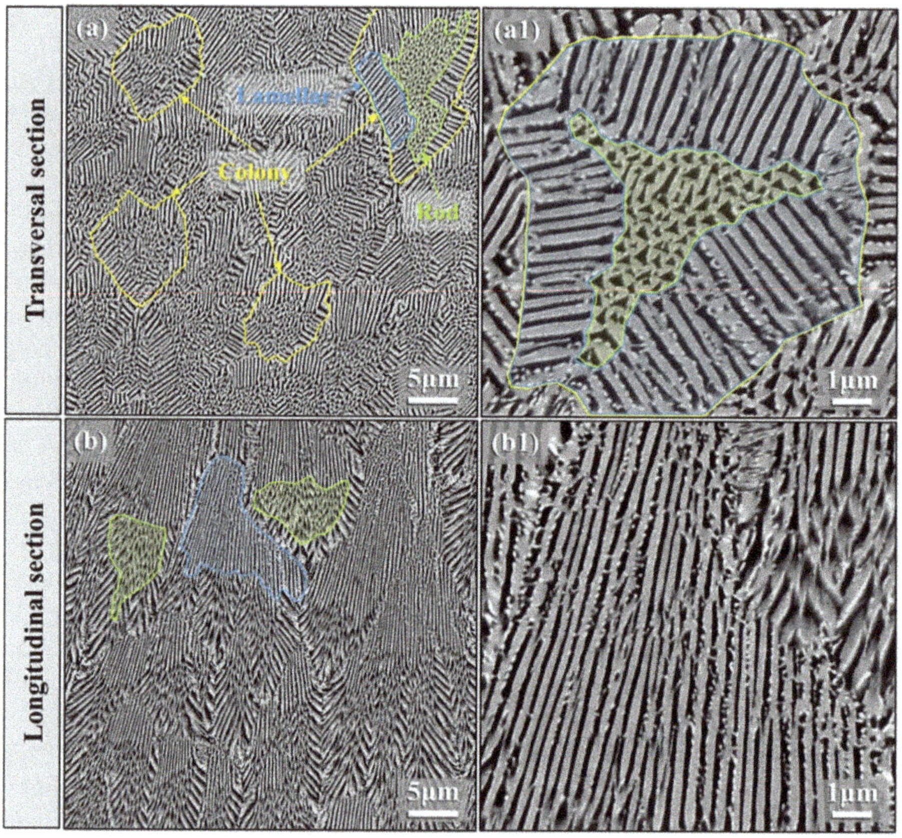

Figure 7.36. SEM images of the directionally solidified Al_2O_3/YAG/ZrO_2 ternary eutectic ceramics: (a) transversal section, (a1) colony structure, (b) longitudinal section and (b1) lamellar structure. Reprinted from [75], Copyright (2025), with permission from Elsevier.

eutectic ceramics were dependent on the pre-sintering temperature, as presented in figures 7.37(e) and (f).

Figure 7.38 presents microstructural variation in the lamellar and rod-like structures of the Al_2O_3/YAG/ZrO_2 ternary eutectic ceramics as a function of pre-sintering temperature, involving three stages. The first stage was the collapse of the ZrO_2 layers. Because the ultrafine ZrO_2 layers had the highest surface energy, they tended to fracture at relatively low temperatures, as illustrated in figure 7.38(a2). At the second stage, the Al_2O_3 and YAG layers started to deform. Meanwhile, ZrO_2 grains gradually became to be equiaxed. The YAG layer was squeezed into pillar-like structure with varying widths, whereas the Al_2O_3 layer appeared as three toothed comb structure, as described in figure 7.38(a3). Finally, in the third stage, both the Al_2O_3 and YAG layers collapsed, resulting in the formation of isolated grains, as seen in figure 7.38(a4) and confirmed in figure 7.37(f). The morphology evolution of the grains is schematically shown in figures 7.38(b1) and (b2).

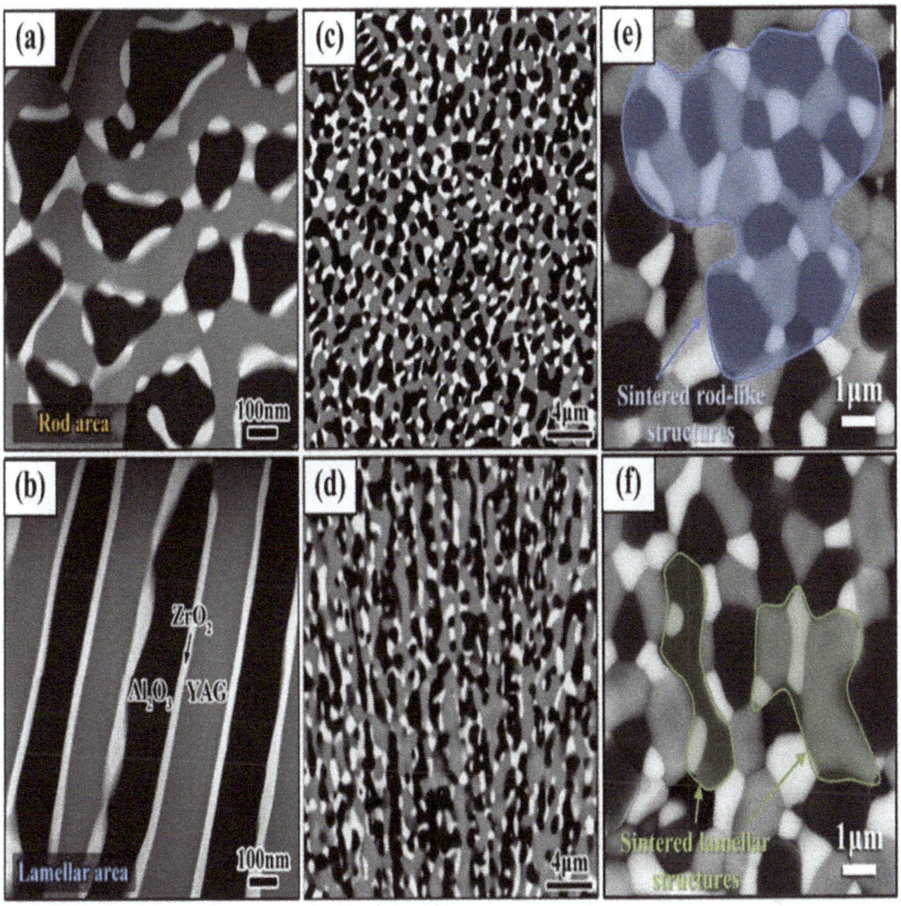

Figure 7.37. TEM images of the directionally solidified Al_2O_3/YAG/ZrO_2 ternary eutectic structures: (a) rod-like region and (b) lamellar region. BSD images after annealing at 1500 °C for 100 h heat treatment: (c) rod-like structure and (d) lamellar structure. BSD images of the Al_2O_3/YAG/ZrO_2 eutectic ceramics sintered at 1670 °C for 2 h: (e) rod-like structure and (f) lamellar structure. Reproduced with permission from [75], Copyright © 2025, Elsevier.

7.3 Concluding remarks

Several applications of ceramic materials processed using 3D printing technology have been summarized and discussed, including biomedical, piezoelectric, microwave, transparent ceramics, and ceramic matrix composites. The two main advantages of 3D printing for ceramic and composite fabrication are design freedom and reduced material waste. On one hand, it enables the fabrication of complex shapes, geometries, and structures limited only by human imagination. On the other hand, the process consumes only the required amount of material, generating minimal waste, which is especially beneficial for expensive ceramic materials.

However, 3D printing technology also has its disadvantages, such as low production rate, relatively low quality of final ceramics, and limited materials

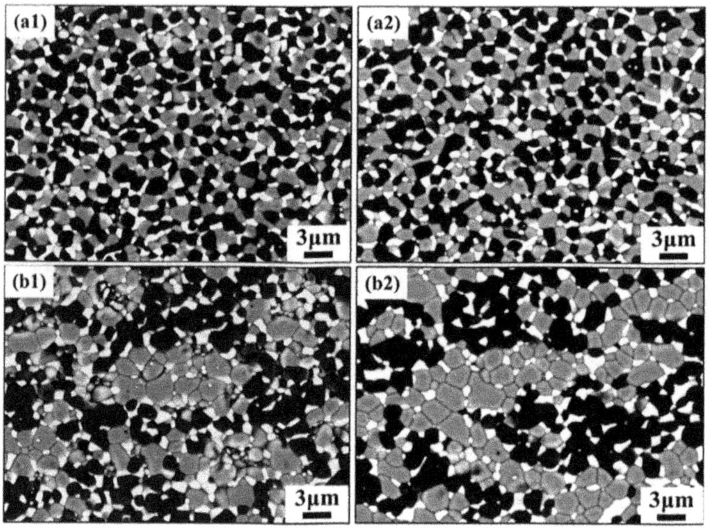

Figure 7.38. SEM images of the sintered Al_2O_3/YAG/ZrO_2 eutectic ceramics: (a1) 1670 SEC, (a2) HIP-ed SEC, (b1) 1685 SPC and (b2) HIP-ed SPC. Reprinted from [75], Copyright (2025), with permission from Elsevier.

options. Due to the additive nature, it is currently suitable only for small-scale sample fabrication at low production rate. Large-scale production remains in its infancy. Meanwhile, the range of printable ceramic materials is still limited compared to traditional ceramic manufacturing methods, as not all ceramic compositions can be readily printed.

Future research should focus on several key areas, including the development of more advanced printing systems, expansion of printable material options, and optimization of printing processes. Expanding the printable ceramic material range and improving their properties and performance will be of crucial importance. This includes developing new ceramic inks or slurries, minimizing the use of organic additives, increasing the solid loading levels, and enhancing material sinterability. Efforts should also be made to improve the printing speed and resolution of 3D printers to promote commercialization and industrial adoption of the printing technology. Additionally, exploring novel applications of 3D printable ceramic materials deserves continues attention. Overall, the applications of 3D printable ceramics are expected to continue expanding and to be widely reported in future research.

References

[1] Bandyopadhyay A, Traxel K D and Bose S 2021 Nature-inspired materials and structures using 3D printing *Mater. Sci. Eng.* R **145** 100609

[2] Dehurtevent M, Robberecht L, Hornez J C, Thuault A, Deveaux E and Béhin P 2017 Stereolithography: a new method for processing dental ceramics by additive computer-aided manufacturing *Dent. Mater.* **33** 477–85

[3] Galante R, Figueiredo-Pina C G and Serro A P 2019 Additive manufacturing of ceramics for dental applications: a review *Dent. Mater.* **35** 825–46
[4] Lu Y Q, van Steenoven A, Dal Piva A M D, Tribst J P M, Wang L, Kleverlaan C J and Feilzer A J 2025 Additive-manufactured ceramics for dental restorations: a systematic review on mechanical perspective *Front. Dent. Med.* **6** 1512887
[5] Wang G Q, Wang S R, Dong X S, Zhang Y J and Shen W 2023 Recent progress in additive manufacturing of ceramic dental restorations *J. Mater. Res. Technol.* **26** 1028–49
[6] Chaudhary S, Avinashi S K, Rao J T D and Gautam C 2023 Recent advances in additive manufacturing, applications and challenges for dentistry: a review *ACS Biomater. Sci. Eng.* **9** 3987–4019
[7] Su G Y, Zhang Y S, Jin C Y, Zhang Q Y, Lu J R, Liu Z Q *et al* 2023 3D printed zirconia used as dental materials: a critical review *J. Biol. Eng* **17** 78
[8] Xiao X D, Yi C H, Xiao Z H, Qiu R L, Huang X H, Dong H B *et al* 2024 Research progress and applications of high performance lithium disilicate glass-ceramics *J. Ceram* **45** 1098–111
[9] Ebert J, Özkol E, Zeichner A, Uibel K, Weiss Ö, Koops U, Telle R and Fischer H 2009 Direct inkjet printing of dental prostheses made of zirconia *J. Dent. Res.* **88** 673–6
[10] Aboras M, Muchtar A, Azhari C H, Yahaya N and Mah J C W 2019 Enhancement of the microstructural and mechanical properties of dental zirconia through combined optimized colloidal processing and cold isostatic pressing *Ceram. Int.* **45** 1831–6
[11] Kumagai T 2018 Isostatic compaction behavior of yttria-stabilized tetragonal zirconia polycrystal powder granules *Powder Technol.* **329** 345–52
[12] Moin D A, Hassan B and Wismeijer D 2017 A novel approach for custom three-dimensional printing of a zirconia root analogue implant by digital light processing *Clin. Oral Implants Res.* **28** 668–76
[13] Osman R B, van der Veen A J, Huiberts D, Wismeijer D and Alharbi N 2017 3D-printing zirconia implants; a dream or a reality?—An in-vitro study evaluating the dimensional accuracy, surface topography and mechanical properties of printed zirconia implant and discs *J. Mech. Behav. Biomed. Mater.* **75** 521–8
[14] Yao Y X, Cui H B, Wang W Q, Xing B H and Zhao Z 2024 High performance dental zirconia ceramics fabricated by vat photopolymerization based on aqueous suspension *J. Eur. Ceram. Soc.* **44** 116795
[15] Mohammed M K, Alahmari A, Alkhalefah H and Abidi M H 2024 Evaluation of zirconia ceramics fabricated through DLP 3d printing process for dental applications *Heliyon* **10** e36725
[16] Jung J M, Kim G N, Koh Y H and Kim H E 2023 Manufacturing and characterization of dental crowns made of 5-mol% yttria stabilized zirconia by digital light processing *Materials* **16** 1447
[17] Su C Y, Wang J C, Chen D S, Chuang C C and Lin C K 2020 Additive manufacturing of dental prosthesis using pristine and recycled zirconia solvent-based slurry stereolithography *Ceram. Int.* **46** 28701–9
[18] Hoffmann M, Schubert N H, Günster J, Stawarczyk B and Zocca A 2025 Additive manufacturing of glass-ceramic dental restorations by layerwise slurry deposition (LSD-print) *J. Eur. Ceram. Soc.* **45** 117235
[19] Shen W, Wang G Q, Wang S R, Zhang Y J, Kang J F, Xiao Z *et al* 2025 Designing and vat photopolymerization 3D printing of glass ceramic/zirconia composites functionally gradient ceramics for dental restorations *Ceram. Int.* **51** 4441–52

[20] Wang G Q, Fu K, Wang S R and Yang B B 2020 Optimization of mechanical and tribological properties of a dental SiO_2–Al_2O_3–K_2O–CaO–P_2O_5 glass-ceramic *J. Mech. Behav. Biomed. Mater.* **102** 103523

[21] Lopez C D, Diaz-Siso J R, Witek L, Bekisz J M, Cronstein B N, Torroni A *et al* 2018 Three dimensionally printed bioactive ceramic scaffold osseoconduction across critical-sized mandibular defects *J. Surg. Res.* **223** 115–22

[22] Szpalski C, Nguyen P D, Vasiliu C E C, Chesnoiu-Matei I, Ricci J L, Clark E *et al* 2012 Bony engineering using time-release porous scaffolds to provide sustained growth factor delivery *J. Craniofac. Surg* **23** 638–44

[23] Fierz F C, Beckmann F, Huser M, Irsen S H, Leukers B, Witte F *et al* 2008 The morphology of anisotropic 3D-printed hydroxyapatite scaffolds *Biomaterials* **29** 3799–806

[24] Hamza M, Kanwal Q, Hussain M I, Khan K, Asghar A, Liu Z Y *et al* 2025 Recent progress in 3D printed piezoelectric materials for biomedical applications *Mater. Sci. Eng. R-Rep.* **164** 100962

[25] Zeng Y S, Jiang L M, He Q Q, Wodnicki R, Yang Y, Chen Y *et al* 2022 Recent progress in 3D printing piezoelectric materials for biomedical applications *J. Phys. D: Appl. Phys.* **55** 013002

[26] Park J, Lee D G, Hur S, Baik J M, Kim H S and Song H C 2023 A review on recent advances in piezoelectric ceramic 3D printing *Actuators* **12** 177

[27] Chen Z Y, Song X, Lei L W, Chen X Y, Fei C L, Chiu C T *et al* 2016 3D printing of piezoelectric element for energy focusing and ultrasonic sensing *Nano Energy* **27** 78–86

[28] Cheng J, Chen Y, Wu J W, Ji X R and Wu S H 2019 3D printing of $BaTiO_3$ piezoelectric ceramics for a focused ultrasonic array *Sensors* **19** 4078

[29] Liu C L, Du Q P, Wu J M, Zhang G Z and Shi Y S 2024 Novel 3D printed PZT-based piezoceramics for piezoelectric energy harvesting via digital light processing *Chem. Eng. J.* **492** 152004

[30] Zheng K, Quan Y, Ding D F, Zhuang J, Wang Y K, Wang Z *et al* 2024 3D printed piezoelectric focused element for ultrasonic transducer *Ceram. Int.* **50** 51863–9

[31] Benford J 2008 Space applications of high-power microwaves *IEEE Trans. Plasma Sci.* **36** 569–81

[32] Chiao J C, Li C Z, Lin J S, Caverly R H, Hwang J C M, Rosen H *et al* 2023 Applications of microwaves in medicine *IEEE J. Microw* **3** 134–69

[33] Raveendran A, Sebastian M T and Raman S 2019 Applications of microwave materials: a review *J. Electron. Mater.* **48** 2601–34

[34] Wang T T, Lu X F and Wang A 2020 A review: 3D printing of microwave absorption ceramics *Int. J. Appl. Ceram. Technol.* **17** 2477–91

[35] Holkar R R, Umarji G G, Shinde M D and Rane S B 2025 3D printing of microwave materials, components and their applications—a review *J. Manuf. Process.* **137** 280–305

[36] Lou Y H, Wang F, Li Z J, Zou Z Y, Fan G F, Wang X C *et al* 2020 Fabrication of high-performance $MgTiO_3$–$CaTiO_3$ microwave ceramics through a stereolithography-based 3D printing *Ceram. Int.* **46** 16979–86

[37] Pan Z X, Wang D, Guo X, Li Y M, Zhang Z B and Xu C H 2022 High strength and microwave-absorbing polymer-derived SiCN honeycomb ceramic prepared by 3D printing *J. Eur. Ceram. Soc.* **42** 1322–31

[38] Zhao W Y, Shao G, Jiang M J, Zhao B, Wang H L, Chen D L *et al* 2017 Ultralight polymer-derived ceramic aerogels with wide bandwidth and effective electromagnetic absorption properties *J. Eur. Ceram. Soc.* **37** 3973–80

[39] Li Q, Yin X W, Duan W Y, Cheng L F and Zhang L T 2017 Improved dielectric properties of PDCs-SiCN by *in situ* fabricated nano-structured carbons *J. Eur. Ceram. Soc.* **37** 1243–51
[40] Li Z B, He J B, Jin L H, Liu X Q and Shao B T 2021 Self-assembled core–shell amorphous SiC_x,N_y,O_z,@SiC_xO_y composites with high thermal stability for highly effective electromagnetic wave absorption *ACS Appl. Electron. Mater.* **3** 2589–600
[41] Johnson R, Biswas P, Ramavath P, Kumar R S and Padmanabham G 2012 Transparent polycrystalline ceramics: an overview *T. Indian Ceram. Soc.* **71** 73–85
[42] Tsabit A M and Yoon D H 2022 Review on transparent polycrystalline ceramics *J. Korean Ceram. Soc.* **59** 1–24
[43] Volfi A, Esposito L, Biasini V, Piancastelli A and Hostasa J 2024 Industrial potential of additive manufacturing of transparent ceramics: a review *Open Ceram* **20** 100682
[44] Wang S F, Zhang J, Luo D W, Gu F, Tang D Y, Dong Z L *et al* 2013 Transparent ceramics: processing, materials and applications *Prog. Solid State Chem.* **41** 20–54
[45] Wu J, Wang Z J, Hu Z C, Liu X L, Tan D Q, Dai Y *et al* 2023 Recent progress and challenges of transparent AlON ceramics *Trans. Nonferrous Met. Soc. China* **33** 653–67
[46] Xiao Z H, Yu S J, Li Y M, Ruan S C, Kong L B, Huang Q *et al* 2020 Materials development and potential applications of transparent ceramics: a review *Mater. Sci. Eng. R-Rep.* **139** 100518
[47] Zhang G R, Carloni D and Wu Y Q 2020 3D printing of transparent YAG ceramics using copolymer-assisted slurry *Ceram. Int.* **46** 17130–4
[48] Zhang G R and Wu Y Q 2021 Three-dimensional printing of transparent ceramics by lithography-based digital projection *Addit. Manuf* **47** 102271
[49] Seeley Z, Yee T, Cherepy N, Drobshoff A, Herrera O, Ryerson R *et al* 2020 3D printed transparent ceramic YAG laser rods: matching the core-clad refractive index *Opt. Mater.* **107** 110121
[50] Jones I K, Seeley Z M, Cherepy N J, Duoss E B and Payne S A 2018 Direct ink write fabrication of transparent ceramic gain media *Opt. Mater.* **75** 19–25
[51] Chen Q M, Li H B, Han W J, Yang J, Xu W T and Zhou Y F 2024 Three-dimensional printing of yttrium oxide transparent ceramics via direct ink writing *Materials* **17** 3366
[52] Zhang S, Gal C W, Sutejo I A, Abbas S, Choi Y J, Kim H *et al* 2025 Overcoming transparency limitations in 3D-printed yttria ceramics *J. Mater. Sci. Technol.* **225** 59–71
[53] Wang H M, Liu L Y, Ye P C, Huang Z Y, Ng A Y R, Du Z H *et al* 2021 3D printing of transparent spinel ceramics with transmittance approaching the theoretical limit *Adv. Mater.* **33** 202007072
[54] Carloni D, Zhang G R and Wu Y Q 2021 Transparent alumina ceramics fabricated by 3D printing and vacuum sintering *J. Eur. Ceram. Soc.* **41** 781–91
[55] Ji H H, Chen H T, Zhang B H, Mao X J, Liu Y, Zhang J *et al* 2022 Direct ink writing of aluminium oxynitride (AlON) transparent ceramics from water-based slurries *Ceram. Int.* **48** 8118–24
[56] Chan K F, Zaid M H M, Mamat M S, Liza S, Tanemura M and Yaakob Y 2021 Recent developments in carbon nanotubes-reinforced ceramic matrix composites: a review on dispersion and densification techniques *Crystals* **11** 457
[57] Cho J, Boccaccini A R and Shaffer M S P 2009 Ceramic matrix composites containing carbon nanotubes *J. Mater. Sci.* **44** 1934–51
[58] Karadimas G and Salonitis K 2023 Ceramic matrix composites for aero engine applications-a review *Appl. Sci.* **13** 3017

[59] Porwal H, Grasso S and Reece M J 2013 Review of graphene-ceramic matrix composites *Adv. Appl. Ceram* **112** 443–54

[60] Wang Y, Liu H T, Cheng H F and Wang J 2014 Research progress on oxide/oxide ceramic matrix composites *J. Inorg. Mater.* **29** 673–80

[61] Zhang C J, Wang X Z, Jiao F, Wang J H and Qiu Y X 2025 Advances in the processing of ceramic matrix composites: a review *Int. J. Adv. Manuf. Technol.* **137** 4243–81

[62] Sun J X, Ye D R, Zou J, Chen X T, Wang Y, Yuan J S *et al* 2023 A review on additive manufacturing of ceramic matrix composites *J. Mater. Sci. Technol.* **138** 1–16

[63] Brinckmann S A, Patra N, Yao J, Ware T H, Frick C P and Fertig R S 2018 Stereolithography of SiOC polymer-derived ceramics filled with SiC micronwhiskers *Adv. Eng. Mater.* **20** 201800593

[64] Yang J H, Yu R, Li X P, He Y Y, Wang L, Huang W *et al* 2021 Silicon carbide whiskers reinforced SiOC ceramics through digital light processing 3D printing technology *Ceram. Int.* **47** 18314–22

[65] Xiong H W, Zhao L Z, Chen H H, Wang X F, Zhou K C and Zhang D 2019 3D SiC containing uniformly dispersed, aligned SiC whiskers: printability, microstructure and mechanical properties *J. Alloys Compd.* **809** 151824

[66] Xiong H W, Chen H H, Zhao L Z, Huang Y J, Zhou K C and Zhang D 2019 SiC_w/SiC_p reinforced 3D-SiC ceramics using direct ink writing of polycarbosilane-based solution: microstructure, composition and mechanical properties *J. Eur. Ceram. Soc.* **39** 2648–57

[67] Zhang H, Yang Y, Hu K H, Liu B, Liu M and Huang Z R 2020 Stereolithography-based additive manufacturing of lightweight and high-strength C_f/SiC ceramics *Addit. Manuf* **34** 101199

[68] Taurino R, Martinuzzi S, Padovano E, Caporali S and Bondioli F 2025 Laser additive manufacturing of Al_2O_3 and ZrO_2-based eutectic ceramic oxide: an overview *J. Eur. Ceram. Soc.* **45** 117133

[69] Liu H F, Su H J, Shen Z L, Jiang H, Zhao D, Liu Y *et al* 2022 Research progress on ultrahigh temperature oxide eutectic ceramics by laser additive manufacturing *J. Inorg. Mater.* **37** 255–66

[70] Liu Z, Song K, Gao B, Tian T, Yang H O, Lin X *et al* 2016 Microstructure and mechanical properties of Al_2O_3/ZrO_2 directionally solidified eutectic ceramic prepared by laser 3D printing *J. Mater. Sci. Technol.* **32** 320–5

[71] Fan Z Q, Zhao Y T, Tan Q Y, Yu B W, Zhang M X and Huang H 2020 New insights into the growth mechanism of 3D-printed Al_2O_3-$Y_3Al_5O_{12}$ binary eutectic composites *Scr. Mater.* **178** 274–80

[72] Pastor J Y, Llorca J, Salazar A, Oliete P B, de Francisco I and Peña J I 2005 Mechanical properties of melt-grown alumina-yttrium aluminum garnet eutectics up to 1900 K *J. Am. Ceram. Soc.* **88** 1488–95

[73] Su H J, Zhang J, Deng Y F, Liu L and Fu H Z 2009 A modified preparation technique and characterization of directionally solidified $Al_2O_3/Y_3Al_5O_{12}$ eutectic *in situ* composites *Scr. Mater.* **60** 362–5

[74] Song K, Zhang J, Jia X J, Su H J, Liu L and Fu H Z 2012 Solidification microstructure of laser floating zone remelted Al_2O_3/YAG eutectic *in situ* composite *J. Cryst. Growth* **345** 51–5

[75] Hao S Q, Su H J, Zhao D, Li X, Shen Z L, Liu Y *et al* 2025 Complex shaped Al_2O_3/YAG/ZrO_2 eutectic ceramics with excellent high temperature mechanical properties printed by vat photopolymerization *Addit. Manuf* **101** 104703

[76] Fan Z Q, Zhao Y T, Tan Q Y, Mo N, Zhang M X, Lu M Y *et al* 2019 Nanostructured Al_2O_3–YAG–ZrO_2 ternary eutectic components prepared by laser engineered net shaping *Acta Mater.* **170** 24–37
[77] Liu H F, Su H J, Shen Z L, Jiang H, Zhao D, Liu Y *et al* 2022 Formation mechanism and roles of oxygen vacancies in melt-grown $Al_2O_3/GdAlO_3/ZrO_2$ eutectic ceramic by laser 3D printing *J. Adv. Ceram* **11** 1751–63
[78] Liu H F, Su H J, Shen Z L, Zhao D, Liu Y, Guo Y N *et al* 2021 One-step additive manufacturing and microstructure evolution of melt-grown $Al_2O_3/GdAlO_3/ZrO_2$ eutectic ceramics by laser directed energy deposition *J. Eur. Ceram. Soc.* **41** 3547–58
[79] Liu H F, Su H J, Shen Z L, Zhao D, Liu Y, Guo Y N *et al* 2021 Preparation of large-size $Al_2O_3/GdAlO_3/ZrO_2$ ternary eutectic ceramic rod by laser directed energy deposition and its microstructure homogenization mechanism *J. Mater. Sci. Technol.* **85** 218–23

www.ingramcontent.com/pod-product-compliance
Ingram Content Group UK Ltd.
Pitfield, Milton Keynes, MK11 3LW, UK
UKHW051908100926
8556IPUK00005B/807

9 780750 348294